TRANSPORTATION SYSTEMS PLANNING

METHODS AND APPLICATIONS

New Directions in Civil Engineering

Series Editor
W. F. CHEN
Hawaii University

Published Titles

Published Titles (Continued)

Unified Theory of Reinforced Concrete
Thomas T.C. Hsu

Water Treatment Processes: Simple Options
S. Vigneswaran and C. Visvanathan

Transportation Systems Planning: Methods and Applications
Konstandinos G. Goulias

TRANSPORTATION SYSTEMS PLANNING

METHODS AND APPLICATIONS

Edited by

Konstadinos G. Goulias

CRC PRESS

Boca Raton   London   New York   Washington, D.C.

Library of Congress Cataloging-in-Publication Data

Transportation systems planning : methods and applications / edited by Konstadinos Goulias.
     p. cm. -- (New directions in civil engineering)
     Includes bibliographical references and index.
     ISBN 0-8493-0273-0 (alk. paper)
     1. Transportation and state. 2. Transportation--Planning. I. Goulias, Konstadinos G. II. Series.

HE193 .T733 2002
388--dc21
                                                  2002074026

Visit the CRC Press Web site at www.crcpress.com

© 2003 by CRC Press LLC

No claim to original U.S. Government works
International Standard Book Number 0-8493-0273-0
Library of Congress Card Number 2002074026
Printed in the United States of America 1 2 3 4 5 6 7 8 9 0
Printed on acid-free paper

Introduction

K.G. Goulias
Editor-in-Chief

Transportation engineering and transportation planning are two sides of the same coin aiming at the design of an efficient infrastructure and service to meet our needs for accessibility and mobility. Many well-designed transport systems that meet these needs are based on good understanding of human behavior. For example, in ergonomics — the science for designing tools for humans — models of action–reaction are used to make sure that machines (e.g., cars) are designed to match human characteristics. In a similar way, when we plan cities and the services provided to their inhabitants, understanding human behavior is a key ingredient that drives most decisions in planning and operations. Since transportation systems are the backbone connecting the vital parts of a city, in-depth understanding of human nature is essential to the planning, design, and operational analysis of transportation systems.

Analytical methods that help us understand and predict human (travel) behavior have only partially been developed in the past few decades. In this area, basic and applied research has been very active in quantitative methods since the 1970s, with one notable development: the disaggregate demand models. However, it is only recently that we experienced the benefits in practice from these methods. Unavoidably, assimilation of research findings into decision making for public policy is very slow. Analyses and explanations for this, but not justifications, abound. For example, a comprehensive overview and comparison between markets and governments are given by Sowell (1996), and a few specific examples are reported in the volume edited by Altshuler and Behn (1997).

Recent environmental and transportation legislation initiatives in most industrialized countries, however, have accelerated this process, increasing the pressure on public agencies to introduce major changes to their evaluation tools. Similarly, in private enterprise, where market forces are imposing operational efficiency and customer service quality, we start to observe changes in tool making, e.g., due to a need for improved supply chain management. These trends will continue to press for better decision support systems. This handbook is an integral part of a worldwide effort to design these tools and to disseminate information about principles, theories, methods, models, data, information, and applications. Complex issues are discussed with intellectual clarity and provide innovative solutions, many of which have never been published before. The handbook belongs to the groundbreaking effort by CRC Press to chart new directions in civil engineering with a series of books written by a wide-ranging cast of authors engaged actively in research, education, and consultation in transportation systems worldwide.

The handbook's main objectives are to:

1. Provide a comprehensive and in-depth guide to new theories, methods, and tools for transportation planners — in engineering, economics, geography, anthropology, sociology, psychology, and business, among others.
2. Show with examples details of the tools and their application in the field, allowing for longer exposition and discussion than in the usual peer-reviewed journals, professional conference proceedings, and other encyclopedic presentations.
3. Create a reference that is a collection of methodological developments that have been used recently or can be used in future planning applications. A few examples in this handbook are also unrestrained presentations of experimental methods that may introduce key innovations in planning.

The handbook is published during very interesting times in which a major paradigm shift is being experienced by our field. This is a movement toward an inherently dynamic approach to explaining the world surrounding us. In addition, it is a multivoiced movement in directions that emerge from many different disciplines. To capture these trends, emphasis is given to ideas from a transdisciplinary (as opposed to simply an interdisciplinary or multidisciplinary) viewpoint by recruiting authors that are recognized as innovation leaders in their respective disciplines worldwide, and they have experimented with methods that are situated at the intersection of disciplines. Contrary to other handbooks, which are encyclopedic reviews of fundamental contributions to be stored in the permanent record (e.g., Hensher and Button, 2000; Papageorgiou, 1991), this handbook extends far beyond modeling in engineering and economics, moving along the lines of development described in Gärling et al. (1998).

Implied in all this is an advocacy position: transportation engineers and economists must become very familiar with transportation planning methods, tools, and applications from viewpoints grounded in other disciplines. This will enable veterans and novices to gain access to research and practice in a discipline that is developing its own theories, conceptual models, and communication artifacts that are by far more human centered and realistic. A shift like this will have a profound positive influence on infrastructure and service design that can help us design transportation systems that are not only sustainable, but also green and closer to the needs of the users.

Another objective of understanding human behavior is conceptual integration. Explanation of facts from different perspectives can be considered jointly to form a comprehensive understanding of people, their groups, and their interactions with the natural and built environment. In this way, we may see explanations of human behavior fusing into the same universal principles. These principles eventually will lead to testable hypotheses (using either quantitative approaches, qualitative methods, or a combination of the two). All these different perspectives will yield Wilson's (1998) famous consilience among, for example, psychology, anthropology, economics, the natural sciences, geography, and engineering.

Transportation Planning and Travel Behavior

Travel behavior research aims at understanding how traveler values, norms, attitudes, and constraints lead to observed behavior. Traveler values and attitudes refer to motivational, cognitive, situational, and disposition factors determining human behavior. Travel behavior refers primarily to the modeling and analysis of travel demand, based on theories and analytical methods from a variety of scientific fields. These include, but are not limited to, the use of time and its allocation to travel and activities, methods to study this in a variety of time contexts and stages in the life of people, and the organization and use of space at any level of social organization, such as the individual, the household, the community, and other formal or informal groups. The movement of goods is included in all this because it is motivated by human needs and because it has very strong interfaces and relationships with the movement of persons.

This handbook is the continuation along one possible direction of the behavioral travel demand models of the 1970s (Stopher and Meyburg, 1976; Domencich and Mcfadden, 1975), with developments mainly from the social sciences. In this way, travel behavior is examined from both objective (observed by an analyst) and subjective (perceived by the human) perspectives in an integrated manner among four dimensions: time, geographic space, social space, and institutional context. On a few occasions the chapters in this handbook include and integrate time as conceived in everyday life with perceptions of time and space by humans. This research includes theory formation, data collection, modeling, inference, and simulation methods to produce decision support systems for policy assessment and evaluation.

Handbook Organization

The handbook is organized into three sections. The first contains chapters that are dedicated to describing transportation systems and reviewing travel behavior theories from different perspectives. The second reviews data collection and data analysis methodological innovations and provides a bridge to the large body of literature on the subject. The third contains two chapters on methods with an application slant and a few applications indicative of new developments.

Section I: Transportation Systems and Theories of Human Behavior

The first section of the handbook starts with an overview chapter on transportation systems, issues addressed in planning and design, and trends observed around the world. Emphasis is given to environmental issues and their relationship to transportation. Examples and analyses from the United States and Europe are reviewed, and a comprehensive review of other sources of information is also offered. The complex relationship among the agents and issues in transportation is identified. Then the chapter provides an overview of some elements of taxonomy for the different decision support tools needed. This is followed by a general theoretical framework called activity theory, which can be used as a lens to understand other model systems. It also provides a unique perspective for modeling and simulation that may be the pragmatic approach needed for travel demand forecasting.

In the second chapter, Ram Pendyala targets one of the most important developments in travel demand modeling: the development and implementation of activity-based models of travel demand that explicitly recognize the important role played by time and space in shaping individual and household activity and travel patterns. In this chapter a data-driven theory development approach is followed. The chapter starts with a descriptive analysis of the ideas and issues in time use, introducing many fundamental definitions, data collection methods, and statistics on time allocation by individuals. Then it provides an overview with examples of model development in this context, using stochastic frontier models of time–space prism vertices, structural equation models of household time allocation and activity engagement, econometric joint discrete–continuous models to study maintenance activity episode timing and duration, and utility models of welfare derived from time use and activity patterns. The chapter also includes a comprehensive review of developments in time use, activity analysis, and their relationship to travel behavior.

The third chapter, by Reginald Golledge and Tommy Gärling, gives an overview of theories on the movement of people in space and in time. The focus is on disaggregate spatial behavior and behavioral travel choice. The authors review human spatial behavior by examining concepts such as cognitive maps and wayfinding, activity participation and travel, information and its role in the choice process, and path selection. Travel demand forecasting is then cast in terms of latent variables, and a synthesis of the most recent travel demand forecasting models is provided. The chapter also includes an extensive review of other key references in this subject.

In the fourth chapter, Frank Southworth gives a comprehensive overview of freight transportation issues that goes beyond the traditional focus on mode choice. The chapter is indicative of a worldwide research movement to understand how freight systems work today and how they can be improved in the future. Emphasis in the chapter is given to the agents operating in freight and their agency, costs, demand and supply, productivity and performance measures, and freight's safety and environmental impacts. A case is also made in favor of a freight modeling and forecasting component in metropolitan and regional planning and forecasting.

In the fifth chapter, Eric Miller addresses one of the most important and positive developments in travel demand forecasting in the past 20 years: integrated land use–transportation models. As Miller notes in the chapter, these models capture "both urban system evolution and the associated evolution of urban travel demand in a comprehensive and integrated fashion." In the chapter we find reviews of the most important land use models and the principles underlying them, including the spatial interaction, utility-based, and Lowry models. The chapter then moves to simulation models that attempt to capture the salient characteristics of land use decisions in the real world and reflect market functioning. This is followed by an extensive presentation of limitations and specific solutions for model failures. An overview of the latest land use models is also given, and the major design and implementation issues for developing and using integrated land use–transportation models are presented.

The sixth chapter, by Kevin Krizek, provides a more comprehensive and, at the same time, more focused viewpoint when analyzing land use and transportation: it examines the relationship among land use–urban form, lifestyles, and travel. The chapter provides in sequence a summary of past and current research and casts a net around the key issues. Then it reviews shortcomings of past research and describes strategies addressing these shortcomings. Using examples from Seattle, Washington, the author provides a convincing case for using urban form and lifestyles as key ingredients in building transportation models to examine travel demand.

Section II: Data Collection and Analysis

The second section of the handbook is about data and their analysis. More traditional data collection methods are documented elsewhere (in Chapter 1 a guide to data collection literature is provided, and in Chapter 2 Ram Pendyala gives references for activity-based surveys). For this reason, only Chapter 7, written by Sean Doherty, describes a family of new methods in data collection that is more likely to support the newer theories described in the first part of the handbook. The chapter describes interactive survey methods aimed at understanding the decision processes followed in activity scheduling. Doherty notes: "What is shared by these methods is a desire to interactively observe *how* decisions are made and their dynamics, not just the results of these decisions in the form of static observed activity–travel patterns." The chapter starts with an outline of the basic components of the activity scheduling decision process. Then each component of the framework is discussed in depth, the types of data from the process are described, and data analysis areas are explained. Examples are from projects in Canada, the United States, and Europe, and the chapter includes a discussion on the latest state-of-the-art techniques and technologies in the data collection field.

The next four chapters focus on regression methods — the premier tool in analyzing data succinctly and efficiently. Chapter 8 provides a brief overview of statistical and econometric methods and models and provides a guide to the extensive literature using a taxonomy of the dependent variables (the variation of which we try to understand), leaving three topics for more in-depth scrutiny in Chapters 9 to 11.

Chapter 9 describes multilevel analysis and provides examples of individual choices of time allocation, exploiting the nested data hierarchy of households, persons, and occasions of measurement. The multilevel models in the chapter are first presented in terms of their advantages and then compared to a single-equation regression in econometrics. A review of the different traditions in model building is provided, together with a detailed description of the basic multilevel single-equation model. Then the multivariate version of multilevel regression is provided, with an example from the time allocation to activity and travel in two days, allowing us to study the effects of information and communication technology on time allocation. Other applications of multilevel models are also discussed, and a brief guide to the literature is given.

Discrete choice models have seen a tremendous development in the past two decades, and today severe limitations and restrictions of the original formulations can be relaxed, leading to substantial gain in realism of the choice models thus derived. Chandra Bhat, in Chapter 10, presents an overview of the structure, estimation techniques, and transport applications of three representative and very important model groups. These are the heteroskedastic, generalized extreme value (GEV), and flexible structure models; for each class, alternative formulations are also discussed. The additional flexibility comes at the cost of evaluating one-dimensional integrals (in the heteroskedastic extreme value model) or multidimensional integrals (in the flexible model structures). Bhat demonstrates that effective methods exist and provides a discussion on the most appropriate use of these techniques.

Chapter 11, by Thomas Golob, is an impressive review of methodological developments and applications of structural equations. Golob has introduced structural equations in transportation planning and has argued successfully for many years that structural equations should be used widely. It is only in the past five years that other researchers, some of whom are well known and respected, exhibited an interest in latent variable models (structural equations are one of the most efficient and robust methods handling latent variable models). This chapter is a reprint from a paper published in *Transportation Research* and provides an excellent overview.

Section III: Systems Simulation and Applications

The third section of the handbook is dedicated to transportation system simulation and related applications. Each chapter is an example of a unique modeling tradition, and taken together, they offer a rich toolbox of options available today to the practicing planner and a rich workbench for many researchers that are yet to join us.

Chapter 12 is written by Eric Miller, who is a pioneer in microsimulation methods in transportation. The chapter is written as a tutorial of concepts and methods that have been used in travel-related forecasting research applications and can be used in practice. In sequence, Miller defines the term *microsimulation*; discusses why microsimulation may prove useful, the need for creating synthetic samples, and some of the major operational issues; and concludes with a brief presentation of several microsimulation models from a wide sample of applications. This chapter defines the framework within which the rest of the chapters should be considered.

In Chapter 13, Debbie Niemeier draws yet another framework that serves as a reality check on data, models, and simulation systems. The chapter describes in a comprehensive and exhaustive way the entire regulatory background of mobile source emissions and clearly defines modeling implications. In a big picture Niemeier integrates travel demand, traffic simulation, and vehicle emissions models. She notes, however, that "the separation between the transportation modeling and air quality research communities is still quite vast," and concludes on a hopeful note that potential for innovation exists in the interface between these two fields.

Chapter 14 picks up on a line of thought discussed by Miller in his chapter on microsimulation: the need to provide a synthetic sample with all the social, economic, and demographic characteristics needed by the regression models encountered in the first two parts of the handbook. The chapter provides an example of demographic microsimulation that builds on past experience and uses new data and more flexible programming techniques.

In Chapter 15 a Dutch team provides a brief presentation of the most complete microsimulation model system, using rule-based approaches. This model system was discussed briefly by Golledge and Gärling in Chapter 3, who introduced it as "in line with approaches in cognitive psychology." The model system is the most comprehensive, and it is the only operational computational process model of transportation demand; for this reason, it is becoming a standard of reference for many other model systems that either precede or follow ALBATROSS. In this chapter, Theo Arentze, Frank Hofman, Henk van Mourik, and Harry Timmermans illustrate four policy scenarios and discuss interpretation of the results extensively.

Chapter 16 concludes the handbook by presenting an activity-based approach that targets small metropolitan communities that do not have the capabilities and resources of larger metropolitan areas. It takes the aging four-step model system and transforms it into a dynamic activity-based system. Experiments with this method are then discussed, and results from a validation exercise are presented.

References

Altshuler, A.A. and Behn, R.D., *Innovation in American Government*, Brookings Institution Press, Washington, D.C., 1997.

Domencich, T. and McFadden, D., *Urban Travel Demand: A Behavioral Analysis*, North-Holland, Amsterdam, 1975.

Gärling, T., Laitila, T., and Westin, K., Eds., *Theoretical Foundations of Travel Choice Modeling*, Elsevier, Amsterdam, 1998.

Hensher, D.D. and Button, K.J., Eds., *Handbook of Transport Modelling*, Pergamon, Oxford, 2000.

Papageorgiou, M., *Concise Encyclopedia of Traffic and Transportation Systems*, Pergamon, Oxford, 1991.

Sowell, T., *Knowledge and Decisions*, Basic Books, New York, 1996.

Stopher, P.R. and Meyburg, A.H., Eds., *Behavioral Travel-Demand Models*, Lexington Books, Lexington, MA, 1976.

Wilson, E.O., *Consilience: The Unity of Knowledge*, Vintage Books, New York, 1998.

Editor-in-Chief

Dr. Goulias was born in Athens, Greece, where he completed his pre-university studies. Then he entered the Universita degli Studi della Calabria with a scholarship from the Italian government. He graduated with a Laurea degree (B.S. and M.S.), with specialties in civil engineering and territorial planning. His thesis, in Italian, involved the creation, testing, and application of an algorithm for maximum likelihood estimation of multinomial logit models for mode choice. Subsequently (August 1986 to August 1987), he received an M.S.C.E. from the University of Michigan, Department of Civil Engineering, concentrating on traffic engineering, highway design, and probabilistic models for civil infrastructure systems. Between August 1987 and August 1991, he pursued his Ph.D. at the University of California–Davis, with a brief visit to the University of California–Irvine and two brief visits to The Netherlands. In his Ph.D. dissertation he created a demographic and travel demand microsimulator (MIDAS) for The Netherlands that was funded by the Dutch Ministry of Transport and the California University Transportation Center in 1990 and 1991. During this period he also worked on a variety of projects, including telecommuting studies, panel and longitudinal survey design and analysis, dynamic econometric modeling, and freight simulation models.

Goulias came to Pennsylvania State University in the fall of 1991 as assistant professor and was promoted to associate professor in 1997 and professor in 2002. He is also the director of the transportation operations program at the Pennsylvania Transportation Institute and the director of the Center for Intelligent Transportation Systems, both since 1997. In 2002 Dr. Goulias became the director of the Mid-Atlantic Universities Transportation Center.

At Pennsylvania State, his primary research interests are in the area of quantitative transportation planning. His emphasis is on forecasting the demand for transportation services and on the impact simulation and forecasting of policy actions. He is working on the development of statistical–econometric and computer-based stochastic simulation methods and tools in five research directions: analysis of the dynamics of traveler behavior; development of computerized decision-making tools; analysis of interaction between information, telecommunications, technologies, and transportation systems; e-commerce and transportation; and sustainable and green transportation. In the past 10 years his projects have received funding from local, state, federal, and international private companies and public agencies.

He teaches at the undergraduate, graduate, and continuing education levels. He also co-developed joint courses in business and management at Pennsylvania State and at the Technical University of Lisbon during his Fulbright Senior Chair Award. Goulias also organized and delivered courses, seminars, lectures, keynote speeches, and resource papers at local, national, and international professional meetings and agencies. He is chairing the National Research Council–Transportation Research Board (TRB) committee on traveler behavior and values (A1C04), and he has been the chair of a subcommittee on activity analysis and travel patterns. He has also been a member of other TRB committees, an executive council member in the Institute of Transportation Engineers, and a committee secretary in the American Society of Civil Engineers. Additionally, he is a member of the International Association for Travel Behavior Research.

Goulias is and has been on the Editorial Advisory Board of refereed journals (such as *Transportation Research Part A* and *ITS Journal*) and regularly reviews papers submitted to *Transportation*, *Transportation Research*, *Transportation Research Record*, and *Transportation Science*. He also serves on a variety of advisory boards, expert panels, and engineering and science foundations proposal review committees in the United States, Canada, Europe, and Australia.

Contributors

Theo Arentze
Urban Planning Group/EIRASS
Technische Universiteit Eindh
Eindhoven, The Netherlands

Chandra R. Bhat
Department
of Civil Engineering
University of Texas
Austin, Texas

Sean T. Doherty
Department of Geography
and Environment
Wilfrid Laurier University
Waterloo, Ontario, Canada

Tommy Gärling
Department of Psychology
Göteborg University
Göteborg, Sweden

Reginald G. Golledge
Department of Geography
and Research
University of California
Unit on Spatial Cognition
and Choice (RUSCC)
Santa Barbara, California

Thomas F. Golob
Institute of Transportation
Research
University of California
Irvine, California

Konstadinos G. Goulias
Pennsylvania Transportation
Institute
Pennsylvania State University
University Park, Pennsylvania

Frank Hofman
Ministry of Transportation,
Public Works and Water
Management
Rotterdam, The Netherlands

Kevin J. Krizek
Humphrey Institute
of Public Affairs
University of Minnesota
Minneapolis, Minnesota

JoNette Kuhnau
Pennsylvania State University
University Park, Pennsylvania

Eric J. Miller
Department of Civil Engineering
University of Toronto
Toronto, Ontario, Canada

Debbie A. Niemeier
Department of Civil and
Environmental Engineering
University of California
Davis, California

Ram M. Pendyala
Department of Civil and
Environmental Engineering
University of South Florida
Tampa, Florida

Frank Southworth
Oak Ridge National Laboratory
Oak Ridge, Tennessee

Ashok Sundararajan
AECOM Consulting
Transportation Group
Fairfax, Virginia

Harry Timmermans
Urban Planning Group/EIRASS
Eindhoven University
of Technology
Eindhoven, The Netherlands

Henk van Mourik
Ministry of Transportation,
Public Works and Water
Management
Rotterdam, The Netherlands

Contents

III   Systems Simulation and Applications

I

Transportation Systems and Theories of Human Behavior

1

Transportation Systems Planning

CONTENTS

Konstadinos G. Goulias
Pennsylvania State University

1.1 Introduction

As discussed in the introduction of this handbook, one can identify a mainstream approach to transportation planning and a second that is richer and more concerned with modern issues. This new emergent viewpoint in approaching transportation problems recognizes the presence of complexities, nonlinearities, and uncertainties that were neglected in the past for the sake of simplicity. The nature of policies to be assessed and the realization that interdependent systems need to be studied and modeled in their totality motivates building decision support systems that are increasingly expanded to incorporate processes and ideas from other related fields. For example, the wider acceptance of discrete choice models, which consider the person as a decision unit, motivates the need to provide data about persons. These can be data on demographics (age, gender), economics (employment, income), and social situations and roles (e.g., household type, indicators of the role in the household). Production of these data to be used in forecasting future choices requires one to employ demographic evolutionary methods that produce this information in future years for which an assessment of policy impacts is made. Many more examples, a few of which are included in this handbook, show that we are experiencing an "immigration" of disparate methods from other fields into transportation systems planning. In this way, the resulting model systems are very often the result of a somewhat haphazard amalgamation of methods that have been designed at different levels of scale (person, community, city), based on different behavioral assumptions (e.g., optimizing, satisficing, adaptive, or opportunistic behavior), and estimated with data from different periods or horizons (e.g., a typical day, a given year defined generically, a census decade, and so forth). For these reasons, different models may not be entirely consistent and interoperable, and their predictions are surrounded by large error bands that provide information that is sometimes sufficient for some type of decision making and other times totally inadequate for any analysis. Many of these models, however, share the same motivation and their ultimate aim is to solve specific *transportation problems*. In this chapter overviews of these problems and of the most recent issues in designing transportation system planning models to solve the problems are provided.

The traditional viewpoint of transport experts and policy makers is that transportation systems exist to provide for the safe and efficient movement of people and goods in an environmentally responsible manner. This definition encompasses not only the benefits to society from a well-designed transportation system, but also the critical issues that we have yet to address and resolve. In fact, older transportation planning textbooks and handbooks would define the transportation problem as composed of a few key dimensions: *safety*, including fatalities, injuries, and property damage due to accidents; *efficiency*, optimal allocation of resources in moving people and goods; *access*, provision of enabling technologies and services to people that need to reach and use opportunities; *comfort*, travel in environments without causing unnecessary stress and strain due to noise or other factors; and *environmental pollution*, production of contaminants in the air, water, or soil that are at higher levels than naturally found and that cause harm to animals, plants, and humans. Safety, efficiency, comfort, and access have seen a tremendous improvement over the last 40 years in all industrialized countries, and they have become valued aspects of transportation systems worldwide. Environmental pollution control, however, in spite of the spectacular improvements in internal combustion engines and emission control, appears to be inhibited by an exponential increase in trip making. This is particularly acute in the more urbanized environments and is the motivation behind many policies.

This is expanded today, and more recent analyses examine the role transportation systems play in our society as a whole from more integrated and systemic viewpoints. In fact, transportation in this approach is viewed as another medium for economic and social development, and its evolution needs to be "guided" with regulation, education, and market manipulation to maximize its positive effect on economic development and provide equitable progress and access to opportunities while minimizing its negative impacts on welfare and the environment (see the examples in Doyle and Hess, 1997; TRB, 2000). Under this somewhat more complex position, the private automobile can be examined in a more critical way and contrasted with many other mobility options that may yield the same benefits but at lower social costs. However, for many tangible and intangible reasons and in many situations, the private automobile is the only feasible and available option. This may be viewed as a threat (e.g., to the environment), but it also opens the opportunity to view the (private) automobile not only as a tool for economic development, but also as a technological opportunity to advance our moral duty of protecting the environment while developing regions and countries in a sustainable way.

These considerations, as expected, are also changing transport-related government positions. Past policies, analyses, and actions on transport systems focused on ways to increase the capacity of individual system components such as roads and terminals (ports, airports, stations), with occasional attention to energy and environmental concerns as well as other social impacts. Attention was paid to transportation system components if they could be studied as independent units, and policies would target a small portion of the system. In addition, decisions were reserved to technical experts, and very little, if any, public input was solicited (see Creighton (1970), who has documented planning work in the 1950s and 1960s). Then, in the 1970s, mainly because of the oil crisis, attention was also paid to managing the transportation system as a system of interconnected components. Realizing that increased capacity is not sufficient to satisfy increasing demand for services, with congestion and air pollution in large metropolitan areas as the earlier evidence, policy analysts and policy makers shifted their attention to a more efficient management of facilities (e.g., utilize the capacity of a highway by spreading its use in a day). New construction was reserved for strategic interventions such as the provision of connectivity among existing roadways. Typical examples with their roots in the 1970s approach transportation systems in more systemic ways, e.g., the National Highway System in the United States and the Trans-European Network in the European Union (EU).

The shift of policies away from expanding capacity to managing demand and the introduction of a systems approach to transportation has been advocated since the late 1960s, with a first attempt to develop comprehensive plans that would be continuously updated and organized in such a way that all governmental levels would cooperate in working toward a common vision. One such example is the U.S. Highway Act of 1962 (see Smerck (1968) for a history leading to 1962). Similarly, but to a much lesser extent, the Treaty of Rome in 1957, which constitutes the foundation of today's European Union, identified transport as one of the key sectors for a common European policy, but it did not have the specificity of the U.S. Highway Act because is was too early for the Union. In the 1970s a major oil crisis provided the needed momentum to reconsider transport policies because the dependency on fossil fuels was becoming a weakness for Western economies. Between October 1973 and January 1974, world oil prices doubled due to a 4.2-million-barrel cutback by select oil producers. This led to a new era in policies, including international military (defense) initiatives, giving birth to a wide variety of strategies to curb the ever increasing private automobile use and the dependence of the United States and Europe on imported oil. Most of the ideas, policies, and strategies seen in the field today were defined and tested in the years just after the oil crisis (Rothenberg and Heggie, 1974; Meyer and Miller, 2001).

While methods and approaches in the 1950s and the 1970s have solved many problems, 30 years later policy debates continue to depict a very grim picture of the private car's role in creating the problems we face in major cities (Pucher, 1999). An added problem to the list we saw before is the West's dependency on foreign oil, which is still substantial. To counter this, many new policies and strategies are needed to provide transportation services while mitigating and minimizing the negative consequences of a car-centered transportation service provision. This is particularly important when we cast transportation services in terms of sustainable development and mobility. At the same time and in clear contrast, more pragmatic defenders of the automobile are also emerging to express popular feelings in favor of the automobile's freedom, flexibility, convenience, and comfort, but also to warn that realistically competitive alternatives to the automobile do not exist yet (Dunn, 1998, 2000).

These problems are not the monopoly of the Western industrialized world. Transportation professionals and transportation systems around the world face similar challenges. In its millennium paper, the committee on International Activities of the Transportation Research Board (TRB) (National Academy of Sciences in the United States) lists the following as challenges (Linzie, 2000):

- Operating transport services and facilities will evolve. The type, extent, and quality of service to users will be under continuous evaluation. The interoperability (working together) of transport services will be an issue.

- Transport organizations will continue a strong trend toward more competition in the delivery of transport services and facilities (e.g., deregulation and privatization).

- Financing and subsidies will always be discussed in the transport sector. Electronic innovations will permit more possibilities for efficient user fees, and democratic governments will ensure some perception of equity.

- Environmental effects of transportation will increase in importance as long-term issues of air quality, water quality, noise, land use, and hazardous waste become priorities for quality of life and sustainability.

- Safety and security of passengers and freight will continue to be emphasized as the public moves toward a zero tolerance of accidents and damage.

- Government regulations, now moving toward economic self-regulation, may change from time to time to balance efficiency with equity and fairness.

- Transportation organizations will have to be more prepared to respond to the threat of climatic change, including effects of emergencies such as hurricanes, floods, and earthquakes (and more recently other intentional threats).
- There is a threat of urban congestion and suburban sprawl for sustainable transport.
- Worldwide coordination and cooperation among transport officials and professionals in research and development will continue to increase in the 21st century.

These themes encompass many of the headings in positions taken by national and international organizations under their call for sustainable transport (OECD, 2001). They are also the same themes found in the European transport policy. However, there are key differences between U.S. and EU transport policies. For example, the need to integrate national transport systems in Europe as different phases of unification are progressing is receiving the bulk of attention; the emissions regulations are aggressive and ambitious on paper, but their implementation is still unknown. In contrast, the United States has a mature energy consumption and emission control legislative framework (including rules and regulations that are currently redefined and debated at all levels of government). Whenever possible we will distinguish between typically U.S. vs. European issues. However, many common themes exist, and the aim to develop suitable analytical tools is the same worldwide.

In terms of complete analytical tools that enable us to assess transportation systems for a sustainability viewpoint, we have very little. There are even less tools that approach transportation systems from a sustainable viewpoint and recognize the complexity and dynamic nature of the relationships needed to study impacts for modern-day policy actions. In contrast, the evolution of analytical methods from the 1970s has seen a tremendous improvement in computational capabilities and a variety of modeling and simulation advances that enable the creation of smarter tools. The emerging urgent need for stronger analytical tools and powerful analytical–computational methods is the key reason why we are starting to observe new methodological and practical developments in transportation planning. In this handbook examples of some of these tools are provided, with pointers to other references and book chapters in other recent handbooks. One key motivation behind the development of the tools is the assessment of the environmental impact of transportation; thus, additional emphasis is given to that aspect.

The remainder of this chapter is organized as follows. First, one of the most comprehensive definitions of "sustainable transport" is provided in its three constituent and interacting dimensions as a backdrop for subsequent sections. Within this same section is also contained a description of the basic elements in many policy instruments. Then a section follows on trends in transport systems use that are divided into urban and national categories to point out and illustrate the most critical issues (particularly air pollution and congestion, which are worse in urban environments), but also to show that a few issues are national in character. Next, three sections address the relationship between transportation and energy consumption, transportation and air pollution, and transportation and safety, with more emphasis on the United States, which appears to be the leading nation in a worldwide unsustainable path that is predominantly private car centered. These are examples of problems, solutions, and unresolved issues that motivate many contemporary policy plans and actions worldwide.

The 1970s also gave us policy tools to manage the transportation system and travel demand. Over time, many lessons were learned about the success and failure of these tools. In addition, we have seen increasing discontent with the analytical tools to assess many policies and the aging of the legacy transport model used by many metropolitan planning organizations. For this reason, the last two sections discuss transportation control measures and their more recent versions, which appear to emphasize a balanced carrot-and-stick approach to policy. The chapter closes with a modeling and simulation framework section that is currently emerging in the field.

1.2 Sustainable Transport

The motivation underlying considerations of sustainability is the realization that in our everyday life humans have been and continue to be wasteful. As a result, we are running our transportation system on credit that very soon we will not be able to renew. In the words of economists, we are reaching the limits of economic growth when we exhaust our energy resources, deplete the ozone layer, cause or do nothing to curb global warming, accelerate land degradation, and contribute to the extinction of species (for a comprehensive framework, see Bartelmus, 1994). One much celebrated example is the petroleum fuel we use that is not regenerated by nature (nonrenewable). Therefore, new and better policy initiatives are needed to provide us with safer, cleaner, and less wasteful cars, but also entire systems that promote sustainability (and possibly green engineering). A distinction should be made here between green and sustainable. Sustainable means that we are able to support a function or process by some degree of renewal that sometimes is complete and other times can be renewed with additional effort. Green means that we have eliminated the risk of harming the environment and the resource used is completely renewable.

Bernow (2000) provides one of the most succinct and interesting distinctions between conventional development and sustainable development with focus on transportation policies. Conventional development is characterized by an attempt to foster convergence in solutions, focus on the short-term impacts, and pay attention to physical–material capital. In addition, it fosters competition, consumerism, and individualism. Among its key elements we find faith that technology will provide solutions, and analyses of cause and effects are based on reductionism and assumptions of linearity. Sustainable development, on the other hand, counts on diversity for creative solutions and emphasizes the long-term payoffs. Social capital is deemed more important than physical capital, with emphasis on cooperation, quality of life, and community. Instead of technology in sustainable development, we find attention on ecology, and its analyses are characterized by ideas of emergence and complexity. Most sustainable transport initiatives depart from three basic dimensions of sustainable development, as illustrated in Munasinghe's diagram of the mutually reinforcing pillars of sustainability (World Bank, 1996, Figure 1.5, p. 28). These pillars function as dimensions; they are the economy (economic and financial aspects), the environment (environmental and ecological aspects), and social systems. Each pillar serves a specific objective to support effective policies that (1) provide for continuing improvements in material standard of living, (2) optimize attainment of overall quality of life, and (3) share the benefits from transportation equitably with all segments of the population.

1.2.1 Economic and Financial Sustainability

Recognizing the strategic role played by transport in fostering material growth, sustainable transportation means:

- Increased competition in transport services by privatizing specific aspects of the services (in essence, injecting competition)
- A move toward more efficient financing that charges users for the total costs of their movement
- Direct involvement of all affected communities in the decision process

Particularly important, as the World Bank (1996) notes, is that infrastructure accounts for 25 to 50% of the value of the total capital stock and contributes only 5% to the total cost of transport services. Other aspects of the system requiring a more careful scrutiny are the economic justification on decisions on the purchase and use of vehicle fleets and the organization of the logistic chain. In fact, supply chain management is a field that is entirely dedicated to the enhancement (some say optimization) of producing and selling goods to the consumer, and it spans the entire *chain*, from the extraction of the initial material needed to manufacture goods to the consumer purchasing an item at a store or having it delivered to his or her home or business location.

1.2.2 Environmental and Ecological Sustainability

Time after time, particularly in developed countries, provision of transport has followed a path of auto dependence. This dependence increases energy consumption and, because most autos use a specific type of internal combustion engine, also increases dependence on fossil fuels. In addition, it generates health-damaging air pollution and results in too many road fatalities and injuries. Policies in this dimension attempt to:

- Use technologies that can move our vehicles with a minimum need for oil
- Eliminate fuels that decrease air quality through undesirable emissions
- Minimize the effects of transportation systems on other aspects of the environment (water and soil)
- Develop services and transportation systems designs that are safer

As Gilbert and Nadeau (2001) state, sustainable economic development and growth may be hard to achieve without resorting to slowing economic growth. For this reason, among others, it is very important to consider the third pillar of sustainability, social sustainability.

1.2.3   Social Sustainability

At the core of this dimension is provision of access to activities and services for all (equity principle). Unfortunately, priority in transport policies has been given to higher mobility (ability to move) instead of higher accessibility (ability to reach locations and opportunities). Since mobility is "purchased," specific groups of the urban or rural population do not have adequate access to basic services, yet they are burdened with the costs (e.g., health, taxation) of mobility. Policies in this dimension are designed to provide access to opportunities, but also to empower individuals to participate in policy definition as well as transport project selection.

In addition to the issues above and policy instruments operating within each of these three dimensions, the World Bank (1996) points out that there are policy instruments that reinforce each other. Among these we find improved asset maintenance, technical efficiency of supply, safety initiatives, contract design, public administration, and charges for external effects. For example, transportation system components that are not maintained because they are unsustainable lead to environmental damages and are more likely to harm the less wealthy population segments. Synergy between two pillars does not ensure sustainability. For example, the private automobile fosters economic development and provides accessibility to many population segments but harms the environment. While this is valid for the entire world, North American, European, and Australian situations appear to be converging to similar findings and present many similarities in the proposed strategies. Similarities may also be found and could emerge in Asia, particularly in Japan, for which we have only limited information.

1.2.4   Policy Instruments

Meyer and Miller (2001) provide a comparison between traditional transportation systems planning and its more recent sustainable development orientation. They provide a comprehensive review organized along eleven dimensions in which four aspects are of particular importance: *types of issues*, *types of strategies*, *pricing of transportation services*, and *role of technology*. In the Meyer–Miller review, the label "traditional processes" is used to indicate current transportation planning orientation, while "sustainable development-oriented processes" are rapidly emerging planning activities.

1.2.4.1   Issues

In the traditional process we find congestion, mobility and accessibility, environmental impacts at large scales (regions and states), economic development, and social equity. In the sustainable orientation are concerns about global warming and green house gases (GHGs), biodiversity and economic development, community quality of life, energy consumption, and social equity.

1.2.4.2   Strategies

In the traditional process we find initiatives for system expansion and safety, efficiency improvements, traffic management, demand management, and Intelligent Transportation Systems (a conglomerate of information and telecommunications technologies aiming to resolve specific system management and user needs and problems). In the sustainable orientation we find maintenance of existing systems and their facilities, traffic calming and urban design, emphasis on the connections and relationship among modes (the key words are *multimodal* and *intermodal* aspects of travel), transportation and land use interaction and integration, demand management for reducing motorized transport, demand and increase in nonmotorized travel, and education and public involvement.

1.2.4.3   Pricing

In the traditional process we find subsidies to transportation users and the true total costs to society are not reflected in the price to travel. In the sustainable orientation we find pricing that includes environmental costs, and transportation services are priced as utility services. Litman (2001) demonstrates how most costs of vehicle use are either fixed or external, and therefore do not affect individual traveler trip decisions. Among his suggestions for sustainable transport we find a call for internalizing external costs, shifting fixed costs to variable costs, and implementing revenue-neutral tax shifts. Forkenbrock (1999) offers a similar proposition for trucks. An attempt to quantify many of these costs and to incorporate them into the collective decision-making process is also represented by the more recent Transportation Research Board report on the costs of urban sprawl (TCRP, 2002). Pricing of transportation services, particularly in urban environments (Gomez-Ibanez, 1999), provides an accessible treatment on this subject. Research in Europe appears to be very active in this area; examples can be found at htttp://www.Europa.EO.INT/Comm/Transport/Extra/Home.html

1.2.4.4   Technology

Technology in traditional processes is used to promote individual mobility, meet government-mandated performance thresholds and standards, minimize negative impacts, and improve system operations. In the sustainable orientation technology is used for travel substitution and provision of more options, pollution is minimized by benign technology, a perspective of life cycle cost assessment is embraced, and more efficient use of the transportation system is advocated.

Governments and community groups in recent years are increasingly confronted with problems within each of these four dimensions and exhibit coordinated movements toward sustainable orientations. In parallel, surveys and polls of the population indicate that solutions to these problems will need to be more creative and innovative than in the past. Recent surveys show that transportation system users want freedom of movement, well-maintained transportation systems, more options to participate in their everyday activities, and more reliable transportation systems in all modes. However, they also indicate that environmental issues, in particular air quality and energy conservation and efficiency, should also be addressed (OmniBUS survey of the Bureau of Transportation Statistics, 2001; Goulias et al., 2001a).

1.2.4.5   Bringing It All Together

Balancing such a diversity of human nature needs and wants is a very complex task at any level of government. Given the complexity of transportation systems and the desires of their customers, public agencies are increasingly considering *portfolios of policies* that can address and resolve a few of these problems by mutual strengthening. These portfolios need to be implemented based on a timeline that is dictated by the timescale of their impact and implementation requirements (e.g., some types of regulations need preparatory work and extensive public debate). For this reason, we must consider not only combinations of policy actions, but also a *dynamic path of policy implementation*. Most important, however, is an attempt to endorse and use one key strategic planning approach that can be named performance-based planning. Approaching planning in this way requires communities to decide on a future they would like to achieve (the vision) and to set *performance criteria and targets* to help them know when they have achieved the vision and to guide them in their path toward that vision (Goulias et al., 2001a). Then scenarios are created of possible paths and a continuous monitoring system is put in place to determine progress. Performance criteria and targets can be defined in two parallel and complementary ways: the scientific way, in which evidence about desirable targets emerges from more or less rigorous research (see Banister et al., 2000), and the political process (see an example mentioned in Banister et al., 2000, p. 122). This can be expanded to incorporate explicitly performance criteria from a supply chain viewpoint (Morash, 2000). In the United States the voice of the public is a mandatory and dictated task in transportation planning activities by recent federal legislation under the label "public involvement." In fact, many new long-range transportation plans at the state level contain strong public involvement elements and are performance-based, with the targets derived directly from a public involvement campaign (Goulias et al., 2001b).

Public involvement does not help us realize benefits without instruments for implementing policy actions. There are at least two groups of instruments in our toolbox: *regulation-based instruments*, such as the limits imposed on energy consumption (e.g., the corporate average fuel economy in the United States, dictating that the fuel consumption of passenger cars sold in the United States should not exceed a maximum limit), and *market-based instruments*, such as "feebates," which in essence penalize wasteful transport options through fees and provide rebates (discounts) for the use of more environmentally friendly options. Among the market-based instruments we also find soft policy methods, such as individualized marketing (i.e., customer-specific information provision combined with incentives for environmentally friendly behaviors such as walking and biking). One such success story comes from Western Australia (John and Bröeg, 2000). Other similar approaches to travel behavior change are described and compared in Bradshaw (2000). Within these portfolios of strategies, and given the domination of the automobile in private and public fleets, one promising bundle of solutions is new technology (e.g., new vehicles, new fueling systems, and creative use of information and telecommunication systems).

However, more traditional transportation plans and programs are designed for local communities and regions not possessing the means to influence technology development. In fact, most actions that are included in these programs (roadway and facility supply management, land use management, and travel demand management) assume technology as an exogenous quantity (see the U.S. examples in Meyer and Miller (2001) and the EU examples in Banister et al. (2000)). As expected, new technology development and incentives for implementation are in the realm of national planning (the federal government in the United States, individual member countries in the European Union, and the variety of EU promulgations for common policies); worldwide initiatives such as the Rio Summit, with the stabilization of CO_2 in 1992 (Bartelmus, 1994); and the United Nations Framework Convention on Climate Change (UNFCCC), famous for its December 1997 Kyoto agreement and the controversial listing of the Appendix I countries and targets. Private enterprise, however, particularly in the new and environmentally friendly fuels arena, may prove to be the most promising solution because of the worldwide car ownership trends, as we will see later in this chapter. With strengthening unification and the design of common policies for participant countries in Europe, we also see support for new policies and technologies by the European Union. Emphasis on common new and advanced technology is one of the most positive consequences of unification, and it is already starting to show the benefits of increased global competitiveness (e.g., see the wireless telephony superiority of European networks and telephone manufacturers). As expected, there are barriers, one of which is the fragmentation of jurisdictions to the implementation of new ideas, but there are also solutions outlined in plans.

Policies and plans for action are usually defined at different levels of government. These levels are defined based on ethnic, topographic, and historical reasons, but they seem to exhibit similarities across the Western world. Two key players in this, but at very different geographic scales, are the national governments and the cities. Because of the wide differences in the types of problems faced by each of these governments, common ground and coordination are needed for cooperation. To tie together national policies with local government and regional policy actions, many governments invest in "partnerships." An example, from the long-range transportation planning efforts in the United States is the emphasis on the cooperative planning process. Another example is the grassroots movement that started with help from the United States government in the alternatively fueled vehicles arena. One such case is the Clean Cities program, an 82-community initiative coordinated by the U.S. Department of Energy (http://www.ccities.doe.gov). The key common element among these examples is the cooperation and compatibility (borrowing from another transportation area, we name this property *policy interoperability*) among policies, plans, and programs defined at all levels of government. In fact, we may have examples of government policies that are contradicting each other. At the local level we often find local laws and regulations by municipalities dictating minimum parking spaces per employer or residence. This contradicts policies that aim at the reduction of driving a car to work (such as car- or vanpooling and public transportation) because parking availability (very often at no charge to the car driver) is a strong incentive for car use. At the national level policies aimed at gasoline price reduction contradict environmental policies aimed at the reduction of fossil fuel use. These contradictions, as well as our spectacular economic growth in the past 50 years, underlie the trends we review below.

1.3   An Overview of Trends

Looking at our past is not a pleasant activity because this past is not very flattering for our skills as planners. During the last 25 years and today, Europe and the United States are characterized by a marked (some would say explosive (Banister et al., 2000)) mobility increase. Most indicators, particularly for highway and air travel, have experienced an upward trend, even when within individual countries we do not see significant population increases. As most statistics from government agencies report, there are many contributing factors to this explosion, which can be summarized as follows:

- Economic activity (e.g., measured in terms of gross domestic product) has increased steadily in Europe, the United States, and a few less industrialized countries (see Banister et al., 2000; BTS, 2002; Gilbert and Nadeau, 2001).
- Household size decreased and the number of households increased, creating the need for additional housing units. Household composition, however, is very different, even among industrialized countries (e.g., Spain, Portugal and Ireland have households that are significantly larger than other EU countries with similar wealth per capita).
- Employment experienced many shifts, including increased labor force participation by women, increases in part-time employment, and a shift to the service industry.
- Consolidation of the once geographically dispersed businesses (retail stores) led to a movement to the suburbs, where land was less expensive. This also motivated wider urban sprawl and an overall increase in the distances traveled.
- Provision of high-speed facilities (e.g., autobahn, autostrada, motorway, expressway, freeway) to connect intraurban and extraurban locations further motivate the use of the private automobile.
- Differential evolution of costs favor the private automobile (e.g., gasoline costs decreased while public transportation costs remained constant or increased).
- Practically nonexistent policy to internalize external costs (e.g., costs of air pollution to public health deterioration) favors the use of specific modes, such as private car, truck, and airplane.

Table 1.1 provides an overview and a snapshot in 1999 of a few key demographic and economic indicators for selected countries.

These social and demographic trends, however, cause and are accompanied by different effects and events in different environments. Figure 1.1 provides a comparison between the United States and other countries on an indicator that is often used as benchmark: motorization. As expected, the United States has more automobiles in circulation, but when this is reported per capita, Germany has the highest number, although it has by far higher proportions of public transportation use. Figure 1.2 shows the new vehicles purchased. In the year 2000 alone the global manufacturing output was 59,765,616 vehicles, and the total sales worldwide was 57,629,253 vehicles. The different rates of purchasing new vehicles, with the United States being the consumer champion, are also an indicator of the capability of each country to control air pollution from vehicle use and other types of pollution caused when vehicles reach the end of their life (e.g., not all the material from which vehicles are made can be recycled and a portion of it is harmful to the environment). On the one hand, when new technologies to curb emissions are introduced, the United States appears to have the largest capacity in meeting targets because of the relatively

TABLE 1.1   A Selection of Demographic and Other Characteristics in a Few Countries

	Japan	France	Germany	Sweden	U.K.	Canada	Mexico	U.S.
Population (in millions)	127	59	83	9	60	31	100	276
Land area (in thousands of square kilometers)	375	546	349	411	242	9,221	1,923	9,159
Population density (persons per square kilometer)	337.7	108.7	237.1	21.6	246.3	3.4	52.2	30.1
GDP (billion dollars) (purchasing power parity)	$2,950	$1,373	$1,864	$184	$1,290	$722	$866	$9,255
GDP per capita (purchasing power parity)	$23,400	$23,300	$22,700	$20,700	$21,800	$23,300	$8,500	$33,900
Percent of surface passenger (kilometers by road transport)	58.4	92.1	92.7	93.8	95.2	99.7		99.0
Percent of surface freight (ton kilometers by road transport)	55.8	67.5	62.8	63.2	83.8	22.0		29.7

Source: From Federal Highway Administration, Highway Statistics 2000, FHWA, 2001.

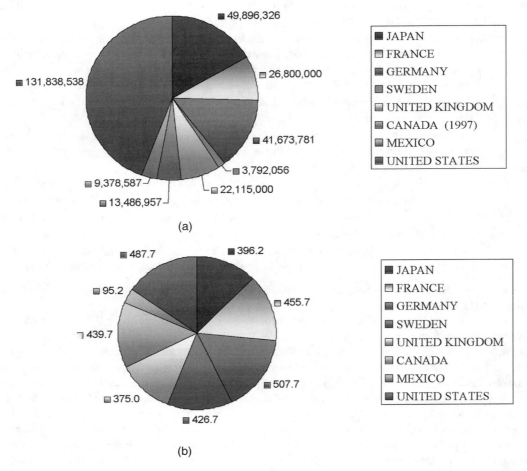

FIGURE 1.1   Car ownership in select countries. (a) Number of automobiles; (b) automobiles per 1000 persons. (From Federal Highway Administration, Highway Statistics 2000, FHWA, 2001.)

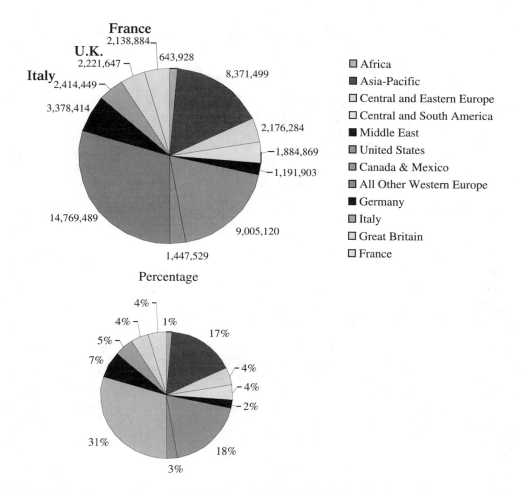

FIGURE 1.2 Car sales around the world in 2000. (From Global Market Data Book, Crain Communications, London, 2001.)

young age of its vehicle fleet, due to the shorter time consumers keep the same vehicle in the household. However, on the other hand, this shorter life of vehicles will contribute the most in pollution at the end of their lives unless higher recycling rates are introduced in the newly manufactured vehicles (exporting used vehicles to less developed countries is not a global solution, but merely a shift of the pollution issues elsewhere). In addition, as one would expect, high car ownership and high car sales lead to equally high levels of car utilization. Table 1.2 provides an overview for a small number of countries for which comparable data exist (compiled by the U.S. Department of Transportation (DOT) every year). In the remaining portion of this section we distinguish between the more urbanized environments and the countries in their entirety to provide additional details about these trends. Rural environments are equally important and face issues that have not received extensive attention in the literature. For convenience, specific focus on them is excluded from the presentation here.

1.3.1 Urban Environments

In a comparative study of major cities around the world, Newman and Kenworthy (1999) and Kenworthy and Laube (1999) examine the patterns of automobile dependence and use, its infrastructure, and land use patterns. These cities exhibit three distinct groups of "typical" developments.

TABLE 1.2   Key Transport Indicators in Selected Countries

Year Reported	Japan (1998)	France (1999)	Germany (1998)	Sweden (1998)	U.K. (1999)	U.S. (1999)
Total vehicle kilometers of travel Automobiles (in millions)	482,551	402,000	528,000	56,992	380,100	2,529,267
Total vehicle kilometers of travel Automobiles (per capita)	3,813	6,776	6,377	6,423	6,387	9,179
Total vehicle kilometers of travel Motorcycles (in millions)	—	4,000	15,400	600	4,600	17,061
Total vehicle kilometers of travel Buses (in millions)	6,520	2,300	3,700	1,172	5,000	11,925
Total vehicle kilometers of travel Trucks (in millions)	256,983	112,000	57,800	8,642	77,300	1,773,035
Average vehicle kilometers of travel per automobile	9,671	14,629	12,475	15,029	17,187	19,099
Average vehicle kilometers of travel per motorcycle	—	1,723	3,377	2,378	6,725	4,109
Average vehicle kilometers of travel per bus	27,429	28,750	43,690	77,616	62,500	16,363
Average vehicle kilometers of travel per truck	12,347	19,512	13,192	25,544	25,025	16,363

Source: From Federal Highway Administration, Highway Statistics 2000, FHWA, 2001.

The first is a group of cities located mainly in the United States and Australia that are very dependent on automobile use. They do not show a relative gain in the percentage of their gross regional product (GRP) spent on commuting; their average trip times to work are similar; transit cost recovery is very low; and their road expenditures are higher than the other two groups. However, the resulting higher costs due to automobile dependence are not charged directly to the automobile users; instead they are spread out to the entire population. The higher levels of per capita car use means higher energy use, higher emissions production, more accidents, and more costly transportation fatalities.

The second group comprises cities with good public transportation systems, mostly European and relatively more affluent Asian cities (e.g., Singapore, Tokyo, and Hong Kong). These cities have the least costs associated with their transportation systems. In this category we also find Toronto, New York, and Sydney.

The third group contains cities in rapidly developing Asian countries that have considerably more costly transportation systems than their relatively more wealthy neighboring countries, probably due to their orientation to car travel. These cities include Bangkok, Seoul, and Beijing, among others.

A key finding in these city reviews is the complex relationship between regional wealth (measured by GRP) and car use and dependency. U.S. and Australian cities use the car the most. Perth in Western Australia had 7023 km/capita and a GRP of U.S. $17,697 in 1990; Phoenix had 11,608 km/capita and a GRP of 20,555; while European cities had an average car use of 4519 km/capita and an average GRP of 31,721 in 1990. The developing Asian cities have low kilometers per capita, but also low GRP, which in essence shows a clearly nonsustainable path of growth and development in which the costs by far outweigh productivity and transportation cost recovery.

Transport-related deaths in 1990 (in deaths per 100,000 persons) as reported in Newman and Kenworthy (1999) were 14.6 for U.S. cities, 12.0 for Australian cities, 6.5 for Toronto, 8.8 for European cities, 6.6 for wealthy Asian cities, and 13.7 for developing Asian cities. Exceptions for developing Asian cities are the Chinese cities, with just 4.8 deaths per 100,000 persons. Their large modal share in nonmotorized transport is one contributing factor to this positive indicator. However, as recent studies and predictions show, this will change rapidly in the future (He and Wang, 2001).

In terms of CO_2 transport (public and private) emissions per capita in 1990 (kilograms per person), U.S. cities produce 4536; Australian cities, 2789; Toronto, 2434; European cities, 1888; wealthy Asian cities, 1158; and developing Asian cities, 837. Public share of emissions is usually a minuscule

TABLE 1.3 International Comparison among Urban Environments

City Type	U.S. Cities	Australian Cities	Toronto	European Cities	Wealthy Asian Cities	Developing Asian Cities
Smog emissions (kilogram per person)	252	233	216	101	30	89
Smog emissions (grams per passenger kilometer)	16	20	23	12	5	21
Urban density (persons per hectare)	14	12	28	50	153	166
Percent workers using public transport	9	14	20	39	60	38

Source: From Newman, P. and Kenworthy, J., *Transp. Res. Rec.*, 1670, 1999.

proportion of the total. When comparing these same cities in terms of smog emissions (NO_x, SO_2, CO, and other volatile hydrocarbons), Newman and Kenworthy (1999) report the following data in Table 1.3. Toronto's higher emission rates are apparently due to less stringent emissions standards in Canada. This is also the case for Australia, which has an older national vehicle fleet than the United States. The United States contributes lower emissions on a per kilometer traveled basis due to its corporate average fuel economy (CAFÉ) and the mandated tailpipe emission standards, currently resulting in major benefits (discussed later), and its much younger fleet of automobiles. These emissions, however, are not lower than those of European cities that are populated with much less vehicles and use public transport at much higher modal split proportions.

On one hand, these comparisons show a disturbing trend for U.S. and Australian cities. These cities continue their pattern of urban sprawl and waste of land and other resources. This example is followed by many developing cities, which exhibit similar patterns. Even worse, however, they are likely to possess a lower capacity to provide solutions, due to their lower economic potential. On the other hand, European cities and Toronto appear to have found at least partial solutions. This may be due to historical and cultural reasons (many European cities have centers with urban forms designed for pedestrian circulation), necessity, and public policy.

1.3.2 National Levels

Pucher and Lefevre (1996) have created a comparison of urban transportation-related trends around the world. Trends provided include absolute traffic by mode of transportation, transportation modal share, measures of congestion, safety statistics, air quality descriptions, and an overview of settlement patterns. The authors also document trends in transportation investment, subsidies, taxation, control strategies of individual behavior, and control strategies of corporate behavior. Among their key findings is that as income increases, people choose to live in lower density settlements (in the suburbs) and consume greater quantities of automobile transportation, predominantly using fossil fuels.

Increases in motorization are also strongly correlated with urban sprawl (but also other types of decentralization), leading to problems of air pollution, congestion, and safety statistics that do not seem to be improving worldwide. These findings are confirmed by Schipper and Marie-Lilliu (1999) with the addition that distance traveled per capita is increasing faster than distance traveled per vehicle because of increases in car ownership. In addition, geographic density of population is a major factor, explaining the amount of travel per person because of choice, necessity, or both. For example, people in the United States, Canada, and Australia with lower densities and higher incomes travel more and use the private automobile the most. Table 1.2 provides partial evidence to support these claims. Schipper et al. (1999) demonstrate similar trends in Asian countries. Given the worldwide projections of an increasingly wealthier world (OECD, 2001), it is reasonable to expect that many of these figures will continue to worsen for some time into the future. One snapshot of this potentially worsening situation is also the predominant use of all types of vehicles in urban environments, as Figure 1.3 shows. As expected, however,

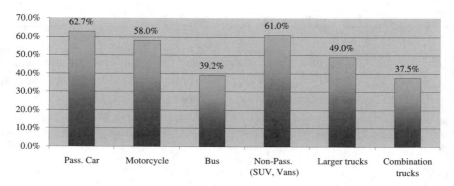

FIGURE 1.3 Distribution in the United States of vehicle use by type. Percent of urban travel over total kilometers (United States, 1999). (From U.S. Department of Transportation, Highway Statistics, Federal Highway Administration, Washington, D.C., 2000.)

the combination trucks (tractor and a trailer) and buses serving longer-distance travel among cities appear to be used mostly outside urban environments.

1.4 Transportation and Energy

Today the demand for oil is 74 million barrels per day and appears to be growing consistently worldwide. Demand growth in developing countries is higher than in the more developed portion of the world, and it is expected to grow as it did in the past for the United States (Schafer and Victor, 2000). The growth is not due exclusively to the growth of gasoline consumption, which appears to be stabilizing at 1990 levels. Instead, aviation and diesel fuel consumption appears to be increasing much more rapidly, as the International Energy Agency claims (see http://www.iea.org/, accessed April 2002).

In spite of this growth, the paradigm of exhaustible resources for energy shortages and transportation oil consumption (as theorized after the 1970s oil crisis) does not seem to be the popular theory in resource economics today. Instead, a more complex and dynamic theory is emerging that examines the roles of technology and social institutions in exploring, researching, defining, testing, implementing, expanding, exploiting, and abandoning energy resources and associated technologies. In fact, many analysts and energy agencies have estimates showing that liquid fuels (conventional and unconventional) may last up to 1500 years.

These resources, however, tend to be concentrated in specific geographic regions where organized cartels control prices that can also create artificial scarcity, contributing to and generating major economic shocks and disruptions. Moreover, energy consumption statistics and forecasts as they relate to the United States show on average over recent years that:

- U.S. oil net imports as a percent of consumption was 49.6% in 1999 (Davis, 2000). This is almost twice its value in 1983 (28.1%) and it is expected to increase further unless something drastic is done.
- OPEC's share of world production is expected to grow to more than 50% by 2020 (CEC, 1999).

In a global market there are no countries that will not feel the effects of supply disruptions from oil-exporting nations. Worldwide transportation as a sector is particularly exposed to these effects, and the United States is even more vulnerable. Our transportation system is almost completely dependent on oil. The energy consumption distribution by source for transportation in the United States for 1999 was 97.4% petroleum and 2.6% natural gas, with all other types of energy being very small. At the same time, the net imports as a percentage of U.S. oil consumption was 50% in 1999. Particularly controversial are the military expenditure estimates for defending oil supplies from the Middle and Near East. Estimates of this are representing an average of $32 billion per year, with a standard deviation of the estimate at approximately $22 million. DeLucchi (1997) provides a comprehensive discussion about automobile costs and raises doubts about the defense-related costs.

In addition to United States, transportation's energy use is growing everywhere. In 1989 world energy consumption was estimated to be 344.83 quadrillion Btu, 39.0% of which was from petroleum. In 1998 total energy consumption was 377.72 quadrillion Btu (a 9.5% increase from 1989), of which 39.6% was from petroleum (Davis, 2000).

In the United States, with an average percent increase in annual rate of energy consumption of 1.3% in the past 10 years, transportation increases are slightly higher than other sectors, at an approximately 1.4% increase per year. Transportation had 16.07% of the total national energy consumption in 1970, 22.57% in 1989, and 28.00% in 1999 (Davis, 2000). Within transportation, highway travel consumes 77.2% of the total energy, with automobiles and trucks the first and second top consumers, respectively. Trucks and automobiles consume almost 97% of the total gasoline used in the United States, and trucks also consume approximately 66% of the diesel fuel. These facts make transportation far more vulnerable to oil price shocks than any other sector of the economy.

One way to disengage the United States and other oil-consuming countries from this dependency is to develop new fuels and associated strategies. However, the benefits from these technology-based strategies will be realized only after they have penetrated the market in substantial quantities to function as stabilizers. As is evident from the air pollution and climate change arguments, transportation energy issues and strategies are defined as a group of policies and actions that address all three aspects simultaneously because they are strictly and inherently related.

1.5 Transportation and the Environment

Transportation impacts the environment not only during facility construction, when the disfigurement of land is most evident, but also after facilities are opened to operation. In this section the atmosphere (air pollution) is emphasized. The impacts of transportation on soil and water contamination and related studies are of equally paramount importance, but are neglected in this presentation (for a more recent study and related references, see Nelson et al., 2001).

1.5.1 Climate Change

Current thinking about transportation and the world's climate change (as expressed by the Intergovernmental Panel on Climate Change, a group of leading scientists) indicates that CO_2 is the most important contributor to climate change and that observations suggest human influence on global climate. In December 1997 in Kyoto, Japan, participants in the United Nations Framework Convention on Climate Change set specific targets for reducing GHG emissions. Since the United States is the world's largest emitter of anthropogenic carbon dioxide (approximately one fourth of the world's total) and its transportation sector produces more than any other country, under the Kyoto Protocol the United States is expected to reduce its emissions by 7% below that of 1990 levels in the 2008–2012 period. Based on data summarized in the annual report by the Oak Ridge National Laboratory (Davis, 2000), transportation contribution to carbon dioxide emissions from fossil fuel consumption has been 30.5% in 1984, 32.2% in 1990, and 32.6% in 1998. Motor gasoline consumption accounts for 60.8% of this 1998 inventory of CO_2.

Given these trends and current and projected use of automobiles, most analysts believe reductions within the Kyoto time frame will not happen because of the amount of time required for technological turnover in the real world, but it may be achieved in the next 30 years (DeCicco and Mark, 1998; Greene and Plotkin, 1997). In addition, when considering CO_2, CH_4, and N_2O in their totality, based on United Nations reports for the countries in what is known as Appendix I, in the period 1990–1997 there have been reductions in man-made emissions in only a few countries by a few percentage points (except for Germany, the third largest overall contributor, which reported a 14% reduction). The United States, as the largest contributor, reports a 10% increase, and Japan, the fourth largest contributor, reports a 9% increase. The Russian Federation, the second largest contributor, provided no data (www.unfccc.de).

Within the United States, California appears to be the leader in environmental policy, technology implementation, and development of strategies, particularly as they relate to GHGs. California's Energy Commission is the designated state agency for analyzing energy-related issues and setting directions; the following is a summary of the strategies of the California Energy Commission (CEC) to decrease CO_2 and other GHG emissions from transportation uses (CEC, 1998):

- Continued development and promotion of clean, alternatively fueled vehicles (AFVs)
- Continued alternative fuel vehicle infrastructure development
- Production and use of biomass to produce transportation fuels
- Pricing measures to reduce vehicle miles traveled (VMT)
- Higher fuel economy standards
- Alternative fuel vehicle incentives, including fuel subsidies and vehicle purchase incentives
- VMT taxes and congestion fees to reduce VMT
- Land use and transportation strategies to reduce congestion, improve air quality, and reduce CO_2 emissions

Schafer (2000), based on a model that forecasts motorization and travel in 11 regions around the world (see also Schafer and Victor, 2000), concludes that GHGs and particularly CO_2 can be inhibited in the medium term (2010 to 2020) and controlled in the long term (by 2050) only with very drastic measures that necessitate technological solutions. The most important findings are:

- Replacement of petroleum fuels by natural gas will have a major positive effect in decreasing CO_2.
- Fuel efficiency improvements are necessary but not sufficient.
- Longer-term reductions can be achieved only by using transportation fuels that are not carbon based.

CO_2 has received most of the attention in publications and targeted GHGs. However, more than 10 years ago, DeLucchi (1991) (see also his bibliography, DeLucchi, 1996) pointed out that emissions of GHGs should also include CH_4, N_2O, and criteria pollutants such as CO, NO_x, and nonmethane organic compounds. It is also very important to examine the carbon content of fuels (grams per British thermal unit) and the efficiency of fuel use (British thermal unit per mile or British thermal unit per kilowatt).

1.5.2 Air Quality

Transportation is also responsible for a large portion of the air pollution emanating from fuel combustion, causing morbidity, mortality, and ecosystem damage. The legislative framework for air quality and transportation in the United States is supported by the Clean Air Act (CAA) of 1970, its amendments in 1990 (CAAA), the Intermodal Surface Transportation Efficiency Act (ISTEA), the more recent Transportation Equity Act for the 21st Century (TEA-21), and Titles 23, 40, and 49 of the U.S. Code and Code of Federal Regulations. CAAA and ISTEA introduced a different philosophy in dealing with air pollution from transportation sources. Regulations from these require an integration of transportation plans and investments with the goal of improving air quality. For example, they mandated the application of transportation control measures (TCMs) — such as strategies and actions to decrease the number of vehicles driven alone by motivating people to carpool — with clear implementation schedules and goals. In addition, they introduced the process of conformity in plan development — in essence a check on compatibility among transportation plans and programs with air quality attainment of specific standards (see Chapter 13 in this handbook). A similar process appears to be evolving in the U.K., as described in Beattie et al. (2001) and the European Framework Directives of 1996, such as the Auto Oil I Programme (for a discussion, see Fenger, 1999; Banister et al., 2000; Commission of the European Communities, 1996, 2000).

The CAAA also focused on technological improvements by mandating progressively tighter vehicle emission standards, cleaner fuels, and vehicle inspection and maintenance programs. Tier I emissions standards have been required for automobiles from model year 1994, and new bus standards were also introduced. In 1997 NO_x reduction standards were also introduced. Air pollution is particularly acute in urban environments, as discussed previously, and motivates many of the stipulations in CAAA, which redefined the National Ambient Air Quality Standards (NAAQS), as maximum concentrations of pollutants that cannot be exceeded. In addition, it also gives authority to the U.S. Environmental Protection Agency (EPA) to impose highway fund sanctions when noncompliance is found (Savonis, 2000; Niemeier's chapter in this handbook). California's air resources board also defines standards that are more localized than the federal standards (http://www.arb.ca.gov/aqs/).

NAAQS specify maximum acceptable concentrations beyond which unhealthy conditions exist and also define the criteria pollutants to be regulated in order to meet the NAAQS on carbon monoxide (CO), nitrogen oxides (NO_x), sulfur dioxide (SO_2), particulate matter with aerodynamic size smaller than 2.5 µm (PM2.5), particulate matter with aerodynamic size smaller than 10 µm (PM10), ozone (O_3), and airborne lead (Pb). Details on the standards, their history, and associated legislation can be found in Wark et al. (1998). CAAA also defined the majority of air pollution controls for transportation.

Transportation (mostly motor vehicles) contributes large quantities to four of the criteria pollutants. Based on data from Davis (2000), in 1998 transportation produced 78% of the CO, of which 56.3% was contributed by highway vehicles. For the same year, transportation contributed 53.4% to the total NO_x produced, more than half of which (31.8% of total NO_x production) was due to highway vehicles. A little less than half (43.5%) of the volatile organic compound (VOC) emissions are from transportation, with 29.7% of the total produced by highway vehicles. VOC with NO_x combine in the atmosphere and, with the help of sunlight, form ground-level ozone, which is the primary component of smog. Lead received 13.1% of its total production from transportation in 1998, in spite of its spectacular decline since the introduction of unleaded gasoline worldwide. Transportation's share of emissions is smaller for PM10 (2.1%), PM2.5 (7.2%), SO_2 (7.2%), and NH_3 (5.2%). Figure 1.4 provides an overview (BTS, 2001).

The CAAA and other transportation legislative initiatives, together with technological advances by the auto manufacturers, have worked very well. Pickrell (1999) (see also Table 1.4) and Davis (2000) show a remarkable downward trend in emissions by the United States in the past 30 years. David Greene (1999), a recognized national expert in energy and transportation, attributes this spectacular gain to advanced

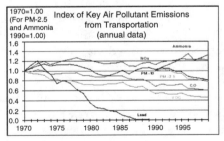

Thousands of Short Tons of Transportation Air Emissions	1997	1998
Carbon monoxide (CO)	55,437	54,170
Oxides of nitrogen (NOx)	10,077	9,975
Volatile organic compounds (VOC)	6,513	6,510
Particulate matter < 10 microns (PM-10)	420	405
Particulate matter < 2.5 microns (PM-2.5)	336	323
Ammonia	250	260
Lead	0.5	0.5

SOURCE: U.S. Environmental Protection Agency, Office of Air Quality Planning and Standards (OAQPS). 1998a. *National Air Pollutant Emission Trends Update: 1970-1997* (Research Triangle Park, NC: December 1998).

Despite rapid growth in vehicle use over the past two decades, emissions of carbon monoxide (CO) and volatile organic compound (VOC) have declined, and lead emissions have been almost eliminated, leading to improved air quality. There have been reductions in particulate emissions (PM) at the 10 micron classification. Only emissions of nitrogen oxides (NOx) remain above 1970 levels. (Ammonia and PM-2.5 were added to the list of regulated pollutants recently.)

With the exception of lead, onroad vehicles contribute the largest share of air pollutants among all models.

FIGURE 1.4 Evolution of emissions in the United States from transportation. (From Transportation Indicators, Bureau of Transportation Statistics, BTS, Washington, D.C., 2000.)

TABLE 1.4 Twenty-Five-Year Evolution of Annual Emissions in Percent Change from 1970 to 1995

	Airborne Lead (Pb)	CO	VOC	NO$_x$
Passenger cars and light trucks	−99%	−35%	−55%	−4%
Trucks	Not available	−14%	Not available	+21%

Source: From Pickrell, D., *Transp. Res. A*, 33, 527–547, 1999.

pollution control technology such as three-way vehicle catalytic converters, multipoint fuel injection, and electronically controlled combustion.

This was also reflected in pollutant concentrations monitored by the U.S. EPA, which shows a decline of 94% in recorded lead concentration and a decline of 54% in CO of the U.S. urban areas between 1975 and 1996. In addition, violations of the NAAQS for lead, NO$_2$, and CO were very few (Pickrell (1999) notes "virtually eliminated"), and violations for ozone decreased by 90%. Encouraged by these results and threatened by the loss of benefits achieved from the continuous increase in travel, the EPA has moved to a second wave of more stringent requirements.

Tailpipe exhaust emissions standards have become increasingly more stringent for each vehicle manufactured model year. These are different for each category of vehicle, based on gross vehicle weight. California went a step further and reduced its emission standards by dramatically lowering HC, CO, and NO$_x$ for its certification of new vehicles sold in California (Davis, 2000). The California Low Emission Vehicle (LEV) program is a set of requirements for new vehicles sales that dictates each major vehicle manufacturer to meet a set of emission standards. Starting from model year 1994, LEV mandates that a minimum percentage of new car sales need to be in a given category of these lower emission standards. This was first introduced for passenger vehicles and then extended to other vehicles. Other states have also considered the LEV program (for an example assessing the environmental and economic effects of these policies in Pennsylvania, see Goulias et al. (1993)).

Recent U.S. policy developments are also pressing pollution control for light-duty gasoline vehicles toward new levels of stringency. California's LEV II rules and federal tier 2 standards are levels set for light-vehicle emissions by regulations that are phased in the market starting in 2004. Key aspects include a greater emphasis on NO$_x$ control (0.07 g/mi NO$_x$ for light cars and trucks < 6000 lb in gross vehicle weight (GVW)) and inclusion of all vehicles in this standard. This is a reduction by factors of five to ten from current levels. There are also rules about extended durability requirements and harmonization of light trucks with passenger car standards. In addition, the rules include reductions for the national average gasoline sulfur levels. An example of standards in the newly defined LEV II program is provided at the California air resources board (ARB) website (http://www.arb.ca.gov/msprog/levprog/levii/factsht.htm), which contains a general overview of the program and the rationale behind these newer and stricter rules.

In its report to Congress, the EPA stated that total vehicle miles traveled (in the United States) grew from 1 trillion in 1970 to 2.5 trillion in 1997, and is expected to grow at the rate of 2 to 3% each year (in fact, the Bureau of Transportation Statistics (BTS) reports 2.56 trillion in 1997 and 2.63 trillion in 1998 for roadway travel alone). In addition, almost half of the passenger vehicles sold in 1998 were higher-polluting light-duty trucks, such as sport utility vehicles. With these as key motivations and the fear that benefits from tier I standards would be lost, the U.S. government launched a new program. At the national level 23 automobile manufacturers, most states (except for New York, Massachusetts, Vermont, and Maine), and the District of Columbia agreed to participate in the National Low Emissions Vehicle (NLEV) program, which parallels the earlier California program (http://www.epa.gov/oms/).

Alternatively fueled vehicles are one possible solution to the air quality and energy problem. In this arena, very important for assessing proposed solutions is the California mandate for zero emitting vehicles, with the 10% requirement for 2003 and its gradual increase to 16% by 2018 being the most notable targets. Additional information may be found at http://www.zevinfo.com/electric/zevchanges.pdf. In addition, a variety of incentive programs exist, as shown at http://www.arb.ca.gov/msprog/zevprog/zevprog.htm. NLEV allows market flexibility for program participants through:

- A market-based credit system for both the auto and oil industries to reward those who lead the way in reducing pollution sooner than required
- An averaging program created to meet both car emission and gasoline sulfur standards
- Strong interim standards for auto manufacturers and refiners while they work toward full compliance of the new standard
- Extra time for small refiners to meet the sulfur standards

In the United States the estimated number of alternatively fueled vehicles grew from about 250,000 in 1992 to about 430,000 in 2000. In addition, consumption of alternative transportation fuels grew from about 230 million gasoline-equivalent gallons in 1992 to about 370 million gasoline-equivalent gallons in 2000 (Joyce, 2001). Government policy has impacted these air pollution trends. Among the policies examined are investment in public transportation, investment in roads, traffic calming, traffic demand management, and traffic supply management, as well as government regulations of auto manufacturing companies that resulted in automobiles that are significantly more fuel efficient, safer, and cleaner than models of several years ago. However, the long-term trend of lower emissions will be very difficult to maintain given the expected increases in automobile travel and its continued and sustained growth. This is a particularly disturbing indicator because as less motorized countries develop their motorization (see, for example, the Chinese scenarios in He and Wang (2001), in which China reaches today's U.S. oil consumption and CO_2 emission in just 30 years), technology may be the only solution.

1.5.3 The Energy Policy Act

Two national goals — to enhance U.S. energy security and to improve environmental quality — are the motivations behind the passage of the Energy Policy Act (EPAct), which encompasses many of the aspects related to:

- Energy supply and demand
- Energy efficiency
- Alternative fuels
- Renewable energy

Several parts, or titles, of the act were designed to encourage use of alternative fuels, not derived from petroleum, which could help reduce dependence on imported oil for transportation. These include methanol, ethanol, and other alcohols, blends of 85% or more of alcohol with gasoline, domestic natural gas and liquid fuels, liquefied petroleum gas (propane), coal-derived liquid fuels, and hydrogen and electricity. For alternative fuels the act includes voluntary and regulatory methods to create a fundamental change and to provide a viable alternative fuel market. One of these initiatives is the Clean Cities program. Figure 1.5 shows a map of the Clean Cities program. The EPAct defines several programs, three of which are of particular interest to transportation planning practitioners. Within the EPAct, one objective is to build an inventory of alternatively fueled vehicles in centrally fueled fleets in metropolitan areas. The fleets targeted are state governments and alternative fuel providers, the federal government, and local governments and private entities.

The EPAct requires state government and alternative fuel provider fleets (fleets with more than 50 light-duty vehicles located in one of 125 designated metropolitan areas) to purchase AFVs as a percentage of their annual light-duty vehicle acquisitions. In addition, fuel provider fleets are required to use alternative fuels whenever feasible (www.ott.doe.gov/epact/state_fleets.html). This includes buses and law enforcement agencies.

Federal fleets are required to purchase 75% of their new light-duty vehicles from the available AFVs. Additional rules effective April 2001 also establish a petroleum reduction goal of 20% by 2005 for federal fleets, compared to their fiscal year (FY) 1999 usage. Agencies are also required to submit strategies for how they will achieve the targets set for them.

FIGURE 1.5 The Clean Cities program. (From http://www.ccities.doe.gov. Accessed January 2000.)

The Department of Energy (DOE) also has the authority to impose AFV acquisition requirements on private and local government fleets. This portion of the EPAct is still under consideration. New fuels may also be added and qualify under the EPAct through a special petition program. One such type of fuel known as P-series and biodiesel have been included in the list of approved alternative fuels. The DOE is estimating that because of its regulatory programs, an annual demand for approximately 30,000 AFVs is created. Estimates from the Energy Information Administration (EIA) show almost 90% of the current 430,000 AFVs to be liquid propane gas (LPG) or natural gas fueled. However, the share of AFVs designed for LPG and methanol is declining, and the share of those designed for natural gas, ethanol, and electricity are increasing (Joyce, 2001).

In the Clean Cities program, the government's objectives, such as energy security, fuel diversity, air quality, and economic opportunity, are combined with commercial objectives and voluntary commitments from fuel suppliers, vehicle suppliers, and fleet owners to form locally based partnerships (U.S. DOE, 2001, www.ccities.doe.gov). The program, organized by the U.S. DOE and mandated in EPAct 1992, is the mechanism used to seek voluntary commitments from suppliers, providers, and fleet purchasers. This is a niche market because it is protected by federal legislation, the U.S. DOE is motivated to make it a success, and it has a national goal. In 1999 it already had 77 participating communities spanning the entire United States. Today the program has 82 coalitions located in almost every state and 11 large corridors.

1.5.3.1 Fleets

From the earlier inception of alternative fuels, fleets appeared to be a good target (Webb et al., 1989; Madder and Bevilacqua, 1989). Some of the reasons are:

High-mileage: Since fleets consume larger quantities of fuel, fleet managers may achieve cost savings when alternative fuels are less expensive.

Centrally located facilities: Many fleets have their own refueling and service facilities and can reach economies of scale easily. In addition, many fleets tend to "garage" their vehicles in one location during the night, which makes electric vehicle recharging easier.

Uncommon uses: There are fleets with a high use of idling stages, such as airports and other major transportation terminals.

Predictable routes and scheduling that can be specialized: Fleet managers can assign vehicles based on their performance characteristics to routes, tasks, and schedules.

Life cycle advantage: Fleet managers are more likely to acquire vehicles with long-term benefits because they tend to optimize based on the entire life cycle of a vehicle's use.

Some typical fleets identified by U.S. DOE for its Clean Cities program are taxis, delivery fleets, shuttle service and transit bus fleets, airport ground fleets, school bus fleets, and national park vehicles.

The following is an excerpt from http://www.ccities.doe.gov/success/ev_rental.shtml:

EV Rental Cars — CNG, Electric

Budget EV Rental Cars offers electric vehicles for rent at Los Angeles International Airport, and recently expanded its selection to include the dedicated compressed natural gas (CNG) Honda Civic GX. The electric vehicles available are the Honda EV Plus, Ford Ranger, GM EV1, Toyota RAV-4, Daimler-Chrysler EPIC minivan, and the Nissan Altra. Electric recharging stations are available to electric-vehicle renters through a partnership with the LA Department of Water and Power. Natural gas vehicle renters receive fueling cards from Pickens Fuel and Southern California Gas Company that give them access to the many natural gas refueling stations in the LA area. Budget EV Rental Cars also is expanding its service locations. An EV Rental Center is now open at the Sacramento airport. The rental fleet includes 20 electric vehicles that are being offered for rental at rates as low as $44 per day. Charging is free at the 100 electric charging stations in the Sacramento area. For more information, call 1-877-EV RENTAL, or check out the EV Rental Web site. For a fun story about a consumer's first experience with renting an electric vehicle, visit My First Day With an EV1 From EV Rental Cars.

1.6 Transport and Safety

A success story in transportation is safety for passengers. Some countries have improved road geometry and imposed mandates to auto manufacturers for safer vehicular design. They have also improved compliance of the population with seatbelt use and increased investment in medical technology (Noland, 2001). This has led to lower fatality and injury rates for motorists. Better sidewalks, separate lanes for bicycles, traffic calming, and other pedestrian protective technologies were also met with considerable success. Table 1.3 provides a brief overview and comparison with other countries (Figure 1.6).

Data from 1975 to 1998 (Davis, 2000) on fatalities by different vehicle sizes and modes show that pedestrian, bicyclist, and motorcycle safety is improving. However, and inevitably by current technologies, pedestrian safety is not improving as fast as desired to reach a zero incidence. As motorization increases, because of interactions among different modes, these gains may not be sustained. To address this, recent legislation, such as TEA-21, has motivated many states to set performance targets in decreasing fatalities. This also creates unique opportunities for auto manufacturers willing to partner with government to improve the safety record of the United States. In addition, fatalities in smaller vehicles (subcompact, compact, and intermediate passenger cars and light trucks) are increasingly following car ownership trends. Some key findings in the safety arena worth noting here (Greene, 1999) are:

- A National Highway Traffic Safety Administration (NHTSA) study shows that a uniform decrease in the weight of all cars in the United States will not improve the fatality record. Instead, a decrease in the light truck size and weight may yield major benefits.
- The fleet distribution of vehicle weight (relative frequency of each vehicle class) is a major contributing factor. Changes in vehicle size and weight that attempt to bring all vehicles on the road closer to each other in terms of mass and weight are more likely to yield safety benefits.

The conclusion from these two points is that if subcompact and compact vehicles are substituted by larger and heavier vehicles and, at the same time, full-size and large trucks are replaced by smaller ones, we will see a decrease in fatalities. These gains will be more pronounced in regions where vehicles tend to exhibit some

FATALITIES AND FATALITY RATES FOR SELECTED COUNTRIES - 1998

OCTOBER 2000		ASIA	EUROPE				AMERICA		TABLE IN-6
		JAPAN	FRANCE	GERMANY	SWEDEN	UNITED KINGDOM	CANADA	MEXICO	UNITED STATES
Fatalities		10,805	8,437	7,776	529	(1997) 3,599	(1997) 3,064	N/A	41,471
Fatality Rate per 100 Million Vehicle Kilometers		1.45	1.72	1.30	0.78	0.88	N/A	N/A	0.98

Source: Prepared from "World Road Statistics," International Road Federation, 2000, "and the 1998 Highway Statistics," FHWA.

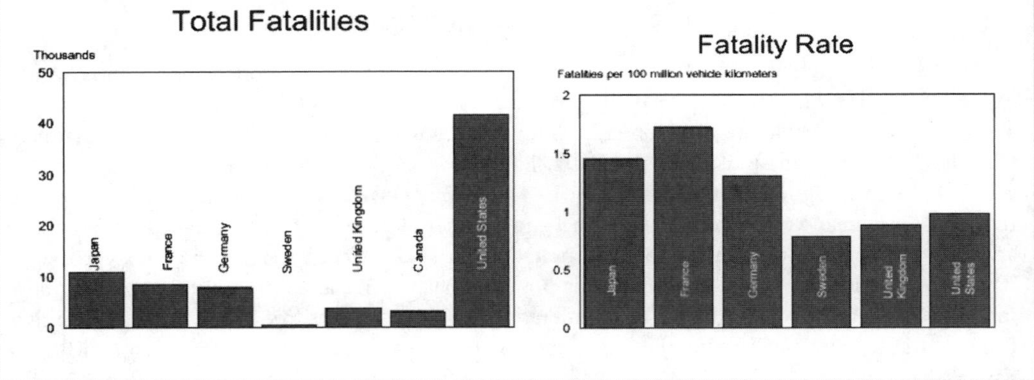

FIGURE 1.6 Safety indicators comparison. (From U.S. Department of Transportation, Highway Statistics 1999, Federal Highway Administration, Washington, D.C., 2000.

sort of weight and size uniformity. No evidence exists on the separate effects of size and weight (Greene, 1999). This increases the uncertainty about safety, energy efficiency, and the market. At the same time, the use of lighter material will also lower vehicle weight, possibly increasing the occupants' risk. However, if technologies that improve other safety features are added, e.g., elimination of welding, electronics and information technology, and vehicle control enhancements, new vehicles may be safer. Figure 1.7 illustrates the safety issue in U.S. fatalities for smaller private cars and trucks. Safety indicators are a sample of the interconnectedness in transportation decision making. Energy efficiency and air pollution controls have influenced vehicle size and weight and, most important, vehicle size and weight diversity — first by making the U.S. fleet more homogeneous, and then less homogeneous as the auto manufacturers discovered and used legislative loopholes. This may have delayed safety improvements, but it also provides a strong momentum to work toward larger vehicles with less polluting engines. The safety example is an illustration of complex interactions among policies that are enacted at the federal level; they are influenced by manufacturers and auto consumers, and they affect local communities as well as the manufacturers and the auto consumers.

1.7 Transportation Control Measures

Federal legislation targeting vehicle fuel economy and exhaust emissions is one of the emphases in curbing the negative effects of the automobile. The other set of policy instruments, as mentioned above and reviewed extensively by Meyer and Miller (2001), are the transportation control measures.

Ten years ago professionals actively involved with transportation planning practice reached a widespread consensus about the inability of *traditional transportation solutions* to solve urban and suburban congestion as well as the concomitant environmental problems. Humphrey's (1990) statement is a testimony to this realization:

Thus, it is timely to think about congestion from a broader perspective; and, further, strategies must be developed that include a combination of innovative organizational approaches and new funding mechanisms as well as a wide range of transportation and land use actions.

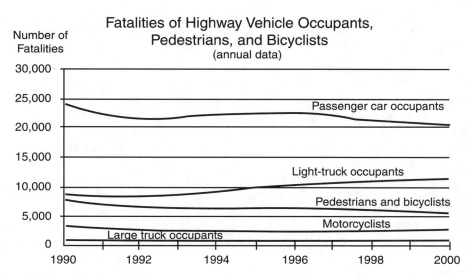

FIGURE 1.7 The remarkable and chilling evolution of safety indicators in the United States. (From Transportation Indicators, Bureau of Transportation Statistics, Washington, D.C., 2000.)

At about the same time the Clean Air Act Amendments of 1990 defined a series of actions and policy initiatives that, in essence, define transportation control measures as:

- Public transit
- High-occupancy vehicle (HOV) roads and lanes
- Employer-based transportation management programs
- Trip reduction ordinances
- Traffic flow improvement programs
- Fringe and corridor parking for multiple occupancy vehicles and transit
- Programs to limit and restrict vehicle use in downtown areas
- Shared ride and HOV programs
- Roads designated for exclusive use by nonmotorized modes
- Bicycle lanes and facilities (storage)
- Control of vehicle idling
- Programs to reduce cold-start vehicle emissions
- Flexible work schedules
- Programs and ordinances to reduce single-occupant vehicle (SOV) travel, including provisions for special events and activity centers
- Construction of paths for exclusive use by pedestrian and nonmotorized vehicles
- Vehicle retirement and replacement (scrappage) programs

TCMs were originally perceived to be regulatory mandates that, as expected, were not popular. Over time, however, the market nature (e.g., incentives) of many TCMs and experience using them started to enable the introduction of increasingly politically stronger approaches using a more balanced carrot-and-stick method. In addition, more comprehensive approaches started to appear in which attention was paid to specific population segments and their needs, as well as the bundling of TCMs into integrated management approaches not only for the larger metropolitan areas but also for smaller cities (Goulias and Szekeres, 1994).

Within this air quality framework a richer set of TCMs is now included in the planners' repertoire (see the examples in Meyer (1998) and the textbook by Meyer and Miller (2001)), and they have been expanded to include a long list of potential programs and actions such as:

1. Congestion pricing and toll programs
2. Emission, VMT, and other fee programs (including carbon taxes and trading)
3. Intelligent Transportation Systems (ITS) — the use of telecommunications and information technology to manage, control, and provide information about the transportation system
4. Accelerated retirement of vehicles programs
5. Telecommunications to substitute or complement travel (home and satellite telecommuting, teleconferencing, and teleshopping)
6. Land use growth and management programs
7. Land use design and attention to neighborhood design for nonmotorized travel
8. Goods movements (freight) programs to improve operations
9. Highway system improvements in traffic operations and flow
10. Alternative transit service improvements and expansion
11. Highway improvements for high-occupancy vehicles
12. System improvements for nonmotorized travel
13. Special event planning and associate traffic management
14. Public involvement and education programs
15. Individualized marketing techniques with improved information and communication with the customer

Most of these initiatives are public investments or are mandated by regulations and local ordinances. For this reason, it is of paramount importance to assess their successes and failures with tools that can be defended when scrutinized. In fact, one notable scrutiny of these tools was done in the Bay Area Lawsuit. Garrett and Wachs (1996) provide an overview on how traditional simulation models are outpaced by the same legislation that defined many of the policies described above, and how this made metropolitan planning organizations particularly vulnerable to legal challenges. Most important, however, the sense of urgency that emerged from the Bay Area Lawsuit is now exacerbated by the evolving trends described in the next section.

A second key aspect of these policies that is different from the federal rules and regulations that target vehicle energy consumption and emissions is the presence of many diverse agents implementing the policies. For example, in Metropolitan Transportation Commission (MTC) Resolution 2131, which approved a variety of TCMs, we find among others the California legislature (responsible for increasing bridge tolls), the California State Department of Transportation (CALTRANS) (responsible for improvements in the HOV system), local cities and transit authority (responsible for improved ferry services), and local employers (responsible for the volunteer employee trip reduction programs). Fragmentation of responsibilities and roles can be considered a threat to any plan, program, and implementation strategy and has been challenged by environmental advocacy groups (Garrett and Wachs, 1996). Depending on the local circumstances, however, this fragmentation could become an opportunity because of the potential to increase the total funding for these programs and the ability to recruit stronger support when a well-coordinated campaign is created (e.g., see the program at http://www.wmin.ac.uk/transport/inphormm/inphormm.htm, accessed April 2002).

1.8 The Future

In spite of the uncertainty of future technological, policy, and market developments, we know transportation all over the world will push its demand for oil to higher levels. The world motorization in the last 50 years has seen an explosive growth, with the U.S. share dropping from 70% of the world's light vehicles in 1950 to just over 30% in 1998 (Birky et al., 2001). In analysis that is based on population projections

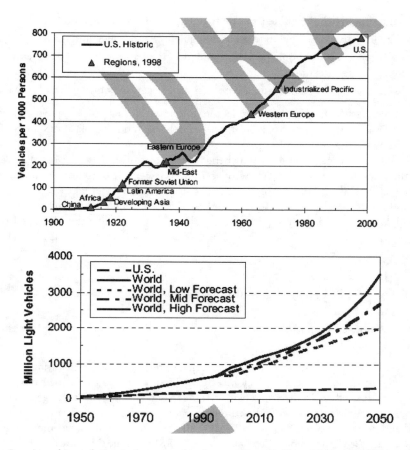

FIGURE 1.8 Car ownership evolution in the past and in the next 50 years. (From Birky, A. et al., Future U.S. Highway Energy Use: A Fifty Year Perspective, Draft Report prepared for Office of Transportation Technologies, Energy Efficiency and Renewable Energy, U.S. Department of Energy, 2001.)

from the World Bank, U.S. government agencies predict the total number of light vehicles around the world to most likely increase by three to five times in the next 50 years. This growth is expected to take place mostly in emerging economies without a decrease in car ownership elsewhere, as shown by the two images in Figure 1.8.

Since alternatives to oil for mass marketing do not exist yet, carbon emissions associated with world transportation oil use will also grow (some forecasts claim exponentially). Vehicles using alternative fuels that are from renewable sources are the only hope to reduce carbon emissions, particularly electric vehicles, hybrid electric vehicles, and later on fuel cell-powered vehicles. In the near term (5 to 10 years) vehicles using ethanol and electricity are potential niche alternatively fueled vehicles. Ethanol, as a liquid fuel, is largely compatible with the present infrastructure and can be blended with petroleum products. Natural gas is another fuel that has been given attention and an embryonic infrastructure exists to support it.

Electric vehicles are supported by electric power companies, so the industry, due to deregulation, has market incentives to pursue growth and penetration in transportation use of electric power. Given the availability of large quantities of low-cost fossil fuels (Maples et al., 2001), stronger policies are needed to accompany this push for electric vehicles in the transition to low-carbon fuels. Professor Sperling (1995), a leader in American thinking for alternative fuels, claims in his book *Future Drive* that "electric propulsion is the key to a sustainable transportation and energy system." His argument and strategy suggestions are (Sperling, 1995):

- Improved technology holds more promise than large-scale behavior modification.
- Technology initiatives must be matched with regulatory and policy initiatives.
- Government intervention should be flexible and incentive based, but should also embrace selective technology-forcing measures.
- More diversity and experimentation are needed with regard to vehicles and energy technologies.

In the longer term, hydrogen has the potential for substantial carbon reduction benefits and applications. The main disadvantage for hydrogen is its cost of delivery (and complication due to the need for carbon sequestration).

To examine potential scenarios and to study benefits and costs of these scenarios, models of the type described in this handbook are used. Below we give an overview of the output from models of this type designed at the national level. First, however, policies considered are reviewed to provide an idea of the background and motivation for many of these models. The Department of Energy in the United States performed and commissioned many studies on market penetration of alternatively fueled vehicles. In a recent summary provided to the U.S. Congress (U.S. DOE, 1999, with amendments in 2000) we find the following requests for strategies:

- Subsidies to vehicle buyers
- Subsidies to fuel providers for infrastructure development
- Subsidies to fuel purchasers
- Other governmental support that includes information programs, research and development, and technology transfer

The proposition of the U.S. DOE is to advance its alternative fuels program by following two parallel paths in which alternative fuels (i.e., natural gas, ethanol, electricity, biodiesel, propane, and methanol) are promoted, while blending fuels (i.e., ethanol, ethers, and biodiesel) into gasoline and diesel is also given an opportunity. Specific examples of an incentive structure based on tax credits, tax deductions, exemptions, and additional mandates and authorities to enforce fuel use in EPAct fleets are included in these recommendations.

In a 1992 study by the Congressional Research Service (CRS) to define the critical path for a 5% market penetration by electric vehicles in the United States, it was estimated that $4 billion is required to subsidize batteries and $3 billion to subsidize the incremental cost of the vehicles for 10 years. In addition, recharging infrastructure is needed and research and development needs to accelerate by additional investments. A portion of this investment has been made in the United States. It is unknown, however, how large the investment is to date.

Using simulation models to study the effect of alternate policy scenarios such as taxation and incentives, the CEC (1998) concludes the following with respect to transportation policy and energy:

- Fuel taxation based on carbon content is the most efficient pricing strategy, since taxes target GHGs directly.
- Given the national and international nature of auto manufacturing, nationwide fuel taxes and feebates would reduce carbon emissions by a greater amount than state-only taxes and feebates of the same magnitude.
- Pricing measures and higher fuel economy standards and feebates appear effective as measures to reduce carbon emissions. HOV lanes and transit use may have to be expanded, and monetary incentives for alternatively fueled vehicles combined with pricing measures, in order for these measures to be truly effective.
- Fuel taxes and congestion fees could offer significant social benefits, since they reduce congestion and other driving effects, in addition to carbon emissions. Studies on whether these pricing measures are regressive are inconclusive.
- Although state-only feebates reduce consumer surplus, they do not appear to affect equity adversely. Further, feebates more effectively promote the demand for alternatively fueled vehicles

than carbon taxes, for a given level of carbon reduction. Feebates do not appear to reduce driving and, therefore, may not offer the high social benefits of pricing measures. In addition, state-only feebates that increase fuel efficiency may not be allowed by the federal government.

- Nationwide feebates and higher fuel economy standards appear to reduce carbon emissions and increase consumer surplus for drivers. These policies, however, may reduce the average costs of driving and, as a result, may actually increase VMT and the external costs related to driving.
- Subsidies for alternative fuels would probably have to be accompanied by increased gasoline taxes in order to show an overall decline in carbon emissions.

With respect to predictions about electric vehicles and barriers to implementation, the CEC reports that:

- Electric vehicles (EVs) would become the most prevalent type of AFV on the state's roads and could reach a population of 1 million vehicles 10 years after the zero emission vehicle (ZEV) regulations are in place (in 2004).
- Non-fossil-fueled electric-generating facilities are expected, possibly as a result of California's recent initiative to ensure R&D funding over the next several years for such technologies.
- More efficient natural gas-fueled generation units are needed to raise the operating efficiency of the electricity supply system enough to increase the CO_2 benefit of EVs.
- Improvements in the operating efficiencies of EV technologies are needed to make EVs more effective in reducing CO_2 emissions.

There are a few key advantages of electric vehicles and hybrid electric vehicles (HEVs) that should be mentioned as a summary:

- EVs and HEVs introduce a nonoil fuel that functions as a "stabilizer" to combat oil price shocks.
- EVs and HEVs can respond to the need for worldwide fuel efficiency for CO_2 production inhibition.
- Only noncarbon fuels can succeed in curbing CO_2 production. Electricity is one of the "fuels" to remove that last portion of CO_2, and EVs may be the only solution to transition toward non-carbon-based fuels.
- Given the current motorization trends, EVs and HEVs may be the only option we have for mitigating air pollution damages.
- Mandatory national and regional programs require increasingly lower exhaust emissions standards. EVs and HEVs are the only technologies that come close to the zero emission vehicle.
- Averaging by auto manufacturers makes EVs and HEVs a good option, so that the more desirable market vehicles (e.g., sport utility vehicles) can continue to be the profit makers.
- EPAct has created a viable market with regulations and a volunteer program that allows a respectable market entry for EVs and HEVs.

For these reasons and in order to examine in more detail the impacts of legislation on energy consumption and air pollution, the U.S. Department of Energy has developed extensive forecasts for the entire country and its major regions (Northeast, Southwest, and so forth). Every year the United States produces a report about nationwide energy consumption from all sectors of the economy. Within this inventory and forecasting exercise, one module focuses on the transportation sector, and within the transportation sector, a more detailed analysis of market trends is performed. This complements an overall energy consumption trend analysis and aids in understanding and assessing the effects of policies on market penetration. Figure 1.9 shows the evolution of the market for alternatively fueled vehicles from 2000 to 2020. The hybrid electric vehicles are expected to dominate the new car sales market after 2003 because of legislative mandates, as discussed earlier in this chapter. As expected, this dramatic increase has a delayed effect in the vehicle stock because of the delay in replacing existing vehicles with environmentally friendlier ones, as depicted in Figure 1.10. The policies in EPAct are in essence imposed to the databases used in this model, and incentives and disincentives are built into the equations used to produce these predictions without a truly behavioral model that allows for possible resistance to EPAct implementation by consumers.

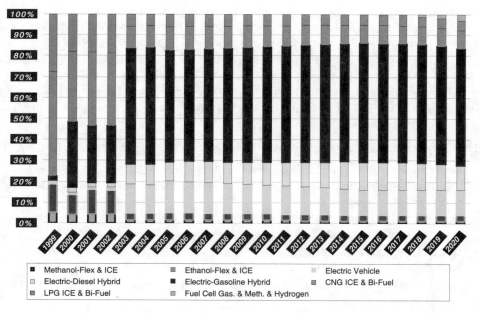

FIGURE 1.9 New light-duty vehicle sales in households and fleets in the United States.

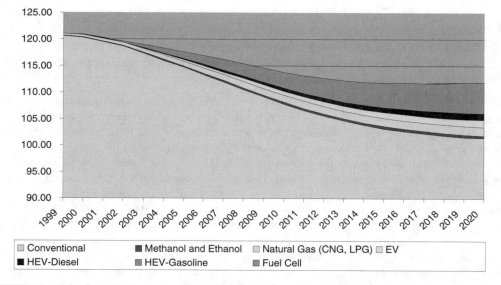

FIGURE 1.10 Light-duty vehicle stock evolution in households and fleets in the United States.

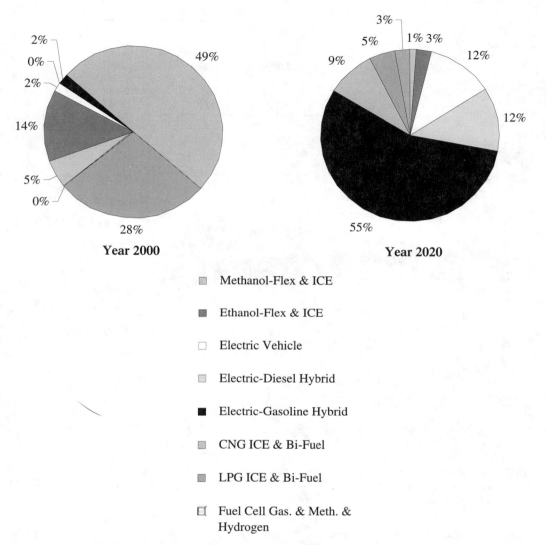

FIGURE 1.11 Composition of alternatively fueled vehicles in the United States. In 2000, AFVs were approximately 2.4% of total (of 8.9 million vehicles, 212,000 were AFVs). In 2020, AFVs will be 10.9% of total sales (of 8.1 million vehicles, 866,000 will be AFVs; HEV-gasoline fueled vehicles are projected to be approximately 480,000).

As a result, the market size is substantial, as demonstrated by Figure 1.11 and Table 1.5, providing two snapshots in 2000 and 2020 from that same U.S. DOE forecast. New car sales for light-duty vehicles (passenger cars and light trucks < 8500 lb) are estimated and shown in Table 1.5, which also displays the vehicles expected to be in organized fleets. Some of these projections are also confirmed by other studies using different methods. Among studies, however, the exact compositions and market sizes differ substantially because they are based on different assumptions (Vyas et al., 1997). Nevertheless, a general agreement exists that under current legislative mandates and gasoline prices, the market depicted in Figures 1.9 to 1.11 is a reasonable expectation, and if oil prices increase, we may even see significantly higher market penetration levels.

Similar projections are found in European projects (e.g., in the UTOPIA project, UR-97-SC-2076 (Moon, 2001)). Within this project a European model called STEEDS was used to provide scenarios of market penetration under a variety of policy scenarios in the EU. These scenarios are somewhat opti-

TABLE 1.5   Sales of Alternatively Fueled Vehicles and Fleet Composition

	2000	2010	2020
Total New Cars	8,886,000	7,505,900	8,112,100
Alternatively Fueled (AFVs)	212,000	6694,600	866,800
Electric and Hybrid	73,000	553,300	689,700
Fleets (>10 veh.)	2,105,000	1,778,700	1,922,420
AFVs in Fleets	124,370	129,020	139,450

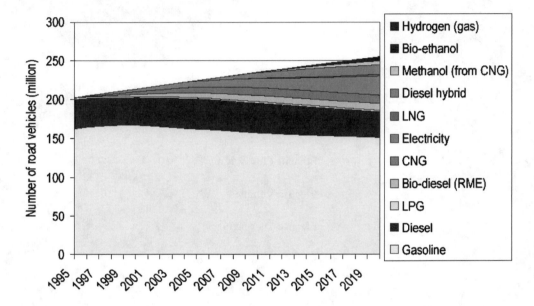

FIGURE 1.12 EU 15 Stock evolution under current legislation based on UTOPIA analysis. (From Moon, D., Urban Transport: Options for Propulsion Systems and Instruments for Analysis, UTOPIA EC, UR-97-SC-2076, Final Report submitted to the European Commission, Brussels, 2001.)

mistic, as the authors of the final report claim. Figure 1.12 displays the roadway vehicle stock by type, and Figure 1.13 shows the composition of market share in 2020. Both figures tell the same story as in the United States: we should expect a rather substantial increase in market penetration by alternatively fueled vehicles under current prevailing political and policy initiatives and policy actions. These predictions, however, are not based on a well-tested behavioral theory; instead, they are projections of the effects legislation may have on the market.

Moreover, many predictions are based on untested assumptions about the behavior of individuals and their households, groups of individuals (firms and fleets), and larger aggregates of individuals (institutions or entire communities). Considering the TCMs described above and the actors identified in the description of the relationship between transportation and the environment, we need tools and models that go beyond the government's simple projection tools and beyond current urban simulation practice. This is particularly important when we study implementation policies. For example, the DOE assessment using simple projection techniques shows the beneficial effect of introducing alternatively fueled vehicles and a target of market penetration for these vehicles. A series of policy actions need to be defined and their effect estimated when trying to meet the targets. Assessment of the success of each these policies can be done with computerized models that are at the heart of our decision support systems for policy analysis in transportation. The following section describes a framework of these types of models.

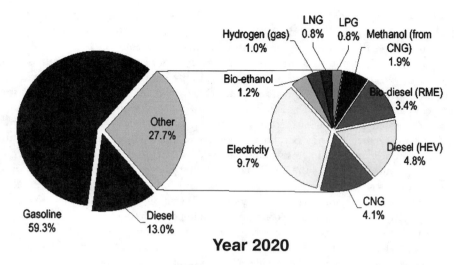

Year 2020

FIGURE 1.13 Market share of vehicles by type. (From Moon, D., Urban Transport: Options for Propulsion Systems and Instruments for Analysis, UTOPIA EC, UR-97-SC-2076, Final Report submitted to the European Commission, Brussels, 2001.)

1.9 The Model Framework

Given the wide range of policies and initiatives described above, we cannot limit the discussion here to one model that is applied to one geographic area for one specific policy. Instead, we need to address multiple model frameworks designed for different policy analyses. This will make the discussion somewhat more general, particularly when compared to the models and applications in subsequent chapters in this handbook.

For very good reasons, past focus of transportation planning and modeling has been on urban transportation. Those model frameworks were targeting issues that major metropolitan (urbanized) areas examined then and now. For example, in the late 1950s Chicago, Detroit, and Pittsburgh were the focus of the first regional models that defined the state of practice in regional long-range comprehensive plans and related simulation models, leading to the well-known four-step model (Creighton, 1970). Later suburban and rural transportation received some attention because the boundaries of many cities were not clearly defined and urban sprawl became increasingly evident. Computational barriers, however, were severe, and for this reason, many models were just approximate, sketchy representations of reality (see the sketch planning models in National Cooperative Highway Research Program (NCHRP) 187). The models were sufficient for public policy interventions, with effects that were by far larger than the models' errors (e.g., building a beltway around major urban centers, such as Detroit). In addition, jurisdictional authority and contractual reasons also motivated past transportation planning studies to be defined and limited to specific cities, regions, states, and so forth, with no formal and clearly defined relationship with other geographic scales. To perform these studies, artificial boundaries are drawn and the behavior of persons and firms within these boundaries is examined in more detail; a summary treatment is given to those outside. On the one hand, travelers' behavior may be totally unrelated to these artificial boundaries, and this artificiality in determining boundaries may be a biasing factor for our models. On the other hand, however, in land use and land use regulations, these boundaries may be the differentiating factor of policies and land use possibilities (e.g., zoning regulations that determine density of development and use in different areas of the city, school taxes that may motivate different location patterns in municipalities that have all other attributes similar, taxation and fee collection for business location and access to main roadways) that need to be accounted for in model building and plan development. Unavoidably, however, these multiple model frameworks are very much related and interacting because the output of one model may be the input of another, in the same way as the outcome of a policy at one level of government determines and conditions the outcome from other levels.

There are four dimensions that one can identify to build a taxonomy of policies and models: (1) *geographic space* and its conditional continuity, (2) the *temporal scale* and calendar continuity, (3) interconnectedness of *jurisdictions*, and, most important, (4) the set of relationships in *social space* for individuals and their communities. These four dimensions very often cannot be disengaged from each other. For example, when one considers issues associated with a change in propulsion fuels and energy consumption, discussed in the previous sections, the time horizons are very often 25 or more years. This also dictates planning that expands beyond the borders of one country (e.g., the oil producers are organized in a common financial interests group named the OPEC, and the consumers are organized in blocks of trading countries such as the European Union that exercise their influence in oil production matters). Issues of this type can be named issues of grand scale because (1) they are long range; (2) they involve larger geographical regions than a single country; (3) economic and cultural relationships among the countries involved often influence the course of action in policy development; and (4) issues in social space are elevated at more abstract levels in which relationships are more often in the realm of international politics and based on historical and cultural relations, as well as on trading and defense. At exactly the other end of the spectrum, we can consider as an example a case of traffic calming — a group of measures and actions aimed at eliminating the traffic that travels through a neighborhood. In this case the time horizon is made by a few years (e.g., a first year of data collection and fact finding, a second year of building facilities and educating the public of "new" ways of traveling, and a third year of verification and final adjustment). The geography is the neighborhood and its topographic as well as other land use characteristics. Control measures are usually designed (through a contractor) by the municipality in which the neighborhood belongs. However, any traffic signals and other signs that are installed, as well as any highway geometric design changes needed, are examined for compliance with local, state, and federal rules and regulations by the associated authorities and agencies. In addition, different agencies have jurisdiction on speed limits, access to highways, and safety standards, depending on the type of facilities that are affected by proposed changes.

The second dimension is the continuity of time and the artificiality of the time period considered in models. For example, models used in long-range planning use typical days (e.g., a summer day to capture worst-case scenarios for some type of emissions and pollutant concentration or a special event day to capture a period during which facilities are used at capacity). In many regional long-range models the unspoken assumption is that we target a typical work weekday in developing models to assess policies. Households and their members, however, do not obey this strict definition of a typical weekday to schedule their activities. They may follow different decision-making horizons in allocating time to activities within a day: spreading activities among many days, including weekends; substituting out-of-home with in-home activities on some days, but doing exactly the opposite on others; and using telecommunications selectively (e.g., on Fridays and Mondays more often than on other days). Obviously, taking into account these scheduling activities is far more complex than existing transportation planning models. A few chapters in this handbook consider this issue, and Chapter 7 provides an example of data collection methods that are promising in eliminating this limitation in regional models.

The third dimension is jurisdictions and their interconnectedness. The actions of each person are regulated by jurisdictions with different and many times, overlapping domains such as federal agencies, state agencies, regional authorities, municipal governments, and neighborhood associations. In fact, in the review above many rules and regulations about environmental protection are defined by a government at a national or supranational level (e.g., the European Union), and they may end up being enforced by a local jurisdiction (e.g., a regional office of an agency within a city). On the one hand, we have an organized way of governance that subdivides jurisdictions and policy domains in clear ways. For example, in the Eastern United States, tax collection belongs to three distinct jurisdictions in decreasing order of percent of income tax collected: federal, state, and municipality. In the majority of situations the federal government collects taxes and then transfers portion of these revenues to the state. In parallel, the state collects taxes and transfers a portion of them to regions and municipalities. In turn, the municipalities will use the revenues from the local taxes and the revenues received from the state for services. In some states municipalities also have the option to collect other fees and taxes that are specific to transportation impacts of businesses. On the other hand, the relationships among jurisdictions and decision making about allocation of resources do not always follow this orderly governance principle of hierarchy and tax purpose. For example, in the most recent transportation legislation, TEA-21, the U.S. Congress allocated a large amount of funds to specific projects that bypassed both the federal and state agencies (these are called earmarks or special projects). Each project is named (e.g., location and purpose, recipient of the grant and purpose) in the legislation, and a governmental entity is assigned to disburse the funds. Decisions about the amount of funds and the purpose are very often between an elected representative's office and a local group or organization. Many of these projects are improvements to the highway system that may increase the incentives for private car use (e.g., when new highways are built for them) and oil consumption, contradicting policies by other levels of government.

In another example the vehicles purchased by households, firms, and fleets are regulated by a variety of overlapping rules and regulations for safety and environmental protection, as well as fiscal responsibility. Each of these rules has an effect on the vehicle performance characteristics, initial price, and life cycle costs. In turn, these attributes influence the household, firm, or fleet in its decision to purchase and operate a vehicle. Because the types of vehicles on the road influence emissions in a significant way, every change in policy at every level of government or jurisdiction may have a significant and substantial effect for emissions.

The fourth and final dimension is social space and the relationships among persons within this space. For example, individuals from the same household living in a neighborhood may change their daily allocation patterns and location visits to accommodate or take advantage of changes in the neighborhood. Depending on the effects of these changes on the highway network, we may also see a shift within the neighborhood social relationships (e.g., the usual outcry when a new road is designed to pass through an existing neighborhood). In contrast, elimination of traffic may increase the use by pedestrian and bicycle facilities, attracting system users from other neighborhoods and thus complicating the relationships, as discussed above.

One important domain and entity within social space is the household. This has been a very popular unit of analysis in transportation planning, recognizing that strong relationships within a household can be used to capture behavioral variation and heterogeneity (e.g., the simplest method is to use a household's characteristics as explanatory variables in a regression model of travel behavior). In this way, any changes in the household's characteristics (e.g., change in composition due to birth, death, children leaving the nest, or adults moving into the household) can be used to predict changes in travel behavior. As other chapters in this handbook describe in more detail, new model systems are created to study this interaction within a household, looking at the patterns of time use in a day and the changes across days and years. One way to visualize these relationships among household members is to look at their use of time within a day and their traveling alone or with others. An example is offered here from two households in South Perth that completed a travel diary based on the New Kontir Design (NKD) travel survey format (Moritz and Bröeg, 1999).

In Figure 1.14, along the horizontal line each hour of a day is reported and along the vertical line the type of activity pursued by each member of a household that contains a male working for pay at a location away from home, a female predominantly working in home duties, and two children that go to school. When one compares the female's time allocation to activity and travel in that day to her children's time allocation, a clear role for her emerges — she accompanies and cares for the children throughout the entire day, while the male appears to allocate his time somewhat independent of the other three members. This may be an apparent phenomenon, and when one changes in minor ways circumstances surrounding the household, a completely different schedule of activities and travel may be seen. The example here shows clearly a few of the limitations in models depicting and used to predict travel behavior that are either summaries of all four persons in Figure 1.14, therefore missing the richness of intersection depicted by this figure, or models of each individual in the figure separately, missing the interdependency among the activity and travel schedules of the four persons in this household. These interdependencies have

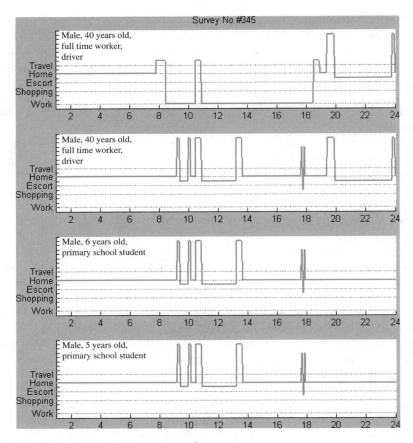

FIGURE 1.14 A family's day in Perth, Western Australia.

profound effects on the modes used by these persons, the fuel consumption of the household, and the emissions produced by each of their vehicles.

It is therefore very important in modeling and simulation, as well as other types of policy analysis, to reflect in the models used for policy analysis not only the interactions described above but also interaction among these four fundamental dimensions. The typical example is long-range planning that is usually defined for larger geographical areas (region, states, and countries) and addresses issues with horizons from 10 to 50 years. In many instances we may find that a large geographic scale means longer time frames applied to wider mosaics of social entities and more diverse jurisdictions. On the other side of the spectrum, issues that are relevant to smaller geographic scales are most likely to be accompanied by shorter-term time frames applied to a few social entities that are relatively homogeneous and subject to the rule of a few jurisdictions. This is not only an important organizing principle, but also an indicator of the complex relationships we attempt to recreate in our computerized models for decision support. In developing the blueprints of these models one can choose from a variety of theories (e.g., neoclassical microeconomics) and conceptual representations of the real world. One such representation, called activity theory, has the potential to become the higher-level theoretical framework needed to think in a systematic way about transportation systems and their planning, and for this reason, it is presented here.

1.9.1 Activity Theory as a Tool for Understanding

Activity theory is a psychological theory that offers a framework that can help us sort out the within-dimensions and among-dimensions relationships among the entities discussed above. It targets human *praxis* and interlinks individual and social levels. It is also a lens through which we can examine travel behavior and transportation (planning) actors from a truly dynamic perspective because this theory was conceived as a system within which humans (and their groups) develop. In fact, activity theory, with activity intended as doing in order to transform an entity into something, is developing into a multivoiced philosophical framework for studying different forms of human praxis as multiple developmental processes involving individuals, households, and their social groups.

The origins of activity theory can be traced to classical German philosophy (Kant, Fichte, Hegel, and later Feuerbach) in which emphasis is given to the active role of humans, their development, and the historicity of their action(s). In addition, activity theory researchers identified formative roots in the concept of practical–critical theorization developed by Marx and Engels. However, the most direct relationship of the activity theory used here is with the Russian (Soviet) cultural–historical school founded by Vygotsky, known in the United States for his contribution to education research, and the later continuation of his work by his collaborators Leontiev and Luria. At the center of Vygotsky's (1978) approach is the utilization of methods and principles of dialectical materialism in studying (psychological and social) phenomena in continuous motion and change. Individual choices in this framework need to be studied in terms of the history in their qualitative and quantitative aspects. Another concept at the center of Vygotsky's approach is the *personal* transformation actions, predominantly work related and work specialization related, have on the individual actor. For example, the tools that a person builds transform nature, and this in turn transforms the person that created the tools in the first place. In travel behavior the dynamic nature of human behavior has been advocated as a key ingredient for more realistic (behaviorally) models that help us to understand and predict phenomena such as motorization; allocation of time, work and play; and the decision structure of activity participation and trip making, alone or with others. These points were raised in a seminal 1988 Oxford conference (selected papers were subsequently published in Jones (1990)), and many key issues identified then guided numerous research projects (many chapters in this handbook provide bibliographies that can be traced back to the Oxford conference). However, the dynamics that have been considered in research models (none of the behavioral dynamic models are currently used in practice) ignore personal transformation. Instead, they focus on household transformation as captured by life cycle (see the microsimulation chapters in this handbook) and lifestyle changes (see Chapter 6). While these are formidable advances over past research and practice, they still need further improvement and adjustments for realism.

Models of personal transformation view actions as artifact mediated, with artifacts defined as material or nonmaterial tools, and object oriented (Vygotsky, 1978, p. 40), where object is intended as objective or goal directed. In fact, Vygotsky's method of thinking about consciousness via the mediation of psychological tools has been named *instrumental method* because both the stimulus and the tool could be considered stimuli affecting the ultimate response. This consideration allowed the incorporation in the explanation of behavior of a more comprehensive treatment of the relationship between a human agent and objects of environment as mediated by cultural means, tools, and signs. Reworking Vygotsky's original ideas, researchers visualize the activity system as the following triangle.

The subject is the central ontological unit whose agency is chosen as the viewpoint of a given analysis. The object is acted upon and can be any type of material entity, conceptual idea, or series of problems to be solved. These objects are transformed into outcomes. Mediation of artifacts is most likely the element making activity theory unique and revolutionary because it maintains control is exercised by humans "from the outside using and creating artifacts" (Engeström, 1999). In addition, the links from subject to object and artifact to object are not separable in the traditional sense of cause and effect because of the artifact mediation and the artifact transformation notions.

In one reformulation of Vygotsky's original idea, Leontiev introduced an emphasis on the division of labor as a fundamental historical process behind the evolution of mental functions. In this conceptualization, work mediated by tools, as described in the activity system above, is also "performed in conditions of collective activity jointly by individuals" (Engeström, 1987). Based on his considerations and with the desire to develop a workable model that can be used in applications, Engeström expanded the original Vygotsky triangle of mediation to incorporate activity context, as shown in Figure 1.15. Before turning to this figure, however, let us review a few basic notions in activity theory.

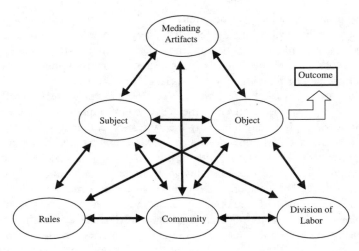

FIGURE 1.15   Activity system as visualized by Engeström.

As mentioned before, *activity theory* is a conceptual system of ideas for understanding praxis. Its basic elements are (1) the hierarchical structure of activity, (2) an object orientedness of the state, (3) the processes of internalization and externalization, (4) tool mediation, and (5) emphasis on development.

1.9.1.1   Hierarchical Structure of Activity

In activity theory the unit of analysis is an activity system. Activity itself is directed at an object (objective) that motivates the activity. Figure 1.15 is the activity system. Leontiev is credited with the following hierarchy:

1. Activities are composed of goal-directed actions.
2. Actions are conscious.
3. Different actions may be undertaken to meet the same goal.
4. Actions are implemented through (automatic) operations.
5. Operations do not have their own goals — they provide an adjustment of actions to current situations (e.g., feedback).
6. The constituents of activity are not fixed and can dynamically change as conditions change.

In addition, Kuutti (1996) and others credit Leontiev for the following levels:

Activity — motive
Action — goal
Operation — conditions

Activities are longer term and become reality through individual and collective actions. These actions share the same object and motive. Groups of actions that have an immediate conscious goal define activity participation, but they cannot be understood without the frame surrounding them. Let us consider an observation you made:

> You are parking your car at the downtown parking lot. The person who just arrived and parked right next to you is wearing a chicken costume. Your mind starts rushing through many ideas: crazy person, advertisement for some restaurant, costume party, and so forth. That *action* alone does not make sense. Then a second car arrives with two other persons wearing the same costume. They are followed by three more cars, and this time a few children come out wearing other costumes and talking about the play at school and how not to forget their lines.

After some patient observation and some reflection about your own children, you find out that this is the end-of-the-year theatrical play at the local school, in which children and their teachers are taking part. The *activity* is the play at the school; the *action* is wearing the costume and arriving by car dressed in it; the *operations* are all the other automatic entities, such as driving the cars, parking them in the parking lot, walking from the parking lot to school, and so forth. Kaptelinin (1996) provides an interesting way to distinguish among the three concepts of activity, action, and operation: when operations face barriers, the agent adapts in an automatic way without noticing; when actions are blocked, the agent resets the goals without negative emotion; "but when the motive is frustrated, people are upset, and their behavior is most unpredictable" (Kaptelinin, 1996). Participation in the theatrical play in the example above is the driving force behind everything else. Wearing costumes and gathering at the parking lot are actions auxiliary to this activity. Driving, finding the parking lot, and parking the car is an automatic process supporting the action(s).

1.9.1.2 Object Orientedness

Humans live in a reality that is objective in a broad sense. For example, the entities that constitute this reality have properties that are considered objective in the sense of natural sciences. A car, a parking lot, and a road are not imaginary and subjective concepts but have a mass associated with them, weight due to gravity, and so forth. However, there are also other objective social and cultural properties associated with them.

1.9.1.3 Internalization and Externalization

Activity theory distinguishes internal from external activities. It also emphasizes that internal activities cannot be understood if they are analyzed independently from external activities. This is mainly due to a process of transformation in which there is a transformation of external activities into internal, and vice versa. Internalization is considered to be social, and the range of actions that a person performs with others is named "zone of proximal development" (Kaptelinin, 1996). Internalization is also mental simulation that provides individuals with a way to try potential interactions with the real world without actually manipulating real objects. Externalization is needed when an internalized action needs to be verified, checked, scaled, calibrated, and so forth. It is also needed when there is collaboration or coordination among individuals.

1.9.1.4 Mediation

Activity theory emphasizes that human activity is mediated by tools in a broad sense. Tools are created and transformed during the development of the activity itself and carry with them the particular cultural and historical meanings embodied in them during their development. These tools through their use represent an accumulation and transmission of social knowledge and influence human external behavior and mental functioning. Examples range from tools that we carry with us for medical or physical enhancement to complex machines that help compute and record information.

1.9.1.5 Development

Fundamental to understanding the phenomena shaping our world is understanding how each phenomenon came to be in the form we observe it today. In activity theory development is the object of study and the general research methodology. For this reason, understanding the formative steps of a given phenomenon using, for example, ethnographic methods that track the history and development of a practice is key.

In activity theory, where activity is intended as behavior, the unit of analysis is an *activity system*, which is composed of *the subject* (e.g., an individual, a household, or a social group), *the object outcome* (e.g., actions and ideas), and *mediating artifacts* (e.g., human material and nonmaterial constructs). Depending on the application, subjects, objects, and mediating artifacts are different entities. For example, in a model predicting travel among places in a region, a subject is a person engaged in work, leisure, shopping, and traveling. In this sense an object is held by the subject and motivates activity, giving it a specific direction and motive. In this case a person's daily schedule in terms of spatial and temporal arrangement of places to visit and amount of time expended at each place is an object aiming at meeting personal and household needs. Nonmaterial objects are also included in this category, such as satisfaction from allocating and

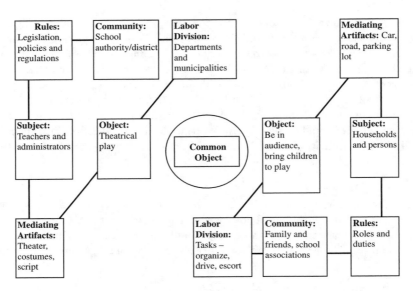

FIGURE 1.16 Activity system interaction for children's theatrical play.

spending time with family and friends or, in the theatrical play example above, the parents' emotions as they observed their children in the play and the children and teachers' emotions as they performed.

In parallel to the modeling of human behavior of transportation system users, activity theory can also be applied to other decision-making units as subjects (e.g., analysts, managers, and policy makers as individuals and groups), with the specific object being the service provided using the transportation system itself. Applications based on these considerations treat the transportation system users and managers as an integrated whole that evolves and cooperates over time. Engeström's (1999) conceptualization is particularly useful in modeling and simulation of time–space behavior by households and their household members, and it has been used in a few recent applications (Goulias, 2001; Yin, 2000) to visualize key modeling aspects. The right-hand-side triangle in Figure 1.16 is the activity system used here as well as in other similar model systems in which the subject(s) and decision-making units are households and the household members. In this conceptual model, however, the true unit of analysis is the entire right-hand-side triangle — the entire activity system of each household. Time allocation as a schedule of actions, such as traveling to the parking lot and participating in the audience at the children's play, is the object. Each household owns vehicles, lives, and moves in a region or city whose facilities, in this framework, are the mediators of behavior. Households and their members consider and have knowledge of many mediating artifacts, such as the distance from home to the play site, location characteristics of the different parking places, and fixed time markers (e.g., prescheduled start and end time of the play, as well as its location). Building a forecasting model system considering households alone, however, is not sufficient to capture the dynamics of the travel behavior you observed at the parking lot because of the intervening roles of other decision-making units (groups and individuals). Figure 1.16 shows one such group, the educators (teachers and administrators that worked to put the play together).

The teachers, in the left-hand-side triangle of Figure 1.16, also use mediators to plan and manage the play (e.g., the theater, the costumes of the actors, the script, and so forth). Their object is mainly the theatrical play. From this viewpoint, there is ongoing cooperation between the two groups (households and teachers/administrators) through the common object of making the play happen. For example, the children need their parents' consent to be in the play and need to be escorted to a prespecified place before the play; the parents need to be at the theater to become the audience.

However, often contradictions emerge between activity systems and within each activity system among its constituent parts. For example, rules and regulations dictating the opening and closing times of the facilities around the theater interfere with the ability and degree of freedom of households to schedule their arrival and departure, as well as other activities before and after the play. The conceptual framework of Figure 1.16 enables identification of these contradictions for modeling interactions and cooperation among activity systems. For example, typical contradictions between the parents' need for parking space and the allocation of parking to teachers and students arriving in costumes lead to new forms of common object development (e.g., new rules and regulations for parking next year may allow the dropping off of children at the entrance and the reservation of parking lots closer to theater). Other contradictions may emerge among each of the entities in each of the triangles, leading to the emergence of new developments (e.g., think of the parents that arrived late and whose children were afraid of not making it to the play). Decision support model systems can and should incorporate these ideas of contradictions between different activity systems because change is determined by these contradictions.

Given the research questions at hand and the new modeling frameworks that are emerging in transportation planning, a more prominent role may be needed for qualitative research methods to understand and influence travel behavior. However, as discussed elsewhere (Goulias, 2001), using qualitative research methods to satisfy this need implies adoption of different theoretical paradigms, such as the activity theory approach here, which may contradict the dominant paradigm in research and practice. The qualitative research methods field, which is very rich in ideas, methodologies, and traditions, has a high potential for innovation. Examples of success stories for the transportation field exist, and some of them have fostered positive change. Suggestions for practice and interesting case studies have been provided by Clifton and Handy (2001).

1.10 Summary

Since the beginning of the past century transportation systems have experienced a remarkable evolution in quantity and quality. Today we have propulsion systems that consume very little compared to their predecessors; our internal combustion engines and their exhaust treatments contain reliable and inexpensive components (e.g., catalysts), emitting very few pollutants; we drive safer vehicles on safer roads; and we have many safety features and options to chose from. In addition, our vehicles are characterized by significant improvements in comfort and reliability, and a wide variety of models are available to consumers. Similar arguments can be made for public transportation in countries such as Germany, Austria, and The Netherlands, but to a lesser degree than the private automobile and the service systems designed around it. In general, however, transportation planners, engineers, and the automotive industry met many past challenges. In this process other problems were created and a few thorny issues remain unresolved, particularly because of the attractiveness of private automobiles.

For example, transportation demand will continue to grow and consumers will continue to move to larger vehicles for comfort and safety reasons. A major implication of all this is increased energy consumption and air pollution. In contrast, communities and their governments continue to mandate lower emissions per vehicle, different propulsion systems to decrease dependency on fossil fuels, and better energy efficiency standards.

To meet the gap between transportation demand trends and desired futures, a variety of policy options have been devised and are currently being tested worldwide. Among these we find:

- The definition of pollutant concentration standards and enforcement, with major emphasis on ozone, CO, NO_x, HC, PM, and CO_2
- Land use changes and controls and an attempt to link optimal land use favoring environmentally friendly modes
- Transportation demand management (TDM) in the form of carpooling, ride sharing, telecommuting and teleshopping, and information provision (the TCMs)
- Intelligent Transportation Systems that make use of information technology for transportation — road telematics

- Development and implementation of new vehicle technology, with emphasis on new fuels and propulsion systems
- City-based integrated plans that combine land use control, TDM, ITS, and clean cars

In Europe and the United States fuel efficiency and emissions are mandated and an expectation exists that larger organized fleets will be the first target groups for implementation of the mandates. In addition, both the European Union and the United States intend to phase into the market cleaner cars, making the manufacturers responsible for introduction. In parallel, safety is not eliminated from car requirements, so size alone will not be sufficient to resolve any safety-related issues. Moreover, legislation is in place to introduce many relatively new fuels (CNG, LPG, methanol, ethanol, hydrogen, electricity), and many new ideas are emerging, such as clear zones in Europe, individual marketing in Australia, and clean cities in the United States.

In the modeling and simulation of these programs and the agents participating in them, activity theory appears to be the only general framework that can be used to incorporate the four dimensions (time, geographic space, social space, and jurisdictional authority) in the same framework and reflect interactions across multiple scales within each dimension. It also incorporates formidable conceptual machinery to model development dynamics for each entity in the domain of a model system. In fact, the evolution of model systems in transportation planning is following a path to more flexible model structures that contain increasingly more complex relationships among decision-making entities. Consideration for the need to address multiple scales and their interaction is also increasing. From this viewpoint, model development is moving toward a more activity theory-based general framework. The following section provides a road map to other resources that contain collections of these models and ideas.

1.10.1 Key References in Planning and Travel Behavior Modeling

Transport systems planning and simulation have experienced a radical change of modeling paradigms in which macroscopic approaches to forecasting are increasingly replaced by microscopic (disaggregate) model systems targeting travel behavior of individuals and their households. This transformation in approaches is particularly pronounced in travel behavior, which is the leader in new model development in this field. Witness to this are the following books, which provide overviews of the journey from aggregate to disaggregate models:

Hensher, D.A. and Button, K.J., *Handbook of Transport Modelling*, Pergamon, Amsterdam, 2000.
Meyer, M.D. and Miller, E.J., *Urban Transportation Planning*, 2nd ed., McGraw-Hill, Boston, MA, 2001.
Ortuzar, J. deDios and Willumsen, L.G., , *Modelling Transport*, 3rd ed., Wiley, Chichester, U.K., 2001.

This evolution is presented in more detail, with examples from the most recent model systems in travel behavior, in three key journals: *Transportation Research Record*, *Transportation Research*, and *Transportation*. In addition, a very important series of edited conference proceedings of the International Association for Travel Behavior Research is an excellent reference for many new developments and the most important research directions in this field. The following five volumes contain many interesting and groundbreaking articles:

Hensher, D.A., *The Leading of Travel Behaviour Research*, Pergamon, Amsterdam, 2002.
Mahmassani, H.S., *In Perpetual Motion: Travel Behavior Research Opportunities and Application Challenges*, Pergamon, Amsterdam, 2002.
Ortuzar, J. deDios, Hensher, D.A., and Jara-Diaz, S., *Travel Behavior Research: Updating the State of Play*, Pergamon, Amsterdam, 1998.
Stopher, P. and Lee-Gosselin, M., *Understanding Travel Behavior in an Era of Change*, Pergamon, Amsterdam, 1997.
Travel Model Improvement Program, Activity-Based Travel Forecasting Conference: Summary, Recommendations, and Compendium of Papers, prepared by Texas Transportation Institute, U.S. DOT, Washington, D.C., 1997.

In parallel to modeling and simulation, data collection has also evolved over time, and a few publications provide summaries that can be used to build a good understanding of the key issues. The latest in a series of data collection publications describing the state of the practice, the state of the art, and future directions is the volume of proceedings from the International Conference on Transport Survey Quality and Innovation in Grainau, Germany, in 1997, published by the Transportation Research Board in collaboration with the Ministry of Transport in The Netherlands, Socialdata in Munich, Germany, and the Transport Research Centre, Melbourne, Australia:

Jones, P. and Stopher, P., Eds., Transportation Research Circular E-C008, TRB, Washington, D.C., 2000.

Two additional documents of particular use are a manual on travel surveys and a case study on data needed for regional transportation models:

Travel Model Improvement Program, Data Collection in the Portland, Oregon Metropolitan Area Case Study, prepared by Cambridge Systematics Inc., U.S. DOT, Washington, D.C., 1996.
Travel Model Improvement Program, Travel Survey Manual, prepared by Cambridge Systematics Inc., U.S. DOT, Washington, D.C., 1996.

1.10.2   Key Websites in Planning and Travel Behavior Modeling

There are also a few useful websites (accessed in April 2002) with rich data content and information about transportation policy and planning:

http://www.trafficlinq.com/ — An excellent site with over 1000 links to the most useful websites for transportation professionals. It contains information on public transport, ITS, research, libraries, software, road safety, traffic, and so forth. It also contains a section on books.
http://www.bts.gov — A gateway to the largest public domain clearinghouse of data and information in transportation in the United States. It also contains links to all other major agencies that collect data and the National Transportation Library.
http://www.tc.gc.ca/ — A Canadian gateway to transport. This is the website of the Ministry of Transport.
http://www.europa.eu.int/comm/eurostat/ — A European clearinghouse for data and other information about the European Union and countries that are related to the EU.
http://www.worldbank.org/html/fpd/transport/ — Data, information, and publications about transport around the world.
http://www.roads.detr.gov.uk/ — Very interesting site in the U.K. about regions and regional planning.
http://www.smarturbantransport.com/ — An electronic journal about transport issues from around the world.

More topic-specific sites are:

http://www.nhtsa.dot.gov/ — The safety transportation agency in the United States
http://www.energy.gov/transportation/index.html — Information and news
http://www-fars.nhtsa.dot.gov/ — Safety data in the United States
http://www.ccities.doe.gov/ — Clean Cities program website
http://www.vtpi.org/tdm/ — Comprehensive reference to transportation demand management
http://www-pam.usc.edu/ — Electronic journal on planning and markets published by U.S.C.
http://vwisb7.vkw.tu-dresden.de/TrafficForum/journal — An e-journal of a peer-reviewed paper on transport dynamics

References

Banister, D. et al., *European Transport Policy and Sustainable Mobility*, Spon Press, London, 2000.
Bartelmus, P., *Environment, Growth and Development: The Concepts and Strategies of Sustainability*, Routledge, London, 1994.

Beattie, C.I., Longhurst, J.W.S., and Woodfield, N.K., Air quality management: Evolution of policy and practice in the UK as exemplified by the experience of English local government, *Atmos. Environ.*, 35, 1479–1490, 2001 (www.sciencedirect.com).

Bernow, S., Transportation in the 21st Century: Contexts, Prospects, and Visions, paper presented at Tellus Institute, Seattle, 2000.

Birky, A. et al., Future U.S. Highway Energy Use: A Fifty Year Perspective, Draft Report prepared for Office of Transportation Technologies, Energy Efficiency and Renewable Energy, U.S. Department of Energy, 2001.

Bradshaw, R., Evaluation of Pilot Programmes of Site Specific Advice on Travel Plans Final Report: Executive Summary, University of Westminster, U.K., 2000, available at http://www.wmin.ac.uk/transport/projects/trplsumm.htm (accessed May 2002).

Bureau of Transportation Statistics, Transportation Statistics Annual Report 2000, BTS 01-02, BTS, U.S. DOT, Washington, D.C., 2001.

Bureau of Transportation Statistics, Transportation Indicators February 2002 and Subsequent Months, 2002, available at http://www.bts.gov/transtu/indicators/ (accessed April 2002).

California Energy Commission, 1997 Global Climate Change Report: GHG Emissions Reduction Strategies for California, CEC, 1998 (http://38.144.192.166/global_climate_change/97_report.html).

California Energy Commission, 1999 Fuels Report California Energy Commission Docket Proceeding 99-FR-1, PN 300-99-001, 1999 (http://38.144.192.166/FR99/index.html).

Clifton, K.J. and Handy, S.L., Qualitative Research in Travel Behaviour Research, paper presented at Workshop B5, International Conference in Transport Survey Quality and Innovation, Kruger National Park, South Africa, August 5–10, 2001.

Commission of the European Communities, Directive 96/69/EC of the European Parliament and of the Council, Brussels, 1996.

Commission of the European Communities, Communication for the Commission: A Review of the Auto-Oil II Program, Brussels, 2000.

Creighton, R.L., *Urban Transportation Planning*, University of Illiniois Press, Urbana, 1970.

Davis, S.C., *Transportation Energy Data Book*, 20th ed., Center for Transportation Analysis, Oak Ridge National Laboratory, Oak Ridge, TN, 2000.

DeCicco, J. and Mark, J., Meeting the energy and climate challenge for transportation in the United States, *Energy Policy*, 26, 395–412, 1998.

DeLucchi, M., Emissions of Greenhouse Gases from the Use of Transportation Fuels and Electricity, Vol. 1, ANL/ESD/TM-22, Center for Transportation Research, Argonne National Laboratory, Argonne, IL, 1991.

DeLucchi, M., The Annualized Social Cost of Motor-Vehicle Use in the US, 1990–1991: References and Bibliography, UCD-ITS-RR-96-3(20), Institute of Transportation Studies, University of California, Davis, 1996.

DeLucchi, M., The Annualized Social Cost of Motor-Vehicle Use in the US, 1990–1991: Summary of Theory, Data, Methods, and Results, UCD-ITS-RR-96-3(1), Institute of Transportation Studies, University of California, Davis, 1997.

Doyle, D.G. and Hess, D.B., Transportation and the Economy, Summary of Proceedings, Annual Symposium Series on the Transportation, Land Use, Air Quality Connection UCLA, Extension Public Policy Program, Los Angeles, CA, 1997.

Dunn, J.A., Jr., *Driving Forces: The Automobile, Its Enemies, and the Politics of Mobility*, Brookings Institution Press, Washington, D.C., 1998, 230 pp. (www.brookings.org).

Dunn, J.A., Jr., The auto, plus: Apolitical framework for assessing automobile-related problems and designing appropriate policy responses, *Transp. Q.*, 54, 7–10, 2000.

Engeström, Y., *Learning by Expanding*, Orienta-Kousultit, Helsinki, 1987.

Engeström, Y., Activity theory and individual and social transformation, in *Perspectives on Activity Theory*, Engeström, Y., Miettinen, R., and Punamaki, R.L., Eds., Cambridge University Press, U.K., 1999, chap. 1.

Federal Highway Administration, Highway Statistics, FHWA, U.S. DOT, Washington, D.C., 2001.

Fenger, J., Urban air quality, *Atmos. Environ.*, 33(29), 4877–4900, 1999.

Forkenbrock, D.J., External costs of intercity truck freight transportation, *Transp. Res. A*, 33, 505–526, 1999.

Garrett, M. and Wachs, M., *Transportation Planning on Trial: The Clean Air Act and Travel Forecasting*, Sage Publications, Thousand Oaks, CA, 1996.

Gilbert, R. and Nadeau, K., Decoupling Economic Growth and Transport Demand: A Requirement for Sustainability, paper presented at Conference on Transportation and Economic Development, Portland, OR, 2001.

Gómez-Ibáñez, J.A., Pricing, in Gómez-Ibáñez, J.A., Tye, W.B., and Winston, C., Eds., *Essays in Transporation Economics and Policy*, The Brookings Institution Press, Washington, D.C., 1999.

Goulias, K.G., On the role of qualitative methods in travel surveys, Workshop Report on qualitative methods Q-5, International Conference in Transport Survey Quality and Innovation, Kruger National Park, South Africa, August 5–10, 2001.

Goulias, K.G. and Szekeres, D., Centre Region Transportation Demand Management Plan, Final Draft Report prepared for the Centre Regional Planning Commission, University Park, PA, 1994.

Goulias, K.G. et al., A Study of Emission Control Strategies for Pennsylvania: Emission Reductions from Mobile Sources, Cost Effectiveness, and Economic Impacts, Final Report submitted to the Pennsylvania Low Emissions Vehicle (LEV) Commission for the Mid-Atlantic Universities Transportation Center, PTI Report 9403, University Park, PA, 1993.

Goulias, K.G. et al., Pennsylvania's Statewide Long Range Transportation Plan (PennPlan): An Overview and Summary, Preprint Paper 01–2799, Transportation Research Board 80th Annual Meeting, CD ROM Proceedings, Washington, D.C., 2001a.

Goulias, K.G., Viswanathan, K., and Kim, T., Pennsylvania's statewide long range transportation plan (PennPlan): Performance based planning in the US, in *Urban Transport VII: Urban Transport and the Environment for the 21st Century*, Sacharov, L.J. and Brebbia, C.A., Eds., WIT Press, Southampton, U.K., 2001b, pp. 43–52.

Greene, D.L., Why CAFE worked, in Transportation Research Circular 492, Transportation Research Board, National Research Council, Washington, D.C., 1999.

Greene, D.L. and Plotkin, S., Transportation sector, in Scenarios of U.S. Carbon Reductions: Potential Impacts of Energy Technologies by 2010 and Beyond, Report ORNL-444, Interlaboratory Report for the Office of Energy Efficiency and Renewable Energy, U.S. DOE, Oak Ridge National Laboratory, Oak Ridge, TN, 1997.

He, D. and Wang, M., Projections of Motor Vehicle Growth, Fuel Consumption and CO_2 Emissions for the Next Thirty Years in China, paper presented at the 80th Annual Meeting of the Transportation Research Board, Washington, D.C., January 7–11, 2001.

Humphrey, T.F., Suburban congestion: Recommendations for transportation and land use responses, *Transportation*, 16, 221–240, 1990.

John, G. and Bröeg, W., Individualized Marketing: The Perth Success Story, paper presented at Conference on Marketing Public Transport: Challenges, Opportunities, and Success Stories, Aotea Centre, Auckland, NZ, August 2001.

Jones, P., Ed., *Developments in Dynamic and Activity-Based Approaches to Travel Analysis*, Avebury, Aldershot, U.K., 1990.

Joyce, M., Developments in US Alternative Fuel Markets, 2001 (http://www.eia.doe.gov/cneaf/alternate/issues_trends/altfuelmarkets.html).

Kaptelinin, V., Activity theory: Implications for human–computer interaction, in Nardi, B., Ed., *Context and Consciousness: Activity Theory and Computer Interactions*, MIT Press, Cambridge, MA, 1996.

Kenworthy, J. R. and Laube, F.B., Patterns of automobile dependence in cities: An international overview of key physical and economic dimensions with some implications for urban policy, *Transp. Res. A*, 33, 691–723, 1999.

Kuutti, K., Activity theory as a potential framework for human–computer interaction research, in Nardi, B., Ed., *Context and Consciousness: Activity Theory and Human–Computer Itteractions*, MIT Press, Cambridge, MA, 1996, pp. 17–44.

Linzie, M., Future of International Activities: Transportation in the New Millenium: State of the Art and Future Directions, TRB, Washington, D.C., 2000.

Litman, T., You Can Get There from Here: Evaluating Transportation Choice, Paper 01-3035, presented at the Transportation Research Board 80th Annual Meeting and included in the CD-ROM proceedings, Washington, D.C., 2001.

Madder, G.H. and Bevilacqua, O.M., Electric vehicle commercialization, in *Alternative Transportation Fuels: An Environmental and Energy Solution*, Sperling, D., Ed., Quorum Books, New York, 1989.

Maples, J.D. et al., Alternative fuels for U.S. transportation, in Transportation in the New Millenium: State of the Art and Future Directions, CD-ROM, Transportation Research Board, National Research Council, Washington, D.C., 2000.

Meyer, M.D., *A Toolbox for Alleviating Congestion and Enhancing Mobility*, Institute of Transportation Engineers, Washington, D.C., 1998.

Meyer, M.D. and Miller, E.J., *Urban Transportation Planning*, 2nd ed., McGraw-Hill, Boston, 2001.

Moon, D., Urban Transport: Options for Propulsion Systems and Instruments for Analysis, UTOPIA EC, UR-97-SC-2076, Final Report submitted to the European Commission, Brussels, 2001.

Morash, E.A., Demand-based transportation planning, policy, and performance, *Transp. Q.*, 54, 11–33, 2000.

Moritz, G. and Bröeg, W., Redesign of the Dutch travel survey: Response improvement, in Transportation Research Circular E-C026, *Personal Travel: The Long and Short of It*, TRB, Washington, D.C., 1999, pp. 365–377.

Nelson, P.O. et al., Environmental Impact of Construction and Repair Materials on Surface and Ground Waters: Summary of Methodology, Laboratory Results, and Model Development, Report 448, National Cooperative Highway Research Program, TRB, National Research Council, National Academy Press, Washington, D.C., 2001.

Newman, P. and Kenworthy, J., Costs of automobile dependence: Global survey of cities, *Transp. Res. Rec.*, 1670, 17–26, 1999.

Noland, R.B., Traffic Fatalities and Injuries: Are Reductions the Result of "Improvements" in Highway Design Standards? Paper presented at the 80th Annual Meeting of the Transportation Research Board, Washington, D.C., January 7–11, 2001.

Organization for Economic Co-operation and Development, Policies to Enhance Sustainable Development, Paper presented at Meeting of the OECD Council at Ministerial Level, Paris, France, 2001.

Pickrell, D., Cars and clean air: A reappraisal, *Transp. Res. A*, 33, 527–547, 1999.

Pucher, J., Transportation trends, problems, and policies: an international perspective, *Transp. Res. A*, 33, 493–503, 1999.

Pucher, J. and Lefevre, C., *The Urban Transport Crisis in Europe and North America*, MacMillan Press, London, 1996.

Rothenberg, J.G. and Heggie, I.G., Eds., *Transport and the Urban Environment*, Wiley, New York, 1974.

Savonis, M.J., Toward a strategic plan for transportation-air quality research, 2000–2010, *Transp. Res. Rec.*, 1738, 68–73, 2000.

Schafer, A., Carbon dioxide emissions from world passenger transport, *Transp. Res. Rec.*, 1738, 20–21, 2000.

Schafer, A. and Victor, D.G., The future mobility of the world population, *Transp. Res. A*, 34, 171–205, 2000.

Schipper, L. and Marie-Lilliu, C., *Carbon-Dioxide Emissions from Travel and Freight in IEA Countries: The Recent Past and the Long-Term Future*, Transportation Research Circular 492, TRB, Washington, D.C., 1999.

Schipper, L., Marie-Lilliu, C., and Lewis-Davis, G., Rapid Motorization in the Largest Countries in Asia: Implication for Oil, Carbon Dioxide and Transportation, paper presented at the 78th Transportation Research Board Annual Meeting, Washington, D.C., 1999.

Smerck, G.M., *Readings in Urban Transportation*, Indiana University Press, Bloomington, 1968.

Sperling, D., *Future Drive: Electric Vehicles and Sustainable Transportation*, Island Press, Washington, D.C., 1995.

Transit Cooperative Research Program, Costs of Sprawl: 2000, TCRP Report 74, TRB, National Research Council, Washington, D.C., 2002.

Transportation Research Board, Current Practices for Assessing Economic Development Impacts from Transportation Investments, National Cooperative Highway Research Program, Synthesis 290, TRB, National Research Council, Washington, D.C., 2000.

U.S. Department of Energy, Replacement Fuel and Alternative Fuel Vehicle Technical and Policy Analysis, pursuant to Section 506 of the Energy Policy Act of 1992, U.S. DOE, Energy Efficiency and Renewable Energy, Office of Transportation Technologies, 1999.

U.S. Department of Energy, Assumptions to the Annual Energy Outlook 2001, U.S. DOE/EIA0554, Washington, D.C., 2001.

Vyas, A.D. et al., Electric and Hybrid Electric Vehicles: A Technology Assessment Based on a Two-Stage Delphi Study, ANL/ESD-36, Center for Transportation Research, Energy Systems Division, Argonne National Laboratory, Argonne, IL, 1997.

Vygotsky, L.S., *Mind in Society: The Development of Higher Psychological Processes*, Cole, M. et al., Eds., Harvard University Press, Cambridge, MA, 1978.

Wark, K., Warner, C.F., and Davis, W.T., *Air Pollution: Its Origin and Control*, Addison-Wesley, Menlo Park, CA, 1998.

Webb, R.F., Moyer, C.B., and Jackson, M.D., Distribution of natural gas and methanol: Costs and opportunities, in *Alternative Transportation Fuels: An Environmental and Energy Solution*, Sperling, D., Ed., Quorum Books, New York, 1989.

World Bank, *Sustainable Transport: Priorities for Policy Reform*, World Bank, Washington, D.C., 1996.

Yin, X., *An Activity Theory Approach in Intelligent Transportation Systems Design*, M.S. thesis, Pennsylvania State University, University Park, 2000.

2

Time Use and Travel Behavior in Space and Time

Ram M. Pendyala
University of South Florida

2.1 Introduction

Transportation systems are planned and designed to provide people with the ability to engage in activities at locations and times of their preference. When people cannot engage in activities at locations and times of their preference, the transportation system is deemed to provide a poor level of service. Transportation models are aimed at modeling and forecasting where and when the demand for travel will occur so that the transportation system can be planned and designed to meet the projected travel demand and ensure a high quality of life for the residents and visitors of a geographical region. Thus the analysis of travel behavior is inextricably linked to the concepts of space and time, and there is a growing body of literature that makes a strong case for the development of transportation models and planning methods that explicitly recognize the role of space and time dimensions in people's travel behavior.

The traditional approach to travel demand analysis and forecasting has relied on models of travel demand that are based on computing four major aspects of travel behavior:

1. How many trips are made? (trip generation)
2. Where are trips made? (trip distribution)
3. By what means of transportation are trips made? (modal split)
4. On what route or path are trips made? (network assignment)

The spatial element of travel plays an important role in all steps of the modeling process, but is mostly captured in the second step, trip distribution. In this step, trip origin and destination locations are identified using measures of zonal attractiveness or activity levels and degree of separation to determine trip interchanges between zone pairs. The spatial characteristics of trips in turn influence the choice of mode and route in the subsequent two steps of the four-step modeling process.

The time element of travel is less explicitly captured in the current transportation modeling process, although it is at least as important as the spatial element. The timing of travel is often addressed through transportation models that are formulated or adjusted to obtain peak hour or peak period travel demand, and the duration of travel is often addressed by the inclusion of different forms of travel time variables in trip distribution, modal split, and network assignment models.

Significant changes in the past few decades in sociodemographic characteristics of households, urban structure, industrial composition, and transportation systems have resulted in increasingly complex activity engagement and travel patterns. Consequently, although infrastructure expansion continues to play a major role in transportation planning and analysis, there is a growing emphasis being placed on transportation systems management and, more recently, on the role and impacts of various travel demand management (TDM) strategies and transportation control measures (TCMs). Although current travel models capture several fundamental aspects of transportation demand, they are based on sets of assumptions and paradigms that do not adequately reflect the spatiotemporal interdependencies that are inherent in the organization of activities and travel. This realization has led to the growing interest in travel forecasting methods that incorporate activity participation and time allocation behavior. It has been increasingly realized that transportation is a derived demand in that the way individuals and households organize their lives dictates when and where they travel. Recent developments in the transportation modeling field have paid considerable attention to the notion of time use with the belief that understanding the mechanisms of activity participation and time allocation will lead to increased capability in forecasting travel demand and evaluating planning options (Kitamura et al., 1997b).

As an example of the importance of recognizing time and space dimensions in models of activity and travel demand, one may consider the case of telecommuting. When a worker telecommutes (from home), the commute to and from the work location is eliminated. Therefore, the worker now has additional time available for pursuing activities. The elimination of the commute trip influences the duration of travel or activity engagement. Besides influencing duration, telecommuting may influence the timing and location of activity engagement. Whereas a worker may have pursued nonwork activities in combination with the commute when traveling to and from work, the worker may now choose to engage in nonwork activities at other times of the day and at locations closer to home. In the absence of the commute trip, the worker no longer has the need or opportunity to link nonwork activities to the commute trip and the work location. Analyzing these spatial and temporal shifts in activity engagement patterns is important for accurately assessing the impacts of telecommuting on travel demand.

The role of time in travel behavior analysis is further amplified by the fact that it is a finite and critical resource that is consumed in the engagement of activities and travel. All activities and trips consume time, and regardless of the time span under consideration, there is only limited time within which an individual can pursue activities and trips. In turn, the spatial dimension is very closely related to the temporal dimension as the distance traversed and the set of possible destination opportunities are dictated by timing and time availability. Thus, there is only a finite spatiotemporal action space in which an individual can pursue activities and travel. Moreover, there may be additional personal, work- and school-related, household, institutional, and modal constraints that limit the size of the spatiotemporal action space of an individual.

In recent years, the state of the art in travel demand modeling has moved in the direction of developing and implementing activity-based models of travel demand that explicitly recognize the important role played by time and space in shaping activity and travel patterns of individuals. In the new planning context where TDM strategies and TCMs are inherently linked to time and space dimensions, activity-based approaches that capture the relationships between time use and travel behavior in space and time offer a stronger behavioral framework for conducting policy analyses and impact studies.

This chapter aims to provide a general overview of the role of time use in analyzing travel behavior in space and time. It includes several specific examples that demonstrate how the explicit recognition of the notion of time can offer valuable insights into human activity and travel behavior.

2.2 Time Use and Travel: A Descriptive Analysis

Activity-based travel analysis is increasingly being recognized as a powerful methodology to model human travel behavior. Activity-based travel analysis recognizes that individuals' activity and trip patterns are a manifestation of their decision to allocate time to various activities during a day. Travel is then derived from an individual's desire to perform an activity at a location away from the previous activity location. Recent research has argued that information on individual activity engagement behavior offers the potential to enrich our understanding of the complex and dynamic nature of travel executed by people. Benefits of the activity-based approach include the determination of (1) spatial and temporal constraints on activity and travel choice, (2) scheduling and sequencing of activities in time and space, (3) interactions between activity and travel decisions, and interactions between individuals, and (4) roles played by members of a household in accomplishing household activities and tasks.

Time use research is playing an increasingly important role in activity and travel behavior research because of the recognition that many travel choices are governed by time, which is a limited resource that is consumed according to one's needs and preferences. Activity data are often derived from time use studies that record all in-home and out-of-home activities and all trips performed during the survey period in a sequential manner. Potentially, the explicit representation of time use in travel demand models will further help to explain people's travel choices over the course of a day.

Time use and activity studies have focused on the examination of various aspects of activity and travel behavior, including:

1. Daily time allocation: In these studies, the total time allocated to various activity categories or purposes is examined or modeled at the day level. In these studies, individual episodes are not explicitly considered. As such, while these studies focus on daily time use and allocation behavior, they are unable to consider issues such as activity timing, frequency, episode duration, or activity scheduling and sequencing. On the other hand, they provide strong insights into daily time allocation and the trade-offs associated with having to allocate time among various activity types in a typical 24-h day.

2. Activity episode duration: Studies of episode duration analysis have focused on modeling the duration of individual activity episodes by purpose or category. These analyses provide a powerful mechanism for understanding the factors that influence individual activity episode durations and the probability that a certain activity will be terminated given that a certain duration has elapsed. Episode duration models have traditionally taken the form of hazard-based duration models and Tobit models that help explain time use in the context of a single activity episode. Within the context of these models, it is often possible to reflect the interdependence among activity episodes and the timing of activities, as the end of one activity episode reflects the beginning of another activity episode. On the other hand, issues associated with daily time allocation to activities, sequencing of activities, and scheduling of activities are more difficult to capture explicitly in models of episode duration.

3. Activity timing and scheduling: Activity timing and scheduling models focus on identifying when a certain activity or trip will be pursued. Hazard-based duration models, time-of-day period-based discrete choice models, and heuristic algorithms have been used to model activity timing and scheduling behavior. Although these studies do not necessarily capture time use behavior, they do examine the role of time in activity–travel behavior, as timing and scheduling decisions are, by definition, temporal in nature.

4. Activity sequencing: Activity sequencing studies are concerned with the sequence in which various activities and trips are linked. Thus, activity sequencing studies directly capture the essence of trip

chaining because trip chaining is simply a manifestation of activity sequencing decisions. Various methods, including sequence alignment techniques, discrete choice models, and heuristic rule-based algorithms, have been used to model activity sequencing decisions. While activity sequencing does occur along the time dimension and is closely related to activity timing and scheduling, these studies have incorporated time use behavior only in a limited way.

5. Activity frequency: Activity frequency models focus on the number of occurrences of various types of activities. These models often take the form of Poisson or negative binomial regression equations, discrete choice models, or other models suitable for representing count phenomena. In general, these models do not explicitly capture the time dimension, as they are exclusively focused on the number of times an activity is pursued, regardless of the durations of the episodes.

Among the five types of studies noted above, the first three are directly related to the notion of time use and its role in travel behavior. Therefore, only examples of models pertaining to these three aspects of time use (i.e., daily time allocation, activity episode duration, and activity timing) are presented in this chapter. The examples presented in this chapter serve as applications of the usefulness of the notion of time use in transportation demand and policy analysis.

Over the past several years, there have been several activity-based time use and travel surveys undertaken in the transportation planning, modeling, and survey research arenas. However, time use research and studies have been undertaken in the social sciences for many years. These surveys have afforded the opportunity to quantify the activity and time use behavior of individuals in the context of their travel. The remainder of this chapter provides descriptive analysis and statistics on activity and time use patterns that have been obtained in some recent surveys.

When examining activity and time use patterns, it is very important to note that activity and time use behavior varies considerably by demographic segment and by survey method. Demographic characteristics that may contribute to differences in time use and activity patterns include employment status, age, sex, education, income, household composition, and land use–transport environment. Besides demographic factors, the survey methodology may also result in differences in activity and time use patterns. For example, the design of the survey instrument may have important implications for the reporting of activities and time. While some instruments are sequential in nature, collecting information on each activity pursued by an individual in a sequential fashion, other instruments utilize the time diary format, where individuals enter their activities in various time intervals, similar to a day planner or personal calendar. Also, whether the survey is self-administered (e.g., mail-out mail-back survey) or interviewer administered (e.g., computer-assisted telephone interview (CATI)) may have an important bearing on the activity and time use data collected in a survey. Within the scope of this chapter, it is not possible to provide a rigorous analysis and description of time use and activity patterns by demographic segment while controlling for survey method. Therefore, the statistics presented here distinguish only between commuters and noncommuters and are derived from CATI surveys.

The data presented in this section are derived from three different surveys conducted in the past decade. All of the surveys may be regarded as activity-based time use and travel surveys administered by CATI techniques. The three surveys include the 1996 San Francisco Bay Area activity survey, the 1998 Miami activity survey of commuters, and the 1994 Washington, D.C., activity survey of commuters. Among these three surveys, only the 1996 San Francisco Bay Area survey includes a sample of noncommuters; therefore, the sample derived from this survey is split into commuter and noncommuter samples for describing time use and activity characteristics. Even though all surveys were administered by similar means, they used different activity categories. As such, any comparison of statistics across the three surveys must be done with caution, recognizing that the activity categories may not be exactly equivalent.

The 1996 San Francisco Bay Area activity survey was a 2-day time use and travel survey conducted in the nine counties of the San Francisco Bay Area. Detailed information on both in-home and out-of-home activities and trips undertaken by an individual was recorded in the survey. While information on all trips and trip segments (in the case of chained trips) was collected, in-home activity information was requested only for those activities that were longer than 30 min in duration. However, many of the respondents provided detailed information on all in-home activities, regardless of duration. On the other hand, information on all out-of-home activities was collected irrespective of their duration.

The CATI survey elicited a favorable response from 14,431 persons residing in 5857 households in the bay area. They provided detailed household and person level socioeconomic and demographic data. The survey intended to collect detailed activity and trip information for all individuals residing in a household. However, not all individuals who provided demographic data furnished complete activity and trip information. Only 8817 individuals residing in 3919 households provided detailed activity and trip information over a 48-h period. After extensive data checking, cleaning, and merging and organizing, the final data set included 7982 persons residing in 3827 households. Among the 7982 persons, 4331 were commuters and the remaining 3651 were noncommuters. Full-time or part-time workers who had at least one work activity outside the home during the survey period were treated as commuters. Individuals reporting activities performed out of the study area and individuals who provided activity trip information for 1 day or less during the survey period were not included in the final sample.

The Miami–Dade County activity-based travel behavior and time use survey was conducted in Florida in 1998. The survey collected detailed information on both in-home and out-of-home activities and on all travel associated with these activities. Unlike the San Francisco Bay Area survey, the Miami survey collected activity and travel behavior data for only a 1-day (24-h) period. In addition, the sample consisted exclusively of commuters who were defined as individuals who commuted to a regular work or school location at least 3 days a week. Only one randomly selected commuter was chosen to participate from each household.

Similar to the Bay Area survey, the Miami survey was administered using the CATI technique. Socio-economic and demographic information about the household and about persons residing in the household was collected first. Information regarding the usual commute to and from work was collected from the randomly selected commuter. Activity and time use data were collected only from eligible commuter respondents. Unlike the Bay Area survey, the Miami survey did not have any duration threshold for reporting of activities. All activities, regardless of their length, were recorded in the data set. Similar to the Bay Area survey, the Miami survey included information on all trips, including individual trip segments of chained trips.

Socioeconomic and demographic data were collected for 2539 persons residing in 1040 households. As mentioned earlier, activity and trip data were collected only from commuters, with the constraint that each commuter must be drawn from a different household. A total of 803 commuters provided detailed information on their usual commute to and from work; of these, 640 provided detailed activity and trip information for the 24-h survey period. The analysis presented here, however, is performed only on a sample of 589 commuters, as the remaining respondents included full-time students with no work. Even though the omitted respondents were considered commuters from a survey standpoint, it was felt that they should not be included here for reasons of compatibility and comparability across the surveys.

Finally, a very detailed activity-based travel survey was administered using CATI techniques to a random sample of 656 commuters in the Washington, D.C., metropolitan area in 1994. This survey was conducted as part of a larger study to develop an activity-based travel demand forecasting system and policy evaluation tool called AMOS — Activity Mobility Simulator. As is typical with most travel surveys, the survey gathered information on the socioeconomic and demographic characteristics of the commuters. In addition, commuters were asked to provide data on their typical travel patterns over the duration of an average week. The survey then collected very detailed and revealing preference information on all out-of-home and in-home activities that one randomly chosen commuter in a household pursued over a 24-h period.

Table 2.1 provides a summary of the socioeconomic characteristics of the households, while Table 2.2 provides a summary of the person characteristics in each of the survey samples. An examination of Table 2.1 shows that the survey samples exhibit both similarities and differences with respect to household characteristics. It should be noted that the Miami and Washington, D.C., samples include only households

TABLE 2.1   Household Characteristics of Survey Samples

Characteristic	San Francisco	Miami	Washington, D.C.
Sample size	3827	640	656
Household size	2.3	3.2	2.7
Income:			
Low (< $30 K)	15.8%	29.4%	n.a.
Medium ($30–$75 K)	44.4%	40.9%	n.a.
High (> $75 K)	26.7%	19.7%	n.a.
Vehicle ownership	1.9	2.1	2.0
% Vehicles ≥ commuters	86.4	64.4	90
Number of workers	1.4	2.5	1.7
Zero worker household	16.5%	n.a.	n.a.
Number of bicycles	1.3	1.4	1.4

Note: n.a. = not applicable or not available.

that have at least one regular commuter who works outside the household. Some of the differences across the survey samples are simply a manifestation of the difference in sampling scheme. In the San Francisco survey sample, 16.5% of the households have no worker who commutes to a workplace outside home. This is reflected in the smaller average household size and number of workers in the household for the San Francisco sample. Auto availability, represented by the percent of households where the number of vehicles is greater than or equal to the number of commuters, is quite high in the San Francisco and Washington, D.C., samples, where about 90% of the households fall into this category. For the Miami sample, the corresponding percentage is only about 65%, reflecting a lower level of auto availability relative to the San Francisco and Miami samples.

The person characteristics summarized in Table 2.2 once again show that there are similarities and differences across the survey samples. Once again, it should be noted that the Miami and Washington, D.C., samples are pure commuter samples. As expected, whereas the commuter samples show relatively strong similarities in person characteristics, the noncommuter sample in the San Francisco survey shows substantial differences in age, license holding, and student status.

Table 2.3 shows the average activity and travel characteristics of the person samples with a view toward providing insights into average time use patterns. Differences and similarities in time use patterns across the survey samples should be viewed in the context of the differences and similarities in their household and person sociodemographic characteristics seen in Tables 2.1 and 2.2.

TABLE 2.2   Person Characteristics of Survey Samples

Characteristic	San Francisco Noncommuters	San Francisco Commuters	Miami	Washington, D.C.
Sample size	3651	4331	589	656
Age (in years):	32.4	41.5	n.a.	40.1
Young (≤ 29)	54.0%	18.8%	25.3%	21.3%
Middle (30–49)	14.5%	53.8%	48.9%	59.7%
Old (≥ 50)	31.5%	27.4%	22.3%	19.9%
Employment status:				
Full-time	n.a.	81.5%	80%	88%
Part-time	n.a.	12.1%	15%	11%
Licensed	48.6%	95.3%	93.0%	98%
Student	44.8%	13.3%	11.1%	n.a.
Work mode choice:				
Single-occupancy auto	n.a.	68%	72%	70%
Car- or vanpool	n.a.	13%	18%	16%
Transit	n.a.	8%	3%	10%
Nonmotorized	n.a.	11%	5%	3%

Note: n.a. = not applicable or not available.

TABLE 2.3 Time Use and Activity Characteristics of Survey Samples

Characteristic	San Francisco		Miami	Washington, D.C.
	Noncommuters	Commuters		
Sample Size	3651	4331	589	656

Daily Activity Durations

Work	00:00 (0%)	06:41 (28%)	07:00 (29%)	07:44 (32%)
Sleep	09:23 (40%)	07:57 (32%)	07:56 (32%)	07:13 (30%)
In-home maintenance	03:43 (15%)	02:28 (11%)	02:29 (11%)	02:25 (10%)
Personal care/child care	01:16 (5%)	01:08 (5%)	01:24 (6%)	n.a.
Out-of-home maintenance	00:47 (3%)	00:44 (3%)	00:45 (3%)	01:00 (4%)
Shopping/personal business	00:34 (2%)	00:23 (2%)	00:24 (2%)	n.a.
In-home recreation	03:46 (16%)	02:12 (9%)	01:51 (8%)	01:47 (7%)
Out-of-home recreation	01:10 (5%)	00:46 (3%)	00:40 (3%)	00:26 (2%)
Eating/meal preparation	01:46 (7%)	01:24 (6%)	01:23 (6%)	n.a.
School	02:21 (10%)	00:07 (1%)	00:00 (0%)	00:00 (0%)
Missing (unaccounted time)	00:05 (0.5%)	00:07 (1%)	00:15 (1%)	00:18 (1%)
Total	00:59 (4%)	01:34 (7%)	01:41 (7%)	02:00 (8%)

Daily Travel Durations

Work	00:00 (0%)	00:29 (32%)	00:34 (34%)	00:45 (38%)
Out-of-home maintenance	00:18 (32%)	00:17 (19%)	00:26 (26%)	00:26 (22%)
Shopping/personal business	00:08 (14%)	00:07 (8%)	00:10 (10%)	n.a.
Child care/serve child	00:01 (2%)	00:01 (1%)	00:06 (6%)	n.a.
Other	00:09 (16%)	00:09 (10%)	00:10 (10%)	n.a.
Out-of-home recreation	00:08 (14%)	00:07 (8%)	00:06 (6%)	00:07 (6%)
Eat meal (out of home)	00:03 (5%)	00:05 (5%)	00:06 (6%)	n.a.
Return home	00:23 (39%)	00:34 (36%)	00:28 (28%)	00:42 (35%)
School	00:06 (10%)	00:00 (0%)	00:00 (0%)	00:00 (0%)

Note: For the San Francisco and Miami samples, the in-home and out-of-home portions of the eating and meal preparation activities are not available. For the Washington, D.C., sample, these portions have been added to the in-home and out-of-home maintenance categories. All durations are represented in hours and minutes in the format hh:mm. Figures in parentheses indicate the percentage of the day (1440 min) dedicated to the activity, except in the case of travel durations, where the figures represent the percent of total travel time dedicated to each travel purpose. n.a. = not applicable or not available.

An examination of the statistics presented in Table 2.3 shows that commuters generally exhibit similar characteristics across the three survey samples. As expected, noncommuters tend to have activity and time use characteristics that are substantially different from those of commuters. While some of the differences in time use characteristics can be related to differences in socioeconomic characteristics, one should be careful in trying to explain differences in time use patterns as a function of differences in socioeconomic characteristics. One may postulate that many sociodemographic factors, often considered explanatory variables of time use, are in fact endogenously determined by an individual's or household's long-term lifestyle choices and short-term activity decisions. Thus, one may be able to infer lifestyle choices by noting time use patterns exhibited by an individual or household.

In addition to the statistics derived from activity-based travel surveys, as shown in Table 2.3, the literature offers additional insights into time use patterns of individuals. Kitamura et al. (1997a) provide a comparative description of time use patterns of survey samples drawn from The Netherlands, California, and the United States (a nationwide sample). Descriptive time use statistics provided in their paper account for sex (male or female), working status (working or not working on survey day), and type of day (weekday or weekend day). Table 2.4 offers a summary of the time use statistics derived from their tabulation.

The Dutch and California data sets represent time use patterns of randomly chosen individuals. The Dutch time use survey included home interviews from a sample of 2964 individuals, with a response rate of 54%. The time use survey conducted in California had a response rate of 62% and yielded a

TABLE 2.4   Activity Durations by Activity Type, Sex, and Working Status for Weekday and Weekend

Activity Category	Day Type	Survey Area	Overall Average	Working		Not Working	
				Male	Female	Male	Female
Paid work	Weekday	Netherlands	02:47	07:49	05:50	n.a.	n.a.
		California	04:07	07:33	07:19	n.a.	n.a.
		U.S.	03:50	07:43	07:18	n.a.	n.a.
	Weekend	Netherlands	00:28	04:47	04:38	n.a.	n.a.
		California	01:35	06:59	05:25	n.a.	n.a.
		U.S.	01:29	06:09	06:08	n.a.	n.a.
Domestic work	Weekday	Netherlands	02:40	00:49	02:16	02:18	04:04
		California	01:50	00:38	01:13	02:26	03:15
		U.S.	01:52	00:58	01:22	02:11	02:43
	Weekend	Netherlands	02:16	01:02	02:00	01:46	02:46
		California	02:04	00:45	01:37	01:54	02:45
		U.S.	02:05	00:58	01:24	02:15	02:28
Child care	Weekday	Netherlands	00:29	00:11	00:19	00:11	00:50
		California	00:18	00:08	00:13	00:08	00:39
		U.S.	00:18	00:10	00:15	00:15	00:28
	Weekend	Netherlands	00:27	00:16	00:21	00:16	00:36
		California	00:15	00:13	00:17	00:12	00:18
		U.S.	00:16	00:08	00:28	00:18	00:19
Shopping, errands	Weekday	Netherlands	00:34	00:13	00:31	00:38	00:46
		California	00:36	00:19	00:30	00:35	01:00
		U.S.	00:32	00:16	00:21	00:35	00:50
	Weekend	Netherlands	00:27	00:16	00:21	00:23	00:31
		California	00:39	00:12	00:16	00:38	00:55
		U.S.	00:36	00:26	00:29	00:36	00:42
Personal care	Weekday	Netherlands	01:09	00:53	00:59	01:24	01:15
		California	01:12	00:56	01:17	01:21	01:20
		U.S.	01:23	01:14	01:16	01:27	01:39
	Weekend	Netherlands	01:14	01:08	01:17	01:12	01:16
		California	01:06	00:57	01:04	00:57	01:15
		U.S.	01:23	01:12	01:27	01:16	01:33
Education	Weekday	Netherlands	00:31	00:16	00:14	01:12	00:28
		California	00:18	00:12	00:14	00:27	00:22
		U.S.	00:43	00:13	00:12	01:37	01:09
	Weekend	Netherlands	00:10	00:06	00:05	00:14	00:08
		California	00:08	00:04	00:02	00:10	00:09
		U.S.	00:12	00:04	00:03	00:13	00:15
Organizational Activities	Weekday	Netherlands	00:19	00:14	00:10	00:25	00:21
		California	00:08	00:05	00:09	00:07	00:11
		U.S.	00:10	00:09	00:07	00:13	00:14
	Weekend	Netherlands	00:21	00:16	00:05	00:26	00:19
		California	00:19	00:02	00:23	00:20	00:22
		U.S.	00:23	00:04	00:09	00:25	00:32
Entertainment	Weekday	Netherlands	01:09	00:41	00:53	01:27	01:24
		California	00:31	00:23	00:23	00:46	00:39
		U.S.	00:36	00:25	00:23	00:51	00:47
	Weekend	Netherlands	01:41	02:08	02:05	02:56	02:37
		California	01:13	00:57	00:42	01:21	01:19
		U.S.	01:10	00:46	00:43	01:20	01:16
Sports, hobbies	Weekday	Netherlands	00:54	00:26	00:39	01:07	01:11
		California	00:34	00:25	00:18	01:09	00:40
		U.S.	00:43	00:24	00:27	01:14	00:56
	Weekend	Netherlands	01:08	00:34	00:46	01:10	01:12
		California	00:50	00:35	00:25	00:57	00:55
		U.S.	00:57	00:23	01:00	01:11	01:05

continued

TABLE 2.4 (CONTINUED) Activity Durations by Activity Type, Sex, and Working Status for Weekday and Weekend

Activity Category	Day Type	Survey Area	Overall Average	Working		Not Working	
				Male	Female	Male	Female
TV, reading	Weekday	Netherlands	03:12	02:41	02:22	04:28	03:14
		California	03:43	02:43	02:22	05:36	05:02
		U.S.	03:18	02:31	02:19	04:06	04:07
	Weekend	Netherlands	03:43	03:20	02:33	04:25	03:31
		California	04:04	02:26	02:50	05:03	04:12
		U.S.	03:56	03:10	02:46	04:27	04:05
Meals	Weekday	Netherlands	01:17	00:59	01:01	01:29	01:27
		California	01:12	01:06	00:57	01:26	01:20
		U.S.	01:11	01:01	00:59	01:24	01:21
	Weekend	Netherlands	01:25	01:14	01:09	01:25	01:27
		California	01:23	01:12	01:07	01:29	01:27
		U.S.	01:17	01:01	00:59	01:24	01:23
Sleep	Weekday	Netherlands	07:50	07:20	07:37	08:14	08:02
		California	07:49	07:25	07:29	08:18	08:17
		U.S.	08:01	07:23	07:32	08:43	08:34
	Weekend	Netherlands	08:32	07:40	07:43	08:37	08:38
		California	08:32	07:36	08:07	09:04	08:37
		U.S.	08:45	08:01	07:42	09:07	08:57
Travel	Weekday	Netherlands	01:09	01:29	01:11	01:07	00:56
		California	01:35	02:05	01:30	01:35	01:10
		U.S.	01:25	01:34	01:29	01:25	01:13
	Weekend	Netherlands	01:04	01:12	00:57	01:11	00:59
		California	01:47	01:56	01:46	01:51	01:40
		U.S.	01:30	01:38	01:36	01:29	01:26
Sample size (Diary days)	Weekday	Netherlands	14,820	3732	2067	2768	6253
		California	1013	301	259	154	299
		U.S.	2206	597	568	481	726
	Weekend	Netherlands	5928	363	223	2237	3105
		California	566	90	50	179	247
		U.S.	841	109	105	307	388

Note: All durations are represented in hours and minutes in the format hh:mm. n.a. = not applicable or not available.

Source: Kitamura, R., van der Hoorn, T., and van Wijk, F., A comparative analysis of daily time use and the development of an activity-based traveler benefit measure, in *Activity-Based Approaches to Travel Analysis*, Ettema, D.F. and Timmermans, H.J.P., Eds., Pergamon, Elsevier Science Ltd., U.K., 1997. With permission.

sample of 1564 individuals. Whereas the Dutch time use survey employed a weekly time use diary format with 15-min time intervals and closed activity categories, the California survey adopted a sequential activity diary format in which all information about in-home and out-of-home activities and travel was collected sequentially for a 24-h period using an open activity category structure. The overall U.S. data set was obtained from a nationwide sample of 3047 individuals residing in 44 states. Time use information is available for a 1-day period for this sample of individuals. All of the surveys utilized similar activity categorization schemes, thus facilitating comparative tabulation of time use patterns across the surveys. In all of the survey samples, about 55% of the individuals are female. With respect to age distributions, the Dutch and California data sets are quite similar, while the U.S. national data set has a larger proportion of elderly individuals. The U.S. and California samples differ from the Dutch sample with respect to marital status; in general, the Dutch sample includes a larger proportion of married persons and persons who have never been divorced or separated.

This chapter has provided a descriptive analysis of time use statistics from recent activity-based and time use surveys conducted in various areas of the United States and the Netherlands. As mentioned in the earlier parts of this chapter, activity and time use studies have focused on various aspects of behavior, including such items as the frequency, scheduling, timing, and sequencing of activities. While some of these items will be addressed in the context of the examples and applications furnished in subsequent sections of this chapter, this section has focused mainly on daily time use behavior for the sake of emphasis and clarity in presentation. Even though a detailed presentation and discussion of the frequency, scheduling, and sequencing of activities is beyond the scope of this chapter, it is very important to note that daily time use patterns and time allocation behavior are inextricably linked to such aspects of activity behavior.

Time use and activity surveys have been conducted around the world over the past several decades. The measurement and research of time use is quite complex, and extreme care must be exercised in the design and administration of time use and activity surveys. In general, a time use survey collecting information on in-home and out-of-home activities should yield a total of about 20 to 25 activities per person per day, with about one fifth of these activities constituting trips (i.e., four or five trips per day per person). These general values may be used as broad guides to ensure that an activity and time use survey is yielding information consistent with past experience. These figures may vary considerably, depending on the nature and composition of the sample, the level of detail regarding activities and trips that is captured in the survey instrument, and the design and administration of the survey. As pointed out by Harvey (2002), there is a merging of traditions between travel and time use studies that bodes well for the transportation planning profession as time use surveys become increasingly amenable to collecting detailed travel information.

2.3   Example Application 1: Modeling Time–Space Prisms

The notion of time–space prisms was introduced by Hägerstrand (1970) to describe the spatiotemporal constraints in which people make activity and travel decisions. Since then, many researchers in the travel behavior arena have addressed or utilized the concept of time–space prisms for modeling activity and travel engagement patterns of individuals. The representation of spatiotemporal constraints in the modeling of human activity and travel behavior is very important. In any given day, a person has only 24 h available and much of that time may be spent on basic subsistence activities, including sleeping, working (to earn a living), and personal and household care. The temporal aspects of these types of activities tend to be rigid and impose constraints on an individual's potential activity–travel engagement pattern. Similarly, in a spatial context, one can postulate that fixed home and work locations (coupled with various temporal constraints) limit the range of spatial choices for a person. Thus, it can be seen that time–space constraints play an important role in shaping people's activity–travel patterns.

The accurate and complete representation of time–space prisms has taken on added importance in the context of the emergence of microsimulation approaches to travel demand forecasting. Whereas in the traditional zone-based four-step travel demand modeling approaches one did not focus on the individual traveler, microsimulation approaches attempt to simulate activity and travel patterns at the level of the individual traveler. When dealing with individual travelers and their potential behavioral responses to evolving transportation policy scenarios, it is imperative that a mechanism be developed by which individual time–space prisms can be accurately modeled.

This section is aimed at developing a methodology by which temporal vertices of time–space prisms of individuals can be effectively represented in a comprehensive framework that encompasses both in-home and out-of-home activity engagement and time use. The approach involves the use of recent activity and time use data to model temporal vertices of time–space prisms for each individual as a function of his or her socioeconomic and demographic characteristics.

Thus, this section presents an attempt to define the beginning and ending point (called a vertex) of Hägerstrand's prism. While a trip is observable and is by definition always contained in a prism, the prism itself can rarely be defined based on observed information. Although the vertices of a prism are often determined by coupling constraints (e.g., one must be at a certain place by certain time), such constraints are often unobserved or not well defined. For example, consider a commuter who must report at work by 9:00 A.M. In this case a prism has one of the vertices located at the workplace at 9:00 A.M. in the space–time coordinates. The other vertex, which designates the beginning point of the prism, is not defined, except that it is located at the home base somewhere prior to 9:00 A.M. along the time axis.

In this section, models are developed to locate prism vertices along the time axis. The models are formulated as stochastic frontier models, which are used to estimate the location of an unobservable frontier (or an upper or lower bound) based on the measurement of an observable variable that is governed by the frontier. In this study, the location on the time axis of a prism vertex is the unobservable frontier, and the starting or ending time of a trip is the observable quantity governed by the frontier. In particular, this section focuses on three items of interest:

1. Formulation and estimation of time vertices using stochastic frontier models
2. Comparison of space–time prism vertices between geographic areas
3. Investigation of day-to-day variability in time vertices

By definition, a trip in a prism always starts at or after the origin vertex of the prism, and ends at or before its terminal vertex. While the beginning and ending times of a trip are almost always available from travel survey data, the origin and terminal vertices of a prism are normally unobserved. A modeling approach, therefore, is adopted in this study to estimate the location of prism vertices using observed variables.

Adopted in the modeling approach are the following inequalities:

$$\text{at origin vertex: } \tau_o \le t_o$$

$$\text{at terminal vertex: } t_t \le \tau_t \tag{2.1}$$

where τ_o is the location along a time axis of the origin vertex of a prism, τ_t is the location of the terminal vertex, t_o is the beginning time of a trip in the prism, and t_t is the ending time of the trip. It is assumed that τ_o and τ_t are unobserved. From the inequalities,

$$t_o = \tau_o + u_o, \, t_t = \tau_t - u_t \tag{2.2}$$

where u_o and u_t are nonnegative random variables.

A possible model that applies to these relationships is the stochastic frontier model, whose general form can be presented as

$$Y_i = \beta' X_i + \varepsilon_i = \beta' X_i + v_i - u_i \tag{2.3}$$

where i denotes the observation; Y_i is the observed dependent variable (in this case a trip beginning or ending time); β is a vector of coefficients; X_i is a vector of explanatory variables; and v_i and u_i are the random error terms, $-\infty < v_i < \infty$ and $u_i > 0$. In the context of this study, $\beta' X_i + v_i$ can be viewed as the location of the terminal vertex of a prism with the random element, v_i. The observed trip ending time (Y_i in the above notation) will not exceed $\beta' X_i + v_i$ because u_i is nonnegative. A model for an origin vertex can be formulated similarly as $Y_i = \beta' X_i + v_i + u_i$.

In the econometric literature on stochastic frontier models, v_i is typically assumed to be normal and a truncated (half) normal distribution is often used for u_i. In this case, the distribution of ε_i is given as (subscript i is suppressed below)

$$h(\varepsilon) = \frac{2}{\sqrt{2\pi}\sigma} \left\{ 1 - \Phi(\varepsilon\lambda/\sigma) \right\} \exp\left[-\frac{\varepsilon^2}{2\sigma^2} \right], -\infty < \varepsilon < \infty \tag{2.4}$$

where $\sigma^2 = \sigma_u^2 + \sigma_v^2, \lambda = \sigma_u/\sigma_v, v \sim N(0, \sigma_v^2)$, and u has the density function

$$g(u) = \frac{2}{\sqrt{2\pi}\sigma_u} \exp\left[-\frac{u^2}{2\sigma_u^2} \right], u \ge 0 \tag{2.5}$$

This formulation is adopted with an observed trip starting or ending time as Y_i and selected attributes of the individual and household, including person commute characteristics, as X_i. Because of the way the model is constructed, the inequalities of Equation (2.1) are always satisfied. Yet, there remains the question of whether $\beta'X_i + v_i$ in fact represents the prism constraint in the strict sense of Hägerstrand. One could argue that $\beta'X_i + v_i$ may represent a threshold that an individual subjectively holds as the earliest possible starting time or the latest possible ending time for a trip, but may not coincide with actual constraints that are governing travel. For example, a commuter may believe that he or she cannot possibly leave home before 6:30 A.M. in the morning; thus the origin vertex of his prism before the work starting time is located subjectively at 6:30 A.M. But it is not likely that this is an objectively defined constraint. In fact, the same commuter may leave home before 6:00 A.M. for a business trip.

Models of prism vertices are estimated in this study with empirical data without any information on the individual's beliefs or perceptions of prism constraints. Yet observed travel behavior is governed by subjective beliefs and perceptions, e.g., "I must return home by midnight" or "I cannot possibly leave home before 6:30 A.M." Thus some ambiguity is unavoidable about the nature of $\beta'X_i + v_i$; it is unlikely that it represents a prism vertex in the strict sense of Hägerstrand. It is yet reasonable to assume that $\beta'X_i + v_i$ is nonetheless a useful measure for the practical purpose of determining the earliest possible departure time or latest possible arrival time for a trip.

It is often considered that workers' daily activities are regulated by their work schedules. It may then be assumed that the work starting time defines the terminal vertex of a worker's morning prism before work, and the work ending time defines the origin vertex of his or her evening prism after work. The prism during the lunch break is determined by the beginning time and ending time of the break. Work schedules that define these prism vertices are determined primarily by institutional factors, and personal or household attributes are expected to have relatively small effects. There is therefore little room to apply such a model as described above to prism vertices that are defined by a work schedule. Therefore, stochastic frontier models are presented in this section for the origin vertex of workers' morning prisms and the terminal vertex of workers' evening prisms. For those prism vertices that are defined by work schedules, different approaches (e.g., using observed frequency distributions of work starting or ending times by industry and occupation) may be more effective.

Commuter samples from the Miami and San Francisco Bay Area surveys (described earlier in this chapter) were used to estimate stochastic frontier models of prism vertices. Pendyala et al. (2002a) present models of the following prism vertices:

- Origin vertex of the commuters morning prism — Miami and San Francisco
- Terminal vertex of the commuters evening prism — Miami and San Francisco

As data are available for a 2-day period in the San Francisco data set, separate models are estimated for each day and for the pooled data set so that day-to-day variability in time vertex locations can be explored.

The dependent variables of the models presented in this section are defined with the time of day expressed in minutes, with 12:00 A.M. (midnight) being 0; so 6:00 A.M. is expressed in the model as 360, and 6:00 P.M. as 1080. All models assume that v_i has a normal distribution and u_i has a half-normal distribution. The expected value of u_i is evaluated as

$$E[u_i] = \left(\frac{2}{\pi}\right)^{1/2} \hat{\sigma}_u \tag{2.6}$$

where $\hat{\sigma}_u$ is an estimate of σ_u.

For the sake of brevity, this section presents two tables that are representative of the model estimation results that can be accomplished using econometric software such as LIMDEP. The model for the origin vertex of the Miami commuter's morning prism is presented in Table 2.5. The model is a cost frontier model and is formulated as $\beta'X_i + v_i + u_i$. The model is found to offer plausible indications. The model shows that a full-time worker has a origin vertex about 86.5 min earlier than a nonworker, while the corresponding

TABLE 2.5 Stochastic Frontier Model of Miami Commuters' Morning Prism Origin Vertex

Variable	Coefficient	t-Stat
Constant	383.4	15.17
Full-time worker	−86.5	−3.69
Part-time worker	−57.1	−1.98
Work at home	86.8	3.38
Student	−47.7	−1.88
Home to work commute time (hours)	−31.4	−1.73
Car availability: vehicles ≥ drivers	19.3	1.65
Driver's license	51.9	3.35
Age ≥ 50 years	−21.7	−1.95
Family with children (5–15 years)	−15.4	−1.62
R^2, adjusted R^2	0.260, 0.248	
L(C), L(β)	−3592.1, −3464.2	
χ_c^2 (df)	255.9 (9)	
Var(v), E[u], Var(u)	1807.6, 141.4, 11404	
N	569	

Note: L(C) is evaluated by setting all coefficients to 0, except constant. df = degrees of freedom.

figure for a part-time worker is about 57 min. On the other hand, those who work at home have origin vertices about 87 min later than those who work outside the home. Similarly, the variable representing students also has a negative coefficient, though not as much as those associated with full- or part-time workers. The origin vertex of the Miami commuters' morning prism is pushed earlier as commute time increases — about 30 min for every hour of commute. Greater car availability and the possession of a driver's license provide for origin vertices that are later in the morning; this is presumably because of the faster travel times and flexibility associated with the ability to drive alone. Older individuals and those in families with children have slightly earlier origin vertices than other groups. E[u] is 141 min, indicating that the first time of departure from home is, on average, about 2 h 20 min after the origin vertex.

Table 2.6 shows the results of the model estimation effort for the terminal vertex of the commuter's evening prism of the San Francisco sample. For the San Francisco commuters, 2 days' worth of data is available. Therefore, model results are shown by day and for the pooled sample as a whole. The work end time and the final time of arrival at home for the San Francisco commuter sample show distributions that have high variance and less well-defined patterns. This is also evidenced in the model estimation results. For example, the effect of a working day for a full-time worker is minimal (and not significant for the second day). On the other hand, the working day of a part-time worker and working multiple jobs shifts the terminal vertex by about 30 min later in the day. Similarly, school also shifts the terminal vertex later in the day. For every hour of commute, the vertex is shifted later by about 25 min. This result is found to be very symmetric with that of the origin vertex of the morning prism, where 1 h of commute shifted the origin vertex 25 min earlier in the day (table not presented in this chapter). Being older or of minority status (Hispanic or Black) is associated with relatively earlier terminal vertices for the commuter's evening prism. On the other hand, being a single person and having a driver's license are both associated with later terminal vertices. This may be because of the greater flexibility for final home arrival that these individuals may have relative to those who have families and those who cannot drive. A greater number of workers or cars is associated with marginally later terminal vertices, once again presumably due to the flexibility afforded by these variables.

E[u] is found to be about 2 h 45 min, indicating that commuters in the San Francisco sample, on average, arrive home about 2 h 45 min prior to the terminal vertex. For both survey samples, it was found that the goodness of fit of the model of the evening prism terminal vertex is substantially poorer than that found for the morning prism origin vertex. The greater variability in the final arrival times at home may be contributing to this poor fit.

TABLE 2.6　Stochastic Frontier Model of San Francisco Commuters' Evening Prism Terminal Vertex

Variable	Pooled		Day 1		Day 2	
	Coefficient	t-Stat	Coefficient	t-Stat	Coefficient	t-Stat
Constant	1292.71	83.80	1286.12	61.80	1299.73	56.70
Working day (full-time worker)	8.94	1.57	22.59	2.92	−6.36	−0.75
Working day (part-time worker)	38.98	4.14	39.58	3.13	39.24	2.78
Multiple jobs	26.11	2.93	30.20	2.51	20.90	1.59
School day	69.05	6.29	92.61	5.74	42.48	2.82
Commute time (minutes)	0.46	4.03	0.49	3.06	0.43	2.58
Age	−2.11	−11.74	−2.09	−8.61	−2.16	−8.04
Hispanic	−19.23	−2.34	−26.44	−2.38	−10.60	−0.87
White	−8.56	−1.39	−14.50	−1.74	−1.72	−0.19
Black	−13.67	−1.40	−27.36	−1.99	0.05	0.00
Single person	41.05	6.25	43.83	4.95	38.98	3.99
No. of workers	6.04	2.06	10.96	2.70	1.15	0.27
No. of cars	5.50	2.50	4.96	1.64	6.02	1.88
No. of bicycles	−5.17	−3.33	−5.32	−2.52	−5.03	−2.20
Driver's license	43.26	4.90	35.00	3.00	52.67	3.93
Low income (< \$30 K)	−15.32	−2.36	−19.71	−2.29	−10.48	−1.07
R^2, adjusted R^2	0.062, 0.060		0.078, 0.074		0.052, 0.048	
L(C), L(β)	−45922, −45631		−24045, −23854		−21875, −21761	
χ_c^2 (df)	583 (14)		381 (14)		228 (14)	
Var(v), E[u], Var(u)	18260, 166.8, 15873		16949, 169.9, 16485		19525, 162.6, 15089	
N	6885		3606		3279	

Note: L(C) is evaluated by setting all coefficients to 0, except constant. df = degrees of freedom.

Overall, it is seen that the stochastic frontier modeling methodology is capable of representing the terminal vertices associated with beginning or ending points of space–time prisms, at least for commuters who tend to have more structured weekdays. Further research is warranted in the context of estimating vertex locations for nonworkers.

Pendyala et al. (2002a) present comparisons for two items of interest:

- Comparisons between Miami and San Francisco commuter samples with respect to origin vertex of morning prism and terminal vertex of evening prism
- Comparisons between first and second days of the San Francisco commuter sample with respect to origin vertex of morning prism and terminal vertex of evening prism

For the sake of brevity, two comparisons are shown in this section. The first comparison, shown in Figure 2.1, pertains to distributions of expected vertex locations and observed final home arrival times. The peak home arrival time for the Miami sample appears to be about 7 P.M., while that for the San Francisco sample appears to be about 30 min earlier at 6:30 P.M. The distributions of expected vertex locations peak for both samples at about 9:30 P.M., with the Miami sample showing a more pronounced peak than the San Francisco sample. Greater household obligations (child care, etc.) associated with larger household sizes in the Miami area may be contributing to this difference. Thus it is found that most individuals arrive home about 2.5 to 3 h prior to their vertex location.

The second comparison, shown in Figure 2.2, examines day-to-day variation in origin vertex of the morning prism and the first time of departure from home for the San Francisco Bay Area survey sample.

The distributions are strikingly similar. The peaks of the observed distributions are shifted about 2 h to the right (later in the day) of the peaks associated with the distributions of the expected vertex locations. The distributions of the expected vertex locations are very similar between the 2 days, as are the distributions of observed home departure times.

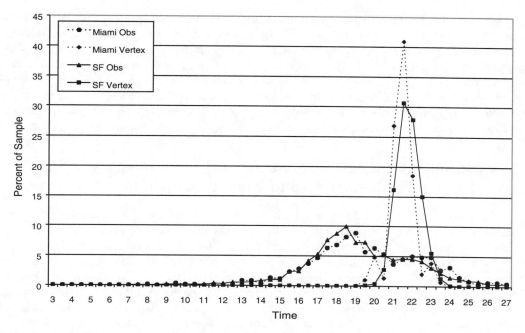

FIGURE 2.1 Distribution of expected vertex locations and final arrival at home.

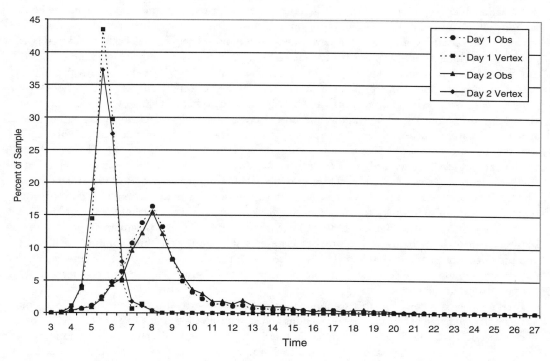

FIGURE 2.2 Distribution of expected vertex locations and first departure from home: comparison between day 1 and day 2 for San Francisco commuters.

Overall, the analysis and model estimation results showed that similarities across geographic areas are more pronounced in the case of origin vertices associated with the morning prism of commuters. The greater variability in home arrival times contributes to greater differences across geographical areas when one considers the terminal vertex locations of the evening prism. Also, comparisons between 2 days of travel show striking similarities between the distributions of expected vertex locations and observed departure and arrival times. The stochastic frontier modeling method is effective for modeling temporal extremities. Estimation results provide strong indications that the temporal vertices associated with space–time prisms are significantly influenced by people's socioeconomic, demographic, and commute characteristics.

2.4   Example Application 2: Structural Equations Modeling of Household Activity and Travel Durations

Recent advances in activity-based approaches involve the microsimulation of individual activity–travel patterns in the space–time continuum. The individual person is typically considered the decision or choice maker, and model system components attempt to represent various aspects of the activity–travel behavior of the individual traveler. These models are becoming increasingly sophisticated in their ability to reflect the effects of various types of constraints on activity–travel patterns. In this context, household interactions and the constraints and opportunities that such interactions bring about can also play a big role in influencing individual activity–travel patterns. Household members allocate tasks among one another, make trade-offs, or join together in activity participation, and often may depend on one another for undertaking activities and travel (particularly in the case of children who depend on adults for their transport). As the use of cell phones, e-mail, and other technology becomes increasingly common, one can only expect that the amount of interaction (especially real-time interaction) will increase over time (Meka et al., 2002).

Considering that there is a wide array of possible interactions among household members that merit investigation, the analysis in this section focuses on nonwork activity and time allocation among household members consistent with the notion of time use discussed in this chapter. For an individual, the amount of work activity and travel may have an effect on nonwork activity engagement and travel. As a person spends more time at work or traveling to work, he or she is likely to spend less time at nonwork activities. These intraperson trade-offs are often clear and well captured in models of activity and travel behavior. Similar trade-offs may occur at the interperson level. As one individual in the household spends more time at work or traveling to work, it is possible that the other individual will spend more time taking care of the household obligations and other nonwork activities. Thus, in modeling household activity and travel behavior, it is important to represent such trade-offs to accurately capture household-level trip-making patterns.

The data set used in this study is derived from a traditional household travel survey conducted in southeast Florida (Broward, Palm Beach, and Miami–Dade Counties) during the 1999 calendar year. The travel survey consisted of three steps, including a computer-assisted telephone interview recruitment, a mail-out of survey instruments and travel diaries, and a CATI retrieval of the survey responses. Of the 7500 households that agreed at first to participate, 5168 households actually responded to the survey. The 5168 households were approximately evenly split among the three counties in the region and provided a respondent sample of 11,426 persons reporting a total of 33,082 trips. In general, the respondent sample exhibited socioeconomic, demographic, and travel characteristics consistent with the population in the region.

For the analysis reported in this section, it was necessary to extract a suitable sample that would facilitate the modeling of interperson interactions and activity allocations in a focused manner. In order to do this, households that had two or more adults (18 years or older) of which at least one adult worked outside the home were extracted to form a multi-adult worker household sample. This sample consisted of 1262 households. All of the analysis and model estimation reported in this section has been conducted on this sample of 1262 households.

In order to make comparisons between two adult household members meaningful and easy to interpret, the adults were numbered (given a person ID) based on work duration and age. The following method was used to assign identification numbers:

- Adults were assigned numbers in descending order of their total daily work activity duration. The adult with the longest work duration is person 1, and the adult with the next longest work duration is person 2.
- If two adults had identical work durations, then the older adult was assigned a lower person identification number.
- If two adults had exactly the same work durations and age, then the identification numbers were assigned randomly between them. This situation, however, never occurred in the sample.

As the sample is a carefully selected sample of households with multiple adults, the average household size is rather large, at 3.4 persons per household. About one third of the households are two-person households. Average vehicle ownership is 2.3 vehicles per household, with the majority having two vehicles. Nearly three quarters live in single-family dwelling units. As expected, the income variable had a poor response rate, with more than one quarter of the households not providing income information. Considering the nature of the selected household sample, the rather high proportion falling in the high-income category ($80,000 and above) is not surprising. On average, the households have about one child per household, with nearly one half reporting no children. About 60% of the households have two workers; once again, this is consistent with the nature of the sample.

With respect to person attributes, the average age is very similar between person 1 and person 2. However, person 2 has a higher proportion of elderly (greater than 60 years) individuals. While only 1% of those classified as person 1 are not employed, the corresponding percentage among those classified as person 2 is 23%. Consistent with the employment pattern, the income distribution shows a higher personal income for person 1. Among those who commute, about 85% choose to drive alone to work, while about 10 to 12% choose to car- or vanpool to work. Very small percentages use transit or other modes.

Table 2.7 provides average activity frequencies, activity durations, and travel durations for persons 1 and 2. The table makes a distinction between the entire sample of 1262 persons and the subset of persons who actually participated in the activity. The latter set is considered the nonzero set, and the sample size for each activity is in parentheses under the respective average. For example, 98 persons (among the 1262 classified as person 1) pursued shopping. The average shopping activity frequency for this set of 98 persons is 1.07.

With respect to activity durations, the average work duration for those classified as person 1 is about 8 h 20 min. As expected, the average for those classified as person 2 is considerably lower because of the higher incidence of noncommuters among the person 2 sample. However, in line with the higher activity frequencies they exhibited, those classified as person 2 spent more time at nonwork activities. Even if one were to focus on the nonzero samples, those classified as person 2 show consistently higher daily average activity durations for nonwork activities.

Average travel durations show trends that are similar to those shown by activity frequencies and activity durations. Those classified as person 1 spend about 40 min traveling to work, while those classified as person 2 spend about 30 min traveling to work. However, among the nonzero observations, those classified as person 2 spend more time traveling to work than those classified as person 1. On average, those classified as person 2 spend more time traveling to nonwork activities than those classified as person 1. One anomaly is found in the context of travel to school. Among those who actually participated in school activity, those classified as person 1 spent more time traveling to school than those classified as person 2. In general, the person samples spend an average of about 110 min traveling to various activities. This is quite high, but consistent with expectations given the larger household size and multiworker, multiadult, and multivehicle nature of the sample.

The modeling of within-household interactions in activity engagement involves dealing with multiple endogenous variables in a simultaneous equations framework. Work and nonwork activity frequencies,

TABLE 2.7   Daily Time Use and Activity–Travel Frequencies on Travel Survey Day

Characteristic	Person 1 (All)	Person 1 (Nonzero)	Person 2 (All)	Person 2 (Nonzero)
Activity Frequencies				
Work	1.19[a]	1.19 (1262)	0.73	1.21 (763)
Shopping	0.08[a]	1.07 (98)	0.20	1.09 (229)
Social recreation	0.07[a]	1.06 (79)	0.13	1.14 (141)
School	0.02[a]	1.0 (24)	0.08	1.03 (95)
Return home (includes final home stay, but not initial home stay)	1.33[a]	1.33[b] (1262)	1.55	1.55 (1258)
Other	0.98[a]	2.07[b] (596)	1.65	2.58 (805)
Activity Durations (minutes per day)				
Work	498[a]	499[b] (1262)	207	386 (678)
Shopping	2[a]	30[b] (89)	8	48 (215)
Social recreation	6[a]	100 (70)	12	121 (125)
School	2[a]	108[b] (21)	14	194 (88)
Total at home (includes initial and final home stay)	755[a]	757[b] (1260)	887	894 (1251)
Other	34[a]	77[b] (551)	71	121 (745)
Travel Durations (minutes per day)				
Work	42.5[a]	42.7 (1255)	29.5	49.6 (751)
Shopping	1.3[a]	17.3 (96)	3.8	20.9 (227)
Social recreation	1.2[a]	20.1 (78)	2.7	24.6 (138)
School	0.75[a]	38.8 (24)	2.1	27.8 (94)
Return home (includes final return home trip)	40.3	43.5 (1170)	41.9	48.6 (1087)
Other	21.7[a]	47.6[b] (575)	38.9	64.2 (765)

[a] Significantly different from person 2 (all) at the 0.05 significance level.
[b] Significantly different from person 2 (nonzero) at the 0.05 significance level.

activity durations, and travel durations are all activity- and travel-related endogenous variables that are interconnected with one another. When modeling the interactions among several interdependent endogenous variables, simultaneous equations systems offer an appropriate framework for model development and hypothesis testing. In this application, the structural equations methodology is adopted for estimating simultaneous equations systems that capture the interdependencies among household members' activity engagement patterns.

A typical structural equations model (with G endogenous variables) is defined by a matrix equation system, as shown in Equation (2.7):

$$
\begin{bmatrix} Y_1 \\ \cdot \\ \cdot \\ \cdot \\ Y_G \end{bmatrix} = \begin{bmatrix} Y & X \end{bmatrix} \begin{bmatrix} B \\ \Gamma \end{bmatrix} + \begin{bmatrix} \varepsilon_1 \\ \cdot \\ \cdot \\ \cdot \\ \varepsilon_G \end{bmatrix}
\tag{2.7}
$$

This equation can be rewritten as

$$
Y = BY + \Gamma X + \varepsilon
\tag{2.8}
$$

or

$$
Y = (I - B)^{-1}(\Gamma X + \varepsilon)
\tag{2.9}
$$

where Y is a column vector of endogenous variables, B is a matrix of parameters associated with right-hand-side endogenous variables, X is a column vector of exogenous variables, Γ is a matrix of parameters associated with exogenous variables, and ε is a column vector of error terms associated with the endogenous variables.

Structural equations systems are estimated by covariance-based structural analysis, also called the method of moments, in which the difference between the sample covariance and the model implied covariance matrices is minimized. The fundamental hypothesis for the covariance-based estimation procedures is that the covariance matrix of the observed variables is a function of a set of parameters, as shown in Equation (2.10):

$$\Sigma = \Sigma(\theta) \qquad (2.10)$$

where Σ is the population covariance matrix of observed variables, θ is a vector that contains the model parameters, and $\Sigma(\theta)$ is the covariance matrix written as a function of θ.

The relation of Σ to $\Sigma(\theta)$ is basic to an understanding of identification, estimation, and assessment of model fit. The matrix $\Sigma(\theta)$ has three components: the covariance matrix of Y, the covariance matrix of X with Y, and the covariance matrix of X.

Let Φ equal the covariance matrix of X and Ψ equal the covariance matrix of ε. Then it can be shown that

$$\Sigma(\theta) = \begin{bmatrix} (I-B)^{-1}(\Gamma\Phi\Gamma' + \Psi)(I-B)^{-1'} & (I-B)^{-1}\Gamma\Phi \\ \Phi\Gamma'(I-B)^{-1'} & \Phi \end{bmatrix} \qquad (2.11)$$

Before estimating model parameters, it is first necessary to ensure that the model is identified. Model identification in simultaneous structural equations systems is concerned with the ability to obtain unique estimates of the structural parameters. The identification problem is typically resolved by using theoretical knowledge of the phenomenon under investigation to place restrictions on model parameters. The restrictions usually employed are zero restrictions where selected endogenous variables and certain exogenous variables do not appear on the right-hand side of certain equations and selected error correlations are specified to be zero. There are several rules that can be used to check whether a structural equations model system is identified.

The unknown parameters in B, Γ, Φ, and Ψ are estimated so that the implied covariance matrix, $\hat{\Sigma}$, is as close as possible to the sample covariance matrix, S. In order to achieve this, a fitting function $F(S, \Sigma(\theta))$, which is to be minimized, is defined. The fitting function has the properties of being a scalar, greater than or equal to zero, equal to zero if and only if $\Sigma(\theta) = S$, and continuous in S and $\Sigma(\theta)$.

Available methods for parameter estimation include maximum likelihood (ML), unweighted least squares (ULS), generalized least squares (GLS), scale-free least squares (SLS), and asymptotically distribution-free (ADF). Each of these methods minimizes the fitting function and leads to consistent estimators of θ. Ideally, one would use the ADF method of estimation to estimate parameters of structural equations models because of its ability to accommodate limited dependent variables with different asymptotic distributions. However, the software used in this research (AMOS version 4.01) does not estimate intercepts when the ADF method is utilized. In this particular modeling effort, it was found that values of model coefficients estimated by the ADF method are virtually identical to those obtained by the ML method of estimation. This is presumably because most of the endogenous variables considered in this analysis are continuous variables. Hence, in this case, the ML method of estimation may be employed without adversely influencing the estimation results.

The fitting function that is minimized in the ML method of estimation of structural parameters is shown in Equation (2.12):

$$F_{ML} = \log | \Sigma(\theta) | + \mathrm{tr}\,(S\,\Sigma^{-1}\,(\theta)) - \log | S | - (G + K) \qquad (2.12)$$

where G is the number of excluded endogenous variables on the right-hand side of the model and K is the number of included exogenous variables on the right-hand side of the model.

The asymptotic covariance matrix for the ML estimator $\hat{\theta}$ is given by

$$\left(\frac{2}{N-1}\right)\left\{E\left[\frac{\partial^2 F_{ML}}{\partial\theta\,\partial\theta'}\right]\right\}^{-1} \tag{2.13}$$

When $\hat{\theta}$ is substituted for θ, an estimated asymptotic covariance matrix that allows tests of statistical significance on parameters of $\hat{\theta}$ is obtained.

A comprehensive structural equations model was estimated on the sample of 1262 households to explore causal linkages between two adult persons in multiadult households. Hypotheses regarding interperson interactions coupled with statistical measures of fit and significance were used to guide the model development process. Ultimately, a model that offered behaviorally sound interpretations and satisfactory statistical indications was obtained. This model is represented in a path diagram format in Figure 2.3. The variables in the left column represent exogenous variables. The variables in the middle column represent endogenous variables. The error terms are represented by circles in the last column.

An examination of the figure shows the types of causal relationships and error covariances that are represented in the model. Work travel and activity durations were taken to be exogenous variables, as events related to the work activity tend to be relatively nondiscretionary and inflexible in nature. Therefore, work travel and activity durations for both persons 1 and 2 are in the left-most column. Endogenous variables include nonwork activity duration, nonwork travel duration, and nonwork trip frequency for both persons 1 and 2, making a total of six endogenous variables. The causal relationships represented in the model are shown by curved arrows between pairs of endogenous variables. Similarly, error covariances included in the model are shown by curved double arrows in the figure.

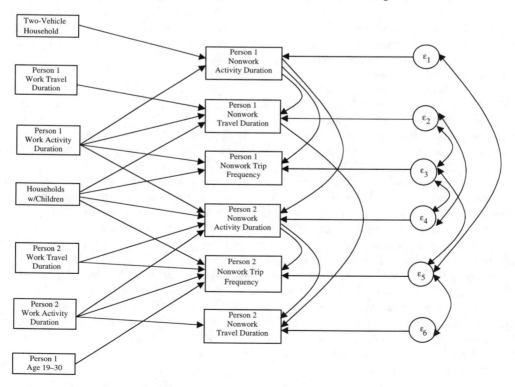

FIGURE 2.3 Causal structure of structural equations model of household time use and activity engagement.

Estimation results are shown in Table 2.8. The table provides estimates of the direct, indirect, and total effects. To understand these different effects, consider the relationships depicted in Figure 2.3. Person 1 work activity duration directly affects person 2 nonwork activity duration (as depicted by the arrow between these variables). However, person 1 work activity duration also affects person 1 nonwork activity duration. In turn, person 1 nonwork activity duration affects person 2 nonwork activity duration. Thus, person 1 work activity duration also indirectly affects person 2 nonwork activity duration through the mediating variable — person 1 nonwork activity duration. The indirect effect of person 1 work activity duration on person 2 nonwork activity duration is the product of the two direct effects that cause the indirect effect. The total effect of person 1 work activity duration on person 2 nonwork activity duration is the sum of the direct and indirect effects.

All of the model coefficients presented in Table 2.8 are statistically significant at the 0.05 level of significance, with a few exceptions that are significant at the 0.1 level. The model χ^2 goodness-of-fit statistic indicates that the hypothesis that the matrix of sample moments is equal to the matrix of model implied moments can not be rejected at the 0.05 level of significance. Thus the model fit is statistically acceptable. The other measures of fit provided at the bottom of the table are also in line with agreeable standards of fit for a structural equations model of this nature. In the discussions that follow, it should be noted that person 1 represents the adult in the household who spent the longest time working and may therefore be considered the primary worker of the household.

The model results provide some very logically consistent findings. First, consider the intraperson activity–travel interactions. As expected, work activity duration negatively affects nonwork activity duration for the same person. For both person 1 and 2, as work activity duration goes up, their respective nonwork activity engagement reduces. Similarly, work travel duration has a negative effect on nonwork travel duration for both persons 1 and 2, although it is an indirect effect for person 2. As work travel duration goes up, time spent traveling for nonwork activities reduces as well, indicating that individuals look for destinations in close proximity to get their nonwork activities done. Work travel directly affects nonwork activity duration for person 2 with a negative direct effect. As person 2 travels longer for work, this person spends less time pursuing nonwork activities. For both persons, the derived nature of travel demand can be seen very clearly. Person 1 nonwork activity duration directly affects the nonwork trip frequency and the nonwork travel duration. The same holds true for person 2. The positive coefficients indicate that larger activity durations are associated with larger travel durations or trip frequencies for nonwork activities. Thus, the model has effectively captured the intraperson relationships.

With respect to interperson interactions, the findings are logically consistent and reflect the special nature of interpersonal interactions in households. Household members may pursue some activities jointly while trading off and allocating others. Here are the main findings:

- The model suggests that, in the context of nonwork activities, household members may be interacting in a complementary manner. This is possibly because household members jointly undertake many nondiscretionary nonwork activities, including shopping, recreation, and personal business. Thus, one notes that person 1 nonwork activity duration has a positive direct effect on person 2 nonwork activity duration. As person 1 spends more time at nonwork activities, so does person 2 because they are engaging in the activity together.

- Person 1 work activity duration has a positive direct effect on person 2 nonwork activity duration. That is, as person 1 works longer, person 2 engages in more nonwork activities. This is consistent with expectations.

- However, the total effect of person 1 work activity duration on person 2 nonwork activity duration is negative. Consider the following: As person 1 spends more time at work, this person spends less time at nonwork activities (coefficient = –0.145). When person 1 spends less time at nonwork activities, so does person 2 (coefficient = 0.325). Then person 1 work activity duration has a negative indirect effect (coefficient = –0.047) on person 2 nonwork activity duration. The net total effect of person 1 work activity duration on person 2 nonwork activity duration is negative. Thus, when person 1 works longer, person 2 may pick up some of the nonwork household obligations,

TABLE 2.8 Structural Equations Model Estimation Results

Endogenous Variable	Intercept	Effect	Two-Vehicle Household	Household w/ Child	Person 1 Age Group (19–30 years)	Person 1 Work Travel Duration	Person 1 Work Activity Duration	Person 2 Work Travel Duration	Person 2 Work Activity Duration	Person 1 Nonwork Activity Duration	Person 1 Nonwork Travel Duration	Person 2 Nonwork Activity Duration
Person 1 nonwork activity duration	122.306	Total	−10.818	0.000	0.000	0.000	−0.145	0.000	0.000	0.000	0.000	0.000
		Direct	−10.818	0.000	0.000	0.000	−0.145	0.000	0.000	0.000	0.000	0.000
		Indirect	0.000	0.000	0.000	0.000	0.000	0.000	0.000	0.000	0.000	0.000
Person 1 nonwork travel duration	15.845	Total	−3.133	12.329	0.000	−0.026	−0.060	0.000	0.000	0.290	0.000	0.000
		Direct	0.000	12.329	0.000	−0.026ª	−0.018ª	0.000	0.000	0.290	0.000	0.000
		Indirect	−3.133	0.000	0.000	0.000	−0.042	0.000	0.000	0.000	0.000	0.000
Person 1 nonwork trip frequency	1.941	Total	−0.088	0.167	0.000	0.000	−0.004	0.000	0.000	0.008	0.000	0.000
		Direct	0.000	0.167	0.000	0.000	−0.002	0.000	0.000	0.008	0.000	0.000
		Indirect	−0.088	0.000	0.000	0.000	−0.001	0.000	0.000	0.000	0.000	0.000
Person 2 nonwork activity duration	131.835	Total	−3.517	−21.849	0.000	0.000	−0.014	−0.088	−0.208	0.325	0.000	0.000
		Direct	−3.517	−21.849	0.000	0.000	0.033ª	−0.088ª	−0.208	0.325	0.000	0.000
		Indirect	−3.517	0.000	0.000	0.000	−0.047	0.000	0.000	0.000	0.000	0.000
Person 2 nonwork travel duration	40.253	Total	−0.750	−1.606	0.000	−0.002	−0.008	−0.011	−0.068	0.069	0.096	0.128
		Direct	0.000	0.000	0.000	0.000	0.000	0.000	−0.042	0.000	0.096	0.128
		Indirect	−0.750	−1.606	0.000	−0.002	−0.008	−0.011	−0.027	0.069	0.000	0.000
Person 2 nonwork trip frequency	2.187	Total	−0.012	0.233	−0.634	0.000	0.000	−0.001	−0.003	0.001	0.000	0.003
		Direct	0.000	0.305	−0.634	0.000	0.000	−0.001	−0.002	0.000	0.000	0.003
		Indirect	−0.012	−0.072	0.000	0.000	0.000	0.000	−0.001	0.001	0.000	0.000

ª Significant at 90% level. All other variables significant at 95% level.

Note: N = 1262; χ^2 = 45.84 with 37 df (degrees of freedom); p = 0.151; goodness-of-fit index = 0.994.

but the total nonwork activity duration goes down because the two persons have less time to jointly spend together at nonwork activities.

- The complementary nature of nonwork activity engagement is also seen in the interaction between the travel duration variables. The person 1 nonwork travel duration has a positive direct effect on person 2 nonwork travel duration (coefficient = 0.096). Again, this signifies joint activity participation where person 1 and person 2 are likely riding together to undertake household nonwork activities.

- Person 1 work travel duration and work activity duration have negative indirect effects on person 2 nonwork travel duration through the mediating variable — person 1 nonwork travel duration. As person 1 works longer, he or she spends less time on nonwork travel (coefficient = −0.026). When person 1 spends less time traveling to nonwork activities, so does person 2 (coefficient = 0.096). Once again, these indirect effects capture the complementary nature of nonwork travel among household members.

The model provided valuable insights into the interactions between two adult household members in multiadult households. The model captured within-person trade-offs between work and nonwork activity engagement. For each person, as the amount of work activity or travel increased, the amount of nonwork activity or travel decreased. Between persons, the model captured the complementary and joint nature of nonwork activity engagement where household members tend to pursue nonwork activities together. Thus, when one person's nonwork activity or travel increases, so does the other person's nonwork activity travel engagement. These findings are very consistent with results reported in the literature and lend credibility to the importance of reflecting these relationships in activity-based travel demand modeling systems. On the other hand, the findings differ from the initial hypotheses presented earlier in this section, where it was postulated that an increase in nonwork activity participation by one member would decrease nonwork activity participation by the other member. It should be noted that this hypothesis may still hold true if one were to separate maintenance trips from leisure trips. Whereas maintenance activities may be allocated between adults (suggesting a trade-off), leisure activities may be conducted jointly (suggesting a complementary effect). Such effects have been found by Golob and McNally (1997). It is important to preserve the disaggregate activity purpose classification; in this particular analysis, nonwork activities had to be grouped together due to the rather low participation rates in maintenance and leisure activities (when treated separately).

The model offers several additional insights that are useful in a transportation policy context. For example, the model may be used to study the potential presence of induced travel effects. If road capacity increases were to result in reduced work travel durations, then what would happen to nonwork activity and travel engagement? For person 1, the primary worker who works for longer durations, work travel duration has a negative effect on nonwork travel duration (coefficient = −0.026). Thus, if person 1 has a shorter commute due to capacity enhancements, his or her nonwork travel duration will increase. But note that this will also result in an increase in nonwork travel duration for person 2 because, presumably, persons 1 and 2 perform nonwork activities together. Thus, while the vehicle miles traveled for nonwork activities will increase as a result of the capacity expansion, the person miles traveled may increase even more. In general, though, for person 1 it appears that the reduction in work travel duration will affect only nonwork travel duration, suggesting a change in destination (choosing to travel to a farther and more preferred destination) with no change in nonwork trip frequency (trip generation) or activity duration.

On the other hand, for the secondary worker in the household, a reduction in work travel duration results in a potential increase in nonwork trip frequency (coefficient = −0.001) and in nonwork activity duration (coefficient = −0.088). Because nonwork activity duration would rise, so would nonwork travel duration for person 2 (coefficient = 0.128). Thus, it appears that the induced travel effects may be larger for the secondary worker in the household, who has the greater flexibility to pursue other non-work-related activities. Person 1 may not have the same level of flexibility as person 2. For person 2, induced travel effects include changes in trip frequency, destination choice (longer travel duration), and activity duration.

The structural equations model also provides a mechanism for quantifying the effects between variables. The model shows that a 10-min reduction in work travel duration for person 1 would bring about a 0.26-min increase in nonwork travel duration for that person. A 10-min increase in work activity duration for person 1 would bring about a 1.5-min decrease in nonwork activity duration for that person. The corresponding values for person 2 are 0.1 and 2 min, respectively. Every 10 min of nonwork activity engagement leads to about 3 min of nonwork travel for person 1; the corresponding value for person 2 is only 1.3 min. A 10-min increase in the nonwork activity duration of person 1 contributes to a 3.25-min increase in nonwork activity duration for person 2. Thus, it appears that person 1 and person 2 would jointly spend 3.25 min together, while person 1 would spend the other 6.75 min performing a nonwork activity outside home alone. These types of quantifications greatly aid in determining the dynamics of household activity travel patterns in response to changes in socioeconomics or transportation level of service.

2.5 Example Application 3: Two Dimensions of Time Use–Activity Episode Timing and Duration

The previous example application focused on relationships underlying daily time allocation to various types of activities. In analyses of daily time use patterns, information about individual activity episodes is often not considered explicitly. The application presented in this section considers temporal aspects of individual activity episodes to demonstrate how time use analysis can be applied at the individual episode level.

There are two key aspects of the temporal dimension that play an important role in activity–travel demand modeling. They are the timing of an activity episode and the duration (time allocation) of an activity episode. In other words, activity-based analysis allows one to answer two critical questions:

- When is an activity pursued?
- For how long is the activity pursued?

The relationship between activity timing and duration is an important component of activity-based travel demand modeling systems that aim to explicitly capture the temporal dimension. On the one hand, one may hypothesize that the timing of an activity affects its duration. Perhaps activity episodes pursued during peak periods are of short duration, while those pursued in off-peak periods are longer in duration. On the other hand, the duration of an activity may affect its timing. Perhaps activities of longer duration are scheduled during the off-peak periods while activities of shorter duration are scheduled during peak periods. This application example attempts to shed light on this relationship by exploring both causal structures in a simultaneous equations framework. By identifying the causal structure that is most appropriate in different circumstances, one may be able to design activity-based model systems that accurately capture the relationship between activity timing and duration.

This section offers a detailed analysis of the relationship between activity timing and duration for maintenance activity episodes. The analysis is performed on commuter and noncommuter samples drawn from the 1996 Tampa Bay Household Travel Survey. A simultaneous equations system approach where activity timing is represented as a discrete time-of-day choice variable and activity episode duration is represented as a continuous variable is developed and estimated for two different causal structures. One causal structure assumes timing as a function of duration, while the second assumes duration as a function of timing. The discrete–continuous simultaneous equations model offers a powerful framework for analyzing such causal structures.

To illustrate the importance of accurately capturing the relationship between activity timing and duration, two different causal structures may be considered in the context of analyzing the potential impacts of a variable pricing (congestion pricing or time-of-day-based pricing) scheme. Such schemes are aimed at changing the time of travel or activity engagement so that trips otherwise undertaken during the congested peak periods would shift to off-peak periods. The two causal structures worthy of examination are briefly described in the following paragraphs.

2.5.1 Causal Structure D → T

In this structure, activity episode duration is assumed to be predetermined. The timing of an activity is determined next and is dependent on the duration of the activity episode. The model system representative of this mechanism may be represented as follows:

$$D_a^* = \beta_a'X + \alpha_a'Z_a + \varepsilon_a$$

$$T_a^* = \delta_a'S + \omega_a'R_a + \theta_a D_a + v_a P + \xi_a$$

where D_a^* is the latent variable underlying episode duration for activity type a; T_a^* is the latent variable underlying activity timing for activity type a; X and S are the vectors of socioeconomic characteristics; Z_a and R_a are the vectors of characteristics of activity type a; D_a is the observed or measured counterpart of D_a^*; P is the variable pricing (amount charged); ε_a and ξ_a are the random error terms that may be correlated; and β_a, α_a, δ_a, ω_a, θ_a, and v_a are the model coefficients.

Thus, in this model structure, activity episode duration is modeled as a function of socioeconomic characteristics (that do change based on the activity type) and activity characteristics (different activity types may have different characteristics). The time-of-day choice is then modeled as a function of socioeconomic characteristics, activity characteristics, the variable pricing cost, and the duration of the activity episode. Thus, in this scheme, the duration of the activity is predetermined and the timing is determined as a function of the duration. As variable pricing is aimed at merely shifting time of travel, it appears as an explanatory variable only in the timing equation.

2.5.2 Causal Structure T → D

In the second causal structure, the time-of-day choice for an activity episode is determined first and the duration of an activity episode is determined second. The simultaneous equations system representative of this causal scheme is as follows:

$$T_a^* = \delta_a'S + \omega'R_a + v_a P + \xi_a$$

$$D_a^* = \beta_a'X + \alpha_a'Z_a + \theta_a T_a + \varepsilon_a$$

All of the symbols are as described previously. In this scheme, activity timing is a function of socioeconomic characteristics, activity characteristics, and variable pricing. After the timing has been determined, the activity episode duration is determined as a function of socioeconomic characteristics, activity characteristics, and activity timing. Once again, variable pricing appears only in the timing equation.

Now, suppose one is interested in determining the potential impacts of variable pricing on travel demand by time of day. The implications of using the two different structures for impact assessment are very significant. In causal structure D → T, duration is predetermined and is not sensitive to time-of-day choice. In the presence of variable pricing, the extent to which a shift in activity timing may take place is dependent on the activity duration. In causal structure T → D, timing is sensitive to variable pricing and is determined first. The activity episode duration is then adjustable in response to the timing of the activity episode. Thus, the duration is no longer fixed and does not affect the potential shift in timing.

In other words, if one used causal structure D → T to assess variable pricing impacts when in fact structure T → D is the correct one, then one might underestimate the potential shift in traffic. This is because timing is a function of duration and the duration itself is not responsive to variable pricing. So, even though the variable pricing cost may motivate an individual to shift time of travel for an activity, the duration of the activity may preclude the person from doing so. Thus causal structure D → T may inhibit the potential shift in timing. On a similar note, if one used causal structure T → D to assess variable pricing impacts when in fact structure D → T is the correct one, then one might overestimate the potential shift in traffic.

The above example shows the critical importance of identifying the appropriate causal structure that should be employed under different circumstances. It is possible that different causal structures apply to different market segments, activity types, and urban contexts. The analysis in this section attempts to control for some of these aspects by considering activity timing and duration relationships only for out-of-home maintenance activities. Models are estimated for commuters and noncommuters separately to control for the significant influence that work and commute episodes may have on activity timing and duration decisions.

The data set is derived from a comprehensive household travel survey that was administered in 1996 in the Tampa Bay region of Florida. The survey was a traditional trip diary survey — not an activity or time use survey — of the mail-out mail-back type. The survey collected household and person socioeconomic and demographic characteristics, together with detailed information about all trips undertaken over a 24-h period. Households were asked to return one complete diary for every household member (including children); however, as expected, many households returned fewer diaries than household members. The survey instrument was mailed to about 15,000 households, and over 5000 households returned at least one trip diary, resulting in a response rate close to 35%. Given the mail-out mail-back nature of the survey, this response rate may be considered quite reasonable and consistent with expectations.

After extensive checking and data integrity screening, a final respondent sample of 5261 households was obtained. From these 5261 households, a total of 9066 persons returned usable trip diaries. The 9066 persons reported information for a total of 31,459 trips (through the 24-h trip diary). The trip file was used to create an out-of-home activity file where individual activity records were created from the trip records. This activity file included information about activity type, activity timing, activity duration, and other variables pertinent to each activity episode. The analysis presented here focuses on the relationship between activity timing and duration for maintenance activities. Maintenance activities included shopping, personal business, errands, medical and dental visits, and serving passenger or child activities.

These activity records were extracted from the original file to create two maintenance activity record files, one for commuters and one for noncommuters. Commuters were defined as driving-age individuals who commuted to a workplace on the travel diary day, while noncommuters were defined as driving-age individuals who did not commute to a workplace (made zero work trips) on the travel diary day. Also, children under the age of 16 were excluded from the analysis completely. Maintenance activity records that had full information (no missing data) were extracted to create commuter and noncommuter data files for the modeling effort.

Maintenance activities were pursued by 2904 individuals residing in 2386 households. Of these individuals, 1023 were commuters; they reported 1351 maintenance activities. The remaining 1881 individuals were noncommuters; they reported 2899 maintenance activities. The commuter and noncommuter maintenance activity episode data sets included complete socioeconomic and activity information for the respective samples. The samples represent self-selected samples of individuals who actually participated in a maintenance activity on the travel survey day. Thus, in modeling the relationship between activity timing and duration for these data sets, one needs to account for self-selectivity arising from the activity record selection and extraction process.

The average household size for the sample of 2386 households is 2.3 persons per household. More than one half of the households are two-person households in this particular sample. Average vehicle ownership is about 1.8 vehicles per household, with a little more than 40% of the sample owning two cars. More than three quarters of the sample resides in a single-family dwelling unit. About one third of the sample has an annual income of less than $25,000, while about one quarter of the sample has an annual income of greater than $50,000.

The major differences between commuters and noncommuters are consistent with expectations. Commuters are predominantly in the age groups of 22 to 49 years and 50 to 64 years, while noncommuters are older, with more than 60% greater than or equal to 65 years of age. Similarly, 80% of commuters are employed full-time, while only 7.7% of noncommuters are employed full-time.

Among those who undertake at least one maintenance activity, noncommuters undertake (on average) a higher number of maintenance activities. In addition, they allocate more time to maintenance activities and have longer maintenance activity episode durations than commuters. While the commuter sample that reported at least one maintenance activity spent (on average) 1.5 h for maintenance activities during the day, the noncommuter sample spent nearly 4 h. On average, the commuter sample reported activity episode durations of 70 min, while the noncommuter sample reported activity episode durations twice that amount, at a little over 140 min.

The decisions regarding the timing and duration of maintenance activity episodes are modeled using a joint discrete–continuous econometric framework. In such joint systems, logical consistency considerations require certain restrictions to be maintained on the coefficients representing the causal effects of the dependent variables on one another. Specifically, in the context of the joint time of day of activity participation and activity duration model of the current section, the restrictions imply a recursive causal model in which time of day of activity participation affects activity duration or vice versa (but not both). The following discussion describes the restrictions in more detail for the case of a simple binary choice for time of day, as it simplifies the presentation. However, the same restrictions extend to the case of a multinomial choice situation for time of day. This is followed by a presentation of the structure and estimation technique for a multinomial time-of-day and continuous activity duration model.

Let s^* be a latent continuous variable that determines an observed binary variable s representing the time of day of activity participation; s may take the value 0 (say A.M. participation) or 1 (say P.M. participation). Let a be the logarithm of the duration of activity participation (the logarithm form guarantees the nonnegativity of duration predictions). Consider the following equation system, where the index for observations has been suppressed:

$$s^* = \beta'z + \delta_1 a + \varepsilon, \quad s = 0 \text{ if } s^* \leq 0; \; s = 1 \text{ if } s^* > 0$$
$$a = \theta'x + \delta_2 s + \omega, \tag{2.14}$$

where z and x are vectors of observed variables, ε and ω are random error terms assumed to be normally distributed, and β, θ, δ_1, and δ_2 are coefficients to be estimated. Using the second equation to replace a in the first equation, we obtain:

$$s^* = \beta'z + \delta_1\theta'x + \delta_1\delta_2 s + \delta_1\omega + \varepsilon, \quad s = 0 \text{ if } s^* \leq 0; \; s = 1 \text{ if } s^* > 0 \tag{2.15}$$

From the above equation, one can write the following:

$$\text{Prob}\,[s = 0] = 1 - \Phi(\beta'z + \delta_1\theta'x)$$
$$\text{Prob}\,[s = 1] = \Phi(\beta'z + \delta_1\theta'x + \delta_1\delta_2) \tag{2.16}$$

where Φ is the cumulative normal distribution function of $\delta_1\omega + \varepsilon$. The sum of the above two probabilities is 1 only if $\delta_1\delta_2 = 0$, that is, only if either $\delta_1 = 0$ or $\delta_2 = 0$ in Equation (2.14). Intuitively, the logical consistency condition $\delta_1\delta_2 = 0$ states that s^* cannot be determined by s if it also determines s (see Equation (2.15)).

The application of the logical consistency condition leads to a recursive model system. If $\delta_1 = 0$, then the time of day of participation affects the logarithm of activity duration (but not vice versa). If $\delta_2 = 0$, then the logarithm of activity duration affects time of day of participation (but not vice versa). A natural question then is "Which of the two assumptions ($\delta_1 = 0$ or $\delta_2 = 0$) should be maintained?" In this analysis, both recursive systems are estimated and the two alternative systems are tested empirically to provide guidance regarding the causal direction to maintain in a joint time of activity participation and activity duration model system.

The preceding discussion used a binary choice structure for time of day to discuss the need to maintain a recursive structure in the time-of-day–duration model system. The same arguments are applicable even for a multinomial choice context for time of day.

2.5.3 Time of Day Affects Activity Duration

Let i be an index for time of day of activity participation (i = 1, 2, ..., I) and let q be an index for observations (q = 1, 2, ..., Q). Consider the following equation system:

$$u_{qi}{}^* = \beta_i' z_{qi} + \varepsilon_{qi}, \quad \varepsilon_{qi} \sim \text{IID Gumbel }(0,1)$$

$$a_q = \theta' x_q + \delta' D_q + \omega_q, \quad \omega_q \sim N(0, \sigma^2) \tag{2.17}$$

where $u_{qi}{}^*$ is the indirect (latent) utility associated with the ith time of day for the qth observation; D_q is a vector of the time-of-day dummy variables of length I; δ is a vector of coefficients representing the effects of different times of the day of activity participation on activity duration; ε_{qi} is a standard extreme value (Gumbel) distributed error term assumed to be independently and identically distributed across times of the day and observations; and other variables are as defined earlier in Equation (2.14), with the addition of appropriate subscripts. The error term ω_q is assumed to be identically and independently normal-distributed across observations with a mean of zero and variance of σ^2.

In Equation (2.17), the time-of-day alternative i will be chosen (i.e., D_{qi} = 1) if the utility of that alternative is the maximum of the I alternatives. Defining

$$v_{qi} = \left\{ \begin{array}{c} \max \qquad u_{qj}{}^* \\ j = 1, 2, ..., I, \ j \neq i \end{array} \right\} - \varepsilon_{qi}, \tag{2.18}$$

the utility maximizing condition for the choice of the ith alternative may be written as D_{qi} = 1 if and only if $\beta_i' z_{qi} > v_{qi}$. Let $F_i(v_{qi})$ represent the marginal distribution function of v_{qi} implied by the assumed IID extreme value distribution for the error terms ε_{qi} (i = 1, 2, ..., I) and the relationship in Equation (2.18). Using the properties that the maximum over identically distributed extreme value random terms is extreme value distributed and the difference of two identically distributed extreme values terms is logistically distributed, the implied distribution for v_{qi} may be derived as

$$F_i(y) = \text{Prob}(v_{qi} < y) = \frac{\exp(y)}{\exp(y) + \displaystyle\sum_{j \neq i} \exp(\beta_j' z_{qj})}. \tag{2.19}$$

The nonnormal variable v_{qi} is transformed into a standard normal variate using the integral transform result:

$$v_{qi}{}^* = \Phi^{-1}[F_i(v_{qi})] \tag{2.20}$$

where $\Phi(.)$ is the standard cumulative distribution function. Equation (2.17) may now be rewritten as

$$D_{qi}{}^* = \beta_i' z_{qi} - v_{qi}{}^*, \quad D_q = 0 \text{ if } D_{qi}{}^* < 0, \quad D_q = 1 \text{ if } D_{qi}{}^* > 0$$

$$a_q = \theta' x_q + \delta' D_q + \omega_q. \tag{2.21}$$

A correlation ρ_i between the error terms $v_{qi}{}^*$ and ω_q is allowed to accommodate common underlying unobserved factors influencing the time-of-day choice for activity participation and the duration of the participation. For example, individuals who are physically challenged or take things slowly may prefer to participate in activities during the midday periods (rather than early in the morning) and may also have a long duration of participation.

The parameters to be estimated in the joint model system are the β_i parameter vectors in the time-of-day choice model, the θ and δ parameter vectors in the activity duration model, the standard deviation

σ of the ω_q random term, and the correlation parameters ρ_i. The likelihood function for estimating these parameters is

$$= \prod_{q=1}^{Q} \left\{ \prod_{i=1}^{I} \left[\frac{1}{\sigma} \phi(l_q) \Phi(b_{qi}) \right]^{D_{qi}} \right\}, \tag{2.22}$$

where $\phi(.)$ is the standard normal density function, and l_q and b_{qi} are defined as follows:

$$l_q = \left(\frac{a_q - \theta' x_q - \delta' D_q}{\sigma} \right), \quad b_{qi} = \left(\frac{\Phi^{-1} F_i(\beta'_i z_{qi}) - \rho_i l_q}{\sqrt{1 - \rho_i^2}} \right) \tag{2.23}$$

2.5.4 Activity Duration Affects Time of Day

The equation system in this case may be written in the following form:

$$u_{qi}^* = \beta'_i z_{qi} + \gamma a_q + \varepsilon_{qi}, \quad \varepsilon_{qi} \sim \text{IID Gumbel } (0,1)$$

$$a_q = \theta' x_q + \omega_q, \quad \omega_q \sim N(0, \sigma^2). \tag{2.24}$$

Using the same procedures as in the previous section, the above system can be rewritten as

$$D_{qi}^* = \beta'_i z_{qi} + \gamma a_q - v_{qi}^*, \quad D_{qi} = 0 \text{ if } D_{qi}^* < 0, \quad D_{qi} = 1 \text{ if } D_{qi}^* > 0$$

$$a_q = \theta' x_q + \omega_q. \tag{2.25}$$

Assuming a correlation ρ_i between v_{qi}^* and ω_q, the likelihood function for estimating the parameters β_i ($I = 1, 2, \ldots, I$), γ, θ, σ, and ρ_i is exactly the same as in Equation (2.22), with the following alternative definitions for l_q and b_{qi}:

$$l_q = \left(\frac{a_q - \theta' x_q}{\sigma} \right), \quad b_{qi} = \left(\frac{\Phi^{-1} F_i(\beta'_i z_{qi} + \gamma a_q) - \rho_i l_q}{\sqrt{1 - \rho_i^2}} \right). \tag{2.26}$$

Both of the causal structures were estimated on the noncommuter and commuter sample activity episodes to identify the appropriate causal structure for each sample group. Detailed estimation results of joint timing–duration models for each causal structure and sample group are presented in Pendyala et al. (2002b).

Table 2.9 presents a summary of model performance based on goodness-of-fit measures.

For the noncommuter sample, it was found that the model in which activity duration is assumed to be determined first and then influence time-of-day choice offered superior statistical measures of fit compared to the model in which activity timing was assumed to precede and determine activity duration. In addition, the noncommuter model showed a significant error correlation between midday activity participation and activity episode duration, suggesting that noncommuters who do not like to be rushed prefer to engage in longer activities during the midday (avoiding peak periods). These findings suggested that activity timing and duration are closely related for the noncommuter sample and that activity duration precedes the choice of time of day.

For the commuter sample, on the other hand, it was found that both causal structures offered virtually identical statistical measures of fit and that the fits were substantially poorer than those obtained for the noncommuter samples. In addition, all of the error correlation terms were found to be statistically insignificant, suggesting that activity timing and duration could be modeled in an independent and sequential framework. These findings suggest that activity timing and duration are only loosely related

TABLE 2.9   Measures of Fit for Joint Timing–Duration Models

	Noncommuter Sample		Commuter Sample	
Summary Statistic	Time of Day Affects Duration	Duration Affects Time of Day	Time of Day Affects Duration	Duration Affects Time of Day
Log-likelihood at zero, $L(0)$[a]	−7072.00	−7072.00	−3216.60	−3216.60
Log-likelihood at sample shares, $L(C)$[b]	−5756.11	−5756.11	−3058.38	−3058.38
Log-likelihood at convergence, $L(\beta)$	−5584.14	−5316.82	−2990.86	−2990.60
Number of parameters, k[c]	22	21	21	22
Number of observations, N	2899	2899	1351	1351
Adjusted likelihood ratio index at zero, $\bar{\rho}_0^2$	0.207	0.245	0.064	0.063
Adjusted likelihood ratio index at sample shares, $\bar{\rho}_c^2$	0.027	0.073	0.016	0.016

[a] The log-likelihood at zero corresponds to the likelihood function value of the joint model with no variables in the multinomial logit (MNL) time-of-day model, and with only the constant and variance (standard deviation) terms in the log-linear duration equation. All correlation terms are zero.

[b] The log-likelihood at sample shares corresponds to the likelihood function value of the joint model with only alternative specific constants in the MNL time-of-day model, and with only the constant and variance terms in the log-linear duration equation. All correlation terms are zero.

[c] The number of parameters (k) does not include the constant and variance terms in the log-linear duration model.

from a causal decision-making standpoint, although they are correlated with one another. Commuters, who have relatively more constraints imposed by work schedules and commute trips, may not have the ability to exercise a decision process that is characterized by choices, alternatives, and causal relationships. In such a context, it is very difficult to identify a causal structure or relationship underlying activity timing and duration.

The identification of such causal relationships between temporal aspects of activity engagement phenomena is very important from several key perspectives. First, the identification of appropriate causal structures will help in the development of accurate activity-based travel demand model systems that intend to capture such relationships at the level of the individual traveler and activity episode. Second, a knowledge of the true causal relationships underlying decision processes will help in the accurate assessment and impact analysis of alternative transportation policies such as variable pricing, parking pricing, and telecommuting.

2.6   Example Application 4: Time Use Patterns and Quality-of-Life Measures

Transportation investments are increasingly being seen as directly influencing the quality of life in a neighborhood, city, or region. In many surveys of area residents around the world, transportation is consistently rated very highly as an important determinant of quality of life. The level of satisfaction that residents exhibit with respect to the performance and level of service of the transportation system is very often directly correlated with their perception of the quality of life that they enjoy (Pendyala et al., 1998).

The notion of time use can be used to evaluate the impact of transportation investments on people's quality of life. In general, it may be postulated that people's activity–travel patterns are a manifestation of their desire to pursue activities that are distributed in time and space. These activities (i.e., the time spent at these activities) provide a positive utility; otherwise the activities would not be undertaken. Thus, by analyzing the time use associated with an activity–travel pattern, one may be able to measure the level of satisfaction or welfare that a person is deriving from his or her activity-travel pattern. As the activity–travel pattern is often directly affected by the level of service provided by the transportation system, one may then conjecture that the impact of a transportation system improvement or travel demand management strategy on a person's quality of life can be measured through activity-based time use analysis. This section illustrates how the notion of time use can be used to address quality-of-life issues.

The utility or welfare derived from a daily activity–travel pattern is viewed primarily as a function of the amounts of time allocated for out-of-home and in-home activities. Two other important dimensions are monetary expenditures and the quality of time for each activity. The latter is determined by the location, the co-participants, the amounts of nonmonetary resources devoted to the activity, and other contributing factors. The formulation presented in this section combines the concept of the intervening opportunities model, which embodies the satisficing concept and the asymptotic theory of extreme value distributions.

Consider the activity–travel pattern of individual i on day t, and let T_{itk} be the time spent on the kth activity episode, Y_{itk} the monetary expenditure for the kth activity episode, and R_{itk} the location attributes of, and nonmonetary resources devoted to, the kth activity episode. Let $Z_{it} = (T_{it}, Y_{it}, R_{it})$ denote the daily pattern, where $T_{it} = (T_{it1}, T_{it2}, ..., T_{itn})$, and the utility of this pattern be $U(T_{it}, Y_{it}, R_{it})$.

Now consider an individual activity episode and let

$$U_q = B_{k(q)} \ln t_q = \{\beta_{k(q)} [\ln (\eta r_{k(q)}) + \gamma_{k(q)} \ln S_q] + \varepsilon_q\} \ln t_q, \ t_q > 0 \quad (2.27)$$

where t_q is the activity duration of episode q, k(q) is the activity type of episode q, $\beta_{k(q)}$ and $\gamma_{k(q)}$ are unknown coefficients, $r_{k(q)}$ is the density of opportunities for activity k(q), η is the scaling constant, S_q is the travel time expenditure for episode q, and ε_q is the random error term (independently and identically distributed).

The coefficient $B_{k(q)}$, $B_{k(q)} > 0$, may be viewed as the modifier of the basic time utility $\ln t_q$. The modifier is assumed to vary by activity type. As shown below, $B_{k(q)}$ represents the location attributes of activity episode q. In this formulation, both dU_q/dt_q and d^2U_q/dt_q^2 are greater than zero.

The term $\ln (\eta r_{k(q)}) + \gamma_{k(q)} \ln S_q$ reflects the consideration that the utility of an opportunity chosen for the activity increases (on average) with the number of opportunities out of which that opportunity has been chosen. It may be reasonably assumed that an opportunity chosen after traveling S_q is better than those opportunities closer than S_q; otherwise that distance will not be traveled. In a hypothetical feature-less plain, the number of opportunities within S_q may be represented as $n = \eta r_{k(q)} S_q^2$. The utility of the chosen opportunity is maximum ($U_1 ... U_n$), which is asymptotically proportional to $\ln n$, if the U values are independent and identically distributed.

Therefore

$$\ln n = \ln (\eta r_{k(q)} S_q^2) = \ln (\eta r_{k(q)}) + 2 \ln S_q \quad (2.28)$$

Substituting $\gamma_{k(q)}$ for 2, a generalized form, $\ln (\eta r_{k(q)}) + \gamma_{k(q)} \ln S_q$, which is part of the utility expression given above, is obtained.

In applying the above, appropriate zonal density measures may be selected for $r_{k(q)}$ considering the type of activity. Determining S_q for linked trips is not straightforward. One approach is to use a measure of the deviation of the opportunity location from the line obtained by connecting the previous location and the next location (including the home base), for example, max $(t_{iq} + t_{qj} - t_{ij}, 0)$, where i is the previous opportunity, j is the next opportunity, and t_{ij} is a measure of spatial separation between opportunities i and j.

Assuming that the total utility of the series of activities pursued during a day is the sum of the utilities of the respective activities, let

$$U(T_{it}, R_{it}) = \Sigma U_q = \Sigma B_{k(q)} \ln t_q \quad (2.29)$$

where the summation is for all nontravel activities. This form of the utility function may be used to evaluate the quality or level of satisfaction derived from alternative activity–travel patterns. It is noteworthy that the same formulation can be used even if the total utility is considered a product of individual utilities.

This basic utility expression warrants two extensions: (1) incorporation of monetary expenditures, and (2) incorporation of differential effects of travel mode on the quality of travel time. Monetary expenditures or the stock of instruments and devices available for activity engagement do affect the quality of time spent for the activity. For example, the same 2-h dinner may yield different levels of utility depending on the quality of the restaurant, which will be reflected in the monetary expenditure there. Unfortunately, such information is usually not available in travel behavior data sets. Because of this, the formulation presented in this section assumes that such differences can be represented by incorporating measured socioeconomic attributes of the individual into the utility function, and by the random error term, ε_q. This calls for the following modification of U_q:

$$U_q = \{\beta_{k(q)} [\ln (\eta r_{k(q)}) + \gamma_{k(q)} \ln S_q] + B_{k(q)} X_i + \varepsilon_q\} \ln t_q, \, t_q > 0 \tag{2.30}$$

where $B_{k(q)}$ is the coefficient vector and X_i is the attribute vector describing individual i. The unknown coefficients in the utility function may be estimated by using information contained in typical travel survey data sets.

Simplifying the expression in Equation (2.30) and taking expectation, one obtains

$$E(U_q) = [\Delta_{k(q)} M + \Phi_{k(q)} \ln S_q] \ln t_q \tag{2.31}$$

where M is a vector of explanatory variables (including the effects of both X_i and $r_{k(q)}$) and $\Delta_{k(q)}$ and $\Phi_{k(q)}$ are unknown coefficients to be estimated. Formulating the problem as one of utility maximization (subject to the constraints that a day is limited to 24 h) and assuming independence of error terms across activity episodes, linear regression methods may be employed to estimate model coefficients (Kitamura et al., 1997a). A summation of the utilities of individual activities yields the total utility associated with a 24-h activity–travel pattern. That is,

$$U = \Sigma_q [\Delta_{k(q)} M + \Phi_{k(q)} \ln S_q] \ln t_q + \ln t_h \tag{2.32}$$

where t_h is the time spent at home.

Table 2.10 presents sample utility models of activity durations for various activity types that were estimated on the California time use data set by Kitamura et al. (1997a). In general, the model coefficients have the expected sign and offer plausible interpretations. For example, the coefficient of $\ln S_i$ is positive (except in the case of social activities, where it is insignificant at the 95% confidence level), indicating that travel to a farther destination entails a higher level of utility (otherwise a closer destination would be chosen). Similarly, social and recreational activities offer higher utilities toward the end of the week. Plausible interpretations are also offered by the demographic variables.

These results may be used to estimate the utility that a person derives from an activity–travel pattern and the impact of a transportation policy or improvement on the utility or satisfaction derived from an activity–travel pattern.

To illustrate this, consider an individual living in an urban household with two adults and a teenager. Consider an adult with the following pattern on a Friday:

7:00–8:00	Travel to work
8:00–12:00	Work at workplace
12:00–13:00	Lunch break (recreational)
13:00–17:00	Work at workplace
17:00–17:50	Travel from work to shop
17:50–18:20	Shopping
18:20–18:45	Travel to home

Travel, by itself, is assumed neither to produce utility nor to involve any nontime cost. As work is assumed to be fixed and independent of nonwork activities, it may be placed outside the analysis here (inclusion would merely shift the scale of the utility measure by a constant). Then, for this individual, the set of activities that produce utility may be summarized as follows:

Home sojourn = 735 min

TABLE 2.10 Utility Models of Activity Durations

Explanatory Variable	Shopping		Personal Business		Social		Recreational	
	Coefficient	t-Stat	Coefficient	t-Stat	Coefficient	t-Stat	Coefficient	t-Stat
Constant	−0.0198	−1.12	0.0093	0.33	0.1656	1.85	0.0037	0.10
ln (t)	0.0297	4.89	0.0113	1.13	−0.0108	−0.41	0.0407	3.33
Urban household with teenagers	0.0648	2.49	—	—	—	—	—	—
Single person under 35 years	—	—	0.0979	3.41	—	—	—	—
Suburban household with two or more adults under 35 years	—	—	—	—	0.5179	2.74	—	—
Suburban single person under 35 years	—	—	—	—	—	—	0.0653	2.65
Thursday or Friday	—	—	—	—	0.1046	2.03	0.1733	2.96
N	140		83		23		176	
R²	0.182		0.146		0.348		0.158	
Adjusted R²	0.170		0.125		0.245		0.144	
F (df)	15.20 (2, 137)		6.83 (2, 80)		3.38 (3, 19)		10.78 (3, 172)	

Note: F = F statistic; df = degrees of freedom.

Recreational activity = 60 min (travel time S_i is 0)
Shopping activity = 30 min (travel time S_i is 15 min)

The travel time to shopping is 15 min because the travel time between home and work is 60 min and only the additional travel time attributable to shopping enters the utility equation. The utility of this pattern is computed as follows:

Home sojourn:	ln (735) = 6.60
Recreational:	$(0.0037 + 0.0407 \times 0 + 0.1733)$ ln (60) = 0.725
Shopping:	$(-0.0198 + 0.0297 \times$ ln $15 + 0.0648)$ ln (30) = 0.427
Total utility:	6.60 + 0.725 + 0.427 = 7.752

Now suppose traffic conditions are improved by capacity additions (transportation investments) and commuting travel time is reduced from the current 60 min to 45 min. With two commute trips, this would imply a total travel time saving of 30 min per day. If all of the travel time reduction is allocated to home activities and no changes are made to nonhome time allocation (except for starting and ending times), then the home sojourn utility will become ln (765) = 6.640 and the total utility increases from 7.752 to 7.792.

Structural equations models of time use and travel (such as that presented in the second application example of this chapter) have generally found that, for every 10 min of time savings, 7 min would be allocated to in-home activities, 2 min would be allocated to out-of-home nonwork activities, and 1 min would be allocated to additional travel. These values vary widely by demographic group, but these average indicators may be used to gauge the impact of a capacity improvement (travel time reduction).

In the example, of the 30 min saved, if the person allocates 21 min to in-home activities, 6 min to additional shopping activity, and 3 min to additional shopping travel, then the utility calculations are modified as follows:

Home sojourn:	ln (756) = 6.63
Recreational:	0.725 (unchanged)
Shopping:	$(-0.0198 + 0.0297 \times$ ln $18 + 0.0648)$ ln (36) = 0.469
Total utility:	6.63 + 0.725 + 0.469 = 7.824

This example shows how the transportation improvement increases the utility or welfare of the individual from a baseline value of 7.752 to a new value of 7.824. This example shows how time use analysis can serve as a powerful tool for assessing the benefits and welfare that can be derived from transportation improvements and investments. Any additional travel demand (e.g., induced travel) that is brought about by the transportation system improvement should be viewed in the context of the improved quality of life that such transportation improvements provide the residents of a region. Thus, time use analysis serves as a powerful framework for conducting policy analysis.

2.7   Summary and Conclusions

This chapter has provided an overview of the key role played by time use in activity and travel behavior analysis. The representation of time and space in models of activity and travel demand has been greatly facilitated through the development of new analytical methods and technological tools. Some of the methods that are being used to address these dimensions of activity and travel behavior have been presented and discussed in this chapter through four different application examples that clearly demonstrate the power of time use analysis. The methods and applications explicitly covered in this chapter include:

1. Stochastic frontier models of time–space prism vertices
2. Structural equations models of household time allocation and activity engagement
3. Econometric joint discrete–continuous models of causality between maintenance activity episode timing and duration
4. Utility model of welfare or quality of life derived from time use and activity pattern

For each application example, the chapter provides an overview of the concept addressed, the methodology adopted, and the results of the modeling effort. The practical implications of the model results are emphasized in order to demonstrate how time use concepts can be used in the context of activity and travel behavior analysis and modeling.

It must be noted, however, that there are many more methods that have been used to analyze time use patterns and activity durations. For example, Goulias (2002) applies multilevel modeling methods to study daily time use and time allocation to activity types while accounting for complex covariance structures using correlated random effects. Bhat and Zhao (2002) perform a detailed spatial analysis of activity stop generation while accounting for spatial interdependencies among activity–travel choices. They develop spatial mixed ordered response logit (MORL) models and compare their performance against aspatial ordered response logit models that do not account for spatial aspects of activity–travel choices. They find that ignoring spatial interdependencies and aspects of activity stop generation may adversely affect model estimation results. The additional references provided under the "Further Information" section of this chapter provide descriptions of alternative methodologies and their application to time use and activity pattern analysis.

New technological tools are beginning to play an important role in facilitating detailed time–space analysis of activity–travel patterns. Within the scope of this chapter, the key relationship between time and space has been recognized, but the application examples and descriptive statistics have placed a greater emphasis on the concept of time. The analysis of the space dimension has generally lagged the analysis of the time dimension, partially because of the difficulty associated with representing and measuring space. However, Geographic Information Systems (GIS), Global Positioning Systems (GPS), and other technologies are being increasingly used to represent, measure, and model the action space of individuals in the context of time constraints. Interdisciplinary research efforts involving the fields of geography, sociology, planning, and engineering will continue to yield advances in the modeling of the space dimension in the time–space continuum.

The chapter has included numerous descriptive statistics about time use patterns of individuals in different geographic contexts. While these measures and statistics were largely obtained from time use and activity-based surveys, it must be noted that time use and activity-based analysis can also be performed on traditional travel survey data sets that include information about trips. While these data sets do not offer detailed information about in-home activities, they do offer information about out-of-home activities. In fact, two of the four example applications presented in this chapter involved the analysis of time use patterns derived from traditional trip-based travel diary data sets. The structural equations analysis of household interactions utilized the southeast Florida trip diary data set, while the causal analysis of maintenance activity timing and duration utilized the Tampa Bay (Florida) area trip diary data set.

The merging of traditions between travel research (and travel data collection) and time use research (and time use data collection) constitutes the core of the new wave of transportation planning and modeling tools that are being developed around the world. The merging of these traditions has made it possible for travel researchers to obtain a richer understanding of the role played by time and space dimensions in shaping human activity and travel behavior.

References

Bhat, C.R. and Zhao, H., The spatial analysis of activity stop generation, *Transp. Res. B*, 36, 557–575, 2002.

Golob, T.F. and McNally, M.G., A model of activity participation and travel interactions between household heads, *Transp. Res. B*, 31, 177–194, 1997.

Goulias, K.G., Multilevel analysis of daily time use and time allocation to activity types accounting for complex covariance structures using correlated random effects, *Transportation*, 29, 31–48, 2002.

Hägerstrand, T., What about people in regional science? *Pap. Proc. Reg. Sci. Assoc.*, 24, 7–24, 1970.

Harvey, A.S., Time Space Diaries: Merging Traditions, paper presented at International Conference on Transport Survey Quality and Innovation, Kruger Park, South Africa, August 2002.

Kitamura, R., van der Hoorn, T., and van Wijk, F., A comparative analysis of daily time use and the development of an activity-based traveler benefit measure, in *Activity-Based Approaches to Travel Analysis*, Ettema, D.F. and Timmermans, H.J.P., Eds., Pergamon, Oxford, U.K., 1997a, pp. 171–188.

Kitamura, R., Fujii, S., and Pas, E.I., Time use data, analysis, and modeling: toward the next generation of transportation planning methodologies, *Transp. Policy*, 4, 225–235, 1997b.

Meka, S., Pendyala, R.M., and Kumara, M.A.W., A Structural Equations Analysis of Within-Household Activity and Time Allocation between Two Adults, paper presented at CD-ROM Proceedings of the 81st Annual Meeting of the Transportation Research Board, National Research Council, Washington, D.C., January 2002.

Pendyala, R.M., Kitamura, R., and Reddy, D.V.G.P., Application of an activity-based travel demand model incorporating a rule-based algorithm, *Environ. Plann. B*, 25, 753–772, 1998.

Pendyala, R.M., Yamamoto, T., and Kitamura, R., On the formulation of time space prisms to model constraints on personal activity–travel engagement, *Transportation*, 29, 73–94, 2002a.

Pendyala, R.M., Bhat, C.R., Parashar, A., and Muthyalagari, G.R., An Exploration of the Relationship between Timing and Duration of Maintenance Activities, paper presented at CD-ROM Proceedings of the 81st Annual Meeting of the Transportation Research Board, National Research Council, Washington, D.C., January 2002b.

Further Information

There is a vast and growing body of literature dedicated to time use research and activity-based analysis of travel behavior. Several excellent review articles include:

Axhausen, K. and Gärling, T., Activity-based approaches to travel analysis: conceptual frameworks, models, and research problems, *Transp. Rev.*, 12, 324–341, 1992.

Bhat, C.R. and Koppelman, F., A retrospective and prospective survey of time use research, *Transportation*, 26, 119–139, 1999.

Kitamura, R., An evaluation of activity-based travel analysis, *Transportation*, 15, 9–34, 1988.

Pas, E.I. and Harvey, A.S., Time use research and travel demand analysis and modeling, in *Understanding Travel Behaviour in an Era of Change*, Stopher, P.R. and Lee-Gosselin, M., Eds., Pergamon, Oxford, 1997, pp. 315–338.

In addition, there are several books that contain a wealth of information about activity-based approaches to travel analysis and time use research. These books include:

Ettema, D.F. and Timmermans, H.J.P., Eds., *Activity-Based Approaches to Travel Analysis*, Pergamon, Oxford, U.K., 1997.

Jones, P., Ed., *Developments in Dynamic and Activity-Based Approaches to Travel Analysis*, Oxford Studies in Transport, Avebury, Aldershot, U.K., 1990.

Robinson, J.P. and Godbey, G., *Time for Life: The Surprising Ways Americans Use Their Time*, Pennsylvania State University Press, University Park, PA, 1997.

Ver Ploeg, M. et al., Eds., *Time Use Measurement and Research*, National Research Council, National Academy Press, Washington, D.C., 2000.

There are two special issues of the journal *Transportation* that contain papers on time use and activity perspectives in travel behavior research. These two issues contain selected papers presented at the 1997 and 2000 International Association for Travel Behaviour Research Conferences. The first issue is in Volume 26, published in 1999, while the second issue is in Volume 29, published in 2002. This journal is published by Kluwer Academic Publishers and printed in The Netherlands.

The Transportation Research Board (a unit of the National Research Council, Washington, D.C.) has a subcommittee, A1C04(1), dedicated to research on time use and activity patterns in travel behavior. Many of the activities and papers of this subcommittee would be of direct interest to time use professionals. Many of the papers presented at the Annual Meeting of the Transportation Research Board appear on the CD-ROM proceedings and in the *Transportation Research Record, Journal of the Transportation Research Board*. There are several other committees and subcommittees in the Transportation Research Board structure that are involved in advancing the state of the art and the state of the practice in time use and activity-based travel behavior analysis research.

Structural equations modeling methods have been used extensively in the analysis of activity patterns and time use. *Structural Equations with Latent Variables* by K. Bollen (John Wiley & Sons, New York, 1989) provides a detailed description of these methods and their application. There are several papers by T.F. Golob that use the structural equations methods to analyze various aspects of time use and activity behavior. His papers have appeared in *Transportation Research A* and *B*. Besides T.F. Golob and the author of this chapter, others who have used the structural equations approach include R. Kitamura (e.g., *Transportation Research B*, 34, 339–354, 2000) and the late E.I. Pas (e.g., *Transportation Research A*, 33, 1–18, 1999).

Several groups of researchers, including T. Arentze, A. Borgers, H.J.P. Timmermans, M.-P. Kwan, R. Golledge, and T. Gärling, have used Geographic Information Systems and other spatial analysis tools to study the effects of accessibility on spatial patterns of activities and travel. Another important topic, activity scheduling, has been studied by researchers such as E.J. Miller, S. Doherty, and T. Gärling. Papers by these authors and others can be found in the journals identified in this section and are cited in most of the papers already mentioned here.

New statistical methods, including both parametric and nonparametric methods, are being increasingly applied to the analysis and modeling of time use and activity patterns in space and time. J.P. Kharoufeh and K.G. Goulias (*Transportation Research B*, 36, 59–82, 2002) provide an example of a nonparametric kernel density estimation method applied to daily activity durations. C.R. Bhat, through numerous papers appearing in *Transportation Research A* and *B*, has contributed substantially to the application of advanced econometric methods, including mixed logit methods and hazard-based duration models for the analysis of time use and activity patterns. Other researchers (e.g., F. Mannering and D.A. Niemeier) have also applied duration models to the analysis of activity time allocation. R. Kitamura has applied the Tobit modeling approach to studying time allocation behavior among activity types.

Finally, the U.S. Department of Transportation, through its Travel Model Improvement Program (TMIP), has been developing a clearinghouse of information, including references to activity and time use research. A 1996 conference on activity-based methods produced a conference proceedings that has several articles and workshop reports covering time use and activity-based approaches to travel behavior analysis. Information may be obtained at http://tmip.fhwa.dot.gov.

3

Spatial Behavior in Transportation Modeling and Planning

CONTENTS

Reginald G. Golledge
University of California

Tommy Gärling
Göteborg University

3.1 Introduction

The demand for transportation services is a derived demand based on the needs of people to perform daily and other episodic activities. There have been two dominant approaches to investigating this derived demand: (1) studies focused on the spatial behavior of people, that is, the recorded behavior of people as they move between origins and destinations (e.g., Hanson and Schwab, 1995); and (2) an examination of the decision making and choice processes that result in spatially manifest behaviors (e.g., Ben-Akiva and Lerman, 1985; Ortúzar and Willumsen, 1994). The former approach has been typified by the development of methods for describing and analyzing activity–travel patterns. The latter is typified both by structural models that involve modeling the final outcomes of decision processes, but paying little attention to the cognitive processes involved in determining the final decision concerning movement in space, and by behavioral process models paying particular attention to the cognitive factors involved in decision making, as well as to the final choice act (Golledge and Stimson, 1997).

Structural models are built on assumptions such as utility maximization, complete knowledge, optimality, and lack of individual differences among the population. Behavioral models have been built on assumptions of satisficing principles, nonoptimal behavior, constrained utility maximization, and individual differences across the population. The structural models usually represent the aggregate movement activities of populations, while the behavioral models are disaggregate representations of the behaviors of individuals or households. Another chapter in this book focuses on structural models; in this chapter we review research on disaggregate spatial behavior as the source of information about behavioral travel choice models.

Transportation modelers and planners need knowledge of travel behavior, including route choice, mode choice, destination choice, travel frequency, activity scheduling, commuting behavior, and pretravel and en route travel decision making. Since the 1970s, most modeling emphasis has been based on random utility theory. Different travel options are assumed to have an associated utility, which is defined as a function of the attributes of the alternative and the decision maker's characteristics. Ben-Akiva (1995) and Ben-Akiva et al. (1997) provide a recent summary of the state of the art in modeling individual travel choices. They claim that there are few satisfactory existing structural models and that there is a need for behavioral realism, which involves considering heterogeneity of travel preferences, a variety of decision strategies, differentiation between individual and joint decision making for travel, improved consideration of information, and traveler's states of knowledge (e.g., their cognitive awareness or cognitive maps of the travel environment). Many of these concerns have been the focus of the activity-based approach, which emphasizes both travel and the spatial decisions that influence movement behavior (Jones et al., 1983; Kitamura, 1988; Axhausen and Gärling, 1992; Ettema and Timmermans, 1997; Bhat and Koppelman, 2000).

One concern with many of the structural models derived from random utility theory has been their unrealistic behavioral assumptions. Foremost among these has been the assumption of utility maximization, which has allowed the development of models in an optimization framework. But, as part of the activity-based approach, the growing concern with the cognitive demands of travel has led to substantial research into human spatial behavior. This research has included a search for simple measures of spatial ability, individual differences within populations, attitudes toward risk and uncertainty, and variability in path selection criteria. In addition, it is now commonly recognized that decision processes are often dependent on the time of day that travel is to take place and the type of information about network and traffic conditions that is available at that time. To understand day-by-day variability in traffic volumes and network usage, research has been undertaken on the episodic intervals needed for pursuing different types of activities (Recker et al., 1986a, 1986b; Zhou and Golledge, 2000). It has also been recognized that many travel decisions are secondary effects of the choice of locations for home and work.

In the contemporary information technology-dominated society of the 21st century, it has become more widely accepted that the quality, quantity, and timing of information will critically affect travel choices. Travelers can choose only from options of which they are aware, so information affects choice set generation and is instrumental in defining feasible opportunity sets for each trip purpose (Kwan, 1994). Sources of information include the learning that takes place with environmental experience as well as information obtained from secondary sources, such as mass media. To date, considerable research has focused on the task of predicting travelers' use of information sources (Polydoropoulou and Ben-Akiva, 1998; Abdel-Aty and Jovanis, 1996, 1998; Liu and Mahmassani, 1998; Polydoropoulou et al., 1996; Khattak et al., 1995; Mannering et al., 1995; Adler et al., 1993b). Limited research has examined how travelers' perception and memory of the transportation environment (i.e., travel experience) influence activity and travel choice (but see Jha et al., 1996; Kaysi, 1992; Iida et al., 1992; Gärling et al., 1994). A paucity of material at this stage also relates to the issue of spatial abilities (but see Stern and Leiser, 1988; Deakin, 1997; Khattak and Khattak, 1998). In addition, Svenson (1998) and Gärling and Golledge (2000) summarize theories related to the cognitive base of decision-making processes. They point out that humans have limited information processing capabilities, must represent information from long-term memory in a limited-capacity working memory to solve spatial tasks, and often apply heuristic rules to simplify decision making rather than attempting to determine optimal behaviors.

What has been of concern to researchers on spatial behavior (with its implication for transportation modeling and planning) is an understanding of the different regimes for using spatial information. Following ideas offered by psychologists such as Piaget and Inhelder (1967) and Siegal and White (1975), Freundschuh (1992) (see also Gärling and Golledge (2000) for a similar analysis) identifies three different stages or conditions of environmental knowing. The first consists of persons with landmark knowledge (called declarative knowledge or geographical facts). This is fundamentally place knowledge and consists of location-specific factual information. Persons who develop route knowledge are able to link landmarks in sequences and develop routes. The second type of spatial knowledge includes information on distances and directions from their navigation and is sometimes referred to as procedural knowledge. The third condition involves comprehending the layout of landmarks and understanding the integration of routes into networks. It is variously referred to as map knowledge, survey knowledge, or configurational knowledge. Freundschuh's (1992) analysis of the relative ease with which people can travel through regular grid networks as opposed to irregular networks indicated that the most critical factor influencing this type of behavior is spatial ability. He concluded that the use of models assuming homogeneous spatial abilities is unrealistic. His findings have focused considerable ongoing research to determine the nature of spatial abilities, which appear to be most influential in travel behavior (Golledge, 1992; Gärling et al., 1998c). Thus, it has become a matter of record that people have different methods of encoding spatial data, and that their knowledge of physical space and built environments is organized in identifiable ways.

The results of this research tend to indicate that travelers with landmark knowledge can recognize familiar surroundings but are not able to use this knowledge to complete a trip to a new location. These travelers must rely on ancillary information such as maps or directions from others, are captive to the route that is provided for them, and have limited ability to substitute route segments or to take shortcuts. On the other hand, travelers with route knowledge learn a specific set of rules for navigating from any given point to any other given point following a set of landmarks in strict order. Such travelers can recall routes from memory, but usually only one route at a time. Travelers with configurational knowledge have an understanding of the nature of the network and are able to mentally compute spatial relations required to link landmarks and develop routes, even to destinations that have not been previously visited. They are more likely to be able to construct new routes in response to changing travel conditions and are likely to have the greatest number of feasible alternative destinations and routes stored in memory. They have a dynamic understanding of the transport environment, can take shortcuts or select alternative routes when faced with congestion or other adverse travel conditions, and are the most self-confident travelers in the population.

As detailed in other chapters of this book, developments entailing such a detailed analysis of individuals' spatial and nonspatial knowledge have made necessary a transition from a focus on secondary data (i.e., aggregate travel, usually between arbitrarily defined spatial zones and collected by traffic counts or simplified driver interviews), as opposed to the use by behavioral modelers of primary data, much of which is unobservable except through stated preferences, stated attitudes, or behaviors predicted from knowledge of personal information bases and personal (or household) activity patterns. In practical applications, this has meant a shift from the gravity–entropy models that dominated transportation modeling and planning in the 1960s and 1970s to the variety of formats amenable to disaggregate modeling, including logit models, computational process models, and microsimulation models. In the balance of this chapter we will explore the nature of spatial behavior processes and how components of it have been operationalized in such a way that they can be incorporated into modeling and planning activities by processes of contemporary transportation scientists, engineers, and planners.

3.2   The Nature of Spatial Decision Making

Human decision making does not take place in a vacuum. As people age and develop psychologically and intellectually, they accumulate a store of information about environments — the cultural, social, economic, political, legal, and other constraints that limit freedom of choice and freedom of movement — and they develop different levels of spatial abilities and knowledge. Thus, we accept that decision making is influenced by prior knowledge based on experience and learning of the environments and sociocultural systems in which individuals reside and carry out their activities. For any given problem situation one can assume either that there is stored experience in memory that can be called on to help solve any given problem or that knowledge transfer can take place based on experiencing similar situations or based on generalized schemata that people carry over from one environment to another. For example, although a person may never have visited a specific shopping mall before, he or she usually has a generic template or schema of what a shopping center is supposed to be, and this is of help in defining locations for entrances and exits and means of traveling from one level to another, and even in obtaining an understanding of how shops are organized on each level. The same type of schemata may develop in different cultural environments. As another example, U.S. travelers entering different U.S. cities will carry schemata of the transportation network (involving freeways, highways, arterial roads, neighborhood streets, lanes, and alleys) that allow them to categorize parts of the unfamiliar network and to use this network in a manner similar to that which they have experienced in other environments (Kwan et al., 1998). This state of prior knowledge and transferable schemata is derived from the personal experiences of traveling through different environments, by examining representations of environments in the form of maps, images, photographs, and slide or video presentations, or by developing a configural understanding of an environment from a bird's-eye view (e.g., from a lookout or by looking through the window of an airplane).

A person has to be motivated to travel. Examples of travel motives include the feeling of hunger or the need to earn a living, or exposure to an advertisement for a job or for a location at which particular wants and needs can be satisfied. The end result is that an individual, acting either for oneself or for a group, is motivated to move between an origin and destination. Usually the first step in this motivation process is a search for relevant information. This search will include an attempt to familiarize the individual with selected aspects of the environment. This may include the transportation network and the location of different land uses. The motivated person may also have to collect information about traffic volumes and the daily temporal cycles of movement undertaken by the population as a whole. Some of this information can be obtained from secondary sources such as the Yellow Pages telephone directories, printed or televised ads, communication with neighbors, or examination of printed or electronic maps.

Once information is collected, it is encoded and stored in long-term memory. Thus, each individual builds a cognitive map of his or her unique internal representation of the world around him or her (Downs and Stea, 1973b). These cognitive maps are simply encoded databases, and there is no evidence that they are actually stored in cartographic format. For the most part, the term either is accepted as a hypothetical construct or is used metaphorically (Kitchin, 1994). Nevertheless, it is assumed that, when faced with a task involving spatial movement, people are — within the limits of their spatial abilities — able to bring previously encoded information from long-term memory into a working memory and potentially arrange it in map-like or other spatial form so that critical movement decisions can be made (Kuipers, 1978). The essence of these decisions is that potential travelers are able to define a behavior space in which their movements will be located. This behavior space consists of a subset of the total environment, which may be confined to a particular segment or corridor. Information relevant to the movement process is evaluated in this behavior space as part of the spatial decision-making process (Golledge, 1997b). For example, given a particular need (e.g., food) the behavior space will include a set of feasible alternatives at which the desired food could be obtained. The creation of this behavior space is temporally and locationally dependent. The behavior space for food purchase may, for example, be quite different when viewed from the perspective of a home base as the source of a trip, as opposed to the perspective that would be appropriate if another location, such as work or an educational institution, was the origin of the trip. In each case, the feasible opportunity set might change. For example, a potential traveler at a home base may choose a feasible alternative that lies in the opposite direction to the workplace; such an alternative would usually not be considered part of the feasible set if viewed from the perspective of the workplace.

Once the behavior space has been determined, the traveler focuses on movement imagery. In this case, a potential route between the current location and the chosen destination will have to be worked out. This will involve making a choice of travel mode; estimating the time, cost, and distance of travel to the proposed destination; integrating this particular trip into a multiple-stop trip chain if that is the intent of the decision maker; developing travel plans that include optional activities if the desired route is blocked by congestion, hazard, or construction; and assessing or evaluating the likely outcomes of making such a trip.

The final stage of the decision-making process involves implementing the desired behavior and traveling through space between an origin and destination via a particular mode over a segment of the transportation network. At the end of any transaction that is involved with this trip, feedback occurs in that the traveler evaluates and assesses whether the derived behavior satisfies the original demand condition. If it does, then this particular trip may be stored in memory as a potential solution in future task situations of the same type. If not, then evaluation of which part of the constructed process led to failure to meet anticipated levels of aspiration might dictate the necessity for a change in behavior on the next trial (Golledge and Stimson, 1987). This represents part of a spatial learning process. Successful trials can quickly lead to the development of a habitual behavior that then becomes relatively persistent and invariant over time. It is also difficult to extinguish so that, even when a potential trip is temporarily restricted by external events such as congestion, construction, weather, or other form of hazard, the traveler may return to the original spatial behavior once the intervening problem has been surmounted or disappears.

Travel habits represent behaviors that require little conscious decision-making activity prior to their performance (Gärling et al., 2001; Gärling and Garvill, 1993). They represent a significant part of the total trip patterns undertaken. The journey to work is often characterized as a travel habit. In particular, it lends itself to structural modeling and successful prediction of travel. Many other behaviors, however, are not as well entrenched as this type of travel habit. They represent more variable behaviors and may be less easily modeled and predicted by a conventional structural model. Behavioral models have been specifically developed to deal with these variable behaviors that are not easily categorized into a repetitive format. Many types of consumer behavior (apart from food shopping), social behavior, and recreational behavior fall within this latter category.

To briefly summarize this section, studies of spatial behavior have contributed significantly to understanding the decision-making process that goes on prior to the actual selection and implementation of a route choice. Rather than just trying to model revealed behaviors (i.e., the actual traces of movement over the network), models based on spatial behavior attempt to incorporate processes associated with cognitive demands. As we will see later in this chapter, the use of cognitive information carries with it error and belief baggage that biases information stored in memory and may result in inefficient, inaccurate, or unpredictable behaviors.

3.3 Cognitive Maps and Travel Behavior

The focus of this section is to examine the relationship between cognitive maps and travel behavior in urban environments. We do this incrementally, beginning with clarifications of terms relating to cognitive mapping and wayfinding, with an emphasis placed on selecting paths to destinations by using existing transport networks (particularly road hierarchies). We also introduce concerns relating to the role of trip purpose in path selection and discuss how different purposes spawn different path or route selection strategies. Finally, we examine in detail how environmental structures and considerations impact the interaction between cognitive maps, route selection, and activity choice (Golledge, 1999).

Cognitive maps are our internal representations of experienced environments. These environments can be real or imaginary, but they emphasize place ties with objects or interactions and relate nonspatial characteristics to spatially referenced places. There is as yet no clear evidence that cognitive maps have any formal cartographic structure. However, place cell analysis (Nadel, 1999) suggests that environmentally experienced objects are coded in specific place cells and that, upon repeated exposure to images or representations of specific objects or places, neurons in the same cells at specific places in the brain repeatedly fire. There appears to be insufficient evidence about the internal arrangement of place cells, so we do not know if they are randomly distributed throughout the brain or selectively clustered according to some identifiable spatial criteria. Cognitive maps, thus, are the conceptual manifestations of place-based experience and reasoning that allow one to determine where one is at any moment and what place-related objects occur in that vicinity or in the surrounding space. As such, the cognitive map provides knowledge that allows one to solve problems of how to get from one place to another, or how to communicate knowledge about places to others without the need for supplementary guidance, such as might be provided by sketches or cartographic maps.

Little research has been completed on the creation of network knowledge and the relationship between network knowledge systems and real-world transportation systems. We all realize from personal experience that our knowledge of existing networks is partial. But, if we have an overall anchoring structure or general layout understanding of on-route and off-route landmarks, we can — either by using a travel aid such as a map or by independently accessing cognitively stored information — find our way between specific origins and destinations in urban environments. Sometimes this task is simple, with minimal feasible alternative path structures to be considered. At other times the task is complex and substantial and requires meticulous planning and implementation.

In many countries, the use of the household car (or cars) represents an important form of movement. To satisfy economy of movement, minimize air and noise pollution, achieve door-to-door delivery of drivers and passengers, and guarantee independence in route choice, networks of surface roads have been developed. Usually these are differentiated into freeways, highways, arterials (major and minor), local streets, and lanes or alleys. When making a trip, each individual must consider how to use the local road hierarchy. These decisions can be made a priori (as in a travel plan) or en route (as in real-time wayfinding). The mere existence of the hierarchy, combined with individual memories of travel experience, leaves the way open for different route selection strategies to be developed and for different paths to be followed. Thus, one next-door neighbor might try to maximize use of a freeway for, say, a trip to work and maximize use of local streets to facilitate a trip chain on the way home, while another neighbor might use the reverse strategy. Thus, two spatially adjacent householders, going to the same destination, can choose completely different paths. By doing this, their environmental experiences may differ and their cognitive maps may, likewise, be quite different.

In many urban environments, traffic control measures such as one-way streets and limited on-street parking can also influence path selection and, consequently, the nature of the detail that is georeferenced in the cognitive map. Apparently, to facilitate communication and development of a general understanding of complex environments, people tend to define common anchors — significant places in the environment that are commonly recognized and used as key components of cognitive maps — and idiosyncratic or personalized anchors that are related to a person's activities (e.g., specific workplace or home base) (Golledge, 1990). These anchor the layout or structural understanding of an environment (regardless of its scale). Objects and features in an environment compete for a traveler's attention, with the most successful reaching the status of common anchor — recognized by most people and, consequently, incorporated into all their cognitive maps. Other features and objects are less successful in general, but might achieve salience for a specific trip purpose (e.g., the odd-shaped building where I park in order to go to my favorite restaurant). Minor pieces of information are attached to anchors and act as primers and fillers — the second, third, or lower orders of information experienced but used only in selected ways and with varying frequencies.

Individual differences exist in the degrees of knowledge about places, locations, or landmarks and other components of a route or network (Allen, 1999). There is also evidence that there are developmental changes in the ability of humans to learn both route and survey information (Piaget and Inhelder, 1967). Recent researchers have criticized the strict Piaget type sequential–developmental theory of spatial knowledge acquisition, particularly as interpreted by Siegel and White (1975) (e.g., Liben, 1981; Montello, 1998). Still, there appear to be recognizable differences between preschool, preteen, teenage, and adult spatial abilities, both in terms of environmental learning and success in navigating or wayfinding. There is also some evidence that males and females acquire different types of knowledge and use different types of strategies in their wayfinding tasks. In particular, it has been suggested that women use more landmarks and are more likely to use piloting strategies (i.e., travel from landmark to landmark in succession), while males tend to use more orientation and frame-related processes for wayfinding (e.g., Self and Golledge, 2000) and "head out first in the general direction" of a destination. What complicates things even further is that humans do not all behave the same way in the same environments, partly because of different levels of familiarity, partly because of different spatial abilities, partly because of different motivations to travel, partly because of different trip purposes that require them to give different saliencies to environmental features, and partly because people react differently to considerations of geographic scale and its impact on the comprehension of environments (see Bell, 2000).

Allen (1999) suggests that the most widely recognized spatial abilities from psychometric analyses are visualization, speeded rotation, and spatial orientation. Visualization concerns the ability to imagine or anticipate the appearance of complex figures or objects after a prescribed transformation, such as occurs during a paper-folding task. Speeded rotation, sometimes called spatial relations, involves the ability to determine whether one stimulus is a rotated version of another. Orientation is the ability of an observer to anticipate the appearance of an object from a prescribed perspective, such as being able to point to an obscured object in a real or imagined space.

These spatial abilities appear to fall into one of three families: (1) the stationary individual and manipulable objects, (2) a stationary or mobile individual and moving objects, and (3) a mobile individual and stationary objects. Wayfinding appears to be more related to the last of these groupings. Spatial abilities, therefore, are an important component of making and using cognitive maps, as well as playing a critical role in human wayfinding.

Sholl (1996) suggests that travel requires humans to activate two processes that facilitate spatial knowledge acquisition — person-to-object relations that dynamically alter as movement takes place, and object-to-object relations that remain stable even when a person undertakes movement. The first of these is called egocentric referencing; the second is called layout or configurational referencing. Given this conceptual structure, it is obvious that poor person-to-object comprehension can explain why a traveler can become locally disoriented even while still comprehending in general the basic structure of the larger environment through which movement is taking place. Error in encoding local and more general object-to-object relations can result in misspecification of the anchor point geometry on which cognitive maps are based.

Although there are many electronic, hard-copy, and other technical aids that can be used as wayfinding tools, humans nevertheless most frequently tend to use their cognitive maps and recalled information as travel guides. There are three different types of knowledge usually specified with relation to travel behavior. One is route knowledge (or systematic encoding of the route geometry by itself). A second is route-based procedural knowledge acquisition that involves understanding the place of the route in a larger frame of reference, thus going beyond the mere identification of sequenced path segments and turn angles. A third type is survey or configural knowledge, implying the comprehension of a more general network that exists within an environment and from which a procedure for following a route can be constructed.

An individual need not have a correctly encoded and cartographically correct "map" stored in memory to be able to successfully follow a route. Route knowledge by itself requires that a very small section of general environmental information is encoded. In its pure form, the route is completely self-contained, anchored by choice points and on-route landmarks and consisting of consecutive links with memorized choice points and turn angles between the links. The integration of specific routes is a difficult task, but apparently not an impossible one, for many people develop either skeletal or more complete representations of parts of urban networks through which their episodic travel takes place.

Finding and following a route usually also entails many stages of information processing on the part of the traveler. Due to the working of these processes, errors or omissions in the cognitive map are compensated for by the acquisition of relevant information from the environment that helps solve wayfinding problems.

3.4 Human Wayfinding

Many animals, birds, and insects, after controlled or random searches for food or water, return to their home base using a procedure called path integration. This involves constant updating of one's position with respect to home base. After achieving a goal (e.g., finding food), they can return directly home via a shortcut. There is no need to recall a route just traveled or to retrace it. Called dead reckoning by human navigators (e.g., pilots), this strategy *can* also be used by humans, but, because of travel mode and transport network requirements, usually is *not* used.

It is becoming more common to differentiate between navigation and wayfinding. Navigation implies that a route to be followed is predetermined, is deliberately calculated, and defines a course to be followed between a specified origin and destination. Wayfinding is taken more generally to involve the process of finding a path between an origin and a destination that has not necessarily previously been visited. Wayfinding can thus be identified with concepts such as search, exploration, and incremental path segment selection during travel.

Navigation seems to imply that a distinct process is used to define a specific course, either to get to a predetermined known or unvisited destination or to allow the traveler to return home without undue wandering or error. The principal types of navigation include piloting (or landmark-to-landmark sequencing of movement) and path integration (dead reckoning) that allow direct return to the origin without the need for storage and recall of the route being traveled.

Navigation is usually dominated by criteria such as shortest time, shortest path, minimum cost, and least effort, or with reference to specific goals that should be achieved during travel. Wayfinding is not as rigidly constrained, is purpose dependent, and can introduce emotional, value and belief, and satisficing constraints into the travel process. Whereas navigation usually requires the traveler to preplan a specific route to be followed, wayfinding can be more adventuresome and exploratory, without the necessity of a preplanned course that must be followed. While for some purposes travel behavior will be habitualized (thus lending itself to the navigation process), for other purposes variety in path selection may be more common (indicating more of a wayfinding concern).

Whether predetermined or constructed while traveling, a route can be said to have a certain legibility. This is the ease with which it can become known and traversed. This is based on the number and type of relevant cues or features both on and off the route that are needed to guide the movement decisions. It also reflects the ease with which these cues can be organized into a coherent pattern. Legibility influences the rate at which an environment is learned. Most human travelers in urban environments seek to gain legibility for the routes they travel on both a regular (habitual) or intermittent basis.

Human wayfinding is very dependent on trip purpose. The question as to whether specific purposes are better served by certain types of wayfinding strategies remains to be researched. For example, journey to work, journey to school, and journey for convenience shopping may be best served by quickly forming travel habits over well-specified routes. Such an action would minimize en route decision making, and often the resulting route conforms to shortest-path principles. However, journey for recreation or leisure may be undertaken as a search-and-exploration process requiring constant locational updating and destination fixing. Thus, as the purpose behind activity changes, the path selection criteria can change, and, as a result, the path that is followed (i.e., the travel behavior) may also change. Recent work on Intelligent Highway Systems (IHS) and Advanced Traveler Information Systems (ATIS) has shown that humans sometimes respond to advance information on congestion or the presence of obstacles by substituting destinations, changing departure times (particularly early morning), delaying or postponing activities, or selecting alternate routes (particularly in the evenings) (Chen and Mahmassani, 1993). All these produce different travel behavior in response to changing environmental circumstances. Cognitive maps must be very versatile to allow such behavioral dynamics.

3.5   Travel Plans and Activity Patterns

Activity patterns consist of a sequence of activities carried out at different locations in space. In the activity-based approach (Jones et al., 1983; Kitamura, 1988; Axhausen and Gärling, 1992; Ettema and Timmermans, 1997; Bhat and Koppelman, 2000), the tenet is that such activity–travel patterns are the outcome of predetermined interrelated choices sometimes referred to as activity scheduling (Doherty and Miller, 1997). The cognitive representation of choices of destination, mode, departure time, and route contingent on choice of activity has been termed a travel plan (Gärling et al., 1984, 1997; Gärling and Golledge, 1989). Wayfinding is usually controlled by a travel plan.

Understanding activity choice has a long history. Different approaches have been offered by:

1. Chapin (1974), the pioneer of activity-based approaches whose work concerned characteristics of activity patterns and their relationship with sociopsychological propensity factors
2. Hägerstrand (1970), who emphasized which activity patterns can be realized in particular spatial–temporal–functional settings
3. Burnett and Hanson (1982), who advocated a constraints approach, suggesting that utility-maximizing models such as discrete choice models and stated preference–choice models were all based on the unrealistic assumption that individuals were free in choosing the alternatives they liked the best
4. Smith et al. (1982), suggesting the development and use of computational process models based on choice heuristics rather than utility-maximizing behavior, and acknowledging imperfect information and suboptimal choice making
5. Miller and Salvini (1997), who proposed microsimulation models that are used to aggregate the behavior of each individual in a population via simulation processes

The simplest of all behavioral models are single-facet models, usually based on panel or diary data and addressing specific characteristics such as trip chaining, departure time decisions, and activity time allocation. Activity frequency analysis and activity association have been examined by Ma and Goulias (1999), who used a Poisson model to predict the frequency of activities related to subsistence, maintenance, and out-of-home leisure. Other models of this class include those of Kockleman (1999), Lu and Pas (1997, 1999), Golob (1998), and Lawson (1999). An innovative contribution is to use structural equations to simultaneously estimate the relationships between sociodemographics, activity participation, and travel behavior, including the number of stops, time of travel, mode of travel, and the number of trip chains. Golob and McNally (1995) used a structural equation model to analyze activity participation in the travel behavior of couples, using the dominant categories of work, maintenance, and out-of-home discretionary activities.

Activity duration and time allocation modeling can be found in the work of Kitamura et al. (1988, 1992) and Robinson et al. (1992). The emphasis here was on log-linear models examining the commuter duration and work duration as opposed to time allocated to other activities. Kitamura et al. (1998) incorporated activity duration into a model of destination choice. The systematic variation of activities across the days of the week has been examined by Hanson and Huff (1982), Koppelman and Pas (1984), Huff and Hanson (1986), Pas and Koppelman (1987), Bovy and Stern (1990), Pas and Sundar (1995), Ma and Goulias (1997), and Zhou and Golledge (2000).

3.6   Pretravel and En Route Decisions

The past decade or so has seen a paradigm shift in transportation modeling and planning to focus attention on more effective management of travel. The major incentive has been an obvious need for the development of traffic control strategies, rather than strategies focused on providing more infrastructure. As societal changes such as flex time working hours, telecommuting, and in-car dynamic, real-time reception of advance travel information have become more important, modeling and planning attention have been focused on understanding travel behavior. Achieving such a goal is hypothesized to help reduce travel demand by the suppression or selective elimination of redundant, unnecessary trips, by targeting single-occupant vehicles at peak periods of commuting, and by reducing driver frustration, stress, and road rage by providing in-car, en route, or pretravel information about routes and traffic conditions. As more data have been collected by survey research, travel diary, and interview procedures, a more comprehensive understanding of the reasons for trip making and route selection has evolved. In association with this knowledge accumulation has come more detailed examination of the decision-making characteristics of potential drivers, their spatial abilities, and their individual differences with respect to travel preferences. In general, this has produced a body of research designated Intelligent Transportation Systems (ITS), which covers the more effective control of traffic and more efficient transmission of information to actual or potential travelers. Much of this concern has drawn on the activity-based approach described in the last section.

A major goal of ITS is the reduction of congestion and accidents or hazards that are associated with surges in traffic volume. A significant part of ITS is the ATIS. This consists of in-vehicle information and ex-vehicle guidance systems that aid in pretrip planning and en route decision making. Information obtained in advance about current traffic conditions on routes that have been selected as part of travel planning assists the potential traveler in making important decisions such as at what time to begin a trip. Research on individual differences makes us aware that drivers will respond in different ways to the same set of information. For example, advance information on the congested state of a particular route segment may encourage some drivers to delay departure times, others to choose different routes, and yet others not to change their travel plans on the assumption that the congestion will have cleared by the time they have reached the critical spot. Thus, reactions will range from ignoring the advance information to accepting it and changing part of a travel plan. In this way, the ATIS acts as a decision support system — an integrated set of tangible and intangible information that is designed to supplement personal knowledge during problem activities (Densham and Rushton, 1988).

A decision support system does not replace individual decision making, but rather acts as an additional source of information that must be evaluated and integrated into the regular decision-making process. Much of the research in psychology and cognitive science on conflict resolution and decision making has emphasized the importance of offering more than a single solution to a problem. Advance information serves a similar purpose by giving an early warning of potential impediments to travel, allowing a potential traveler to develop a set of alternate strategies that could be implemented in order to achieve the original goal (Adler et al., 1993a, 1993b).

While the nature of travel information has been explored extensively over the last decade and a half, much less research has been undertaken on the most appropriate way for people to receive this information (e.g., by visual signals or graphic map displays in the car, by special radio broadcasts, by voice command interfaces with in-car computers, by dynamic highway traffic signs, and so on). Behavioral research tells us that the probability of ignoring or accepting information provided may vary significantly between sexes and among age groups. Behavioral researchers at this point have therefore generally adopted a multimodal approach in order to reach the greatest number of people in these different response groups. Perhaps the most significant factor emerging from this research, however, is that advance information will be acted on only if it is provided to potential travelers in a realistic time frame (Jayakrishnan et al., 1993, 1994).

One common scenario involves a potential traveler receiving information before the trip is actually initiated. We have already seen that trips for different purposes require different amounts of preplanning. Trips to work, for example, often become more or less habitual, encouraging stereotyped behavior and repetitive travel over a well-defined route. Trips for other purposes may be more variable, in terms of both the times of departure and the times of travel (often varying considerably during the day), and whether the proposed trip will be part of a trip chain. Axhausen and Gärling (1992) emphasized the importance of access to information in the pretrip planning phase. Jou and Mahmassani (1998) and Mahmassani and Jou (1998) undertook diary surveys of commuters in two different environments — the north central expressway corridor in Dallas and the northwest corridor in Austin — to examine dynamics of commuter decisions. In particular, they focused on departure times and selection of the routes to be followed for both the morning and evening commuters. They modeled pretravel decision making concerning route selection, departure time, and route-switching patterns to other factors such as time of day of travel, normal time of departure, trip length, path selection criteria, nature of the route to be followed, and expectations as to the likelihood that pretrip planning would have to be changed. Significant results included evidence of greater route-switching activity in the evening commute and a later frequency of time switching in the morning commute. Mahmassani and Herman (1990) previously reviewed the evolution of approaches focused on traveler information from models that were microeconomics-based analyses of idealized situations to elaborate simulation studies and critical observation work in laboratory and real-world conditions. Certainly, manipulation of departure times appears to be a first-order response to advance traveler information that specifies congestion or other problems along preselected routes.

En route decisions require additional information other than personal evaluations of traffic conditions. For example, if information is given en route to a driver about congestion or other impediments to travel, along with the time or distance along the route to the location of these barriers, the traveler must evaluate *in situ* the potential impact of the warnings on his or her travel plans. The driver must integrate at the same time the perception of the current speed of traffic, traffic volume, time lapses associated with completing designated sections of the route, familiarity with the network on which he or she is traveling, and familiarity with adjacent neighborhoods through which he or she may have to travel if departing from the preset route, while at the same time reevaluating travel goals and expectations associated with the specific trip. The traveler may also have to review his or her knowledge of landmarks and other important reference nodes on and off an alternative route and evaluated conditions of safety and uncertainty that may go along with a change in travel plans.

While en route, a traveler has a number of alternative strategies that are available in response to the receipt of negative information about the route being followed. Recent studies focusing on the nature of these choice alternatives have been undertaken by Bonsall and Parry (1991), Allen et al. (1991), Ayland and Bright (1991), Ben-Akiva et al. (1991), Khattak et al. (1993), and others. This early research examined the en route travel behavior change pattern in both laboratory experiments and in real-world conditions. Adler et al. (1993a) characterize en route driver behavior as an integrative process through which they assess the current state of a system and adapt travel behavior in response to the severity of their perceptions. They suggest that possible strategies would include route diversion, new information acquisition, revision of travel objectives, delay of travel, substitution of routes, substitution of destinations, and reordering of scheduled priorities. Factors that influence which of these are likely to be chosen include estimates of delay; estimates of travel time involved in waiting or clearance or by taking new routes; perception of the ease of travel and safety of alternative routes; the amount of prior experience with congested conditions on the original route; the risk-taking propensity of individual drivers; their tolerance thresholds with respect to delay; expectations of meeting the original travel goals, objectives, mode of travel, focus of trip, and time of day of trip; and the potential for rescheduling an activity.

Adler et al. (1993a, 1993b) devised a simulation method (FASTCARS) that allowed participants to make choices resulting in road changing, lane changing, and information acquisition while traveling between a given origin and destination. Information was provided through highway advisory radio (HAR) and In-Vehicle Navigation Systems (IVNS). The HAR system provided real-time traffic incident and congestion information for the freeways in the network. The IVNS calculated the shortest time path from the driver's current position to the destination of choice. Both these types of information were fed to participants, and the consequent activities and choices were evaluated after relating behavior profiles to trial event data. The results thus incorporated current traffic conditions with behavioral profiles to examine the role of spatial behavior in travel choice. Most studies assume that drivers' responses reflect their perceptual and cognitive processing ability, both of which are temporally and spatially dependent. The recording of physiological or psychological changes in driver behavior in real time, however, is still lacking. It is likely, because of safety conditions associated with these types of studies, that microsimulation, virtual immersive, or virtual desktop environments are likely to be the most effective way of examining driver responses to changing traffic conditions.

3.7 Path Selection Criteria

Human wayfinding can thus be regarded as a purposive, directed, and motivated activity that may be observed and recorded as a trace through an environment. The trace is usually called the route or course. A route results from implementing a travel plan (Gärling et al., 1984; Gärling and Golledge, 1989) that consists of predetermined choices defining the sequence of segments and turn angles that comprise the course to be followed or the general sector or corridor within which movement should be concentrated.

The criteria used in path selection vary significantly with trip purpose. Traditionally, the major types of path selection criteria include shortest path, shortest time, shortest distance, least cost, turn minimization, longest leg first, fewest obstacles (such as traffic lights or stop signs), congestion avoidance, minimization of the number of route segments, restriction to a known corridor, maximization of aesthetics, minimization of intermodal transfers, optimization of freeway use, avoidance of known hazardous areas, least patrolled by authorities, and minimization of exposure to truck or heavy freight traffic.

Most studies of travel behavior have adopted the assumption that travelers desire to minimize time, cost, or distance. Such assumptions facilitate the development of tractable, mathematical models that can use simple network structures to provide optimal route selection solutions to different types of movement problems. This has been the strength of traditional microeconomic models. Over the past decade, however, psychological and behavioral geographic studies have indicated that rational optimizing behavior is not widespread among individual travelers (Pas and Koppelman, 1986, 1987; Gärling and Golledge, 2000). So what criteria *are* used? Golledge (1997a) conducted a variety of laboratory experiments in regular and irregular networks. For about half the population, shortest-path trips were chosen regularly. However, that same path was often not chosen when individuals were asked to retrace the route from the destination to the origin (e.g., 60% retraced it in a simple grid network environment, but only 20% retraced it in a more complex irregular network). Thus, depending on the nature of travel and the traveler's location at which to start a trip, different path selection criteria might be used. Criteria that have been found in both empirical and laboratory studies include fastest time, minimizing left turns, minimizing total turns, driving the longest leg first, driving the shortest leg first, trying to approximate a straight-line shortcut route between an origin and a destination, always heading in the direction of the destination, and defining a travel corridor beyond whose boundaries travel would not take place (Golledge, 1997a).

Apparently, people use different criteria for different purposes. Since much of the research has focused on the dominant home–work–home trip (usually without intermediate stops), the tendency has been to accept an assumption that drivers will minimize time, distance, or cost. An analysis of travel behavior, however, has shown that the trip home is not always a simple reversal of the trip to work. This is partly because of the increased probability of a trip chain being undertaken on the way home, partly because of the perceptions of the ease or difficulty of retracing the route (Mahmassani et al., 1997). Thus, as the trip purpose changes from shopping to recreational or health- and professional-related needs or purposes, to education, or to religious purposes, the reasons for choosing a particular route may also change. At times, maximizing the aesthetic value of a particular route (e.g., on a recreational trip) may be more important than minimizing travel. Suddenly one cannot assume that all the people, say, traveling on a freeway at 5:15 P.M. on a weekday, are going directly home. Thus, while it may be expected that the bulk of them may be doing this, it is not necessarily a good assumption to build into a planning strategy for travel behavior at that time of day. Usually there are a number of feasible route selection criteria that are imbedded in daily activity patterns.

3.8 Behavioral Models for Forecasting Travel Demand

In the preceding sections we have reviewed research on human spatial behavior. How can the findings of this research be used in transport modeling and planning? In this section we briefly review some modeling approaches that build on behavioral assumptions and whose purpose is to forecast travel demand in such a way that it can be used in transportation planning.

The standard travel demand forecasting procedure consists of a household base, a cross-classification model for trip production, a regression-based model for trip attraction, a gravity model for trip distribution, a multinomial logit model for mode choice (often focused largely on home and work trips only), and a network assignment procedure for highway or transit travel. Among these, only the multinomial logit model has been based on behavioral principles, although it is usually made operational at an aggregate rather than disaggregate level.

Ben-Akiva et al. (2000) suggest that it is possible to identify a model with limited latent variables using only observed choices. To use maximum likelihood estimation, we need the distribution of the utilities, $f(U \mid X, X^*; \beta)$. An additive utility is a common assumption in the transportation literature:

$$U = V(X, X^*; \beta) + \varepsilon \tag{3.1}$$

That is, the random utility is decomposed into the sum of a systematic utility, $V(\cdot)$, and a random disturbance, ε. The systematic utility is a function of both observable and latent variables. β values are utility coefficients to be estimated.

Choice can then be expressed as a function of the utilities. For example, assuming utility maximization:

$$y_i = \begin{cases} 1, & \text{if } U_i = \max_j \{U_j\} \\ 0, & \text{otherwise} \end{cases} \tag{3.2}$$

where i and j are index alternatives. From Equations (3.1) and (3.2) and an assumption about the distribution of ε, we derive $P(y | X, X^*; \beta)$, the choice probability conditional on both observable and latent explanatory variables.

$$P(y_i = 1 | X, X^*; \beta) = P(U_i \geq U_j, \forall j \in C)$$

$$= P(V_i + \varepsilon_i \geq V_j + \varepsilon_j, \forall j \in C)$$

$$U_i = V_i + \varepsilon_i \quad \text{and} \quad V_i = V_i(X, X^*; \beta), \ i \in C$$

where C is the choice set. The most common distributional assumptions result in logit or probit choice models. For example, if the disturbances, ε, are independent and identically distributed (i.i.d.) standard Gumbel, then

$$P(y_i = 1 | X, X^*, \beta) = \frac{e^{V_i}}{\sum_{j \in C} e^{V_j}} \qquad \text{(logit model)}$$

or, in a binary choice situation with normally distributed disturbances,

$$P(y_i = 1 | X, X^*, \beta) = \Phi(V_i - V_j) \qquad \text{(binary probit model)}$$

where Φ is the standard normal cumulative distribution function.

Choice indicators could also be ordered categorically, in which case the choice model may take on either ordered probit or ordered logistic form. Finally, to construct the likelihood function, an assumption about the distribution of X^* is needed. Assuming X^* is independent of ε and its distribution can be described by a vector of parameters γ, the result is

$$f(y | X; \beta, \gamma) = \int_{X^*} P(y | X, X^*; \beta) f(X^*; \gamma) dX^*$$

Ben-Akiva et al. (2000) further argue that, although the likelihood of a choice model with latent explanatory variables is easily derived, it is quite likely that the information content from the choice indicators will not be sufficient to empirically identify the effects of individual-specific latent variables. Therefore, indicators of the latent variables are used for identification, which leads to more elaborate model systems that combine choice models with latent variable models. When the complexity increases even further, other approaches are needed.

The fact that many trips are routine or repetitive (usually representing more than 50% of the total trips made on any given weekday in particular) has provided the basis for successful modeling and planning using structural models (McFadden, 2002). However, to forecast demand for more variable types of travel (e.g., weekend or leisure travel), it may be necessary to completely understand the decision-making *process*, more than is possible purely on the basis of building a successful structural model. At the same time, predictive validity of behavioral process models is not likely to be equally good (Gärling et al., 1998a).

It may be questioned whether an increased understanding of the underlying travel choice process parallels the progress that has been made with respect to the development of applications. The term *activity scheduling* is used to refer to the choice process resulting in a travel plan that eventually is implemented in an activity–travel pattern. Limits on human information processing capacity render optimal activity scheduling generally infeasible unless the task is very simple (Gärling, 2001). An important goal of research is, therefore, to specify the kinds of simplification people are likely to make. To this end, behavioral process models have used a formalism called production systems, which are sets of conditional rules that can be encoded in computer programs (Smith et al., 1982; Gärling et al., 1994).

The development of process models has focused travel choice research on important issues (Gärling et al., 1998c). With reference to Table 3.1, it has been a shift of focus from time-invariant determinants of single choices with no learning (upper left corner) to the *process* of making multiple choices (concerning multipurpose multistop trips) in which learning takes place (lower right corner). At the same time, the tractability of mathematical–statistical models decreases. Yet behavioral process models allow modeling of more complex activity–travel choice. For instance, it is now realized that utility maximization is an unrealistic assumption. In response to this, process models based on bounded rationality assumptions and employment of noncompensatory decision rules have been developed (e.g., Arentze and Timmermans, 2000). This development may influence structural models in the future. To this end, Ben-Akiva et al. (1999a,b) have extended a conceptual framework as a basis for (travel) choice modeling that includes affective factors. In addition, it is also essential to model how information is searched, perceived, and remembered.

If encoded in computer programs, it may be possible to make exact predictions from production system models, for instance, in simulating the outcome of policies on individuals or households (e.g., Gärling et al., 1998b; Pendyala et al., 1997, 1998). In contrast to statistical–mathematical structural models, estimating free parameters of process models is nevertheless considerably more difficult. With some success (Ettema, 1996), structural and process models have been combined to this end. In fact, both types of models should be compatible. Still, the validity of process assumptions is not easily judged from estimates of the parameters of structural models. Thus, there are problems to be solved concerning data and methods of data collection with reference to process models. Some such solutions appear to be forthcoming (Doherty, 1998; Ben-Akiva and Bowman, 1998).

Three systems of models of activity–travel scheduling (Bowman and Ben-Akiva, 2000; Kitamura and Fujii, 1998; Arentze and Timmermans, 2000) have recently been proposed. These models are operational, so they can forecast activity–travel patterns. They also aspire to make realistic behavioral assumptions.

In Ben-Akiva and Bowman (1997), a system of nested discrete choice models for travel demand forecasting is described. It is assumed that decisions with different time frames are hierarchically organized. Mobility and lifestyle decisions (e.g., choosing to purchase an automobile, residential choice) condition longer-term activity and travel scheduling, which in turn conditions daily activity and travel *re*scheduling. The latter is the major focus of the model. A primary daily activity pattern is assumed to exist. Interrelated choices are then assumed to be made for tours, including a primary activity (out of or in the home), the type of tour for the primary activity (the number, purposes, and sequence of activity stops), and the number and purposes of secondary tours. Timing and mode are chosen for tours. A hierarchy of choices is again postulated, this time on the basis of priority. Choices are assumed to maximize utility at each level. A hierarchical organization of interrelated choices seems reasonable to assume, because it restricts the size of the choice sets. Still, the empirical examples indicate that, from a behavioral point of view, the choice sets may be unrealistically large. It would therefore be reasonable to make the additional assumption that people, instead of maximizing utility,

TABLE 3.1   Different Foci of Past and Current Research on Travel Choice

	Structure	Process
Single choice	No learning	No learning
	Learning	Learning
Multiple choice	No learning	No learning
	Learning	Learning

use some simplifying choice heuristics. Furthermore, it is also assumed that the choices are made sequentially (not taking into account subsequent choices). A drawback is that the basis for the hierarchical organization (priority) is not defined. For instance, it may not be realistic to assume that priority does not change over time (Doherty et al., 2000).

A similar system of discrete or continuous choice models is reported in Kitamura and Fujii (1998). It is labeled the prism-constrained activity–travel simulator (PCATS) because it takes as a starting point the time–geographical concept of a prism that defines the maximal range of possible travel within a certain time period (Hägerstrand, 1970). Thus, it is assumed that the choice sets are restricted, but that each choice maximizes utility. In an open period (no activities chosen), a two-stage choice of an activity (out of home vs. in home, followed by type) conditions choice of location, which in turn conditions choice of mode. At the lowest-level activity, duration is chosen. In summary, the model system is similar to that proposed by Ben-Akiva and Bowman (1997) in that it may realistically describe activity–travel rescheduling that forms part of a routine activity–travel pattern.

ALBATROSS (Arentze and Timmermans, 2000) is a third model system. Like that proposed by Ben-Akiva and Bowman (1997), several time horizons are assumed. The detailed model concerns short-term activity–travel scheduling and rescheduling. In this respect, the model is similar to PCATS in assuming relatively fixed sets of constraints on choices. An important difference is that, using a decision table formalism, choices are modeled as the application of rules selected from hierarchies of condition–action pairs. This is clearly in line with approaches in cognitive psychology (Payne et al., 1993). Furthermore, although activity- and trip-related choices are assumed to be made sequentially in a fixed order, they are strongly interconnected by means of the condition–action rules. Thus, not only prior choices but also *subsequent* choices or *expectations* influence a particular choice. The model specifies the constraint rules and a base of preference rules. The actual preference rules that people use are determined by fitting the model to diary data on actual activity–travel patterns. In this way the model is adjusted to the data.

Any process model is incomplete if it does not include statements about how travelers learn and adapt to the transportation environment. Gärling (2001) points out that the fact that people are able to solve complex scheduling problems is in large part due to their eminent ability to learn how to simplify information processing, for instance, by chunking information or retrieving ready-made action plans called scripts. A promising development in this respect is the model system proposed by Pendyala et al. (1997) focusing on behavioral adaptation. Furthermore, work is in progress (Arentze and Timmermans, 2001) to augment ALBATROSS with a model of how choice rules change as a function of the outcome of previous choices.

3.9   Summary and Conclusions

Because of individual differences in spatial abilities, differences in the content and structure of cognitive maps, different motivations or purposes for travel, and different preferences for optimizing or satisficing decision strategies, human travel behavior is difficult to understand or predict. If we add to that the unexpected barriers and obstacles to traffic flow that occur spontaneously and intermittently (e.g., from congestion, accidents, construction, or other obstacles that impede movement over a selected path or over a network), then problems of intelligently modeling travel behavior in the real world become substantial. Yet some success has been achieved in doing this, using simplified assumptions about human behavior (e.g., assuming that, knowingly or unknowingly, travelers adopt shortest-path optimizing practices). But models like this and the predictions they make can be very inadequate. The problem facing future research is that of combining travel demand (considering people's activities) with network supply (considering the tracks, corridors, or transport systems available) with an understanding of how humans decide on where they prefer (or have) to go and how they prefer (or have) to get there. A gap thus still exists between knowledge of spatial behavior and the practice of modeling travel choice with the aim of forecasting demand for travel. As argued by Simon (1990), it is unlikely that the behavioral sciences will ever be able to make exact quantitative predictions of behavior. The laws will most likely remain qualitative. However, practitioners should realize that this does not necessarily make the theories less useful. An example is the germ theory and its highly successful applications to fight infection and diseases. A challenge to practitioners is how they can use qualitative behavioral principles in transportation planning — for instance, in making quantitative predictions of travel demand.

Acknowledgment

This work was funded in part by grants to R.G. Golledge from National Science Foundation (grant BCS-0083110) and University of California Transportation Center (grant DTRS99-G-0009).

References

Abdel-Aty, M.A. and Jovanis, P.P., Investigating effect of advanced traveler information on commuter tendency to use transit, *Transp. Res. Rec.*, 1550, 66, 1996.

Abdel-Aty, M.A. and Jovanis, P.P., Modeling Incident-Related Routing Decisions Including the Effect of Traffic Information Using a Nested Logit Structure, paper presented at 77th Annual Meeting of the Transportation Research Board, Washington, D.C., January 1998.

Adler, J.L., Recker, W.W., and McNally, M.G., A conflict model and interactive simulator (FASTCARS) for predicting en route driver behavior in response to real-time traffic condition information, *Transportation*, 20, 83–106, 1993a.

Adler, J. L., Recker, W.W., and McNally, M.G., Using interactive simulation to model driver behavior under ATIS, in *Proceedings of the ASCE 4th International Conference on Microcomputers and Transportation*, ACSE, Baltimore, 1993b, pp. 344–355.

Allen, G.L., Spatial abilities, cognitive maps, and wayfinding: bases for individual differences in spatial cognition and behavior, in Golledge, R.G., Ed., *Wayfinding Behavior: Cognitive Mapping and Other Spatial Processes*, Johns Hopkins University Press, Baltimore, 1999, pp. 46–80.

Allen, R.W. et al., A human factor simulation investigation of driver route diversion and alternative route selection using in-vehicle navigation systems, in *Proceedings of the Vehicle Navigation and Information Systems Conference*, Vol. 1, Society of Automotive Engineers, Dearborn, MI, 1991, pp. 9–26.

Arentze, T. and Timmermans, H.P.J., *ALBATROSS: A Learning Based Transportation Oriented Simulation System*, European Institute of Retailing and Services Studies, Technical University of Eindhoven, Netherlands, 2000.

Arentze, T. and Timmermans, H.P.J., Modeling learning and adaptation processes in activity–travel choice: a framework and numerical experiments, *Transportation*, 2001, submitted.

Axhausen, K. and Gärling, T., Activity-based approaches to travel analysis: conceptual frameworks, models, and research problems, *Transp. Rev.*, 12, 323–341, 1992.

Ayland, N. and Bright, J., Real-time responses to in-vehicle IVHS technologies: a European evaluation, *Transp. Res. Rec.*, 1318, 111–117, 1991.

Bell, S., Children's Comprehension of Location in Different Spaces, unpublished doctoral dissertation, Department of Geography, University of California, Santa Barbara, 2000.

Ben-Akiva, M., de Palma, A., and Kaysi, I., Dynamic network models and driver information systems, *Transp. Res. A Gen.*, 25, 251–266, 1991.

Ben-Akiva, M. and Lerman, S., *Discrete Choice Analysis: Theory and Applications*, MIT Press, Cambridge, MA, 1985.

Ben-Akiva, M.E., Ramming, S., and Golledge, R.G., Collaborative Research: Individuals' Spatial Behavior in Transportation Networks, National Science Foundation Grant Proposal BCS-0083110, University of California–Santa Barbara and Massachusetts Institute of Technology, Cambridge, MA, 2000.

Ben-Akiva, M.E., Ramming, M.S., and Walker, J., Improving Behavioral Realism of Urban Transportation Models through Explicit Treatment of Individual's Spatial Ability, paper presented at European Science Foundation/U.S. National Science Foundation Conference on Social Change and Sustainable Transport, Berkeley, CA, March 11, 1999.

Ben-Akiva, M.E. et al., Modeling methods for discrete choice analysis, *Mark. Lett.*, 8, 273–286, 1997.

Ben-Akiva, M.E. et al., Behavioral Realism in Urban Transportation Planning Models, paper presented at Transportation Models in the Policy-Making Process: Uses, Misuses and Lessons for the Future, a symposium on the problems of transportation analysis and modeling in the world of politics, in memory of Greig Harvey, Pacific Grove, CA, March 4–6, 1998.

Ben-Akiva, M.E. et al., Extended framework for modeling choice behavior, *Mark. Lett.*, 10, 187–203, 1999.

Bhat, C.R. and Koppelman, F.S., Activity-based modeling for travel demand, in *Handbook of Transportation Science*, Hall, R.W., Ed., Kluwer, Dordrecht, 1999, pp. 35–62.

Bonsall, P.W. and Parry, T., Using an interactive route-choice simulator to investigate drivers' compliance with route guidance advice, *Transp. Res. Rec.*, 1036, 59–68, 1991.

Bovy, P.H.L. and Stern, E., *Route Choice: Wayfinding in Transport Networks*, Kluwer Academic Publishers, Dordrecht, Netherlands, 1990.

Bowman, J.L. and Ben-Akiva, M.E., Activity based disaggregate travel demand model system with daily activity schedules, *Transp. Res.*, 35, 1–28, 2000.

Burnett, P. and Hanson, S., The analysis of travel as an example of complex human behavior in spatially-constrained situations: definition and measurement issues, *Transp. Res. A Gen.*, 16, 87–102, 1982.

Chapin, F.S., *Human Activity Patterns in the City: What People Do in Time and Space*, John Wiley, New York, 1974.

Chen, P.S.T. and Mahmassani, H.S., A dynamic interactive simulator for studying commuter behavior under real-time traffic information supply strategies, *Transp. Res. Rec.*, 1413, 12–21, 1993.

Deakin, A.K., Procedural Knowledge and Its Potential for Enhancing Advanced Traveler Information Systems, paper presented at 76th Annual Transportation Research Board Meetings, Washington, D.C., January 12–16, 1997.

Densham, P.J. and Rushton, G., Decision support systems for locational planning, in *Behavioral Modelling in Geography and Planning*, Golledge, R.G. and Timmermans, H., Eds., Croom Helm, London, 1988, pp. 56–90.

Doherty, S.T., The Household Activity–Travel Scheduling Process: Computerized Survey Data Collection and the Development of a Unified Modeling Framework, doctoral dissertation, Department of Civil Engineering, University of Toronto, Canada, 1998.

Doherty, S.T., Axhausen, K., Gärling, T., and Miller, E.J., A conceptual model of the weekly household activity–travel scheduling process, in *Travel Behaviour: Spatial Patterns, Congestion and Modelling*, Stern, E., Salomon, I., and Bovy, P., Eds., Edward Elgar, Cheltenham, U.K., 2002.

Doherty, S.T. and Miller, E.J., Tracing the Household Activity Scheduling Process Using One-Week Computer-Based Survey, paper presented at Eighth Meeting of the International Association of Travel Behavior Research (IATBR), Austin, TX, September 1997.

Downs, R.M. and Stea, D., Cognitive maps and spatial behavior: process and products, in *Image and Environment*, Downs, R. and Stea, D., Eds., Arnold, London, 1973a, pp. 8–26.

Downs, R.M. and Stea, D., Eds., *Image and Environment: Cognitive Mapping and Spatial Behavior*, Aldine, Chicago, 1973b.

Ettema, D., Activity-Based Travel Demand Modeling, doctoral dissertation, Department of Architecture, Technical University of Eindhoven, Netherlands, 1996.

Ettema, D.F. and Timmermans, H.P.J., Theories and models of activity patterns, in *Activity-Based Approaches to Travel Analysis*, Ettema, D.F. and Timmermans, H.J.P., Eds., Pergamon, Oxford, 1997, pp. 1–36.

Freundschuh, S.M., Is there a relationship between spatial cognition and environmental patterns? *Lect. Notes Comput. Sci.*, 639, 288–304, 1992.

Gärling, T., The feasible infeasibility of activity scheduling: an unresolved issue in travel-choice modeling, 2001, submitted.

Gärling, T., Boe, O., and Fujii, S., Empirical tests of a model of determinants of script-based driving choice, *Transp. Res. F*, 4, 89–102, 2001.

Gärling, T., Böök, A., and Lindberg, E., Cognitive mapping of large-scale environments: the interrelationship of action plans, acquisition, and orientation, *Environ. Behav.*, 16, 3–34, 1984.

Gärling, T. and Garvill, J., Psychological explanations of participation in everyday activities, *Behavior and Environment: Psychological and Geographical Approaches*, in Gärling, T. and Golledge, R.G., Eds., Elsevier/North-Holland, Amsterdam, 1993, pp. 270–297.

Gärling, T., Gillholm, R., and Gärling, A., Reintroducing attitude theory in travel behavior research: the validity of an interactive interview procedure to predict car use, *Transportation*, 25, 147–167, 1998a.

Gärling, T. and Golledge, R.G., Environmental perception and cognition, in *Advances in Environment, Behavior, and Design*, Vol. 2, Zube, E.H. and Moore, G.T., Eds., Plenum Press, New York, 1989, pp. 203–236.

Gärling, T. and Golledge, R.G., Cognitive mapping and spatial decision making, in *Cognitive Mapping: Past, Present, and Future*, Kitchin, R. and Freundschuh, S., Eds., Routledge, London, 2000, pp. 44–65.

Gärling, T., Kwan, M.-P., and Golledge, R.G., Computational-process modeling of household activity scheduling, *Transp. Res. B*, 25, 355–364, 1994.

Gärling, T., Laitila, T., and Westin, K., Eds., *Theoretical Foundations of Travel Choice Modeling*, Pergamon, Oxford, 1998c.

Gärling, T. et al., Interdependent activity and travel choices: behavioral principles of integration of choice outcomes, in *Activity-Based Approaches to Travel Analysis*, Ettema, D. and Timmermans, H.P.J., Eds., Pergamon, Oxford, 1997, pp. 135–150.

Gärling, T. et al., Computer simulation of a theory of household activity scheduling, *Environ. Plann. A*, 30, 665–679, 1998b.

Golledge, R.G., The conceptual and empirical basis of a general theory of spartial knowledge, in *Spatial Choices and Processes*, Fischer, M.M., Nijkamp, P., and Papageorgiou, Y.Y., Eds., Elsevier/North-Holland, Amsterdam, 1990, pp. 147–168.

Golledge, R.G., Place recognition and wayfinding: making sense of space, *Geoforum*, 23, 199–214, 1992.

Golledge, R.G., Defining the criteria used in path selection, in *Activity-Based Approaches to Travel Analysis*, Ettema, D. and Timmermans, H.P.J., Eds., Elsevier, New York, 1997a, pp. 151–169.

Golledge, R.G., Dynamics and ITS: Behavioral Responses to Information Available from ATIS, paper presented at 8th Meeting of the International Association for Travel Behavior Research (IATBR), Austin, TX, 1997b.

Golledge, R.G., *Wayfinding Behavior: Cognitive Mapping and Other Spatial Processes*, Johns Hopkins University Press, Baltimore, 1999.

Golledge, R.G., Kwan, M.-P., and Gärling, T., Computational-process modelling of household travel decisions using a geographical information system, *Pap. Reg. Sci.*, 73, 99–117, 1994.

Golledge, R.G. and Stimson, R., *Analytical Behavioural Geography*, Croom Helm, London, 1987.

Golledge, R.G. and Stimson, R.J., *Spatial Behavior: A Geographic Perspective*, Guilford Press, New York, 1997.

Golob, T.F., A model of household choice of activity participation and mobility, in *Theoretical Foundations of Travel Choice Modeling*, Gärling, T., Laitila, T., and Weston, K., Eds., Elsevier, Oxford, 1998, pp. 365–397.

Golob, T.F. and McNally, M.G., A Model of Household Interactions in Activity Participation and the Derived Demand for Travel, paper presented at International Conference on Activity Based Approaches: Activity Scheduling and the Analysis of Activity Patterns, Eindhoven University of Technology, Netherlands, 1995.

Hägerstrand, T., What about people in regional science? *Pap. Proc. N. Am. Reg. Sci. Assoc.*, 24, 7–21, 1970.

Hanson, S. and Huff, J.O., Assessing day-to-day variability in complex travel patterns, *Transp. Res. Rec.*, 891, 18–24, 1982.

Hanson, S. and Schwab, M., Describing disaggregate flows: individual and household activity patterns, in *The Geography of Urban Transportation*, Hanson, S., Ed., Guilford Press, New York, 1995, pp. 166–187.

Huff, J.O. and Hanson, S., Repetition and variability in urban travel, *Geographical Analysis*, 18, 97–114, 1986.

Iida, Y., Akiyama, T., and Uchida, T., Experimental analysis of dynamic route choice behavior, *Transp. Res. B*, 26, 17–32, 1992.

Jayakrishnan, R., Mahmassani, H., and Hugh, T.Y., An evaluation tool for advanced traffic information and management systems in urban networks, *Transp. Res. C*, 2, 129–147, 1994.

Jayakrishnan, R., Mahmassani, H.S., and Rathi, U., User-friendly simulation model for traffic networks with ATIX/ATMS, in *Proceedings of the 5th International Conference on Computing in Civil and Building Engineering*, Anaheim, CA, June 1993.

Jha, M., Madanat, S., and Peeta, S., Perception updating and day-to-day travel choice dynamics in traffic networks with information provision, *Transp. Res.*, 6C(3), 189–212, 1998.

Jones, P. et al., *Understanding Travel Behavior*, Gower, Aldershot, U.K., 1983.

Jou, R.-C. and Mahmassani, H.S., Comparative analysis of day-to-day trip-chaining behavior of urban commuters in two cities, *Transp. Res. Rec.*, 1607, 163–170, 1998.

Kaysi, I., Framework and models for the provision of real-time driver information, unpublished Ph.D. dissertation, MIT, Cambridge, MA, 1992.

Khattak, A.J. and Khattak, A.G., A Comparative Analysis of Spatial Knowledge and En-Route Diversion Behavior across Chicago and San Francisco: Implications for ATIS, paper presented at 77th Annual Meeting of Transportation Research Board, Washington, D.C., January 11–15, 1998.

Khattak, A.G., Schofer, J. L., and Koppelman, F.S., Commuters' en-route diversions and return decisions: analysis and implications for advanced traveler information systems, *Transp. Res. A*, 27, 101–111, 1993.

Khattak, A.J., Schofer, J.L., and Koppelman, F.S., Effect of traffic information on commuters' propensity to change route and departure time, *J. Adv. Transp.*, 29, 193, 1995.

Kitamura, R., An evaluation of activity-based travel analysis, *Transportation*, 15, 9–34, 1988.

Kitamura, R., Chen, C., and Narayanan, R., Effects of Time of Day, Activity Duration and Home Location on Travelers' Destination Choice Behavior, paper presented at 77th Annual Meeting of the Transportation Research Board, Washington, D.C., January 11–15, 1988.

Kitamura, R. and Fujii, S., Two computational process models of activity–travel choice, in *Theoretical Foundations of Travel Choice Modeling*, Gärling, T., Laitila, T., and Westin, K., Eds., Elsevier, Amsterdam, 1998, pp. 251–279.

Kitamura, R., Nishii, K., and Goulias, K., Trip Chaining Behavior by Central City Commuters: A Casual Analysis of Time–Space Constraints, paper presented at Oxford Conference on Travel and Transportation, Oxford, 1988.

Kitamura, R. et al., A Comparative Analysis of Time Use in The Netherlands and California, paper presented at Proceedings of the 20th PATRC Summer Annual Meetings, London, 1992.

Kitchin, R.M., Cognitive maps: what are they and why study them? *J. Environ. Psychol.*, 14, 1–19, 1994.

Kockelman, K.M., Application of a Utility-Theory-Consistent System-of-Demand Equations Approach to Household Activity Travel Choice, paper presented at 78th Annual Meeting of the Transportation Research Board, Washington, D.C., January 10–14, 1999.

Koppelman, F.S. and Pas, E.I., Estimation of disaggregate regression models of person trip generation with multi-day data, in *Proceedings of the Ninth International Symposium on Transportation and Traffic Theory*, Volmuller, J. and Hamerslag, R., Eds., VNU Science Press, Utrecht, Netherlands, 1984, pp. 513–531.

Kuipers, B.J., Modelling spatial knowledge, *Cognit. Sci.*, 2, 129–153, 1978.

Kwan, M.P., A GIS-Based Model for Activity Scheduling in Intelligent Vehicle Highway Systems (IVHS), unpublished Ph.D. dissertation, University of California at Santa Barbara, 1994.

Kwan, M.-P., Golledge, R.G., and Speigle, J., Information representation for driver decision support systems, in *Theoretical Foundations of Travel Choice Modeling*, Gärling, T., Laitila, T., and Westin, K., Eds., Pergamon, Oxford, 1998, pp. 281–303.

Leiser, D. and Zilberschatz, A., The TRAVELLER: a computational model of spatial network learning, *Environ. Behav.*, 21, 435–463, 1989.

Liben, L.S., Contributions of individuals to their development during childhood: a Piagetian perspective, in *Individuals as Producers of Their Development: A Life-Span Perspective*, Lerner, R. and Busch-Rossnagel, N., Eds., Academic Press, New York, 1981, pp. 117–154.

Liu, Y.-H. and Mahmassani, H., Dynamic Aspects of Departure Time and Route Decision Behavior under Advanced Traveler Information Systems (ATIS): Modeling Framework and Experimental Results, paper presented at 77th Annual Meeting of the Transportation Research Board, Washington, D.C., January 1998.

Lu, X. and Pas, E.I., A Structural Equations Model of the Relationships among Socio-Demographics, Activity Participation, and Travel Behavior, paper presented at 76th Annual Meeting of the Transportation Research Board, Washington, D.C., 1997.

Lu, X. and Pas, E.I., Socio-demographics, activity participation, and travel behavior, *Transp. Res. A*, 33, 1–18, 1999.

Ma, J. and Goulias, K.G., A dynamic analysis of person and household activity and travel patterns using data from the first two waves in the Puget Sound Transportation Panel, *Transportation*, 24, 309–331, 1997.

Ma, J. and Goulias, K.G., Application of Poisson Regression Models to Activity Frequency Analysis and Prediction, paper presented at 78th Annual Meeting of the Transportation Research Board, Washington, D.C., January 10–14, 1999.

Mahmassani, H.S., Hatcher, S., and Caplice, C., Daily variation of trip chaining, scheduling, and path selection behavior of work commuters, in *Understanding Travel Behaviour in an Era of Change*, Stopher, P.R. and Lee-Gosselin, M., Eds., Elsevier Science Ltd., New York, 1997, pp. 351–379.

Mahmassani, H.S. and Herman, R., Interactive experiments for the study of tripmaker behaviour dynamics in congested commuting systems, in *Developments in Dynamic and Activity-Based Approaches to Travel Analysis*, Jones, P.M., Ed., Avebury, Aldershot, U.K., 1990, pp. 272–298.

Mahmassani, H. and Jou, R.-C., Bounded rationality in commuter decision dynamics: incorporating trip chaining in departure time and route switching decisions, in *Theoretical Foundations of Travel Choice Modeling*, Gärling, T., Laitila, T., and Weston, K., Eds., Elsevier, Oxford, 1998.

Mannering, F. et al., Travelers' preferences for in-vehicle information systems: an exploratory analysis, *Transp. Res. C*, 6, 339, 1995.

McFadden, D., Disaggregate behavioral travel demand's RUM side: a 30-year retrospective, in *The Leading Edge in Travel Behavior Research*, Hensher, D.A. and King, J., Eds., Pergamon Press, Oxford, 2002, in press.

Miller, E.J. and Salvini, P.A., Activity-Based Travel Behavior Modeling in a Microsimulation Framework, paper presented at IATBR Conference, Austin, TX, September 1997.

Montello, D.R., A new framework for understanding the acquisition of spatial knowledge in large-scale environments, in *Spatial and Temporal Reasoning in Geographic Information Systems*, Egenhofer, M.J. and Golledge, R.G., Eds., Oxford University Press, New York, 1998, pp. 143–154.

Nadel, L., Neural mechanisms of spatial orientation and wayfinding: an overview, in *Wayfinding Behavior: Cognitive Mapping and Other Spatial Processes*, Golledge, R.G., Ed., Johns Hopkins University Press, Baltimore, 1999, pp. 313–327.

Ortúzar, J.D. and Willumsen, L.G., *Modelling Transport*, 2nd ed., Wiley, Chichester, U.K., 1994.

Pas, E.I. and Koppelman, F., An examination of day-to-day variability in individuals' urban travel behavior, *Transportation*, 14, 13–20, 1986.

Pas, E.I. and Koppelman, F., An examination of the determinants of day-to-day variability in individuals' urban travel behavior, *Transportation*, 13, 183–200, 1987.

Pas, E.I. and Sundar, S., Interpersonal variability in daily urban travel behavior: some additional evidence, *Transportation*, 22, 135–150, 1995.

Payne, J.W., Bettman, J.R., and Johnson, E.J., *The Adaptive Decision Maker*, Cambridge University Press, New York, 1993.

Pendyala, R.M. et al., An activity-based micro-simulation analysis of transportation control measures, *Transp. Policy*, 4, 183–192, 1997.

Pendyala, R.M., Kitamura, R., and Reddy, D.V.G.P., Application of an activity-based travel demand model incorporating a rule-based algorithm, *Environ. Plann. B*, 25, 753–772, 1998.

Piaget, J. and Inhelder, B., *The Child's Conception of Space*, Norton, New York, 1967.

Polydoropoulou, A. and Ben-Akiva, M.E., *Effect of Advanced Traveler Information Systems (ATIS) on Travelers' Behavior*, paper presented at 77th Annual Meeting of the Transportation Research Board, Washington, D.C., January 1998.

Polydoropoulou, A. et al., Modeling revealed and stated en-route travel response to advanced traveler information systems, *Transp. Res. Rec.*, 1537, 38, 1996.

Recker, W.W., McNally, M., and Root, G., A model of complex travel behavior: part I: theoretical development, *Transp. Res. A*, 20, 307–318, 1986a.

Recker, W.W., McNally, M., and Root, G., A model of complex travel behavior: part II: an operational model, *Transp. Res. A*, 20, 319–330, 1986b.

Robinson, J.P., Kitamura, R., and Golob, T.F., *Daily Travel in The Netherlands and California: A Time Diary Perspective*, HCG, The Hague, Netherlands, 1992.

Self, C.M. and Golledge, R.G., Sex, gender, and cognitive mapping, in *Cognitive Mapping: Past, Present, and Future*, Kitchin, R. and Freundschuh, S., Eds., Routledge, London, 2000, pp. 197–220.

Sholl, M.J., From visual information to cognitive maps, in *The Construction of Cognitive Maps*, Portugali, J., Ed., Kluwer Academic Publishers, Dordrecht, Netherlands, 1996, pp. 157–186.

Siegel, A.W. and White, S.H., The development of spatial representation of large scale environments, in *Advances in Child Development and Behavior*, Vol. 10, Reese, H.W., Ed., Academic Press, New York, 1975, pp. 9–55.

Simon, H.A., Invariants of human behavior, *Annu. Rev. Psychol.*, 41, 1–19, 1990.

Smith, M.E., Design of small-sample home-interview travel surveys, *Transp. Res. Rec.*, 701, 29–35, 1979.

Smith, T.R., Pellegrino, J.W., and Golledge, R.G., Computational process modelling of spatial cognition and behavior, *Geogr. Anal.*, 14, 305–325, 1982.

Stern, E. and Leiser, D., Levels of spatial knowledge and urban travel modeling, *Geogr. Anal.*, 20, 140–156, 1988.

Svenson, O., The perspective from behavioral decision theory on modeling travel choice, in *Theoretical Foundations of Travel Choice Modeling*, Gärling, T., Laitila, T., and Westin, K., Eds., Elsevier, Amsterdam, 1998, pp. 141–172.

Zhou, J. and Golledge, R.G., An Analysis of Household Travel Behavior Based on GPS, paper presented at the IABTR, Gold Coast, Australia, February 26–March 1, 2000.

Selected Annotated References

Arentze, T. and Timmermans, H.P.J., *ALBATROSS: A Learning Based Transportation Oriented Simulation System*, European Institute of Retailing and Services Studies, Technical University of Eindhoven, Netherlands, 2000.

Based on the findings of a workshop on changing modeling needs, the Ministry of Transportation, Public Works and Water Management commissioned EIRASS at the end of 1996 to develop a prototype of a rule-based system for predicting transportation demand. This project reflected a desire to explore the potential of a new generation of transport demand models that should circumvent some limitations of the existing models. The model should allow one to better assess the likely consequences of flexible work hours, longer opening hours of shops, and similar trends. This book reports the development of this rule-based system, which was given the acronym ALBATROSS. Model development, data collection, and performance of the model are described.

A team of researchers, all members of the Urban Planning Group of the Eindhoven University of Technology and associates of EIRASS, worked on different components of the model system.

Axhausen, K. and Gärling, T., Activity-based approaches to travel analysis: conceptual frameworks, models, and research problems, *Transp. Rev.*, 12, 323–341, 1992.

Abstract: The recent policy discussions about information technology in transport and traffic demand management have increased the interest in activity-based approaches to the analysis of travel behavior, in particular in the modeling of household activity scheduling, which is at the core of many of the required changes in travel behavior. The paper is a state-of-the-art review of conceptualizations and models of activity scheduling with special regard to issues raised by the new policy instruments. In the course of the review, the validity of behavioral assumptions is critically examined and several needs for future research are identified.

Bhat, C.R. and Koppelman, F.S., Activity-based travel demand analysis: History, results, and future directions, Paper presented at the 79th Transportation Research Board Annual Meeting, Washington, D.C., 2000.

Abstract: Since the beginning of civilization, the viability and economic success of communities have been, to a major extent, determined by the efficiency of the transportation infrastructure. To make informed transportation infrastructure planning decisions, planners and engineers have to be able to forecast the response of transportation demand to changes in the attributes of the transportation system and changes in the attributes of the people using the transportation system. Travel demand models are used for this purpose; specifically, travel demand models are used to predict travel characteristics and usage of transport services under alternative socioeconomic scenarios, and for alternative transport service and land use configurations.

The need for realistic representations of behavior in travel demand modeling is well acknowledged in the literature. This need is particularly acute today as emphasis shifts from evaluating long-term investment-based capital improvement strategies to understanding travel behavior responses to shorter-term congestion management policies, such as alternate work schedules, telecommuting, and congestion pricing. The result has been an increasing realization in the field that the traditional statistically oriented trip-based modeling approach to travel demand analysis needs to be replaced by a more behaviorally oriented activity-based modeling approach.

Bowman, J.L. and Ben-Akiva, M.E., Activity-based disaggregate travel demand model system with daily activity schedules, *Transp. Res. A*, 35, 1–28, 2000.

Abstract: They present an integrated activity-based discrete choice model system of an individual's daily activity and travel schedule, intended for use in forecasting urban passenger travel demand. The system is demonstrated using a 1991 Boston travel survey and level-of-service data.

The model system represents a person's choice of activities and associated travel as a daily activity pattern overarching a set of tours. The daily activity pattern includes (a) the primary activity of the day, with one alternative being to remain at home for all the day's activities; (b) the type of tour for the primary activity, including the number, purpose, and sequence of activity stops; and (c) the number and purpose of secondary tours. Tour models include the choice of time, destination, and mode of travel, and are conditioned by the choice of a daily activity pattern. The choice of daily activity pattern is influenced by the expected maximum utility derived from the available tour alternatives.

Downs, R.M. and Stea, D., Eds., *Image and Environment: Cognitive Mapping and Spatial Behavior*, Aldine, Chicago, 1973b.

A concern with the relationship between human behavior and environment has always been at least an implicit claim of social scientists and planners, in theory as well as practice. But never has this concern been manifested so vocally and forcibly as in the very recent past.

Image and Environment addresses itself to this concern by considering how people acquire, amalgamate, and remember all the bits of information necessary to form a comprehensive picture of their environment, and how they then formulate a strategy that enables them to overcome two central behavioral problems: where things are and how to get there from here. The book introduces and gives coherence to the many approaches to this new field of study, and provides an understanding of cognitive mapping as a crucial aspect of the more general process whereby individuals cope with information from and about their total environments.

The approach of the editors — one trained as a psychologist, the other as a geographer — is necessarily interdisciplinary. Two dozen authors from such diverse disciplines as psychology, geography, sociology, neurophysiology, anthropology, biology, and urban design and planning bring an extraordinary richness of viewpoint to this innovative book. An introduction by the editors provides the first genuine attempt to integrate a comprehensive array of papers, which deal with such topics as cognitive representations, spatial preference, developmental sequences, spatial orientation, and cognitive distance. Several of the papers are classics in the field, but three quarters of them have never before appeared in print, and more than half were especially commissioned for this volume. The book also includes the first exhaustive bibliography of work in the field, as well as comprehensive author and subject indexes.

Image and Environment is a major effort to set forth and illustrate a conceptual framework that will unify the contributions of such diverse research areas as cognitive and developmental psychology, human and animal learning, urban sociology, behavioral geography, psychophysics, education, and neurophysiology, as well as the spatial decision-making techniques of architects, designers, and planners. For teachers, students, researchers, and practitioners in all these fields and more, the book will serve as a benchmark of what has been achieved to date and will open up broad new vistas of thought.

Ettema, D. and Timmermans, H.P.J., *Activity-Based Approaches to Travel Analysis*, Elsevier, New York, 1997.

Societal trends have made the need for better travel demand forecasts more urgent, while at the same time making people's travel and activity patterns far more complex. Traditional traffic flow models are no longer sophisticated enough to cope.

Activity analysis is seen by many as the solution. It has had a short but intense history in geography, urban planning, time use research, and, more recently, transportation. Pioneering activity-based models have now been developed to the point where, some argue, it is time to abandon the traditional four-step model for transportation demand forecasting and to adopt activity-based approaches instead. Others claim that the complexity of such approaches and their tremendous data requirements prevent them from having a significant impact.

This book explores these claims and the issues associated with them. An introductory section outlines the debate. The body of the work is organized in four sections: modeling developments, theories and empirical analyses, data needs and data representation, and policy analysis. The final section discusses future research directions.

The work presented here will be of value to researchers, lecturers, and students of transportation, geography, and urban planning; legislative and public policy analysts; and transport planners and consultants.

Gärling, T. et al., Computer simulation of a theory of household activity scheduling, *Environ. Plann. A*, 30, 665–679, 1998.

Abstract: An operational model of household activity scheduling is proposed. The model is based on a theory entailing behavioral principles of how persons acquire, represent, and use information from and about the environment. Choices of destinations and departure times are consequences of the scheduling of a set of activities to be executed in a given time cycle. Illustrative computer simulations of the operational model show realistic effects of work hours, central and decentral living, and travel speed. Several needed improvements of the theory and operational model are discussed, such as incorporating learning effects and choice of travel mode for home-based trip chains. Strategies outlined for empirical tests include comparisons with existing models, psychological experiments illuminating basic assumptions, and use of geographical information systems to process travel diary data for single cases.

Gärling, T., Laitila, T., and Westin, K., *Theoretical Foundations of Travel Choice Modeling*, Pergamon, Oxford, 1998.

This volume fulfills a long-felt need for a single text that documents the theoretical foundations of travel choice modeling. With contributions from a good cross section of the leading researchers in the field, the work provides a valuable reference that will be of lasting interest and value.

Divided into three parts, microeconomic theory, behavioral decision theory, and statistical theory, the book extends approaches to travel choice modeling beyond the consumer theory developed in economics by applying theories from the fields of geography, psychology, and statistics, and in doing so addresses two fundamental questions: What are the theoretical foundations of travel choice modeling? What should they be?

Containing 20 specially commissioned chapters, this book represents the latest and best thinking in this rapidly expanding field. Activity-based and dynamic approaches are fast emerging as the state of the art in transport modeling and are replacing trip-based models. This book tackles the key theoretical foundation that underpins these new approaches by asking:

Are there developments in the traditional microeconomic theory that make it usable?

Is behavioral decision theory a more appropriate theoretical foundation?

Which are the statistical data analytical issues in each case and how can they be solved?

Golledge, R.G., Kwan, M.-P., and Gärling, T., Computational-process modelling of household travel decisions using a geographical information system, *Pap. Reg. Sci.*, 73, 99–117, 1994.

Abstract: Household travel behavior entails interdependent deliberate decisions, as well as the execution of routines not preceded by deliberate decisions. Furthermore, travel decisions are dependent on choices to participate in activities. Because of the complexity of the decision-making process in which individuals are engaged, computational process models (CPMs) are promising means of implementing behavioral principles, which, unlike other disaggregate modeling approaches, do not rely on a utility-maximizing framework. A conceptual framework is proposed as the basis of a CPM interfaced with the geographical information system Arc/Info. How to model households' travel behavior is illustrated in a case study of a single household in which one member started telecommuting.

Golledge, R.G. and Stimson, R.J., *Spatial Behavior: A Geographic Perspective*, Guilford Press, New York, 1997.

How do human beings negotiate the spaces in which they live, work, and play? How are firms and institutions, and their spatial behaviors, being affected by processes of economic and societal change? What decisions are made about the natural and built environment, and how are these decisions acted out? Updating and expanding concepts of decision making and choice behavior on different geographic scales, this major revision of the authors' acclaimed *Analytical Behavioral Geography* presents theoretical foundations, extensive case studies, and empirical evidence of human behavior in a comprehensive range of physical, social, and economic settings. Generously illustrated with maps, diagrams, and tables, the volume also covers issues of gender, discusses traditionally excluded groups such as the physically and mentally challenged, and addresses the pressing needs of our growing elderly population.

Hanson, S. and Huff, J.O., Assessing day-to-day variability in complex travel patterns, *Transp. Res. Rec.*, 891, 18–24, 1982.

Abstract: Recent questioning of assumptions underlying current theory and practice in studies of urban travel behavior is continued. The focus here is on the assumption that the individual's day-to-day travel is habitual and therefore that a one-day record of behavior constitutes a sufficient database for theory and for model building. A rationale for examining the day-to-day variation in an individual's travel is established; then some of the field procedures that can contribute to making longitudinal data suitable for studying this issue are discussed. By using the Uppsala Household Travel Survey data as an example, the efficacy of these procedures is tested. Next, several techniques are described for measuring travel patterns so that day-to-day variability can be detected, and an approach to the measurement problem is outlined with illustrative examples from the Uppsala data, which consist of travel diaries collected over 35 consecutive days. The results of the empirical analysis are preliminary, but they indicate that (a) the quality of longitudinal travel diary data need not deteriorate over the survey period, (b) both employed men and nonworking women exhibit a great deal of repetition in their daily travel–activity patterns, so that (c) days with similar travel patterns can be identified and grouped.

Kitamura, R., An evaluation of activity-based travel analysis, *Transportation*, 15, 9–34, 1988.

Abstract: This paper is a review and assessment of the contributions made by activity-based approaches to the understanding and forecasting of travel behavior. In their brief history of approximately a decade, activity-based analyses have received extensive interest. This work has led to an accumulation of empirical evidence and new insights and has made substantial contributions toward the better understanding of travel behavior. However, practical applications of the approach in transportation planning and policy development have been scarce. Based on an analysis of the inherent characteristics of the activity-based approach, a review of recent (after the 1981 Oxford conference) developments, and a synthesis of the findings from past empirical studies, this study attempts to evaluate the contribution made by activity-based analyses and determine the reasons for the limited practical application. Recommendations are made for the future development of activity-based analysis as a science of travel behavior and as a tool in the practice of transportation planning and policy development.

Kitchin, R.M., Cognitive maps: what are they and why study them? *J. Environ. Psychol.*, 14, 1–19, 1994.

Abstract: It is often implicitly assumed by researchers that their readers understand what a cognitive map and cognitive mapping are, and their justification for study. This paper differs in this respect by explaining explicitly the what and why questions often asked, demonstrating cognitive mapping's multidisciplinary research worth. First, it examines questions concerning what cognitive maps are, the confusion inherent from the use of the term "map," and the usage and reasons for alternative expressions. Second, it examines the theoretical applications or conceptual research concerning the cognitive map's role in the influencing and explaining of spatial behavior; spatial choice and decision making; and wayfinding and orientation; as well as the cognitive map's utility and role as a mnemonic and metaphorical devise; a shaper of world and local attitudes and perspectives; and for creating and coping with imaginary worlds. Third, it discusses cognitive mapping's practical and applied worth concerning the planning of suitable living environments; advertising; crime solving; search and rescue, geographical educational issues, cartography and remote sensing; and in the designing and understanding of computer interfaces and databases, especially Geographic Information Systems (GIS).

Kuipers, B.J., Modelling spatial knowledge, *Cognit. Sci.*, 2, 129–153, 1978.

Abstract: A person's cognitive map, or knowledge of large-scale space, is built up from observations gathered as he or she travels through the environment. It acts as a problem solver to find routes and relative positions, as well as describing the current location. The TOUR model captures the multiple representations that make up the cognitive map, the problem-solving strategies it uses, and the mechanisms for assimilating new information. The representations have rich collections of states of partial knowledge, which support many of the performance characteristics of commonsense knowledge.

Kwan, M.-P., Golledge, R.G., and Speigle, J., Information representation for driver decision support systems, in *Theoretical Foundations of Travel Choice Modeling*, Gärling, T., Laitila, T., and Westin, K., Eds., Pergamon, Oxford, 1998, pp. 281–303.

Abstract: Intelligent Transportation Systems (ITS) utilize advanced communication and transportation technologies to achieve traffic efficiency and safety. There are different components of ITS, including Advanced Traveler Information Systems (ATIS), Automated Highway Systems (AHS), Advanced Traffic Management Systems (ATMS), Advanced Vehicle Control Systems (AVCS), and Advanced Public Transportation Systems (APTS). Development of a system for ITS depends on our ability to deal with a vast amount of information about the locations of places, as well as with the complex representation of the transportation network linking those places, and to incorporate these into a geographic database. The system therefore needs to be constructed based upon the foundation of an integrated and comprehensive Geographic Information System (GIS). As compared to the simplified node-link graph theory representations of transport networks used by current ITS, GIS are able to provide more realistic representations of elements of the complex environment.

Transportation science has an expressed goal of increasing accessibility for all groups of people with regard to the environments in which they live and interact. A significant component of these goals is to further develop ITS through multilevel and multimodal research and testing. This includes contributing to research and transportation system architecture, technology development, policy formation, and operational tests of various systems, including ATMS, ATIS, and APTS. In this paper we focus on ATIS.

McFadden, D., Disaggregate behavioral travel demand's RUM side: a 30-year retrospective, in *The Leading Edge in Travel Behavior Research*, Hensher, D.A., Ed., Pergamon, Oxford, 2002.

Abstract: This resource paper is intended to give a historical account of the development of the methodology of disaggregate behavioral travel demand analysis and its connection to random utility maximization (RUM). It reviews the early development of the subject and major methodological innovations over the past three decades in choice theory, data collection, and statistical tools. It concludes by identifying some topics and issues that deserve more work, and fearlessly forecasts the future course of research in the field.

Pas, E.I. and Koppelman, F., An examination of day-to-day variability in individuals' urban travel behavior, *Transportation*, 14, 13–20, 1986.

Abstract: Day-to-day variability in individuals' travel behavior (intrapersonal variability) has been recognized in conceptual discussions, yet the analysis and modeling of urban travel are typically based on a single-day record of each individual's travel. This paper develops and examines hypotheses regarding the determinants of intrapersonal variability in urban travel behavior.

Two general hypotheses are formulated to describe the effects of motivations for travel and related behavior and of travel and related constraints on intrapersonal variability in weekday urban travel behavior. Specific hypotheses concerning the effect of various sociodemographic characteristics on intrapersonal variability are derived from these general hypotheses. These specific hypotheses are tested empirically in the context of daily trip frequency using a 5-day record of travel in Reading, England.

The empirical results support the two general hypotheses. First, individuals who have fewer economic and role-related constraints have higher levels of intrapersonal variability in their daily trip frequency. Second, individuals who fulfill personal and household needs that do not require daily participation in out-of-home activities have higher levels of intrapersonal variability in their daily trip frequency.

Pendyala, R.M., Kitamura, R., and Reddy, D.V.G.P., Application of an activity-based travel demand model incorporating a rule-based algorithm, *Environ. Plann. B*, 25, 753–772, 1998.

Abstract: In this paper an activity-based travel demand model called AMOS is described. The model system is capable of simulating changes in individual activity and travel behavior that may be brought about by a change in the transportation system. These simulations may then be used to predict the impacts of various transportation policies on region-wide travel characteristics. A rule-based activity-scheduling algorithm is at the heart of AMOS. The algorithm simulates changes in activity and travel patterns while recognizing the presence of constraints under which travelers make decisions. Operationally, the algorithm reads the baseline activity and travel pattern of an individual and then determines the most probable adjustments that the individual may make in response to a transportation policy. In this paper, the scheduling algorithm is described in detail and sample results from a case study in the Washington, D.C., metropolitan area are provided.

Stopher, P.R. and Lee-Gosselin, M., Eds., *Understanding Travel Behaviour in an Era of Change*, Elsevier Science Ltd., New York, 1997.

Travel behavior research has a pivotal role to play in informing the current worldwide debate over the degree to which the growth in personal travel, notably by private motor vehicle, should be encouraged or controlled. At stake are complex public interests concerning air quality, energy, lifestyle, economic development, and the built environment.

This international collection of papers on current methodological and substantive findings from the analysis of personal travel is written by leading travel behavior researchers from the social and engineering sciences. It is organized in four sections: traveler activity and perception, stated preference methods, dynamic behavior, and improvement of behavioral travel models.

The work presented here will be of value to researchers, lecturers, and students of transport planning and engineering; legislative and public policy analysts; transport planners and consultants; and public interest groups.

4

Freight Transportation Planning: Models and Methods

Frank Southworth
Oak Ridge National Laboratory

4.1 Introduction

Freight transportation encompasses the movement of a wide variety of products, from raw materials to finished goods, from comparatively low value-to-weight commodities such as coal, grain, and gravel to high value-to-weight items such as computer parts and pharmaceuticals. It includes easily perishable items such as fresh fruit and vegetables, a wide range of refrigerated items, and a growing number of time-sensitive items for which on-time delivery is crucial to business success. This freight needs to be moved safely and at reasonable cost. It must also be moved in an environmentally sound and socially acceptable manner. The purpose of this chapter is to review the principal issues involved in analyzing freight movements and to describe the analytical methods currently in use or under development for doing so. This includes a review of the data sources and methods for measuring and forecasting freight traffic volumes, as well as their economic, social, and environmental impacts. It also includes methods for measuring the carrying capacity of freight systems and the effects of freight volume-to-capacity ratios on the productivity of the freight industry.

At the beginning of the 21st century most cities and nations find themselves moving more freight than ever before, a good deal of it over long distances and across national borders. On an average day in 1997 some 41 million tons of freight, valued at over $23 billion, was transported within the United States. This represented an average daily freight flow of 310 lb, moving an average distance of 40 mi, for each U.S. resident. In total, this represented some 14.8 billion tons and $8.6 trillion dollars of merchandise, requiring almost 3.9 billion ton-miles of freight activity (BTS, 2001). Much of this freight is a direct result of the growth in population and economic activity, while technological developments have also contributed to a greater reliance on transportation in the production process. The world is also engaging in more trade than ever before. Worldwide merchandise trade (exports) is estimated to have grown from $58 billion in 1948 to $6168 billion in 2000. Between 1960 and 2000, while the worldwide production of merchandised goods grew more than threefold, the volume of international trade increased by a factor of almost 10 (WTO, 2002). Recent projections call for increases in both U.S. and worldwide trade and associated freight volumes well into the current century. Significantly, these growth rates are well in excess of the historical growth rates in freight handling infrastructures and vehicle fleets. With many of these infrastructures already under stress, and suffering from costly traffic congestion, freight planners have an important role to play in the future of the world's transportation and economic systems.

Adding to this professional challenge, these growing demands on today's freight transportation systems come at a time of significant change in both the freight industry itself and in the methods being used to analyze it. Perhaps the most influential of these changes is the rapid evolution and adoption of real-time, telecommunication-based information technologies, the so-called *IT revolution* (Golob and Regan, 2000; Hilliard et al., 2000). This technology has allowed the widespread adoption of electronic commerce (*e-commerce*) as a means of placing contracts, tracking costs, and checking product availability, much of it via the Internet and World Wide Web. This, in turn, has led to new types of business partnerships, including new business arrangements between freight shippers, freight carriers, and a growing variety of third-party freight logistics agents. It has also enabled the rapid adoption of real-time vehicle and cargo tracking and inventory monitoring technologies, which are now encouraging the adoption of *just-in-time* (JIT) freight delivery systems that substitute reliable transportation services for a customer's inventory carrying costs. Since the mid-1950s there have also been some significant advances in freight handling and transport, including the double stacking of trains (Manalytics Inc. et al., 1988); the use of trailer-on-flatcar technology, roll-on roll-off systems, and automated stacking cranes (Ballis and Stathopoulos, 2002); the development of megaships (Bomba et al., 2001); and the use of standardized containers to more easily transfer goods between ship and shore, truck and rail, and truck and plane. The result of all this innovation is that we have today a rapidly evolving freight transportation industry. This industry is currently in need of better data and better methods for tracking, analyzing, and forecasting the potential impacts, financial and otherwise, of both current and newly emerging forms of freight activity.

In addressing the above issues, the rest of this chapter is organized around the following topics: freight agents, freight costs, freight demand (estimation and forecasting), freight supply (capacity issues), productivity and performance measures, and freight's safety and environmental impacts. Much of this discussion treats freight transportation as a clearly identifiable component of metropolitan and statewide transportation planning. A final section of the chapter notes the growing difficulty of doing so. This section focuses on the increasingly close ties between information-rich business logistics and freight transportation operations. These are ties that question the applicability of existing methods for modeling and forecasting many new forms of freight movement. In particular, the pivotal role of freight transportation logistics in the broader arena of supply chain management (Brewer et al., 2001) is considered from the perspective of more efficient freight movement planning. Future developments in freight planning are likely to adopt some combination of these current and newly emerging approaches to freight movement modeling. And as with all forms of planning, data availability is likely to prove a key to its eventual success (Meyburg and Mbwana, 2002).

4.2 Freight Agents: Movers and Shakers

Freight's role in the economy is a central one. It may include moving a raw material from a production site (mine, farm, etc.) to a manufacturing plant, moving processed products from the plant to a distribution center or directly to a retailer, and moving the finished product from the retailer to the final customer. Linking a freight producer to a freight consumer, or customer, can vary from the simple to the complex. On the simple end we have a product being transported directly from manufacturer A to consumer B with no other stops and no transformation of the product en route. A common example is coal transported directly from the mine to a coal-burning power plant. Even in this case, however, a third party in the form of a for-hire freight carrier, such as a railroad, trucking, or barge company, is usually involved. In freight transportation it is usual to refer to the creator or originator of a product to be transported as the *shipper*, and to the receiver of the product as the customer. The transporter of the product is usually referred to as the *carrier*. In cases where the shipper is also the carrier, it is common to refer to this as *private carriage*. Where the carrier is a transportation firm that moves the freight under contract to the shipper, we refer to this as *for-hire carriage*. A third important agent in the freight movement business is the freight broker, or freight *forwarder*, who acts as a go-between in assigning a producer's shipments to for-hire carriers.

The major carriers of freight in the United States and in most of the rest of the world are trucking firms, railroads, airfreight carriers, inland barge operators, seaborne vessel operators, and pipeline operators; there is limited overlap in the ownership and operation of these different modes of transportation at the present time. This in turn has led to a good deal of competition between modes for freight business, but with a degree of cooperation in recent years that reflects the needs of an increasingly demanding marketplace for fast, flexible, low-cost goods delivery. Such cooperation translates in physical terms into *intermodal* transportation, defined here as the end-on transfer of freight between two or more different modes of transport in the process of getting a consignment of freight from its origin to its destination. Common examples of intermodal freight transportation are truck–rail and truck–water shipments of bulk commodities such as coal and grain, as well as truck–air inclusive deliveries of high-value and often time-sensitive commodities such as computer parts and medical supplies (see, for example, Premius and Konings, 2001). A very successful example of truck–air intermodalism is the overnight small package delivery industry, pioneered by companies that have been leaders in a JIT freight delivery revolution that puts a growing premium on speed of transport (Taylor, 2001).

An additional and important player in the freight transportation game is the freight forwarder. These forwarders act as brokers who negotiate deals between shippers and carriers of freight, thereby taking the burden of the shipment logistics away from the shipper (for a price, of course). With the advent of the Internet a new generation of freight forwarders now offers a growing range of services to shippers and carriers, including the use of intermodal transportation. These include a growing number of companies known as third-party logistics (3PL) service providers. Whether starting out as a freight forwarder, freight carrier, or shipper or producer, these 3PLs have become key players in both using and marketing increasingly comprehensive and increasingly information technology-based freight handling services. As a result, a growing number of shippers are turning to 3PLs and to other forms of IT-based logistics companies and freight intermediaries (Song and Regan, 2001) to handle their freight, a condition often referred to as *outsourcing* of transportation management services. The largest of these logistics providers employ hundreds of workers at locations across the country and continent, have arrangements with dozens of carriers to move both air and ground freight, and do annual business in the multimillion dollar range. Types of freight handled can be specialized or varied, depending on company size. (A trip to the World Wide Web identified one firm that handles shipping and other logistical services for companies needing to move food ingredients or additives, paper stock, bottled beverages, plastic and glass containers, and pharmaceutical, health and fitness, video, and printed matter.)

Such 3PLs may offer a range of services, everything from order processing to the carriage, warehousing, and tracking of goods, payments, complaints, and even credit card processing. Within the past decade a newer term, the fourth-party logistics (4PL) service provider, has also found its way into this literature. These are organizations that may themselves include one or more 3PL companies, moving businesses toward increasingly global integration of freight-cum-warehousing-cum-electronic commerce-based order handling systems: systems that link together many different companies to form multienterprise logistics management concerns involved in worldwide trading systems. The number of carriers and shippers associated with these sorts of multifaceted logistics enterprises may be in the hundreds or even thousands in the near future.

Finally, with huge investments of public funds required to build, maintain, expand, and renovate the nation's seaports, airports, highways, and waterways, many publicly elected officials are involved in different aspects of freight transportation. These include regulators; local, metropolitan, regional, and national freight planners; construction engineers; customs agents; statisticians; economists; and lawyers — all with a need to understand what freight is being moved, who moves it, and what the public safety and environmental, as well as economic, impacts of such movements are likely to be. Add organizations such as labor unions, chambers of commerce, and other public interest groups, and it becomes clear that the way we move freight has broad implications for society as a whole. Many of the concerns these people deal with require the ability to derive aggregate (daily, seasonal, annual) estimates and forecasts of the tons as well as the dollar value of the goods moved between places. This in turn requires data collection by public, usually transportation planning, agencies. The modeling of freight flows discussed later in this chapter is based on these public agency data collection efforts.

4.3   Freight Costs

The costs of moving freight include the costs of the labor and the operation and maintenance of vehicles and containers, as well as the costs of the roadways, storage facilities, and terminals required to support pickups and deliveries. Vehicle operating costs include fuel and maintenance as well as insurance, licensing, and related taxes. Over time they also include the costs of vehicle and vehicle parts replacement. Container costs may include cleaning and other special storage needs such as refrigeration or humidification. Hazardous materials movement requires additional precautions in terms of packaging and handling, as well as additional paperwork, including permissions to transport over specific routes. Damaged goods mean lost profits. Accidents en route mean lost goods, lost time, and potentially costly lawsuits (not to mention the potential for lost lives). Each mode has its own particular set of costs to deal with. In the case of trucking and barge transportation, highways and waterways, respectively, are funded out of user taxes on fuels and from vehicle operator licenses. In the case of U.S. railroads, who own their tracks and rights-of-way, there are the costs of company-owned track development and maintenance, including the costs of building and operating stations and some rather large railcar switching yards. Oceangoing transporters must pay port and dock utilization fees. Airfreight operators must pay airport gate access and utilization fees. All modes incur storage and within-terminal handling fees of one sort or another.

To understand why specific modes and mode combinations move certain goods requires an understanding and accurate quantification of these various freight logistics costs, just as remaining competitive in the freight business requires the ability to keep such costs down. It is equally important to understand who is paying these costs: the shipper, carrier, forwarder, or customer. In particular, reliability of service (or in cost terms, the lack of it) is often as important, if not even more important, to shippers and receivers as obtaining the lowest cost of carriage per ton. One reason for this is the trade-off between transportation costs and inventory holding costs. The value of guaranteed on-time delivery is especially important in cases where retention of a high-demand perishable commodity (e.g., milk) requires additional warehousing costs (e.g., refrigeration) in order to ensure that the product is always available to customers (Allen et al., 1985; Vilain and Wolfrom, 2000). This last topic is taken up below under the discussion of freight mode choice.

Freight cost functions are most usefully given in terms of a specific origin-to-destination (O-D) movement, sometimes called a movement channel or a traffic lane, for a specific mode of travel and class of commodity. They may also be time-dependent, varying in some cases by season, as well as by precontracted speed-of-delivery agreements (e.g., overnight, 3-day delivery, delivery by a specified date). In practice, shippers are increasingly contracting for a specific type of service rather than a specific mode of delivery. Hence a shipper may not always know how his cargo got to its destination: only that the carrier or broker he used got it there on time at a given price. This price, usually based on a per unit (e.g., ton, mile, ton-mile) freight rate, may be negotiated for a single shipment or for a contractual period covering weeks or months. For example, it is usual for electric power companies to contract for regular railroad or barge deliveries of what is termed utility coal at a particular rate and for an extended period. In doing so both the customer and the carrier incur risks associated with changes in the market price of the product shipped, as well as changes in the costs of carriage as a result of bad weather or traffic congestion en route. Damage costs are often covered, at least in part, by taking out insurance on both the goods moved and on the vehicle fleet and laborers used to move them. Freight delayed significantly en route can also incur demurrage costs: charges resulting from the need to hold a consignment of goods in storage longer than expected due to late arrival of transportation equipment. Late delivery of such goods can also lead to lost value due to shifts in market price or the perishable nature of the goods. Such delays may be unusual accidents or occurrences, or the result of more generic transportation system problems associated with traffic congestion. Removing or alleviating such congestion is today a major goal of many freight transportation planning studies undertaken by government agencies.

Finally, freight that is moved across international borders is usually subject to trading tariffs, as well as to delays for customs inspections. Additional costs may result from the need to transfer cargoes between foreign and domestic carriers where the latter are the only ones legally allowed to transport certain goods within their national boundaries.

Collecting data on freight costs can be an expensive activity. These costs may be expressed in terms of the resources (fuel, driver time, etc.) needed to move a given volume of freight a given distance, or they may be the resulting freight rates charged by carriers or forwarders for doing so. Getting individual rate quotes for specific shipments has been much simplified by the Internet. Getting representative freight rates of resource costs for industry-wide or region-wide planning studies is a much larger challenge, often requiring sample surveys of shippers or carriers, many of which are less than keen to share proprietary business information. Where such cost data have been collected in the past they are usually oriented toward answering a specific policy question. For examples of freight logistics costs, some listed by individual component, see Cambridge Systematics Inc. (1995), Roberts et al. (1996), and Musso (2001).

4.4 Freight Demand: Estimation and Forecasting

Effective freight movement requires effective freight planning, which in turn requires sound methods and models for forecasting how the demands for freight transportation services will change over time. Past modeling efforts have either focused on the growth in specific commodities, using time series data to project future growth or decline in specific commodity movements, or emulated the traditional four-step urban transportation planning model (TRB, 1997; Cambridge Systematics, 1995). This latter approach appears to be the most popular with metropolitan and statewide planning agencies. It involves linking methods for estimating and forecasting the volume of freight produced by specific industries (*freight generation and attraction*) with methods for estimating the volumes of freight moving between different industries or consumers at different locations (*freight flow modeling*) and with the technological means of transporting this freight (*mode and route choice*).

Figure 4.1 shows the principal freight planning submodels and their key inputs in what is a computationally and data-intensive process. Note that when the planning process calls for commodity flows to be translated into vehicle movements a fifth step is required: the modeling of vehicle load factors. This may occur as step 4 in the modeling process, as shown in Figure 4.1. Alternatively, it may occur at the trip generation stage, producing truck trip forecasts that are suitable for direct application to the subsequent traffic route assignment step. At this assignment step a range of route selection models may be employed. Where truck traffic is concerned it is usual to carry out mixed freight–passenger travel assignments to capture the effects of traffic congestion on shipment times and hence freight delivery costs. These congestion-inclusive costs can then, in theory, be fed back through the freight flow modeling, mode selection, and vehicle loading steps, and iterated until the system of model equations stabilizes on

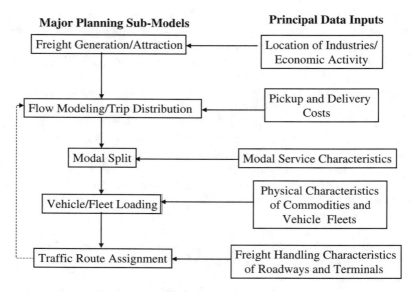

FIGURE 4.1 Multi-step freight planning model: major submodels and data inputs.

a set of transportation costs and flows (see Southworth et al., 1983). Variations on such a process have been used to analyze corridor-specific (Holguin-Veras and Thorson, 2000), metropolitan areawide (Ogden, 1992), and even statewide freight movement systems (Pendyala et al., 2000), although to date with much less frequency and attention to detail than has been put into passenger transportation modeling. For the most part, this modeling has also focused on truck transportation, with multimodal freight modeling receiving limited attention outside of high-volume traffic corridor studies.

4.4.1 Freight Generation and Attraction Models

Methods for estimating the amount of freight generated or received by a specific location, or within a specific geographic area (e.g., a traffic zone, a county), face a nontrivial data collection challenge. Unlike passenger traffic generation models that are based on the number and types of people and vehicles within an area, the freight analyst usually has to deal with difficult-to-obtain data on the number of tons or dollars of economic activity associated with one or more business enterprises, and these are often enterprises that vary a good deal in size and mode of operation, as well as in product mix. Making matters difficult, business data are often guarded as proprietary. Unless the analyst is fortunate enough to be able to survey and obtain the cooperation of a representative sample of the businesses located within an area, he must resort to less direct methods of estimation. This usually means using data on average dollars per ton and average tons per vehicle, as reported by nationally or regionally based sample surveys.

Fortunately, a number of publications and databases now exist to help freight planners with this data issue. A recent synthesis by Fischer and Han (2001) lists the major sources and types of truck trip generation data and provides numerous tables of truck trip generation rates broken down by commodity or vehicle type. The principal data collection methods in use today can be listed as:

- Vehicle classification counts (using in-the-roadway traffic loop counters or video and other types of traffic monitors and sensors)
- Vehicle intercept and special traffic generator surveys (counting, classifying, or surveying vehicles as they enter and leave a specific geographic area over a period of time)
- Truck trip travel diaries (driver- or dispatcher-completed daily travel surveys)
- Carrier activity surveys (typically regulated surveys related to safety or user fee legislation)
- Commodity flow surveys (shipper- or establishment-completed shipment inventories)

Each of these methods has its strengths and weaknesses. Vehicle classification counts and intercept surveys are especially useful for roadway capacity and associated traffic congestion studies. They usually offer the only cost-effective means of capturing truck traffic crossing the major routes into and out of a geographic area. In contrast, special traffic generator surveys focus on high-volume freight generating or attracting locations such as seaports, airports, truck and rail transfer terminals, large industrial parks, and warehousing complexes. Traffic monitoring in such cases may last for a period of days or weeks, depending on the type of equipment used (e.g., video cameras, manual counting). Twenty-four-hour monitoring can yield trip generation rates by time of day, producing peak and off-peak rates. Intercept surveys, where drivers are questioned at selected checkpoints, can also be used to collect additional data on vehicle characteristics (including size and weight, axle configuration, commodity carried) as well as to help identify the volumes of traffic into, out of, and through the area. Similarly, travel diaries can provide an additional wealth of information about not only vehicle characteristics but also where the truck is going and what is being carried. However, diaries can be expensive and difficult to collect, with concerns by truck owners and operators over survey impacts on driver productivity, and dispatchers and drivers may have different knowledge bases when surveyed. Response rates can vary considerably when used to capture wide-area freight activity, causing the added problem of establishing a proper sampling frame (Lawson and Riis, 2001).

Carrier-specific activity surveys offer the most readily available data on barge, railcar, pipeline, oceangoing vessel, and aircraft traffic generators and attractors (see Meyburg and Mbwana, 2001). *Commodity flow surveys* are typically applied to large geographic areas, such as complete metropolitan, statewide, or nationwide surveys, with the emphasis on trade flows and their resulting economic impacts. They tend to be multimodal in nature. They can be especially useful in the estimation of cross-border or external freight flows, in which the volume of freight coming into, moving out of, or passing through a region from or to other regions is of interest. In the United States the Commodity Flow Survey (CFS), carried out in 1993 and 1997 and scheduled for 2002, is the largest of these surveys, with a mandatory response requirement for all shippers included in its sample (U.S. Census Bureau, 1997a). This survey provides national, statewide, and major metropolitan area estimates of the annual tons and ton-miles of freight moved, as well as the dollar value of this freight, broken down by major mode (and mode sequence) with quite detailed commodity classification. This can be useful data when trying to estimate within-state, notably county-based, freight activity totals for use in freight flow modeling (see below), since dollar valued economic activity data by industry types can be obtained at the county level from other sources within the economic census. Translating dollars or tons of commodity movement into annual or daily shipments or vehicle trip rates requires additional data on the distribution of tonnages between vehicle size classes and the average loads carried by vehicles in each size class. In the United States the most widely available source of this type of data for truck trip generation modeling is the Vehicle Inventory and Use Survey (VIUS) (U.S. Census Bureau, 1997b). A common problem for freight traffic generation modeling is the mismatching of industrial classifications used in surveys such as the VIUS and CFS or other national economic and industrial activity data sets. Such problems are further exacerbated when trying to study transborder freight, using data classifications from other countries.

In developing commodity-based or vehicle-based freight trip generation rates the above data sources have for the most part been used in two ways. The first is to combine data on vehicle traffic counts or tons moved with employment or land use data to develop simple trip rates or estimates of tons moved per employee or per unit of land (Cambridge Systematics, 1995; Fischer and Han, 2001). It is questionable how transferable these rates are in any given application. One means of averaging to obtain more robust rates for use in forecasting future freight generations and attractions is to fit least squares regression models to traffic count or commodity tonnage data. The *Quick Response Freight Manual* (Cambridge Systematics, 1997) and Fischer and Han (2001) report a range of past truck trip regression models. Rates are for the most part based on daily truck trips per employee, per acre, or per square feet of floor space given to a particular land use or broad industrial classification. Some studies produce separate rates for trucks in different size classes. In the case of major freight generators such as ports and intermodal terminals, truck traffic can also be estimated from data on the other modes using these same facilities. For example, the Delaware Valley Regional Planning Commission (reported by Fischer and Han, 2001) used the following simple linear regression model for seaport trips:

$$\text{Truck trips/day} = (2.02 \times \text{ship arrivals/year}) - 20$$

and for rail terminals:

$$\text{Truck trips/day} = (0.0095 \times \text{rail cars/year}) + 24$$

In the case of containerized freight, Holguin-Veras and Lopez-Genao (2002) provide a third way of standardizing truck trip rates, by linking the number of daily one-way (inbound or outbound) truck trips to the number of 20-ft equivalent (TEU) containers and, after some further data processing, to the number of container boxes handled annually (from a sample of 21 U.S. container ports). Additionally, separate regression formulas were developed for what are termed "typical" and "busy" days. The rapid growth in container traffic worldwide has increased interest in seaports at which containers are transferred in their thousands from very large oceangoing vessels onto both truck and rail modes (for an example, see Al-Deek et al., 2000).

It should be clear from the above discussion that the volume of freight and the number of vehicle trips required to handle it may be estimated using a number of different data sources.

Ideally, time series data would help tremendously to establish reliable rates as well as assist in forecasting future generation and attraction levels. Little of this data exists at the present time. One reason for using data such as the number of TEUs passing through a seaport or the number of employees engaged in a specific industry within a specific traffic zone is to make such forecasting easier. One of the problems with this approach, however, is the speed with which the relationship between freight volumes and some of these more readily obtained independent variable forecasts can change. For example, higher productivity per employee means more tons moved per labor force in the future. Similarly, changes in container sizes (e.g., from 20-ft to 40-ft containers) can alter the number and perhaps also the type of vehicles used to move them in the future.

4.4.2 Freight Flow–Freight Trip Distribution Models

Freight by its nature is spatial. The *pattern of freight movements* refers to the distribution of an aggregate freight volume between different origin-to-destination pairs of places. Volume here is usually measured in terms of tons or the monetary value of goods transported during a given time period. Operationally, the volume of goods moved per day is important to those either moving the freight or charged with ensuring sufficient transportation system capacity for doing so. For longer-range planning purposes the volumes of freight moved per month and per year are also important data items that need to be collected.

A popular method for modeling (i.e., estimating, forecasting) commodity flows is to develop commodity-specific spatial interaction (SIA) models (for an example, see Black, 1997). If we let V_i refer to the volume of freight (e.g., the annual tonnage, the annual dollar value of production, or output) of a particular commodity in region i, then this freight can be allocated to destinations $j = 1, 2, ..., J$ using the following general SIA model (see Wilson, 1970):

$$T_{ij} = V_i \ A_i \ W_j \ B_j \ f(c_{ij}) \tag{4.1}$$

where T_{ij} is the volume of freight (or value of economic activity) allocated from origin i to destination location j; W_j is the volume of freight (or dollar valued demand) for the commodity of interest by industries located in region j; $f(c_{ij})$ is an inverse function of the costs, c_{ij}, of transporting a unit of the commodity of interest from i to j; and A_i and B_j are the balancing factors that ensure a compliance to the empirically observed or otherwise generated (i.e., trip generation model generated) production $\{V_i\}$ and consumption $\{W_j\}$ totals. Specifically,

$$A_i = [\Sigma_j W_j\, B_j\, f(\, c_{ij}\,)]^{-1} \quad \forall i \tag{4.2}$$

and

$$B_j = [\, \Sigma_i\, V_i\, A_j\, f(\, c_{ij}\,)]^{-1} \quad \forall j \tag{4.3}$$

That is, these two sets of balancing factors are solved using an iterative proportional fitting procedure that ensures that

$$\Sigma_j S_{ij} = V_i \text{ for all } i \text{ and } \Sigma_i S_{ij} = W_j \text{ for all } j \tag{4.4}$$

This sort of model is termed a *doubly constrained* SIA model (Wilson, 1970). Setting all B_j values equal to 1.0 produced a *supply or production constrained* model, in which the constraints on model generated demand totals are relaxed. Setting all A_i values equal to 1.0 produces a *demand or attraction constrained* SIA model, in which the freight shares exactly match the amount of commodity demanded in each region, W_j, but in which the region-specific production totals are allowed to vary from the SIA model estimated values for V_i.

The origin-to-destination freight costs, c_{ij}, in such a model should be derived either directly from empirical data or via econometric modeling from sampled data on observed freight rates, or using observed data on the resource costs involved in transportation (i.e., the fuel, vehicle operation and maintenance costs, driver wages, etc.). Constructing such cost matrices can be an expensive proposition, especially where more than one mode of transportation is used to move such freight. Example freight cost functions include:

$$f(c_{ij}) = \exp(-\beta c_{ij}) \text{ and } f(c_{ij}) = 1/\beta c_{ij} \tag{4.5}$$

SIA models such as that represented by Equations (4.1) to (4.3) above are most often applied to zonally aggregated freight data, where such traffic zones represent anything from a block group area within an urban freight study to a county area within an intercity or statewide freight movement study.

More detailed analysis of freight movements between specific facilities can also be modeled using similar destination choice models and using shipment-specific data coupled with detailed reporting or estimation of shipment costs. In such cases the popular logit choice model can be used, i.e.,

$$T_{ij} = V_i P_{j/i} = V_i\, \exp(u_{ij}) \Sigma_j\, \exp(u_{ij}) \tag{4.6}$$

where V_i is the volume of freight shipped from location i, $P_{(j/i)}$ is the probability of shipping to market j from production location i, and u_{ij} represents a market attractiveness function.

For example, reproducing the production constrained SIA model form introduced above, but applied to shipment specific data, u_{ij} might have a linear additive form such as

$$u_{ij} = -\beta c_{ij} + f(W_j) = -\beta(\alpha 1 + \alpha 2.d_{ij} + \alpha 3.m_{ij} + \alpha 4.t_{ij+} \ldots\,) + (\lambda_1.\ln D_j + \lambda_2.\ln G_j) \tag{4.7}$$

Here the cost of freight movement, c_{ij}, may be made up of specific cost components discussed earlier in this chapter, e.g., driver's time (d), vehicle operating costs (m), and other en route costs, such as highway tolls (t); and W_j is the potential for serving market j, based on the dollar size of the market (D) for the commodity being shipped and possibly other factors (G), such as zonal employment or number of establishments. Alternatively, the above model might use carrier quoted freight rates to represent the c_{ij} values. The key to such models is to find a suitable functional form for u_{ij} that can be fit to the available data, with model calibration involving selection of best-fitting values for β, the various α values, and λ.

Logit models may be applied to either disaggregate, shipment-specific data or to more spatially aggregated data sets. Southworth (1982) provides an example of the former for urban truck freight movements in Chicago. A recent study by Sivakumar and Bhat (2002) describes the latter approach, predicting commodity-specific intercounty and external freight flows for the state of Texas.

A problem with applying traditional logit and SIA models to freight movements is that there are significant differences in the methods used, both within and especially between modes, for routing freight over networks. For example, a good deal of urban truck transport is multistop in nature, with the resulting problem of linking individual cargo movement costs to the volume of goods moved. Airfreight poses a similarly tricky problem. While the goods may be moved from A to B, the aircraft often operates within a well-defined hub-and-spoke system that routes aircraft into and out of major airports on one or both ends of a multistop (often termed a *multileg*) movement (O'Kelly, 1998). With a good deal of freight moving in the belly of passenger aircraft, there is also the problem of costing the freight component of a move. In all modes there is also the issue of capturing any empty *backhauling* costs. In such cases it may be easiest to resort to freight rate data in order to understand current movement patterns. Forecasting future freight movement patterns then depends heavily on the evolution of these hubbing systems. This topic is taken up again below under the traffic assignment discussion.

Where more than mode of transportation may be used to move a commodity, the expense involved in estimating such shipment costs can become that much more resource intensive. This applies to situations involving both multimodal, in the sense of competitive, and intermodal, in the sense of linked or cooperative (e.g., truck–rail) freight movements. In the case of modal competition this requires a method for capturing the combined effects of the available modal cost options on the probability of different suppliers being able to cost-effectively deliver freight to specific markets. This topic is discussed below under mode choice modeling.

In the case of intermodal transportation the analyst needs to consider the costs of transferring the freight from one mode to another. Again, obtaining carrier-quoted freight rates is often an option here for getting around the need to model terminal transfer cost. Choice of one method over the other depends on a study's resources as well as its objectives. If built to analyze policies involving the efficiencies of intermodal transfer terminals, for example, resource-based freight movement costs may need to be computed for each major freight handling activity involved in a source-to-market movement. Collecting shipment rate data for large study areas covering many types of commodity movements usually requires the analyst to construct more or less approximate resource cost-based estimates of c_{ij}, or to develop them around a sample of freight rates for which a relationship between distance or time of transport to rate charged can be established (see Roberts et al., 1996).

Before turning to this issue of capturing the appropriate modal costs within freight flow models, an additional line of development in freight flow modeling is worth describing. This method extends Leontief's classical interindustry input–output (I-O) model of economic activity (Leontief, 1967) to consider spatial interactions (Wilson, 1970, chapter 3). In doing so, it also offers an efficient method for combining available data on both the freight generation and distribution steps in the planning model process shown in Figure 4.1. Starting with the familiar I-O model, let X^m equal the total dollar valued output in industrial sector m, for m = 1, 2, ..., N sectors in the economy of interest. Then we have the following matrix of interindustry relationships between production and consumption of products:

$$X^1 = a^{11}X^1 + a^{12}X^2 + ... + a^{1N}X^N + Y^1$$

$$\vdots$$

$$X^m = a^{m1}X^1 + a^{m2}X^2 + ... + a^{mN}X^N + Y^m \qquad (4.8)$$

$$\vdots$$

$$X^N = a^{N1}X^1 + a^{N2}X^2 + ... + a^{NN}X^N + Y^N$$

where the a^{mn} values are *technical coefficients* that define the *dollar valued* amount of product m required to produce a unit of product n. We now introduce geography into the picture. First, define X_{ij}^{mn} as the amount of m from traffic analysis zone i that is used in sector n in destination zone j, and define Y_i^m to be the final demand for the output of sector m in zone i. Further, if for the moment we define a_{ij}^{mn} to be a set of spatially explicit technical coefficients, we have the following identities:

$$X_i^m = \sum_{jn} X_{ij}^{mn} + Y_i^m = \sum_{jn} a_{ij}^{mn} X_j^n + Y_i^m \qquad (4.9)$$

What we need, then, is a method for computing the a_{ij}^{mn} values. This constitutes a great deal of data for which the information is rarely available. Such data can, however, often be constructed for specific freight generating or freight attracting zones. For example (after Wilson et al., 1981, chapter 10), if we let z_j^{mn} be a set of destination j specific technical coefficients, we can introduce spatial interaction modeling, as described above, explicitly into the process, i.e.,

$$a_{ij}^{mn} = z_j^{mn} \exp(-\beta^m c_{ij}^m) / \sum_i \exp(-\beta^m c_{ij}^m) \qquad (4.10)$$

adopting here the popular negative exponential form of travel cost function. This now lets us restate the combined product (freight) activity generation and distribution model as:

$$X_i^m = \sum_{jn} X_i^{mn} z_j^{mn} [\exp(-\beta^m c_{ij}^m) / \sum_i \exp(-\beta^m c_{ij}^m)] + Y_i^m \qquad (4.11)$$

which is an interindustry, attraction-constrained spatial interaction model in the popular logit form, with similar forms also possible for production (origin), as well as both production and attraction-constrained coefficients.

In the United States this and similar I-O model-based approaches are currently most suitable to intercounty or larger interregional flow modeling, focused on the statewide or multistate regional scale of economic activity. At this level of analysis planners can take advantage of national- and region-specific interindustry coefficients constructed by the U.S. Department of Commerce or by private sector companies who specialize in this sort of analysis. A number of interregional I-O model-based approaches to freight flow estimation and forecasting exist in the open literature. Recent examples of U.S. studies include Vilain et al. (1999) and Sorratini and Smith (2000). Zlatoper and Austrian (1989) also review some earlier econometric studies, including input–output studies. Note that these I-O models produce dollar-valued commodity flows, and therefore represent trade flows rather than freight movements per se. Translation to tons moved or to mode-specific vehicular trips requires additional modeling (cf. Figure 4.1 and see below).

4.4.3 Modeling Freight Mode Choice

While a good deal of freight moved today is largely captive to one mode or another, notably short-distance hauling by trucks, there remains a significant volume of longer-distance freight for which a very real choice is offered by more than one mode of transportation. This includes 1) a good deal of bulk freight where rail, water, or pipelines compete directly, 2) a large volume of freight for which truck and rail compete, and 3) a good deal of high-valued freight, including parcel freight, for which truck and air transportation are both viable options (often in cooperation as well as in direct competition).

While the primary modes of freight transportation are readily identifiable (i.e., truck, rail, water, air, pipeline, and intermodal combinations of these), once we start to analyze freight movements with a specific question in mind, mode choice is seen to be increasingly associated with type of service as well as purely technological attributes. For example, if we are interested in the movements of different types of oceangoing cargo vessels, then we can divide the U.S. merchant fleet in at least two ways (USACE, 2000):

A.	Ocean-Going Freight by Vessel and Cargo Types	B.	Ocean-Going Freight by Vessel and Service Types
1.	Containerships	1.	Liners
2.	Dry bulk vessels		1.1 Containerized
3.	Tankers (liquid bulk)		1.2 Dry bulk/other
4.	Roll-on/roll-off	2.	Tankers
5.	Others	3.	Nonliners
	5.1. Break-bulk		
	5.2. Partial containerships		
	5.3. Barge carriers		
	5.4. Specialized cargo ships.		

Liner vessels as defined under heading B above are vessels operated between scheduled, advertised ports of loading and discharge cargo on a regular basis. In contrast, nonliner cargo vessels do not operate on fixed schedules or itineraries. Both of the breakdowns shown might then be further disaggregated, for example, by distinguishing on the basis of major transoceanic routes or by more exact commodity types (e.g., grains, coal, petroleum). The key point here is that freight planners need to be familiar with a range of definitions and with the attributes that give rise to them when it comes to analyzing the relative merits of different modal freight service types.

With the above in mind, selecting the most appropriate *modal service* may involve a difficult-to-reproduce decision-making process (as more than four decades of freight mode choice modeling can attest to). The most common approach to modal share analysis is to fit some form of discrete choice model, such as a logit model, with mode selection usually based on the least-cost modal option (see McGinnis, 1989; Cambridge Systematics Inc., 1997). One benefit of using logit models is the comparative ease with which they can then be used to create an average transportation cost for all modes selected within a given (i-to-j) transportation corridor. Specifically, if we estimate the probability of selecting mode k from the set of 1, 2, ... K modes available to a specific i-to-j movement, as

$$P_{k/ij} = \exp(-\lambda c_{ijk}) / \sum_k \exp(-\lambda c_{ijk}) \qquad (4.12)$$

then for a given modal cost sensitivity parameter, λ, we have an averaged modal cost of

$$\tilde{c}_{ij} = -1/\lambda \ln(\sum_k \exp(-\lambda c_{ijk})) \qquad (4.13)$$

for use in freight flow models such as those discussed above.

How well such models work is closely tied to the specification of the c_{ijk} cost terms and to the context in which they are applied. McGinnis (1989), reviewing past modal choice models, found that freight rates, service reliability, in-transit time, and condition of the cargo (loss or damage) can all affect mode choice significantly, with responses varying a good deal based on individual shipper, carrier, and commodity characteristics. Recent empirical studies by Wynter (1995), Kawamura (2000), and Wigan et al. (2000) provide examples of the use of logit regression models to quantify responses gathered from shipper or carrier (trucker) stated preference surveys. Such studies offer one means of putting monetary values on such terms as the value of the driver's time (cf. Equation (4.7)). The study by Wigan et al. (2000) also provides quantitative insight into the tradeoffs taking place, from the shipper perspective, between freight rates, transit time, reliability (i.e., proportion of delivery that was late), and damage costs. The very different values placed on travel time by different types of freight delivery service (intercity truckload, intracity truckload, and intracity multiple drop services) demonstrate the need to identify carefully the segment of the freight industry being studied. Given the difficulty of quantifying some of these level-of-service variables and the ways in which they trade off against each other, McGinnis (1989) suggested modeling mode selection as a freight rate minimization problem subject to constraints imposed through the other remaining variables.

Noting the difficulties involved in modeling shipper choice of mode, Roberts et al. (1996) describe the use of a mode choice model based on the *receiver* (customer) of the goods as decision maker. This choice is based on the receiver's total delivered cost, made up of transportation plus logistics plus product purchase cost. This includes the costs to order, ship, load or unload, and store a commodity in transit or on site prior to its use in manufacturing or wholesale operations. While these costs are computed from the shipper perspective, it is the receiver's choice of shipper and preferred shipment size that affect the eventual mode selection. The approach is based on the use of individual shipment records data, taken from such sources as the U.S. Railcar Waybill Sample. This allows detailed modeling of specific transportation options within specific transportation corridors. At this level of analysis it becomes possible to eliminate unlikely modes from consideration due to characteristics of geography, technology, or even carrier policy (e.g., railroads will move as much cargo on rail as possible if a rail siding exists from which to load or unload). Example cost components are presented for truckload, less-than-truckload, long combination vehicle, and private multistop truck transport submodes, as well as from traditional and double-stack rail, Roadrailer, and truck–rail intermodal options. This freight receiver focus meshes well with the growing emphasis being placed on customer demand-driven supply chain analysis, discussed later in this chapter.

Jiang et al. (1999) provide a recent study of freight mode choice in France. They used logit models to estimate the importance of a variety of cost terms on mode selection, while distinguishing between what they call long-term and short-term factors influencing mode choice. Long-term factors include a shipping or receiving firm's type of operation (e.g., a factory, warehouse, shopping center), its size (number of employees), structure (local, national, international), fleet size, geographic location and access to rail branch lines and local highways, and its type of information processing system (reflecting firm logistics practices). Short-term factors include the physical attributes of the good moved (e.g., commodity class, weight, value, packaging), as well as spatial and physical flow attributes (i.e., the origins, destinations, distances, and frequencies of shipments). They produce a *nested logit* model that first selects between for-hire rail, road, and combined or intermodal transportation options, and then they use a log-sum inclusive cost term similar to Equation (4.13) above to model the selection between private and for-hire transport.

Tsamboulas and Kapros (2000) further highlight the complex nature of the decision-making process involved in this mode selection issue, again involving the selection of intermodal (truck–rail, truck–air, truck–water) transportation options. Based on a survey of large shippers, shipping companies and pan-European freight forwarders and road haulers, the study used the technique of factor analysis to first place these companies into three groups based on the importance they each place on a set of 14 decision-making variables. The result was to identify a cost-oriented group with 50% or more of all goods traffic moved intermodally, a quality–cost-oriented group moving from 10 to 50% of their goods intermodally, and a group whose intermodal shares were on the order of 10% or less and whose major criteria for intermodal selection included service reliability, contract duration, and the use of intermodal transportation options for exceptional or unprogrammed shipments. Regression models relating the share of goods moved intermodally were then derived for each group. The selection of the explanatory variables for the study is also of interest, with variables collected under five different headings: transportation cost factors, internal-to-company factors (e.g., commodity types), quality of service factors (reliability, flexibility, safety), external supply side factors (e.g., frequency of rail service), and policy factors (transborder, regional, and local policies).

What all of these modal choice studies demonstrate is the importance in freight planning of understanding the nature not only of the freight, but also of the types of firms and the nature of the geography involved in goods movement. They also indicate that there is not currently, and may not prove to be, a single established method for grouping shippers, carriers, brokers, commodity classes, or types of freight service when it comes to forecasting freight demands.

4.4.4   Converting Tons to Vehicle Loads

Unless the number of vehicular trips is estimated directly, an estimate of the tons or dollars shipped between places may need to be converted into vehicular equivalents for the purpose of assigning traffic to specific infrastructures (routes and terminals). These assignments are an integral component of studies measuring the economic and environmental impacts of traffic volumes on fleet utilization and infra-structure operation and maintenance costs, as well as on delays due to traffic congestion.

Estimating vehicle volumes from aggregate tonnages or dollar values can prove a challenging task, given the variety of vehicle sizes used by each mode of transport and the often variable size of the loads involved. Of particular interest to both freight operators and planners is the percentage of freight carrying capacity devoted to *partial and empty loads*. Often the desire to maximize vehicle carrying capacity runs at odds with required delivery locations and schedules, leading to complex backhauling logistics exercises in order to get the highest productivity out of a vehicle and container fleet, as well as out of the workforce assigned to operate it. A recent estimate of backhauling practices puts empty vehicle miles at 15 to 50% of all truck miles operated in the United States over the course of a year (BTS, 2001).

Interest in the vehicle load problem also stems in the case of trucking from the differential impacts that truck loads of different sizes can have on highway maintenance costs. A topic of considerable public policy interest in the United States, both within and outside the freight industry, is the effect of truck size and weight regulations on safety, modal competition, and freight industry productivity (see Hewitt et al., 1999; FHWA, 2000). Fischer et al. (2000) describe the sort of data sources and data manipulations currently required to convert a set of annual O-D-specific commodity flow estimates (generated by an interregional input–output model) to route-specific daily truck traffic counts in southern California. They used data from both weigh-in-motion (WIM) stations and cordon counts to obtain averaged weekday truck counts by major highway and vehicle axle configuration. To convert from tons to trucks, these axle configurations were then mapped into loaded vehicle weight classes using data from the Vehicle Inventory and Use Survey (U.S. Census Bureau, 1997b). The result is a set of truck trip matrices, including empty trucks, suitable for use in traffic-to-route assignment models. Similarly involved steps are usually required to convert tons of cargo, or numbers of container units, into train, barge, ship, and aircraft loads.

4.4.5   Freight Traffic Assignment Models

Traffic assignment is the term used for allocating vehicle traffic volumes to specific transportation routes. When dealing with multiple origins and destinations, past freight assignments have used a variety of models, including simple all-or-nothing assignments of traffic to a single least cost, least travel time, or shortest distance route, as well as logit-based and other nonlinear programming forms of multipath assignment. There are also significant differences in the ways in which shipments are routed by different modes.

4.4.5.1   Truck Traffic Assignments

To date most metropolitan planning agencies, in the United States and elsewhere, have dealt for the most part with urban truck movements when engaged in freight simulation modeling. In doing so, they have adopted simplifying assumptions that associate trucks in specific size classes with passenger car equivalents (PCEs). For example, a large, single-unit truck may be treated as 1.5 PCEs, while a semitrailer may represent 3 PCEs in terms of its impacts on highway traffic speeds. The principal interest is usually in the effects of freight traffic on roadway damage and replacement costs, highway and neighborhood safety, and traffic congestion. The popular Wardrop equilibrium assignment model is the one most often applied to passenger and mixed passenger and freight highway traffic (see Southworth et al., 1983). Under this approach all routes used between any origin–destination pair have the same travel cost (in terms of travel time or a more generalized cost function), while unused routes are those which cost more. Logit route choice models may also be used for modeling a smaller number of origin–destination pairs, or where the number of available routes through the physical transportation network is limited and readily identifiable. For intercity and long-distance transport, a single least-cost routing model may be appropriate in many cases, and given the strong preference drivers have for interstate routes. Validating the results of such assignments can be problematic, however, given that available, route-specific traffic count data contain a mix of local as well as through traffic and do not always capture the type (e.g., number of axles) of trucks involved (see Black, 1997). Given the multistop nature of a good deal of truck transportation, especially within urban areas, once such routes have been selected the analyst may still have to resort to cost adjusting trip circuity factors (Southworth, 1982, 1983) to the modeling of multistop trip chains (Holguin-Veras and Thorson, 2002), or to microsimulation of individual multistop shipments in order to come up with representative average costs that can be input to traditional highway transportation planning (mode and destination choice) models.

The ability to validate a model generated set of truck traffic assignments depends on the availability of sufficient truck traffic counts against which to compare the resulting model flows. A benefit of the growing volume of truck traffic count data being collected both within and between U.S. cities is not only its potential for examining the resulting model generated assignments, but also its usefulness for adjusting the O-D commodity flow matrices used to generate them. This usually requires a mathematical programming solution. See List and Turnquist (1994) and List et al. (2001) for example applications of such a procedure, often termed *link O-D* modeling in the transportation planning literature.

Rail and intermodal route selection adds further technical challenges. The costs of long distance, notably east–west rail transportation in the United States, often involves the costs of *interlining* between different, privately owned railroad company tracks. With railroads traditionally wanting to keep as much traffic on their own tracks as possible, additional routing subtleties here include recognition of *trackage rights* agreements (by which one railroad allows rolling stock from another railroad to use its tracks) and the presence of small but numerous (around 500 currently) *gateway railroads* that operate short connector lines between the main lines of the major railroad companies. Logit models can again be used to select between a limited number of rail routes serving major intercity corridors. Where intermodal transportation is involved the costs of transferring freight at intermodal terminals becomes a significant element in route choice. Figure 4.2 shows an example truck–rail–truck intermodal route highlighting the various links that may be involved in computing the resource costs associated with traversing a real-world transportation network (see Southworth and Peterson, 2001).

In the case of *air transport*, a large volume of freight travels as belly freight within passenger aircraft. It is therefore important to distinguish this sort of freight from that moved by dedicated airfreight carriers such as Federal Express, United Parcel Service, DHL, and Airborne. It is also important to distinguish the type of delivery service being offered, e.g., overnight vs. 3-day vs. lower priority deliveries. Today a significant amount of this airfreight is in the form of mail and small package delivery. Traffic assignment models must recognize the hub-and-spoke nature of each air carrier's operations in computing likely shipment destinations as well as costs. Most O-D shipments involving air transport involve a truck trip at the beginning or end of the trip (as do many waterway shipments), making them intermodal shipments. Ground access for trucks is therefore an important planning issue around busy hub airports.

For most cases of *inland waterborne commerce* there is rarely more than one route to choose from. Here the principal issue in estimating movement volumes is computation of the potentially significant delays at locks, and the impacts of these delays on fuel consumption and other inland barge operating costs (see Bronzini et al., 1997). Transoceanic and intracoastal shipping concerns often involve channel depths as well as land-side seaport access concerns, both of which may affect port selection, which in turn determines a typically multimodal land–sea pattern of O-D shipments (TRB, 1992). Similar issues affect utilization of the North American Great Lakes. Economies of scale are important here. Sometimes a channel depth increase of just 1 or 2 in. can mean hundreds of thousands of additional tons transportable on a single deep-sea or lakewise vessel.

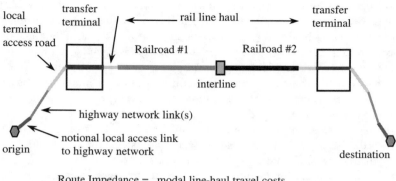

Route Impedance = modal line-haul travel costs
+ intra-terminal transfer costs
+ inter-carrier (interlining) costs
+ local network access and egress costs
+ network-to-terminal local access costs

FIGURE 4.2 Example components of an intermodal (truck–rail–truck) route.

Where empty backhauling of freight is involved, as is often the case, estimating route-specific costs can be problematic, no matter which mode we are dealing with. In such cases it may not be so easy to identify what is meant by a true round-trip cost. Even then the problem of route selection for a specific cargo may be influenced as much by carrier as by shipper or receiver logistics. Modeling load consolidations to maximize carrier productivity (or profit) can play a significant role here. More generally, models that simulate the interplay between carriers and shippers and its ultimate expression as flows on networks are complex, and currently in their early stages of development (Friesz et al., 1985; Harker and Friesz, 1986). Network-based simulation software now exists with which to develop strategic multimodal as well as multiproduct freight traffic assignments, capable of incorporating traffic congestion effects for the modes involved (Crainic et al., 1990; Guelat et al., 1990). Nevertheless, this remains an important and challenging area for further research.

4.5   Freight Supply: Capacity Issues

In general terms *freight capacity* is a measure of the volume of freight that a particular facility, organization, or system can handle during a given time period. That is, it is the number of vehicles, containers, or tons of cargo that can be moved successfully from point to point during that time. The rapid growth in freight traffic on all major modes of transportation has brought the issue of adequate freight system capacity to the attention of national governments, and is currently a major public policy concern in the United States. The key policy issue associated with capacity is whether there is enough of it. These capacity evaluations are usually based on in-transit speeds and any associated traffic delays that result when the volume of traffic passing through a facility approaches or exceeds its designed capacity. In highway engineering, for example, this capacity is reached when a facility's volume-to-capacity ratio (*v/c ratio*) reaches or exceeds 1.0, with different levels of service associated with different ratios (TRB, 2000).

There are now a number of technical references available to help freight planners and engineers compute mode-specific link, intersection, and terminal capacities. Detailed example formulas and references covering each mode, including terminal, airport, and seaport handling capacities, can be found in NCHRP Report 399 (Cambridge Systematics Inc., 1998). This manual also describes the extension of capacity concepts to the transportation corridor level, with an emphasis on the multimodal nature of many freight, passenger, and mixed freight–passenger corridors, involving both parallel and intermodal operations. Recalling Figure 4.2 above, each link and node on the intermodal route shown has an upper limit on its practical carrying capacity. The element with the lowest maximum throughput therefore determines the capacity of the route. Where two or more routes exist in parallel, the capacity of the corridor is the sum of these routes. When different routes share common links, intersections, or terminals, corridor capacity may be a little more challenging to compute, although the solution can be found even in quite complex multiroute cases (as in dense highway networks or within complex intraterminal facilities) by using the well-known maximum flow–minimum cut problem described in operations research textbooks (see also Leighton and Rao, 1999).

The problem with all of the above measures of facility- or corridor-specific capacity, as pointed out by Morlok and Riddle (1999), is that they cannot simply be added up to provide a measure *of freight system capacity*. Where traffic volumes exceed the designed traffic handling capacity of specific facilities, such locations are termed *traffic bottlenecks*. These bottlenecks occur when a high volume of traffic, typically from a number of different origins and destinations, converges on a limited geographic area, resulting in costly traffic delays. Often, these delays occur not on the line-haul portion of a trip but at a traffic terminal such as an airport or seaport, or at some other form of transfer point, such as a railcar switching yard. Effective analysis requires an approach to impacts assessment that looks at all of the major freight flows entering and leaving the region in question, as well as at the condition of the alternative routes available to such freight — not simply at the traffic conditions and capacity immediately surrounding the so-called bottleneck.

A key input to such freight system capacity studies is therefore the development of good aggregate, origin-to-destination, mode- and route-specific freight traffic forecasts. Other required inputs (Morlok and Riddle, 1999) are (1) a consistent measure of transportation system output, and (2) a method for recognizing all of the principal resource limitations on system performance, including vehicle fleet, labor pool, and fuel concerns, as well as constraints on physical infrastructures. Despite some early treatment of this issue, this remains an area in need of additional research and development. Morlok and Riddle (1999) offer a promising approach. They use a reasonably generic mathematical programming formulation to define the potential maximum system output in terms of the cargo moved over a set of origin-to-destination-specific *traffic lanes*. These outputs are subject to physical capacity constraints on transportation network links, terminals, vehicle and container loads, and fleet size. They are also optionally subject to a minimum level of service standards. Finally, plans for supplying effective or practical freight handling capacity must also incorporate environmental and safety constraints, and be able to do so in a quantifiable manner (see Nijkamp et al. (1993) and discussion later).

A good deal of the traffic congestion experienced by today's freight transportation systems occurs at or near *freight terminals*, notably around large airports, seaports, and truck-to-rail and truck-to-truck transfer terminals (including large break-bulk terminals where long-haul freight carried on large, single-, double-, and triple-trailer trucks is transferred to or from a number of smaller, typically intraurban panel or other small-capacity trucks). Planning new terminal sites is therefore a congestion-sensitive issue, requiring good freight generation and attraction forecasts. It also requires careful analysis of current terminal operations. Here the unique nature of many of these operations, and the difficulty of collecting data about them, makes it difficult to develop quantitative models of their operations.

The U.S. Department of Transportation has recently taken steps to come to grips with this idea of transportation system capacity, funding the development of a prototype multimodal freight bottleneck analysis tool for measuring the extent of congestion-induced traffic delays around major transportation terminals (ORNL, 2000). This study adopted the three-stage congestion assessment model shown in Figure 4.3.

The three stages are (1) terminal access, (2) within terminal operations, and (3) terminal egress. Where seaports and airports are concerned, these stages take the form of land-side truck or rail access or egress, within-terminal operations, and port-side (i.e., waterside or air-side) access or egress. Delays at each stage in transferring freight between modes can have a number of different causes. Land-side access and egress delays often result because there is not enough local road space to accommodate large volumes of peak-period truck traffic, especially if this traffic is competing for this same road space with high volumes of passenger traffic. Within-terminal operations include storage as well as throughput, requiring timely

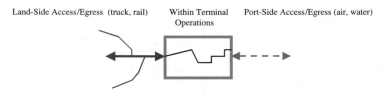

FIGURE 4.3   Three-stage terminal congestion analysis concept.

access to the often limited number of cranes and pieces of handling equipment required to identify, unstack, and load or unload vehicles. Air-side and ocean-side access and egress problems similarly result from too few gates or berths at which to load or unload craft, sometimes causing long delays during peak shipping seasons as well as at certain times of the day. At seaports an additional source of delay results from physical restrictions on channel depths, which in turn limit vessel size, and hence aggregate cargo throughput. At airports the number, length, and arrangement of runways impacts takeoff and landing times during busy periods. Ideally, we could sum up each of the above causes of delay to get a more accurate measure of total delays within a terminal area. However, such simulations taken down to a more operational level can soon become quite complex multistep processes (see Weigel (1994) for a railroad intermodal facility example). They can also become quite data intensive.

A key feature of the ORNL analysis tool is its use of Geographic Information Systems (GIS) technology to not only map the location of such bottlenecks, but also assist in the identification of where the traffic causing and impacted by such bottlenecks originates and terminates. This in turn has the potential to improve the economic evaluation of bottleneck mitigation measures, recent examples of which in the United States have led hundreds of millions of dollars to be invested in infrastructure improvements (see Port of Long Beach, 1994; Cambridge Systematics Inc., 1998).

An important planning issue related to the above concerns over congestion is the siting of freight terminals. Segregation and consolidation of freight transfer activities has long been an economic as well as environmental planning issue. The goal here is provision of good access to major highways, rivers, airports, and rail or pipelines, while dealing with issues of environmental compliance (air pollution, ground water pollution, noise, etc.) and passengers' and local residents' safety. Past solutions include the use of designated industrial parks, possibly encouraged by the creation of specially empowered free trade zones. An interesting recent concept is that of geographically segregated *global freight villages,* formed by clustering a number of industrial–intermodal distributional and logistical companies within a secure perimeter. These villages would usually be located on the outskirts of an urbanized area and, where possible, serve the purpose of rejuvenating abandoned and unsightly urban "brownfield" sites: hence gaining additional environmental cleanup benefits in the process (Weisbrod et al., 2002). These strategies for separating concentrated freight activity from other land uses are often in response to the considerable problems currently experienced within large central business districts and suburban activity centers, where on-street double parking of trucks and poorly accessed and underdeveloped receiving docks and freight elevators result in highly inefficient freight handling practices (for an example, see Morris and Kornhauser, 2000).

4.6   Freight Productivity and Performance

The more cost-effective a region's freight movement system, the better it is for business. Knowing how well the freight sector is doing, and whether it is becoming more or less efficient over time, is of considerable interest to government officials involved with transportation and commerce. How to measure this performance is not always clear, and no single measure may suffice. Today measures of industrial sector performance are often discussed in terms of that sector's *productivity,* i.e., how much it produces at what cost in terms of the resources required to do so. In practice, both performance and productivity can mean different things to different people. At the fully national level, the U.S. Bureau of Transportation

Statistics reports annually on the performance of the nation's transportation system: the miles traveled, the number of people and tons of freight moved, times spent in transit, percentage of on-time service, time lost to traffic delays, and the associated energy, safety, environmental, and economic costs and benefits (BTS, 2001). Productivity measures can be seen here as offering a subset of performance indicators that tie measures of output directly to measures on input (i.e., productivity = output/input). As such, they include measures of something per something, such as ton-miles moved per dollar or per employee, operating costs or fuel consumption per ton-mile, and percent of truckloads or vehicle miles involving empty moves.

A recent study by Hagler Bailly Services (2000) recommends the following six measures as useful indicators of highway freight system performance: (1) costs of highway freight per ton-mile, (2) cargo insurance rates, (3) point-to-point travel times and hours of delay per 1000 vehicle miles on selected freight-significant highways, (4) crossing times at international borders, (5) condition of connectors between the National Highway System and intermodal terminals, and (6) customer satisfaction. While different performance measures may better apply to more localized applications, this list of indicators also serves as a useful pointer to some of today's more pressing freight-related policy questions. A notable member of this list is the customer satisfaction item, involving the polling of those shippers, receivers, and carriers operating on the nation's highways. With traditionally low response rates to survey instruments, approaches to gathering such information are now being actively researched (see Lawson and Riis, 2001). Such surveys can be an important source of information on system- or corridor-specific productivity and performance, as demonstrated by Middendorf and Bronzini (1994). In general, freight industry productivity measures are likely to be most useful when they mirror the measures used by individual carriers and shippers in determining where best to put their individual fleet, labor, operating capital, and other resources.

In terms of improving freight system productivity, a good deal is expected from the adoption of state-of-the-art information technology. The more both suppliers and their potential customers know about available product inventories, including in-transit inventories, the more effectively demands for goods ought to be met. The next generation of computer-based decision support tools will likely tie in-vehicle tracking systems even more closely to the latest shifts in customer demands. They will do this by accessing real-time information on the traffic conditions between customers and available fleet vehicles, and by tracking cargo, container, and vehicle status (e.g., percent empty), allowing rescheduling of vehicle pickups and deliveries "on the fly." Matching vehicle and container sizes to load sizes is one important component of this logistics problem, with the potential for a given cargo to either *weigh-out* or *cube-out* a vehicle. (With aircraft, in particular, there is also the problem of balancing the load properly.) Computer software already exists to help shippers and carriers with this sort of fleet utilization problem. Among the most common algorithms in use today are codes that solve dynamic versions of the well-known "traveling salesman problem" (TSP) and its extensions. These algorithms offer rapidly solved heuristic methods for allocating one or more vehicles to a set of predetermined pickup or drop destinations — at the least overall transportation cost (see, for example, Reinelt, 1994). Such algorithms, containing various degrees of sophistication, can now be found within a range of spatial decision support software. Interactive mapping of vehicle routes tied to rapidly solved TSP problems seems to be a worthwhile near-term research problem with many interesting and potentially cost-saving variations.

4.7 Freight Impacts: Safety and Environmental Issues

Moving freight can be a dirty, noisy, and sometimes dangerous business. The operation of large and heavy vehicles, containers, cranes, and other freight handling equipment poses a variety of work safety problems, including the potential for accidents and exposure to hazardous materials. The fact that freight usually moves most efficiently if delivered in large-capacity vehicles also means that where freight and passenger traffic interact there is the potential for serious traffic accidents should something go wrong with either type of vehicle.

Special conditions apply when moving flammable, combustible, radiological, or otherwise hazardous materials. In the United States and many other countries, strict and detailed regulations exist to control the movement of these substances, with planning studies paying particular attention to the *populations at risk* along hazardous materials routes (Raj and Pritchard, 2000; Hancock, 2001; Hwang et al., 2001). Cleanup costs resulting from hazardous materials spills can be costly and include environmental as well as health-related damages (Abkowitz et al., 2001). GIS software can prove especially useful here, helping to not only map but also efficiently compute the number of people at risk at different distances from a proposed shipment's route (see Frank et al., 2000). Recent events in the United States have also focused government attention squarely on the security of freight transportation systems, in particular on transportation infrastructure assurance. See RSPA (1999) for a discussion of some industry best practices.

Freight planning must also pay due attention to the safety and environmental implications of selecting specific packages, vehicles, and times of delivery. Freight planners must understand these issues and be able to evaluate them as part of the existing regulatory and planning process. Significantly increased public agency concern for the environment during the second half of the 20th century means that most projects involving either the construction or relocation of freight infrastructure must undergo some form of (more or less formal) environmental impact assessment. A comprehensive analysis of freight's environmental impacts on society would include measuring the fossil fuel consumed and the production of a number of health-impacting mobile source emissions (including greenhouse gases), as well as any concerns over land consumption, groundwater runoff, and noise pollution resulting from geographically concentrated freight activity. Ports, freight terminals, and large warehousing complexes are prime sites for study. There is now considerable literature on freight transportation safety, energy, and environmental costs, a good deal of it published by government sources (EPA, 1996; FHWA, 1998; FRA, 1999; USCG, 1999).

At the same time, it must also be noted that coming up with accurate impact measures for specific situations remains a very challenging area of research. The devil here, as it is often said, is in the details of each case study. In particular, the author knows of no well-established set of standards or methods for comparing the energy consumption and environmental impacts across different freight modes that have been shown to be applicable under a wide range of conditions. While a number of studies have been carried out (e.g., Newstrand, 1991; Vanek and Morlok, 1998, 2000; Tolliver and Earth Tech Environment & Infrastructure, 2000), great care needs to be taken (1) to capture all of the relevant stages involved in each freight shipment or aggregate commodity movement, not just the line-haul portions of shipments, and (2) to identify the specific freight technologies being used (e.g., unit train vs. traditional rail). Effective safety and environmental analysis will also usually require commodity, O-D, mode (vehicle configuration), and route specific analyses of freight movements. Add in variability caused by different climatic and other conditions of physical geography and this becomes a difficult and challenging area for further research.

As in other areas of transportation and society, some researchers have begun to examine these issues of safety and environment in a more holistic sense, and to ask what we ought to be planning for in terms of *sustainable* freight transportation systems. The key issue here is what it costs society in terms of accidents, health care, land, and fuel consumption to operate freight systems — and what planners and engineers can do to limit these often negative externalities of freight transportation while still moving goods in a speedy and cost-effective manner. Rodrigues et al. (2001) point out the difficulties caused by some recent and widespread developments in freight logistics. These include the adoption of activity concentrating hub-and-spoke distribution systems for air, rail, and waterborne transportation, and the rise of faster, JIT (vs. warehousing based) services, using high-energy consuming modes such as truck and air to carry a wide variety of low-weight, high-valued goods. Matthews et al. (2001) carry out an interesting comparison in this regard, between traditional and e-commerce book retailing. They conclude that the latter can, under appropriate circumstances, be environmentally beneficial. Much depends on the specific nature of the supply chain linking the producer to the final customer (a topic we return to below). Richardson (2001) also discusses this issue of sustainable transportation with respect to trucking. In the same journal issue, Chatterjee et al. (2001) provide an example impacts study using established models for computing truck congestion, safety, and air quality impacts.

4.8   Some Future Research Directions

There are many valuable directions in which to move current freight planning models, methods, and data collection efforts. Of particular interest are issues tied to the use of JIT services, e-commerce, integrated supply chain management, and any other trends linked to the use of real-time information technologies.

4.8.1   Implications of JIT Delivery

Here there are currently far more questions than answers. Will tighter delivery schedules mean greater reliance on truck and airfreight transportation, and therefore the need for greater investments in our highway and air transportation systems? How will fuel consumption in the freight sector be impacted? Will we see more trucks entering residential neighborhoods? To what extent can real-time traffic rerouting and fleet management software be developed to minimize miles traveled and costs incurred? How flexible will firms become in their contracting for product deliveries on the basis of least offered price? How reliable are such e-business dealings in terms of on-time physical product deliveries, and how can private firms as well as governments develop measures that will help them to identify best freight practices? To what extent will these practices involve multishipper and multicarrier coordination of pickups and deliveries through 3PL and 4PL logistics firms, and what economies of scale and opportunities for reducing total vehicle miles traveled exist in such relationships? These are just some of the questions freight planners need to be asking at the present time.

4.8.2   Demand-Driven Product Supply Chains

Just as the personal transportation planning literature has evolved to consider travel within the broader context of household activity scheduling (see other chapters in this handbook), so too must freight modeling place the demands for goods deliveries within the broader context of transportation-plus-other business logistics. Whether using the services of freight brokers or running their own logistics business, many freight producers as well as customers now see themselves as part of a multistage *product supply chain*. Such supply chains at their most general embody all of the activities involved in satisfying an end-user demand for a product. This includes the extraction, manufacturing, transportation, and retailing of the product and its receipt by a final consumer (usually a household or a company). The movement of goods is often required at more than one step in such a supply chain, such as shipment from a mine or farm to a production site, from the production site to a processing plant, from the processing plant to a wholesale distribution or retail center, and from the retail store to the final customer. Figure 4.4 shows a number of different ways that a product may need to be transported through a supply chain (and the different ways that it may be ordered electronically). One anticipated impact of the Internet and e-commerce is the ability of customers to bypass retailers and other intermediaries and deal directly with wholesalers or even producers through web-based product ordering systems. In not much over a single decade this sort of business-to-business (B-to-B) e-commerce has become a multibillion dollar industry, while business-to-customer (household) (B-to-C) and customer-to-customer (C-to-C) direct deliveries have also grown substantially (Golob and Regan, 2000).

Forecasting freight demands is unlikely to be successful unless some understanding of these rapidly evolving supply chain logistics is built into the planning process. Whether this means adapting the traditional multistep planning model represented by Figure 4.1 or evolving entirely new forms of freight demand–supply balancing models remains to be seen.

Two recent examples of freight modeling incorporating supply chains are the GoodTrip model, developed by Boarkamps et al. (2000), and the multilevel spatial price equilibrium modeling of Nagurney

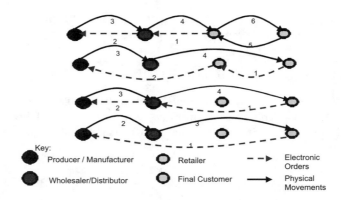

FIGURE 4.4 Example freight and e-commerce supply chains. (Based on Southworth, F., The Digital Economy: Changing the Shape of Transportation Workshop on the New Digital Economy on Transportation, National Academy of Sciences, Washington, D.C., September 14–15, 2000.)

et al. (2002). Both approaches place a strong emphasis on using network-based models to carry out their analyses, and both can be made to handle multimodal transportation systems. Both approaches also place an emphasis on understanding the decision-making roles of shippers, carriers, and receivers of goods. The GoodTrip model can be viewed as an expanded version of the five-step planning model shown in Figure 4.1; it separates out freight productions and attractions by identifying the actors and transactional stages involved in potentially multistep product supply chains of the types shown in Figure 4.4 before computing mode, route, vehicle loading, and destination choice. In doing so, it adds significantly to the behavioral content of the model. Significantly, while freight is moved in these supply chains from producers, through distribution centers, retailers, or other intermediaries, to final customers, these flows are assumed to be *customer demand driven.*

The methodology proposed by Nagurney et al. (2002) reflects this same viewpoint. The various components of a product's supply chain in this case are modeled as a network of commodity flows, information flows, and associated purchase prices that are faced by a particular industry. Given an initial set of customer demands, a set of supplier and retailer or distributor production functions, and a set of shipper or carrier transaction costs for each physical movement component in the supply chain, commodity flows and prices are endogenously solved for and iterated to a form of spatial and economic supply–demand equilibrium *that operates across all of the various supply chains' participants.* This creates a dynamic supply chain model in which *transaction costs* include not only the costs of physically moving the freight, but also the costs of negotiating a price for deliveries and the costs of collecting and using information on available inventories, prices, and supplier or market locations. While still a prototype, this sort of increasingly complex interaction modeling sets the stage for spatial price equilibrium models that can eventually trace the effects of incomplete market information, as well as costly traffic congestion, back to the price paid and the markets selected for goods deliveries. Figure 4.5 shows this sort of analysis as a multilevel network modeling exercise, in which the actual movement of freight over networks is one of four sets of distinctly different, but functionally connected, aspects of freight business logistics. Such ideas and the models to support them are currently in their formative stages. They challenge the limited behavioral content of current freight planning models, but are likely to require additional, or at least new, forms of data collection to become fully realized.

4.8.3   Intelligent Freight Systems and Public–Private Agency Cooperation

To be most successful and universally adoptable, IT-based carrier and shipper solutions to low-cost freight transportation will probably need to be linked to (currently early-stage) publicly funded traffic monitoring and reporting systems. This suggests the need for greater coordination between the private and public sector participants in freight movement, something that has often proved difficult in the past, in

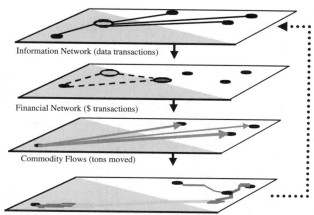

Information Network (data transactions)

Financial Network ($ transactions)

Commodity Flows (tons moved)

Vehicle Routings (traffic counts on multimodal transportation networks)

FIGURE 4.5 Multilevel network analysis of freight logistics.

large part due to the value placed by private companies on the proprietary nature of their customer and product markets and the costs of serving them. However this relationship evolves, freight planners and engineers will need a working knowledge of how individual companies make use of available IT-based technologies for vehicle and product tracking, since these technologies will increasingly affect the when, how, and where freight is moved over specific infrastructures — infrastructures that may themselves eventually become "smart" (in the sense of automation and self-regulation) about handling either dedicated freight or mixed passenger–freight traffic volumes.

4.8.4 Microsimulation of Freight Movements

As with passenger modeling, the opportunity afforded by high-speed computing, coupled with the growing volumes of Internet and other source data on freight costs and volumes, suggests the possibility of simulating freight movements on a shipment-by-shipment or trip-by-trip basis — subsequently cumulating the resulting trips to obtain estimate aggregate freight movement volumes, modal shares, etc.

This *microsimulation modeling* is now being researched as part of at least one ongoing research project (Hancock et al., 2001). Figure 4.6 shows the sort of step-by-step modeling and data needed to simulate and build up a set of freight. Of considerable interest for the future is the ability to create and combine such simulations with real-time product ordering and traffic scheduling information. In the near term this sort of modeling offers some interesting research challenges for analyzing movements within a single company, single supply chain, or single industrial sector.

4.9 Closing Remarks

Two final observations seem useful at this point. First, a conscious effort was made in preparing this chapter to avoid a more traditional mode-by-mode discussion of freight transportation issues. While most freight movement remains dominated by one mode or another, depending in large part on commodity type, it may take a number of modes to get a piece of raw material from its source to its eventual utilization as part of a finished product.

The attention over the past 15 years given to multistage product supply chains has served to reinforce this point. With the emergence of outsourcing of freight and related business logistics, including the rise of 3PL and 4PL service providers, an increasing volume of freight is being handled by organizations not tied to a single mode, but rather driven by the requirements for a particular type of delivery service. Through these companies and the more forward-looking freight producers and consumers, freight

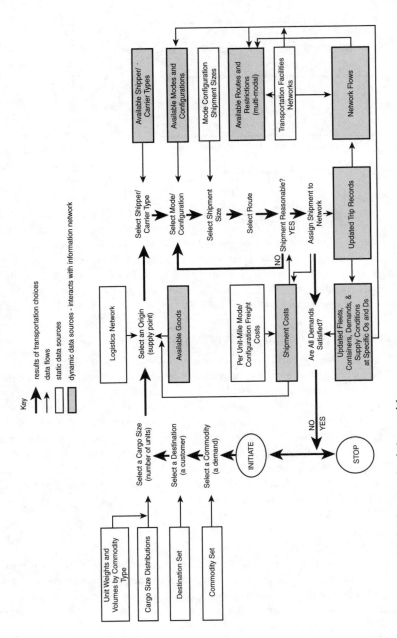

FIGURE 4.6 Example freight microsimulation model.

logistics is now recognized as an integral aspect of most business ventures. It may not be a high priority concern in all industries, but it can be ignored in very few.

Finally, in looking back over the above chapter, the reader will note that in each section there is a brief discussion of, or allusion to, the problems of analyzing freight practices due to data limitations. With the growing realization that cost-effective freight movement is a must in the new global economy, this freight data issue has finally been placed on center stage by a concerned urban and regional planning community (see Meyburg and Mbwana, 2002). Most of these limitations manifest themselves when freight is being analyzed from the perspective of public policy, where a good deal of information is required from a typically large number of companies. Significant concerns over the use of proprietary data are a major reason for this. The sheer variety of ways in which this data could be categorized, and the different data collection and naming conventions already used by the various, typically mode-specific, data collection agencies, can further confound many potentially useful data sets — especially so when international studies involve combining data from different counties. Definitional problems can also arise when trying to compare, in particular, productivity indices across different modes of transportation. The growth of the Internet and the emergence of large 3PL and 4PL service providers means that large volumes of data are being brought together in many new and different ways for individual customers. Yet the very emergence of these out-sourced freight logistics firms is making it difficult to collect even traditional forms of freight data from shippers and carriers who are no longer investing in within-company expertise on shipping costs and logistical practices. As Bronzini (2001) points out, new ways of collecting freight data need to be explored, including less obtrusive data gathering methods focused on administrative records and remote sensing. We cannot, it seems, rely on shipper and carrier surveys alone. This is a very worthwhile area for freight engineers and planners to focus their attention on over the next few years, paying particular attention to the speed at which new data collection and real-time information technologies are becoming available. These same engineers and planners should also make a concerted effort to understand the everyday logistics problems faced by individual carriers and shippers. The single-company vehicle selection, fleet routing, and market targeting strategies that produce the daily on-the-ground (and through the sky) freight movements are after all those that ultimately produce the aggregate freight movement volumes and freight industry productivity levels that planners use to measure progress at the societal level.

Acknowledgments

This work was supported in part by National Science Foundation Grant CMS-0085720: Enterprise-Wide Simulation and Analytic Modeling of Comprehensive Freight Movements.

References

Abkowitz, M.D. et al., Assessing the economic effect of incidents involving truck transport on hazardous materials, *Transp. Res. Rec.*, 1763, 125–129, 2001.

Al-Deek, H.M. et al., Truck trip generation for seaports with container/trailer operations, *Transp. Res. Rec.*, 1719, 1–9, 2000.

Allen, W.B., Mahmoud, M.M., and McNeil, D., The importance of time in transit and reliability of transit time for shippers, receivers and carriers, *Transp. Res. B*, 198, 447–456, 1985.

Ballis, A. and Stathopoulos, A., Innovative transshipment technologies investigated for implementation in seaports and barge terminals, Paper 02-3030, presented at the Transportation Research Board Annual Meeting, Washington, D.C., 2002.

Black, W.R., *Transport Flows in the State of Indiana: Commodity Database Development and Traffic Assignment: Phase 2*, Transportation Research Center, Indiana University, Bloomington, 1997.

Boarkamps, J.H.K., van Binsbergen, A.J., and Bovy, P.H.L., Modeling behavioral aspects of urban freight movements in supply chains, *Transp. Res. Rec.* 1725, 17–25, 2000.

Bomba, M.S., Harrison, R., and Walton, C.M., Planning for megacontainerships: statewide transportation planning approach, *Transp. Res. Rec.*, 1777, 129–137, 2001.

Brewer, A.M., Button, K.J., and Hensher, D.A., Eds., *Handbook of Logistics and Supply Chain Management*, Pergamon Press, New York, 2001.

Bronzini, M.S., Supply chain statistics, in *Handbook of Logistics and Supply Chain Management*, Brewer, A.M., Button, K.J., and Hensher, D.A., Eds., Pergamon Press, New York, 2001, chap. 33.

Bronzini, M.S. et al., Ohio River navigation investment model: Requirements and model design, *Transp. Res. Rec.*, 1620, 17–26, 1997.

Bureau of Transportation Statistics, Transportation Statistics Annual Report 2000, BTS, U.S. Department of Transportation, Washington, D.C., 2001.

Cambridge Systematics Inc., Characteristics and Changes in Freight Transportation Demand: A Guidebook for Planners and Policy Analysts, NCHRP Report 8-30, Transportation Research Board, National Academy Press, Washington, D.C., 1995.

Cambridge Systematics Inc., Quick Response Freight Manual, Report DOT-T-97-10, prepared for the U.S. Department of Transportation and U.S. Environmental Protection Agency, Washington, D.C., 1997.

Cambridge Systematics Inc., Multimodal Corridor and Capacity Analysis Manual, NCHRP Report 399, Transportation Research Board, National Academy Press, Washington, D.C., 1998.

Chatterjee, A. et al., Effect of increased truck traffic on Chickamauga Lock, *Transp. Res. Rec.*, 1763, 80–84, 2001.

Crainic, T.G. et al., Strategic planning of freight transportation: STAN, an interactive-graphic system, *Transp. Res. Rec.*, 1283, 97–124, 1990.

Environmental Protection Agency, Indicators of the Environmental Impacts of Transportation: Highway, Rail, Aviation and Marine Transport, EPA 230-R-96-009, Office of Policy, Planning and Evaluation, U.S. EPA, Washington, D.C., 1996.

Federal Highway Administration, Large Truck Crash Profile: The 1997 National Picture, Analysis Division HIA-20, Office of Motor Carriers, FHWA, U.S. Department of Transportation, Washington, D.C., 1998.

Federal Highway Administration, Comprehensive Truck Size and Weight Study, FHWA-PL-002-029 (multiple volumes), FHWA, Washington, D.C., 2000.

Federal Railroad Administration, *Railroad Accident Incident Reporting System*, FRA, U.S. Department of Transportation, Washington, D.C., 1999.

Fischer, M.J. and Han, M., Truck Trip Generation Data, Synthesis 298, National Cooperative Highway Research Program, Transportation Research Board, National Academy Press, Washington, D.C., 2001.

Fischer, M., Ang-Olson, J., and La, A., External urban truck trips based on commodity flows, *Transp. Res. Rec.*, 1707, 73–80, 2000.

Frank, W.C., Thill, J.-C., and Batta, R., Spatial decision support for hazardous materials truck routing, *Transp. Res. C*, 8, 337–359, 2000.

Friesz, T.L., Viton, P.A., and Tobin, R.L., Economic and computational aspects of freight network equilibrium models: A synthesis, *J. Reg. Sci.*, 25, 29–49, 1985.

Golob, T.F. and Regan, A.C., Impacts of information technology on personal travel and commercial vehicle operations: Research challenges and opportunities, *Transp. Res. C*, 9, 87–121, 2000.

Guelat, J., Florian, M., and Crainic, T.G., A multimodal multiproduct network assignment model for strategic planning of freight flows, *Transp. Sci.*, 24, 25–39, 1990.

Hagler Bailly Services, Inc., Measuring Improvements in Highway and Intermodal Freight, Report DTFH61-97-C-00010 to the Federal Highway Administration, Washington, D.C., 2000.

Hancock, K.L., Hazardous goods, in *Handbook of Logistics and Supply Chain Management*, Brewer, A.M., Button, K.J., Hensher, D.A., Eds., Pergamon Press, New York, 2001, chap. 31.

Hancock, K., Nagurney, A., and Southworth, F., Enterprise-wide simulation and analytic modeling of comprehensive freight movements, National Science Foundation Workshop on Engineering the Transportation Industries, Washington, D.C., August 13–14, 2001.

Harker, P.T. and Friesz, T.L., Prediction of intercity freight flows, I: Theory and II: Mathematical formulation, *Transp. Res. B*, 20, 139–174, 1986.

Hewitt, J. et al., Infrastructure and economic impact of changes in truck weight: Regulations in Montana, *Transp. Res. Rec.*, 1653, 42–51, 1999.

Hilliard, M.R. et al., *Potential Effects of the Digital Economy on Transportation*, Oak Ridge National Laboratory, Oak Ridge, TN, 2000.

Holguin-Veras, J. and Lopez-Genao, Y., Truck Trip Generation at Container Terminals: Results from a Nationwide Survey, paper presented at the Transportation Research Board Annual Meeting, Washington, D.C., 2002.

Holguin-Veras, J. and Thorson, E., Trip length distributions in commodity-based and trip-based freight demand modeling, *Transp. Res. Rec.*, 1707, 37–48, 2000.

Holguin-Veras, J. and Thorson, E., Modeling commercial vehicle empty trips with a first order trip chain model, *Transp. Res. B*, 2002, in press.

Hwang, S.T. et al., Risk assessment for national transportation of selected hazardous materials, *Transp. Res. Rec.*, 1763, 114–124, 2001.

Jiang, F., Johnson, P., and Calzada, C., Freight demand characteristics and mode choice: an analysis of the results of modeling with disaggregate revealed preference data, *J. Transp. Stat.*, 2.2, 149–158, 1999.

Kawamura, K., Perceived value of time for truck operators, *Transp. Res. Rec.*, 1725, 31–36, 2000.

Lawson, C.T. and Riis, A.E., We're really asking for it: Using surveys to engage the freight community, *Transp. Res. Rec.*, 1763, 13–19, 2001.

Leighton, T. and Rao, S., Multicommodity max-flow min-cut theorems and their use in designing approximation algorithms, *J. Assoc. Comput. Mach.*, 46, 787–832, 1999.

Leontief, W., *Input–Output Economics*, Oxford University Press, New York, 1967.

List, G.F. et al., A best practice truck flow estimation model for the New York City region, Paper 02-4128, Transportation Research Board 2002 Annual Meeting CD-ROM, Washington, D.C., 2001.

List, G.F. and Turnquist, M.A., Estimating truck travel patterns in urban areas, *Transp. Res. Rec.*, 1420, 1–9, 1994.

Manalytics Inc. et al., Double-Stack Container Systems: Implication for US Railroads and Ports, Final Report to the U.S. Department of Transportation, Washington, D.C., 1988.

Matthews, H.S., Hendrickson, C.T., and Soh, D.L., Environmental and economic effects of e-commerce: A case study of book publishing and retail logistics, *Transp. Res. Rec.*, 763, 6–12, 2001.

McGinnis, M.A., A comparative evaluation of freight transportation choice models, *Transp. J.*, 29, 36–46, 1989.

Meyburg, A.H. and Mbwana, J.R., Eds., Data Needs in the Changing World of Logistics and Freight Transportation, conference synthesis, prepared by Cornell University for the New York State Department of Transportation, Albany, New York, 2002.

Middendorf, D.P. and Bronzini, M.S., The Productivity Effects of Truck Size and Weight Policies, ORNL Report Number 6840, Oak Ridge National Laboratory, Oak Ridge, TN, 1994.

Morlok, E.K. and Riddle, S.P., Estimating the capacity of freight transportation systems: A model and its application in transportation planning and logistics, *Transp. Res. Rec.*, 1653, 1–8, 1999.

Morris, A.G. and Kornhauser, A.L., Relationship of freight facilities in central business districts to truck traffic, *Transp. Res. Rec.*, 1707, 65–63, 2000.

Musso, A., SOFTICE: survey of freight transport including a cost comparison for Europe, *Transp. Res. Rec.*, 1763, 27–34, 2001.

Nagurney, A. et al., Dynamics of supply chains: A multilevel (logistical/information/financial) network perspective, *Environ. Plann. C*, 29, 2002, in press.

Newstrand, M.W., Environmental impacts of a modal shift, *Transp. Res. Rec.*, 1333, 9–12, 1991.

Nijkamp, P., Vleugel, A.M., and Kreutzberger, E., Assessment of capacity in infrastructure networks: A multidimensional view, *Transp. Plann. Technol.*, 17, 301–310, 1993.

Oak Ridge National Laboratory (ORNL), Intermodal bottleneck evaluation system, prepared by Oak Ridge National Laboratory for the Bureau of Transportation Statistics, U.S. Department of Transportation, Washington, D.C., 2000.

Ogden, K.W., *Urban Goods Movements: A Guide to Policy and Planning*, Ashgate Publishing, Brookfield, VT, 1992.

O'Kelly, M.E., A geographer's analysis of hub-and-spoke networks, *J. Transp. Geogr.*, 6, 171–186, 1998.

Pendyala, R.M., Shankar, V.N., and McCullough, R.G., Freight travel demand modeling: Synthesis of approaches and development of a framework, *Transp. Res. Rec.*, 1725, 9–18, 2000.

Port of Long Beach, The National Economic Significance of the Alameda Corridor, report prepared for the Alameda Corridor Transportation Authority, Long Beach, CA, 1994.

Premius, H. and Konings, R., Dynamics and spatial patterns of intermodal freight transport networks, in *Handbook of Logistics and Supply Chain Management*, Brewer, A.M., Button, K.J., and Hensher, D.A., Eds., Pergamon Press, New York, 2001, chap. 32.

Raj, P.K. and Pritchard, E.W., Hazardous materials transportation on U.S. railroads, *Transp. Res. Rec.*, 1707, 22–26, 2000.

Reinelt, G., *The Travelling Salesman: Computational Solutions for TSP Applications*, Lecture Notes in Computer Science, Springer-Verlag, Berlin, 1994.

Research and Special Programs Administration, *Intermodal Cargo Transportation: Industry Best Security Practices*, RSPA, U.S. Department of Transportation, Cambridge, MA, 1999.

Richardson, B.C., Freight trucking in a sustainable transportation system, *Transp. Res. Rec.*, 1763, 57–64, 2001.

Roberts, P.O., Nanda, R., and Smalley, B., T-RR-T: A Multimodal Freight Diversion Model for Policy Analysis, Paper 960500, presented at the Transportation Research Board Annual Meeting, Washington, D.C., 1996.

Rodrigues, J.-P., Slack, B., and Comtois, C., Green logistics, in *Handbook of Logistics and Supply Chain Management*, Brewer, A.M., Button, K.J., and Hensher, D.A., Eds., Pergamon Press, New York, 2001, chap. 21.

Sivakumar, A. and Bhat, C., A Fractional Split Distribution Model for Statewide Commodity Flow Analysis, Paper 02-2142, Transportation Research Board 2002 Annual Meeting CD-ROM, Washington, D.C., 2002.

Song, J. and Regan, A.C., Transition or transformation? Emerging freight transportation intermediaries, *Transp. Res. Rec.*, 1763, 1–5, 2001.

Sorratini, A.J. and Smith, R.L., Jr., Development of a statewide truck trip forecasting model based on commodity flows and input–output coefficients, *Transp. Res. Rec.*, 1707, 49–55, 2000.

Southworth, F., An urban goods movement model: Framework and some results, *Pap. Reg. Sci. Assoc.*, 50, 165–184, 1982.

Southworth, F., Circuit-based indices of locational accessibility, *Environ. Plann. B*, 10, 249–260, 1983.

Southworth, F. et al., Strategic freight planning for Chicago in the year 2000, *Transp. Res. Rec.*, 920, 45–48, 1983.

Southworth, F. and Peterson, B.E., Intermodal and international freight network modeling, *Transp. Res. C*, 8, 147–166, 2001.

Taylor, S.Y., Just-in-time, in *Handbook of Logistics and Supply Chain Management*, Brewer, A.M., Button, K.J., and Hensher, D.A., Eds., Pergamon Press, New York, 2001, chap. 13.

Tolliver, D. and Earth Tech Environment & Infrastructure, Analysis of the Energy, Emissions, and Safety Impacts of Alternative Improvements to the Upper Mississippi River and Illinois River Waterway System, DACW25-95-D-0006, prepared for the U.S. Army Corps of Engineers, Rock Island, IL, 2000.

Transportation Research Board, Landside Access to U.S. Ports, Special Report 238, TRB, Washington, D.C., 1992.

Transportation Research Board, *A Guidebook for Forecasting Freight Transportation Demand*, NCHRP Report 388, TRB, National Academy Press, Washington, D.C., 1997.

Transportation Research Board, *Highway Capacity Manual*, Special Report HCM2KC, TRB, Washington, D.C., 2000.

Tsamboulas, D.A. and Kapros, S., Decision-making process in intermodal transportation, *Transp. Res. Rec.*, 1707, 86–93, 2000.

U.S. Army Corp of Engineers, Waterborne Commerce Statistics of the United States, USACE, New Orleans, LA, 2000.

U.S. Census Bureau, United States Commodity Flow Survey, Census Bureau, U.S. Department of Commerce, Washington, D.C., 1997a.

U.S. Census Bureau, Vehicle Inventory and Use Survey, Census Bureau, U.S. Department of Commerce, Washington, D.C., 1997b.

U.S. Coast Guard, *Marine Safety Management System*, Coast Guard Marine Safety Directorate, USCG, Department of Transportation, Washington, D.C., 1999.

Vanek, F.M. and Morlok, E.K., Freight energy use disaggregated by commodity: Comparison and discussion, *Transp. Res. Rec.*, 1641, 3–8, 1998.

Vanek, F.M. and Morlok, E.K., Improving the energy efficiency of freight in the United States through commodity-based analysis: Justification and implementation, *Transp. Res. D*, 15, 11–29, 2000.

Vilain, P., Nan Liu, L., and Aimen, D., Estimation of commodity inflows to a substate region: An input–output based approach, *Transp. Res. Rec.*, 1653, 17–26, 1999.

Vilain, P. and Wolfrom, P., Value pricing and freight traffic: Issues and industry constraints in shifting from peak to off-peak movements, *Transp. Res. Rec.*, 1707, 64–72, 2000.

Weigel, M.L., A railroad intermodal capacity model, in *Proceedings of the 1994 Winter Simulation Conference*, Association for Computing Machinery (ACM) Publications, Orlando, Florida,1994.

Weisbrod, R.E. et al., Global Freight Villages: A Solution to the Urban Freight Dilemma, paper presented at the Transportation Research Board Annual Meeting, Washington, D.C., 2002.

Wigan, M. et al., Valuing long-haul and metropolitan freight travel time and reliability, *J. Transp. Stat.*, 3.3, 83–89, 2000.

Wilson, A.G., *Entropy in Urban and Regional Modelling*, Pion, London, 1970.

Wilson, A.G. et al., *Optimization in Locational and Transport Analysis*, John Wiley & Sons, Chichester, England, 1981.

World Trade Organization, *International Trade Statistics, 2001*, WTO, Geneva 21, Switzerland, 2002.

Wynter, L.M., The value of time of freight transport in France: Estimation of continuously distributed values from a stated preference survey, *Int. J. Transp. Econ.*, 22, 151–165, 1995.

Zlatoper, T.L. and Austrian, Z., Freight transportation demand: A survey of recent econometric studies, *Transportation*, 16, 26–46, 1989.

5

Land Use: Transportation Modeling

CONTENTS

Eric J. Miller
University of Toronto

5.1 Introduction

Regional travel demand models typically take as their starting point exogenously determined spatial distributions of population and employment (and any required attributes of these people and jobs) as fixed inputs into the demand modeling system. In so doing, they ignore the fact that these population and employment distributions are the outcome of a dynamic process of urban evolution that is partially determined by the nature and performance of the transportation system. That is, as illustrated in Figure 5.1, a two-way transportation–land use interaction exists, in which transportation is a derived demand from the urban activity system, but also in which the transportation system influences land development and location choice through the provision of accessibility to land and activities. From a policy analysis perspective, the need for a consistent, comprehensive analysis of urban systems is generally well understood, if not always put into practice. In particular, note that many transportation problems such as congestion, air pollution, etc., may have their root causes as much in urban form considerations (e.g., excessive urban sprawl) as in transportation system design per se.

Integrated land use–transportation models are designed to capture most, if not all, of the processes and interactions shown in Figure 5.1. That is, they attempt to model both urban system evolution and the associated evolution of urban travel demand in a comprehensive and integrated fashion. At a

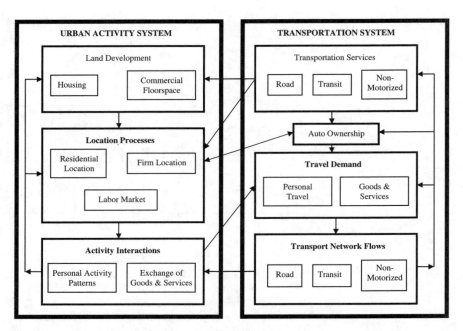

FIGURE 5.1 The urban transportation–land use interaction. (Adapted from Meyer, M.D. and Miller, E.J., *Urban Transportation Planning: A Decision-Oriented Approach*, 2nd ed., McGraw-Hill, New York, 2001.)

minimum, such models are intended to provide the population and employment forecasts required by traditional travel demand models in a more consistent, systematic, and credible fashion than might be possible by other methods. Specifically, they generate these distributions in a way that is consistent with transportation network configurations, congestion levels, etc. Integrated models, however, have the potential to do much more relative to conventional methods, including:

- Providing much more detailed simulation of person and household demographic and socioeconomic attributes, which can be powerful explanatory variables of travel demand
- Providing policy analysis capability for a much wider range of land use, transportation, and other policy measures that might influence travel behavior, either directly or indirectly

Despite the potentially important role that integrated land use–transportation models might play in policy analysis, these models are not currently used in a majority of cities. A number of technical, historical, and resource-related reasons for this state of affairs exist. Integrated models were first built in the early 1960s. These early models represented quite exceptional pioneering efforts, but, on the whole, failed to prove overly useful as policy analysis tools, largely because the computational capabilities, modeling methods, and available data of the day simply were inadequate to support the ambitious requirements and expectations of these models. The weaknesses of these first-generation models were dramatically documented in Lee's seminal paper "Requiem for Large-Scale Models" (Lee, 1973), which had a profound influence on planners' attitudes toward models in general and integrated models in particular for at least the next decade, especially in the United States.

Development work on second- and third-generation integrated models, however, continued around the world, slowly gathering momentum as the computer revolution began to provide modelers with computing capabilities adequate to the task of simulating entire cities, as Geographic Information Systems (GIS) provided computerized databases of sufficient breadth and depth to support such ambitious modeling activities, as our theoretical understanding of urban spatial processes increased, and as our modeling methods for capturing these processes in computerized equations and algorithms improved. The net result of these cumulative advances over the nearly three decades since Lee's requiem is that a considerable variety of integrated models are in operational use around the world, with this number growing steadily.

The purpose of this chapter is to provide an introductory overview of integrated land use–transportation modeling. A single, relatively short chapter such as this one cannot possibly cover the entire contents of such a complex subject. Rather, it has the more modest objectives of sketching the general structure of such models, presenting some of the key modeling principles and methods typically used in current models, and discussing some of the critical design and implementation issues that a planning agency should consider in the development and use of such a model.

5.2 Definitions and Key Concepts

In speaking of land use, travel demand modelers often are simply using this term as a shorthand expression for the zonal population and employment distributions that they require as inputs to their models. More formally, land use refers to the way in which land is used, in terms of the buildings built upon the land (houses, stores, schools, factories, etc.) and the activities housed within these buildings (in-home activities, shopping, education, work, production of goods and services, etc.). As such, land use is essentially synonymous with urban form, which is a term often used by geographers and regional scientists. Given that it is the participation in out-of-home activities that gives rise to the need for travel, transportation system analysts often speak of the urban activity system, which consists of both the physical built form and the spatial–temporal distribution of activities that occur within this built form. In this chapter, as a matter of convenience, we will use the terms *land use*, *urban form*, and *urban activity system* more or less interchangeably, while recognizing the nuances that actually exist among these terms.

From a modeling point of view, the key point to recognize is that, as shown in Figure 5.1, four interrelated but distinct processes define the evolution of the urban activity system over time:

1. Land development, in which the built form changes over time as land is developed and as existing buildings are modified or redeveloped over time. This is the process by which land use, per se, evolves.
2. Location choice, in which households and firms decide where to locate, given the location alternatives (vacant dwelling units or commercial floor space) available at the time the location choice is being made.
3. Activity scheduling and participation, in which households plan their daily lives and then execute these plans in terms of actual participation in activities and the travel associated with this activity engagement.
4. Commercial exchange of goods and services, which includes the full gamut of physical interchanges of persons, goods, and services generated by the urban region's economy, including firm–firm interactions (goods and services exchanges of inputs and outputs), firm–worker labor exchanges (work trips), and firm–household interchanges (shopping, personal business, etc.).

Note that both of the latter two processes occur within a short-run decision-making time frame within which the distribution of residential locations, firm locations, etc., is (temporarily) fixed. Thus, implicit in Figure 5.1 is a temporal dimension in which all four processes are constantly "running," but in which each process tends to operate within a different decision-making time frame and within a different set of constraints. Land development decisions are generally very long run in nature, which for large projects can often play out over literally decades from project conception to final construction. Location choices are also long run in nature, but generally display greater fluidity than land development processes, and certainly are constrained at any given point in time by the supply of available options, as determined by the higher-level land development process. Household activity and travel and commercial economic exchanges are obviously much shorter run in nature, playing out on a daily basis.[1]

---

[1]This is not to say that some of these processes cannot exhibit considerable stability over time. For example, it is not likely that most workers actively re-evaluate their choice of mode to work each and every workday. But they do execute a daily activity schedule that includes the journey to work, and this activity schedule certainly can and does change within shorter time frames than longer-term choices such as residential location.

In parallel to the urban activity system is the transportation system, which similarly evolves over time through a combination of longer-run supply decisions concerning the provision of physical infrastructure and the services operated within this infrastructure, through to the short-run, day-to-day personal travel and goods movements that occur within the system and determine its operating performance levels.

Household auto ownership decision making (e.g., how many vehicles of what types to own or lease) has been included as an explicit box within the transportation system to highlight the important role that it plays in the overall transportation–land use interaction. Auto ownership obviously has a profound influence on travel behavior in terms of mode choice, destination choice, and even trip generation rates. It also, however, affects residential and employment location choices, and in turn is influenced by these location choices (if I don't own a car, my choice of work locations may be limited; households living in suburban locations, on average, own more cars than ones living in central cities; etc.). Although not usually discussed in these terms, household auto ownership is a supply process in which households are able to supply themselves with transportation services that occasionally complement (e.g., commuter rail park and ride) but more usually compete with publicly supplied transit services.

The transportation and urban activity systems interact in three primary ways. First, the activity system drives the transportation system on a daily basis in terms of determining the need for travel. Second, transportation system performance influences this daily activity scheduling process in terms of defining the times, costs, reliability, etc., involved in traveling from one point to another by different modes of travel, thereby influencing the choice of activity location (e.g., shopping at a mall with convenient, free parking vs. downtown, where parking may be expensive and in short supply), activity timing (e.g., shop during off-peak hours to avoid traffic congestion), etc. And third, in the longer run, the accessibility that the transportation system provides to land and activities influences over time both land development and location choice processes. This accessibility is supplied both publicly, through the provision of physical infrastructure and public transit services, and privately, through ownership of personal-use automobiles.

Most of this handbook is devoted to modeling travel demand and to understanding the activity–travel interaction. Therefore, this chapter takes this understanding and associated modeling methods as given and focuses on the longer-run processes of land development and location choice processes, as well as the role that transportation plays in these processes through the provision of accessibility.

Land development, residential and commercial real estate activity, and economic interchanges obviously all occur within the framework of markets, in which demand (consumption) and supply (production) processes interact and determine the exchange of land, floor space, goods and services, etc., as well as the prices at which these commodities are exchanged. Thus, any model of the spatial evolution of the urban activity system should account for both demand- and supply-side processes and should generate the prices (which are both outcomes of and primary inputs into these processes) as endogenous components of the model.

A primary output from an integrated model should be the environmental impacts of the urban activity and transportation systems, including greenhouse gas emissions, mandated air quality emissions, agricultural land consumption, and other environmental indicators of policy importance. Indeed, one of the primary motivations for developing an integrated modeling framework is to better address the short- and long-run environmental implications of transportation and land use policies. Many modelers speak of integrated land use–transportation–environment models (Wegener, 1995) to emphasize this point. The environmental component of such models, however, is in itself a complex modeling problem; this will not be discussed in detail in this chapter.

At least three classes of models exist that deal in a systematic, computerized way with the estimation of future land use distributions. Land accounting and allocation models typically are GIS-based, rule-based systems for estimating future land development on a zone-by-zone basis based on empirically observed past development patterns. Examples of this approach can be found in Landis (1994) and Yen and Fricker (1997). While relatively simply to apply, these models are not discussed further in this chapter since they generally are not sensitive to either the transportation system (and so are not useful in the analysis of transportation policies) or the intraregional economic and market processes that are presumably major determinants of both the location and timing of land development.

Optimization and normative models attempt to generate optimal urban forms, given an assumed objective function that is optimized, subject to a set of system constraints. Examples of optimization-based models can be found in Brotchie et al. (1980), Caindec and Prastacos (1995), and Kim (1989). While useful for exploring what optimal urban forms might look like, as well as how different policies might alter this optimal outcome, such models generally provide little insight into the likelihood that such outcomes will actually occur, or of how the urban area might actually evolve from its current nonoptimal state to the desired optimal future state (given that the actual behavioral processes driving urban evolution are rarely, if ever, inherently system optimizing in nature).

The focus of this chapter is on descriptive and behavioral models that attempt to simulate the evolution of the urban activity system explicitly over time from a known base state to a (most likely or expected) future state. Such models typically move forward in fixed time steps ranging from 1 to 10 years in length, in which the system state at the end of each time step is predicted as a function of the system state at the beginning of the time step, exogenous inputs that are expected to occur during the time step, and the endogenous processes being explicitly simulated within the model. A wide variety of models have been developed to operationalize Figure 5.1; they vary in terms of the comprehensiveness of processes included in the model, treatment of time and space, choice of modeling methods, etc.

For example, Figure 5.2 presents the flowchart for UrbanSim, which is very representative of current practice in operational integrated models. As shown in the flowchart, population and employment distributions in a future year are the outcome of a land development process and move or location choice decisions of households and businesses in the real estate market. These land use distributions drive a conventional travel demand model, which in turn feeds back measures of accessibility to the land use model. Other important components of the modeling system are models that predict how the households and businesses evolve over time, which in turn depend on exogenously forecasted regional control totals (total population growth, total employment growth by sector, etc.).

Miller et al. (1998), Southworth (1995), and Wegner (1994, 1995) all provide detailed reviews of UrbanSim and other operational integrated models. In this chapter, rather than describing in detail these models (all of which differ in many detailed ways from one another), we focus first on discussing some of the key modeling methods that are generally employed in these models, and then on important design and implementation issues involved in the development and application of an integrated land use–transportation modeling system in an operational setting.

5.3 Modeling Spatial Processes

As has been discussed above, the fundamental rationale of integrated land use–transportation models is to model the spatial decision processes that shape the physical form of an urban area over time and determine the physical flows of people, goods, and services within this area. These spatial decision processes include:

- Decisions to develop or redevelop land for various purposes
- Location and relocation decisions of firms
- Residential location and relocation decisions of households
- Labor market decisions of workers (what job to take, where) and employers (what worker to hire)
- Activity–travel decisions of persons and households
- Economic interactions among firms that result in the flow of goods and services among them

In order to deal with this diverse and complex set of processes, a variety of modeling methods and theoretical constructs are required. These include:

- Models of spatial interaction and accessibility
- Models of land and real estate markets
- Models of intraregional economic interaction

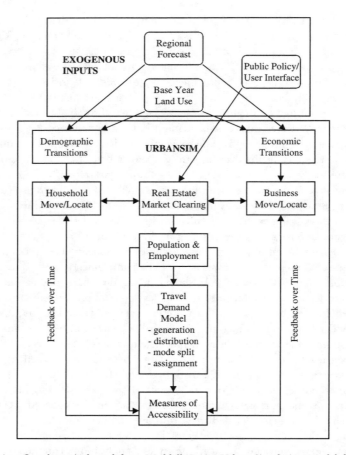

FIGURE 5.2 Urbanism flowchart. (Adapted from Waddell, P., An Urban Simulation Model for Integrated Policy Analysis and Planning: Residential Location and Housing Market Components of UrbanSim, paper presented at the 8th World Conference on Transport Research, Antwerp, Belgium, July 1998.)

Each of these is discussed in some detail in the following sections.

5.4   Spatial Interaction Modeling

The decision of what store to travel to from home to purchase a certain good, the decision of what neighborhood to live in given where one works (or where to work given where one lives), or similar decisions involve the flow or interaction between two points in space (home and store, workplace and residence, etc.), as the outcome of the selection of the destination of the spatial interaction (the store, the residence, etc.), generally given the known location of the interaction origin (the home, the workplace, etc.). Such spatial interactions literally define the transportation–land use interaction, and so it is not surprising that models of spatial interaction play a central role in virtually all integrated models.

Two theoretical approaches dominate the modeling of spatial interactions: entropy maximization (also known as information minimization) and random utility theory. A unique feature of these two approaches is that in the most commonly applied case they result in exactly the same mathematical model. Despite this convergence of the two approaches, it is useful to consider both briefly, since each approach provides its own insights into the fundamental assumptions underlying the operational model. In order to make comparisons between the two approaches more concrete, let us consider as an example the choice of a residential location zone for a one-worker household, given that the place of employment for the worker is known.

5.4.1   Entropy Maximization

The concept of entropy maximization for modeling spatial processes was first developed by Alan Wilson in a seminal paper in the 1960s (Wilson, 1967) as a means of providing a theoretical foundation for gravity-type models of trip distribution. It was later shown that Wilson's model could also be derived from fundamental concepts of information theory (Webber, 1977), which was first developed for applications in communications. The basic notion of entropy maximization is to develop a model that generates the most likely estimates of the spatial interactions of a set of actors (in this case, households looking for a place of residence), given limited information about the actors and the outcomes of their decisions. In particular, it is assumed that this information can be expressed in terms of constraints on feasible outcomes of the actors' decisions. For example, define the following terms:

$H_{i|j}$ = the number of households whose worker is employed in zone j and lives in zone i
H_j = the number of households whose worker is employed in zone j
t_{ij} = the travel time by auto from zone i to zone j during the morning peak period
N_i = the number of housing units in zone i
H = total number of households

The task for the spatial interaction model is to predict $H_{i|j}$, given known values of H_j, t_{ij}, and N_i. Many possible estimates of $H_{i|j}$ might be generated through a variety of models. To be internally consistent, however, all such estimates should satisfy logical constraints defined by the known information. Even in this simple example, many possible constraints might be imposed. In this particular application, the most common set of constraints used are

$$\sum_j H_{i|j} = H_j \text{ for all zones j} \tag{5.1}$$

$$\left\{ \sum_{i,j} H_{i|j} t_{ij} \right\} \Big/ H = t_{avg} \tag{5.2}$$

$$\left\{ \sum_{i,j} H_{i|j} \ln(N_i) \right\} \Big/ H = N_{avg} \tag{5.3}$$

Equation (5.1) simply imposes the logical constraint that the total number of households assigned to all possible residential locations for a given employment zone must equal the number of households associated with this employment zone. In Equation (5.2), t_{avg} is the observed average travel time from home to work (assuming that all work trips occur during the morning peak period by the auto mode), and this constraint imposes the condition that the predicted distribution of worker residential locations should be such that the predicted average travel time to work (i.e., the left-hand side of Equation (5.2)) reproduces the observed average travel time for the system being modeled.

Equation (5.3) imposes a similar sort of constraint, where the left-hand side of the equation is the predicted average value of $\ln(N_i)$, weighted by the number of households choosing zone i for their place of residence, while the right-hand side (N_{avg}) is the observed average value of this term. The rationale for the specification of Equation (5.3) in this particular form is not particularly intuitively obvious. It is chosen primarily because it generates an attractive final model functional form for $H_{i|j}$, as is seen below.

Wilson and others have shown that the most likely equilibrium estimates of $H_{i|j}$ are obtained by maximizing the so-called entropy function:

$$S = -\sum_i \sum_j H_{i|j} \ln(H_{i|j}) \tag{5.4}$$

subject to satisfying the constraints in Equations (5.1) to (5.3). This optimization problem can be solved using the method of Lagrange, that is, by maximizing the function:

$$L = -\sum_i \sum_j H_{i|j} \ln(_{i|j}) + \sum_j \lambda_j \left(\sum_i H_{i|j} - H_j \right) + \beta$$
$$\left(\sum_{i,j} H_{i|j} t_{ij} - Ht_{avg} \right) + \alpha \left(\sum_{i,j} H_{i|j} \ln(N_i) - HN_{avg} \right) \tag{5.5}$$

Differentiating Equation (5.5) with respect to the unknown variable $H_{i|j}$, solving for first-order optimality conditions, and back substituting into Equation (5.1) to ensure that it holds, yields

$$H_{i|j} = \frac{H_j (N_i)^\alpha \exp(\beta t_{ij})}{\sum_{i'} (N_i)^\alpha \exp(\beta t_{i'j})} \tag{5.6}$$

Equation (5.6) is a standard, singly constrained gravity model, variations of which have been used in a wide variety of residential location, employment location, and trip distribution models (with, of course, appropriate redefinition of the variables involved). The choice of the rather odd form for the constraint in Equation (5.3) was motivated by the desire to generate the attraction term $(N_i)\alpha$. If a different functional form for this term is desired, then a different form of the constraint could be written, chosen so that when the new version of Equation (5.5) implied by the new constraint is maximized, the desired term emerges in Equation (5.6). A similar comment holds for the travel impedance term, $\exp(\beta t_{ij})$, or any other term that is included in the model.

Given that entropy maximization seems to merely regenerate a standard gravity model, it is important to note this method makes at least two major contributions to spatial interaction modeling. The first is that entropy maximization provides a formal mathematical and theoretical foundation for gravity models that, prior to Wilson's seminal work, were often criticized for being without any sound theoretical basis. Indeed, information theory shows that, given the problem definition (i.e., the need to predict system behavior given limited information about feasible combinations of that behavior), gravity–entropy models generate the most likely (also sometimes referred to as the least biased) estimates of system behavior achievable given the available information.

Second, entropy maximization provides a formal method for generating model functional forms and estimating model parameters. The functional form of Equation (5.6) can not be arbitrarily chosen; it must be mathematically derivable from a set of logical constraints. Admittedly, freedom exists in the choice of constraints, as illustrated in the example above, but this is no different than the freedom available to modelers in the selection of variables (and their functional form) to include in the systematic utility function of a random utility maximization model. The method to be used in model parameter estimation is also not arbitrary. α and β in Equation (5.6) must be chosen so that the constraints in Equations (5.2) and (5.3) hold for the base calibration data set. These equations can be efficiently solved using the Newton–Raphson root-finding method.

5.4.2 Random Utility Models

If we define $P_{i|j}$ as the probability that a household whose worker is employed in zone j resides in zone i,

$$P_{i|j} = H_{i|j} / H_j \tag{5.7}$$

and if we note that $x^a = e^{a \ln(x)}$, then Equation (5.6) can be rewritten as

$$P_{i|j} = \frac{\exp(\alpha \ln(N_i) + \beta t_{ij})}{\sum_{i'} \exp(\alpha \ln(N_{i'}) + \beta t_{i'j})} \tag{5.8}$$

Equation (5.8) is a standard multinomial logit model, derived from random utility theory, which is discussed in detail in other chapters of this handbook. As briefly sketched here, and as was first shown in detail by Anas (1983), when consistently developed, entropy and multinomial logit models are identical in functional form and estimated parameter values.[2] This is a powerful and perhaps surprising result (given the seemingly quite different theoretical starting points of the two approaches), which seems to be often overlooked by modelers within both the entropy and random utility modeling camps. In particular, the convergence of the two approaches allows modelers to better understand the strengths and weaknesses of spatial interaction models in general.

Starting with Lerman's (1976) seminal application of multinomial logit modeling to the residential location choice problem, random utility models of both the multinomial and nested logit form have been applied to a wide variety of spatial choice processes and, indeed, are the standard tool for modeling these processes in virtually all currently operational integrated models. In general, random utility models have been found to be very flexible and powerful tools for modeling spatial processes for a variety of reasons, including:

1. The explicit tie to microeconomic theory is a powerful one that aids considerably in model specification, validation, and interpretation.

2. Random utility theory is very general and permits a variety of specific models to be developed and applied (multinomial logit, nested logit, probit, generalized extreme value models of various types, etc.). In particular, to the extent that we know (or at least have strong enough insight to hypothesize) that correlations among outcomes exist that cannot be handled within the entropy–multinomial logit framework, we can extend our random utility framework in appropriate ways to accommodate these correlations.[3]

3. The nested logit model structure, in particular, is an extremely attractive and practical method for developing a complex modeling system such as an integrated model, in which many submodels (residential location choice, auto ownership choice, activity–travel decisions, etc.) must coexist and interact in a logical, consistent fashion. The ability to feed back inclusive value terms describing the expected utilities derived from lower-level choices (e.g., travel) into more upper-level decisions (e.g., residential location choice) is an exceptionally efficient and theoretically well-defined method for submodel interfacing. This point is discussed in further detail below in the special and important case of the use of accessibility terms in spatial choice models.

4. The availability of standardized parameter estimation software greatly facilitates the development of operational models.

5.4.3 Accessibility

Closely tied to spatial interaction is the concept of accessibility. Put very simply, accessibility is the raison d'etre of the transportation system: to provide the ability for people and goods to be able to move efficiently and effectively from point to point in space in as unconstrained a fashion as possible. Given this, accessibility obviously must play a central role in the transportation–land use interaction, and measures of accessibility surely must be important explanatory variables in models of spatial decision processes. That is, all else being equal, households presumably will prefer to choose residential locations that provide high access to jobs, stores, good schools for their children, recreational facilities, etc., while businesses will similarly desire locations that provide good access to both their customers and their suppliers.

Before constructing operational measures of accessibility, it is useful to identify the attributes of such a measure implicit in the loose description of the term provided above. These include the following:

1. Accessibility is a point measure, in that each point in space has its own level of accessibility. For example, a point in the downtown of a city will likely have a different level of accessibility to theatres and other cultural facilities than a point on the suburban fringe.

---

[2]Equations (5.2) and (5.3), which must be solved within the entropy formalism to determine the estimates of α and β for a given base set of data, are also the maximum likelihood parameter estimation equations that must be solved within the random utility formalism.

[3]Note that entropy-based models assume that all outcomes are equally likely, except as constrained by the imposed constraints. As a result, the entropy formalism does not readily generalize to allow for the sort of correlations among outcomes accomodated by generalized extreme value, probit, and other random utility models.

2. Accessibility is activity specific. The same point in the downtown will have different accessibility levels to cultural facilities, schools, and big-box building supply centers, to name just a few activity–land use types of possible interest to households considering the downtown as a possible place of residence.
3. Accessibility depends on the ease of travel to potential activity sites. The more activity sites within a convenient travel distance and time, presumably the higher the level of accessibility. Given this, accessibility varies by mode (e.g., the points accessible within a given travel time will be different by car, transit, walk, etc.) and time of day (e.g., peak vs. off-peak).
4. Accessibility depends on the attractiveness of the activity sites available. If there are many restaurants within walking distance of my workplace, but if they are all expensive, serve poor quality food, and provide substandard service, then my accessibility to lunchtime eating establishments is low regardless of the number of sites nominally available.
5. Accessibility is an integrative measure of the potential for spatial interaction. If there are a large number of inexpensive, cheerful, high-quality restaurants close to my workplace, my accessibility to lunchtime eating establishments is high, regardless of whether I actually make a trip to one of those places on a given day or not.

Given these attributes, many measures of accessibility have been developed for a variety of applications, including their use as explanatory variables in integrated models. Probably the simplest such measure involves defining a maximum travel time threshold (e.g., 15 or 30 min) and then adding up all the opportunities for a given type of activity (employment, shopping, etc.) that lie within this travel time threshold for an assumed mode of travel for a given point. For example, in the residential location choice problem discussed above, the accessibility of a given employment zone j to residential housing opportunities within a 30-min drive of zone j during the morning peak period would be

$$A_j = \sum_i H_i \delta(t_{ij}) \qquad (5.9)$$

where $\delta(t_{ij})$ equals 1 if $t_{ij} = 30$ min and 0 otherwise.

Equation (5.9) meets all of the criteria developed above for an accessibility measure. It is defined for a point in space (zone j); it is specified for a particular activity (residential location); it depends on the attractiveness of the activity sites available (in this case, simply measured by the size of the activity site — a very common approach in operational models); it depends on the travel mode (auto) and time of day (morning peak period); and it integrates over the region around the reference point to yield an overall measure of residential location potential for this point. It is also an attractive measure in that it is very easy to compute, especially given modern Geographic Information Systems that readily compute such measures.

The major limitation of Equation (5.9) is that it does not capture the interaction between site attractiveness and location. For example, consider two zones that both have 1000 housing units within a 30-min drive. In the case of zone 1, all 1000 units are actually a 10-min drive away, while for zone 2, the 1000 units are all located 25 min away. Equation (5.9) will return exactly the same level of accessibility for the two zones (i.e., 1000), whereas it is more likely that we would consider zone 1 to have the higher accessibility in this case.

To overcome this difficulty, a common second measure of accessibility that has been used is the denominator of a spatial interaction entropy–gravity model. In the example being considered here, this means using the denominator of either Equation (5.8) or (5.9), that is:

$$A_j = \sum_{i'} (N_{\cdot i})^\alpha \exp(\beta t_{i'j}) = \sum_{i'} \exp(\alpha \ln(N_{i'}) + \beta t_{i'j}) \qquad (5.10)$$

Equation (5.10) obviously also meets all the criteria listed above, with the added advantage that it weights the contribution of the site attractiveness to the overall level of accessibility by the level of difficulty involved in getting there. In the simple numerical example introduced above, if $\alpha = 1$ and $\beta = -2$, then the accessibility of zone 1 would be 10, while zone 2's accessibility would only be 1.6 — reflecting the fact that zone 2 is located more remotely from the housing units than zone 1. Equation (5.10) has another advantage: it directly relates the concept of accessibility (which, recall, describes the potential to interact over space) to actual spatial behavior, through the use of a term taken from a model of spatial choice. Given these advantages, terms such as those in Equation (5.10) have been used as explanatory variables in many land use–transportation models over the years.

Despite the plausible nature of Equation (5.10) as a definition of accessibility, its selection is ad hoc in that it was simply pulled out the air as a reasonable measure. As noted above, Equation (5.9) is a standard logit model, derivable from random utility theory. Given this, the expected maximum utility to be derived by a worker employed in zone j from the choice of a residential location is a known value. It is the so-called inclusive value or log-sum term:

$$I_j = \ln\left\{\sum_{i'} \exp(\alpha \ln(N_{i'}) + \beta t_{i'j})\right\} \tag{5.11}$$

Ben-Akiva and Lerman (1985) convincingly argue that Equation (5.11) is the appropriate definition of accessibility within a random utility framework, since it defines the potential benefit (i.e., expected utility) to be derived at point j from participating in the given spatial process. Equation (5.11) is merely the logarithmic transformation of Equation (5.10), so both measures will generate the same ordinal ranking of accessibilities for a set of zones (given the same model and data), but it is argued that Equation (5.11) is the preferred functional form to be used, since it completes the task of linking the definition of accessibility to the spatial choice processes that underlie and provide meaning to the concept of utility.

In most practical applications, what this means is that a nested logit structure can be used to link travel decisions and location decisions within the integrated model. Figure 5.3 provides a simple extension of the residential location choice case that we have been considering, in which the problem is now defined as one involving the choice of both residential location and mode to work, given a known workplace and a one-worker household. As shown in Figure 5.3, this can be modeled as a nested logit model, in which the longer-run place of residence choice is the upper level and the shorter-run work trip mode choice is the lower level. A simple (but representative) set of equations for this system might take the following form:

$$P_{m|ij} = \frac{\exp(V_{m|ij}/\phi)}{\sum_{m'} \exp(V_{m'|ij}/\phi)} \tag{5.12}$$

$$I_i = \ln\left\{\sum_{m'} \exp(V_{m'|ij}/\phi)\right\} \tag{5.13}$$

$$P_{i|j}\phi = \exp(\gamma' X_{i|j} + \alpha \ln(N_i) + \phi I_i) \tag{5.14}$$

where $P_{m|ij}$ is the probability that mode m will be chosen for a work trip from residence zone i to work zone j; I_i is the inclusive value term for zone I; $V_{m|ij}$ is the systematic utility of mode m for the trip from

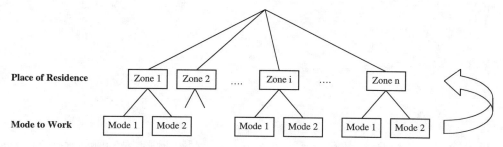

FIGURE 5.3 Nested logit model of residential location and work trip mode choice.

i to j; $X_{i|j}$ is the column vector of explanatory variables influencing the choice of residential zone i for workers employed in j; ϕ is the scale parameter ($0 \leq \phi \leq 1$); and γ is the column vector of parameters.

I_i is an accessibility term defining the access to workplace j from i provided by the transportation system. As this accessibility increases (due, for example, to an improvement in any mode of travel between i and j), the likelihood of the household locating in zone i increases. Note that this decision is based on travel potential, not on actual travel choice. The latter is only made in the lower level of the model, within the day-to-day activity–travel decision making of the household's worker.

5.4.4   Lowry Models

Spatial interaction models define the primary organizing principle of a class of integrated models known as Lowry models, named after the seminal work of Lowry (1964). Figure 5.4 provides a simplified overview of the Lowry modeling framework. Key elements of the Lowry model are the following:

1. Basic vs. retail employment. Lowry divided employment into two fundamental types: basic and retail. Basic employment involves economic activities whose magnitude and location are not a function of a local market, but rather are determined by more macro, extraregional factors. Examples include export-oriented industries, national and international corporate headquarters, major universities, etc. The determination of how much and where such employment activities

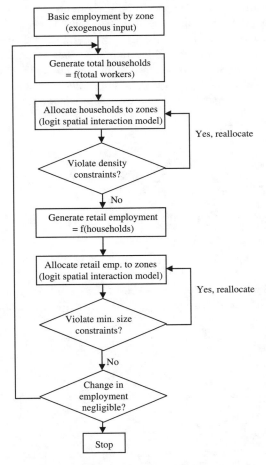

FIGURE 5.4 Lowry model flowchart. (Adapted from Meyer, M.D. and Miller, E.J., *Urban Transportation Planning: A Decision-Oriented Approach*, 2nd ed., McGraw-Hill, New York, 2001.)

are likely to grow, it can be argued, lies outside a model of intraurban processes and might be best handled on a scenario basis as an exogenous input to the model. Retail employment levels and locations, on the other hand, are a function of the size and location of local markets. Examples include retail shopping (hence the label) and other population-serving activities (local services, health care, schools, etc.). Retail employment is modeled as a process endogenous to the model, as a function of local population-based markets, as shown in Figure 5.4.

2. Use of spatial interaction models to determine population and retail employment distributions. Lowry models typically allocate workers' households' residential locations to zones within the urban study area using a spatial interaction model (typically a logit residential zone choice model), given the known workplaces of the employees, with travel time to work typically playing a major role in this spatial allocation process. The allocation of retail employment to zones is similarly accomplished by using spatial interaction models to estimate the amount of shopping interaction that will occur between each residential zone (given the predicted number of households in each zone) and each potential retail employment location.

3. Multiplier effects in the model. In a classic Lowry model, workers are transformed into households through worker-to-household multipliers. The magnitude of retail activity required to serve the population is similarly derived through household-to-retail employment multipliers, typically for several categories of retail activity. Thus, each worker generates a certain number of households. Each household generates a certain number of retail workers to serve it, which in turn generate additional households. As shown in Figure 5.4, this process of employment generating population and population generating employment is iterated until it converges, with the allocation of the households and retail employment to zones in each iteration being accomplished through the spatial interaction models briefly discussed above. This procedure represents a simplified approach to modeling both the household demographics and intraurban economic processes shown in Figure 5.1.

Through the 1960s and 1970s, Lowry-type models were by far the most common type of integrated model, as typified by the work of Wilson, Batty, and others in Britain, and by Lowry, Putman, and others in the United States (Goldner, 1971). To this day, virtually all operational models include Lowry-like elements, including the assumption of at least some exogenous employment components and, often, the use of spatial interaction models in key components of spatial allocation processes. The DRAM–EMPAL (also known as ITLUP) set of models, which are operational in several U.S. cities in particular, is explicitly a Lowry model in its design (Putman, 1996).

5.5 Modeling Land and Real Estate Markets

Two major, interrelated weaknesses exist in traditional Lowry models. The first is that they ignore the land development process. Typically, these models simply allocate households and firms to locations as determined by spatial interaction models without regard to the supply of appropriate building stock to house these activities. Recalling Figure 5.1, such models essentially combine the land development and location choice boxes into a single mixed stage, in which the supply process is essentially ignored. In reality, of course, land is first developed and redeveloped over time in order to supply the houses and commercial buildings demanded by households and firms. That is, markets exist for buildings of different types, with demand and supply processes at work within each market. Location choices are largely made within the context of (temporally) fixed supply, with longer-term land development processes altering this supply over time in response to the behavior of the real estate market.

The second problem is that the price of houses or other buildings is usually absent from these models or, at best, included as an exogenous input variable. Prices, however, clearly are the endogenous outcome of demand–supply interactions. They are strong explanatory variables in both demand and supply functions, and they are what primarily mediate between demand and supply to determine market outcomes (i.e., location choices). In particular, the best location for most households and firms is heavily constrained by affordability. Thus, it is difficult to envision credible models of land use and location choice that do not explicitly include building and land prices as endogenous components of the model. Each of these issues is discussed in the following subsections.

5.5.1   Building Supply and Land Development

Models of building supply and land development are arguably one of the weaker links in most integrated models, which is perhaps surprising given the land use focus of these models. This partially reflects the historical overreliance on spatial interaction models of location choice to act as proxies for land development processes discussed above, but it also reflects the difficulties inherent in modeling this process. Land development decisions can be highly idiosyncratic in nature, can take a long time to play out (e.g., land may be "banked" years in advance of actual development), depend critically on local policies and politics, and often are made by a relative handful of very heterogeneous decision makers (ranging from small individual builders and developers to huge multinational corporations). All of these factors make developing robust, generalized models of building supply difficult, to say the least.

Models of building supply generally make rather simple assumptions about the suppliers being profit maximizers who respond in a lagged fashion to market conditions. Thus, for example, the amount of new housing of a given type supplied in one time period is typically a function of prices and sales levels for this and other housing types in the previous time period, as well as the amount of developable land available in each zone.

5.5.2   Price Determination

Two approaches to the endogenous determination of prices within real estate markets are generally employed in currently operational models. The first involves solving for the set of prices within a given real estate submarket (e.g., the housing market) that balance the demand and supply in this submarket. That is, a set of prices for housing in each zone in the system is found, which results in all households that are active in the market being assigned to a vacant housing unit, while ensuring that no logical constraints are violated (e.g., one cannot assign more households to a given zone than the number of vacancies in this zone). METROSIM (Anas, 1998), MEPLAN (Echenique et al., 1990), and TRANUS (de la Barra, 1989) are all examples of integrated modeling systems that employ this approach to price determination.

The second approach derives from Alonso's (1964) concept of bid rent. In Alonso's theory, households have an amount that they would be willing to pay or "bid" for a house of a given type in a given location, in order to achieve a specific level of utility. Firms are similarly willing to bid a particular amount for locations in order to achieve a given profit level. Building or land owners are assumed to be profit maximizers who will auction their land or buildings to the highest bidder. Thus, whoever values a given location the most will bid the highest for it and thereby receive it. The result of this process is a distribution of activities and prices across the urban area as defined by this bidding process.

Ellickson (1981) extended Alonso's original deterministic model by reformulating it within a probabilistic random bid model in which bids are assumed to be stochastic in nature, and so the probability that a given agent will occupy a given location is equal to the probability that this agent is the highest bidder for this location.

Martinez (1992) has carried this concept further by developing what he calls bid choice theory. If we define the following terms:

I = household income
r_s = rent at location s for a dwelling of type d
T = available time after accounting for compulsory activities
z_s = neighborhood characteristics at location s
acc_s = accessibility to activities at s (as defined by log-sum terms, as discussed above)
P = price of composite good
β_h = parameters for household of type h

then the indirect utility of a dwelling of type d at location s for a household of type t can be expressed as:

$$u_{hs} = V_h(I - r_s, d, T, z_s, acc_s, P, \beta_h) \qquad (5.15)$$

Equation (5.15) can be inverted to obtain the amount which a household would be *willing to pay* at location s for a dwelling of type d to achieve utility level u_h^*:

$$r_{s,max} = V^{-1}(u) = WP_{hs}(I, d, T, z_s, acc_s, P, u_h^*, \beta_h) \qquad (5.16)$$

The household's *consumer's surplus* for a given dwelling or location is the difference between what it would be willing to pay and the actual price of the dwelling:

$$CS_s = WP_{hs} - r_s \qquad (5.17)$$

The optimal location for the household is then the one from within the set of feasible locations, Ω, that maximizes its consumer's surplus:

$$CS^* = \underset{s \in \Omega}{MAX} \{WP_{hs} - r_s\} \qquad (5.18)$$

Conversely, owners (or landlords) will sell (or rent) to the highest bidder. The price of the winning bid will be:

$$r_s^* = \underset{s \in N}{MAX} \{WP_{hs}\} \qquad (5.19)$$

where N is the number of households bidding. Substituting Equation (5.16) into (5.19) and maximizing will yield an equilibrium rent function of the general form

$$r_s^* = r(I_h, d, T, z_s, acc_s, P, u_h^*, w_{hs}, \beta_h, N) \qquad (5.20)$$

which can be empirically estimated from observed housing price data. Finally, if the willingness-to-pay function is assumed to have a random component, and if this random term is assumed to be identically and independently distributed Gumbel, then the probability that household h is the highest bidder for unit s is given by Ellickson's random bid logit model:

$$P_{hs} = \frac{\exp(WP_{hs})}{\sum_{h'} \exp(WP_{h's})} \qquad (5.21)$$

Equation (5.21) can be contrasted with Equation (5.8). Both yield the same result — the probability of a given household occupying a given dwelling unit. In Equation (5.8), however, the decision maker is the household, selecting a residential location from the set of available locations. Equation (5.21) inverts this process: in this equation the decision maker is the dwelling unit (actually, the current owner of the dwelling unit), and it selects the household that will successfully bid for the dwelling unit from among the set of competing households. The majority of currently operational models employ some version of Equation (5.8) (i.e., households choosing dwelling units), with prices being determined so as to match this demand with the available supply in each zone, as has been briefly discussed above. Examples of such models include MEPLAN, TRANUS, and METRO-SIM. However, operational models that employ the bid choice approach (i.e., in which dwelling units/current owners choose households) also exist. These include MUSSA (Martinez, 1996) and UrbanSim (Waddell, 1998).

5.6 Modeling Urban Economies

Over and above urban land and real estate markets, the behavior of the rest of the urban economy has a major impact on transportation network flows of persons and goods, as well as on the land and real estate markets themselves. Most current integrated models explicitly incorporate some representation of the urban economy. One approach (e.g., UrbanSim and MUSSA) is to take as inputs to the land use–transportation model the outputs of an external regional economic model (total employment or change in employment by industrial sector, total population growth, etc.). These regional economic model outputs provide control totals to the land use model and play a similar role to the basic employment inputs of Lowry models.

A second approach involves explicitly modeling the intraurban spatial economy as part of the integrated modeling system (Hunt and Simmonds, 1993). The MEPLAN–TRANUS (Echenique et al., 1990; de la Barra, 1989) family of models is representative of this later approach. Figure 5.5 provides a simplified version of the spatial input–output or social accounting matrix that lies at the heart of a MEPLAN-type model. In this matrix, columns correspond to the production of economic goods and

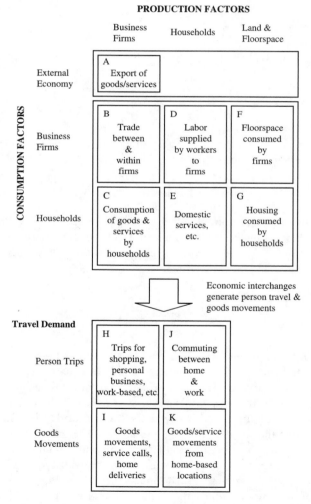

FIGURE 5.5 Social accounting matrix and its mapping into travel demand. (Adapted from Hunt, J.D. and Simmonds, D.C., *Environ. Plann. B*, 20, 221–244, 1993.)

services within the urban area, while rows correspond to the consumption of these goods and services. Some of the goods and services produced within the urban area are exported for final consumption in other regions of the country or the world (box A in Figure 5.5), while the rest are consumed internally within the urban area. This internal consumption occurs both within the urban area's business establishments (as inputs to their production processes, box B) and within the urban area's resident households (as final consumption of consumer goods and services, box C). The labor market (or, equivalently, the determination of place of residence–place of work linkages) can be incorporated into this framework by thinking of households producing labor that is consumed by businesses (box D). Households can also produce and consume labor among themselves (box E) in terms of domestic services and other unofficial interperson economic interactions.

Land enters the model as a third type of entity that is consumed (along with goods and services and labor) by businesses (box F) and households (box G), both of which must occupy land or floor space in order to exist and function.

The economic exchange of goods and services and labor results in the physical flow of goods and people within the urban area. Thus, the derived demand of travel is explicit in this framework, in that both person travel (boxes H and J) and goods movements (boxes I and K) are generated by these economic interactions, given the spatial locations of the businesses and households engaged in these interactions.

5.7 Design and Implementation Concerns

A large number of issues must be considered in the design or implementation of an operational integrated urban model. Different models, of course, will address these issues in a variety of ways, ranging from ignoring them completely to dealing with them in a very computationally detailed or theoretically rigorous manner. No right answer or approach necessarily exists with respect to any one of these issues. As with any model, the right or best design depends on the specific application context (data availability, computational and technical support capabilities, analysis and forecasting needs, etc.). In addition, no one issue or dimension of the problem can be optimized in isolation; it is the overall balance across design dimensions that is important (e.g., very fine spatial resolutions may be difficult, impossible, or unnecessary to maintain within very long range forecasting applications).

Design issues can be grouped into five categories: physical system representation, representation of active agents in the system, representation of decision processes, generic issues (which cut across virtually all physical system, active agent, and process representation considerations), and issues associated with the implementation of the model design within an actual computational environment. The first three categories deal with the substance of the system being modeled: the physical entities, the behavioral entities, and the processes by which these physical and behavioral entities evolve over time. The last two categories are more methodological in nature, dealing with how the representation of these entities and processes is actually implemented within an operational modeling system. Each of these groups of issues is discussed in turn in the following subsections.

5.7.1 Physical System Representation

Fundamental to model design are decisions concerning the representation of the physical elements of the system: time, land (space), buildings, and transportation networks. These decisions fundamentally affect the precision and accuracy of the model, its data and computational requirements, and options for the representation of behavior within the physical urban system.

5.7.1.1 Treatment of Time

All forecasting models must predict how an urban system state in some base year is likely to evolve into the future, typically up to some user-specified forecast horizon year. Choices of model base and horizon years, and the time increment or step used to move the system from the base to horizon year are fundamental design questions.

Also fundamental is the treatment of dynamics within the model. Many models assume that system equilibrium is achieved in each time step and so are able to appeal to the mathematical conditions for equilibrium to solve for the system state at the end of each time step. Ordinary gravity models are a classic example of this approach. Other models do not assume equilibrium. Rather, they explicitly simulate the evolution of the system state from one point in time to another as a function of various assumed processes.

The question of system dynamics is further complicated by the fact that different processes at work within the urban system operate on different time frames. Land development processes operate over time periods of decades or more; many household-level decisions are perhaps made on approximately a yearly basis; many activity–travel decisions change from week to week and from day to day; road network operating conditions (and hence energy consumption and tailpipe emissions) vary from minute to minute and second to second. Reconciling this wide combination of slow (or long-run) and fast (or short-run) dynamics within an overall modeling system is challenging, to say the least.

5.7.1.2 Treatment of Space

The spatial nature of urban systems represents one of the major sources of complexity in the analysis and modeling of these systems. Space enters in terms of both the locations of activities and the flows of people, goods, etc., between these activity locations. Design issues include zone system definition, degree of use of or interface with GIS software, and the degree to which micro neighborhood design attributes are incorporated into the set of spatial attributes maintained within the model.

5.7.1.3 Building Stock

While we often talk rather loosely about land use, most urban activities actually occur within buildings of one type or another, and the built environment, to a large extent, determines the nature of which activities occur where. The extent to which building stock (by amount, type, etc.) is explicitly represented within the model represents an important design decision and is found to vary considerably from one model to another.

5.7.1.4 Transportation Networks

Appropriate representation of both road and transit systems is clearly an essential component of any integrated urban model. Issues here include maintaining consistency in level of detail with the zone system being used, appropriate representation of transit walk access and egress, and appropriate representation of parking supply.

5.7.2 Representation of Active Agents

Active agents are the decision-making units — the people, households, firms, etc., who actually cause the urban area to exist and to evolve over time, through their various activities. People buy and sell homes; participate in the labor market; travel to and from work, school, shopping, etc., every day; (sometimes) get married and have babies; age and (eventually) die; etc. Firms similarly face location–relocation decisions; go through a life cycle process of birth, aging, perhaps with growth, and perhaps eventually "dying"; make land development decisions; supply the goods and services that people buy; provide jobs for workers; etc. Implicitly or explicitly, integrated urban models must address how they are going to represent the two primary active agents within cities: people and firms.

People live within either family or nonfamily units generally referred to as households. For many important activities, such as residential location choice and automobile holdings choice, the household is in most cases the natural decision-making unit, rather than the individual. Thus, the possibility exists that one might wish to explicitly represent both individual persons and households as interrelated but identifiably separable decision-making units within the model.

Other active agents obviously exist within urban areas that have direct impacts on the transportation–land use interaction, notably various government agencies, transportation service providers, etc. The extent to which such agents are explicitly incorporated within an integrated urban model is another design decision, although, in general, such agents are usually assumed to act exogenously to the processes being explicitly modeled.

5.7.3 Representation of Decision Processes

The primary processes that collectively define the transportation–land use interaction have already been discussed in this chapter, along with some of the major approaches used in modeling these decision processes. Probably the most important point to reiterate concerning spatial decision processes is that most or all of these processes are market driven. These include land development and building supply, residential and commercial real estate markets, labor markets, and travel markets. Proper representation of both demand and supply processes within each of these markets is essential to modeling such processes successfully. Implicit in this observation is that prices must be explicitly represented within the model and must be endogenously determined through the demand–supply interaction.

Another important issue that has not been explicitly discussed to this point is the nature and degree of integration between travel demand processes on the one hand and land development and location choice processes on the other. That is, while many integrated models have been mentioned in this chapter, these models vary considerably in how this integration is actually accomplished. In general, two classes of models can be identified. The first are fully integrated models, in that the determination of travel demand is tightly bound with location choice processes. In particular, work trip commuting patterns in these models are directly determined as part of the place of residence–place of work location decision making. Examples of fully integrated models include classic Lowry models and the MEPLAN–TRANUS family of models, in which residential locations are determined given known workplaces. In such models there is no need for the work distribution model of the traditional four-stage travel demand modeling system, since the distribution of work trips is co-determined with the calculation of the population and employment distributions.

This fully integrated approach can be contrasted with connected models in which population and employment distributions are determined within the land use side of the model without explicitly determining place of residence–place of work linkages. These linkages are determined on the transportation side of the model within a traditional work trip distribution model. UrbanSim and MUSSA are both examples of connected models, with the UrbanSim flowchart shown in Figure 5.2 providing a typical example of how such models work. Thus, in connected models, accessibility to employment opportunities (as determined by the travel demand model) enters the location choice model and thereby influences residential location decisions, but these decisions are made without explicitly knowing where household workers are actually employed.

Connected designs have the advantage of allowing land use models to be developed somewhat independently of travel demand models. This may be particularly advantageous when a good travel demand model already exists for an urban area, and one wants to extend the modeling system to include land use–location choice components. It also perhaps facilitates developing more complex, detailed models on both sides of the transportation–land use interaction by partially decoupling the two systems and thereby reducing the overall dimensionality of the problem being modeled. On the other hand, in reality, work trip commuting patterns are the outcome of residential and employment location decisions, and the traditional work trip distribution model is clearly quite an abstract and fairly artificial representation of the actual behavior. Also, as illustrated in the MEPLAN-type approach, theoretically rigorous and internally strongly consistent models can be facilitated by a fully integrated approach.

5.7.4 Generic Design Issues

Integral to the design of the representation of the physical system, the behavioral agents and the processes at work within the system are fundamental choices concerning aggregation level, boundaries between what is endogenous to the model and what is not, and process type. Each of these is briefly discussed below.

5.7.4.1 Level of Aggregation or Disaggregation

Many currently operational integrated models are quite aggregate in both space and time, often using less than 100 zones to represent an entire urban area and working in time steps of 5 or even 10 years. At the other extreme, as is discussed in more detail in Chapter 12, many researchers are experimenting with microsimulation models, in which individual households, building, firms, etc., are the basic model building blocks. Choice of aggregation level will have profound effects on data requirements, options for modeling processes, computation requirements, etc., and represents one of the primary, distinguishing decisions in any model design.

We generally think of the aggregation issue in terms of spatial aggregation (i.e., use of zones instead of individual people as the unit analysis; size of zones used; etc.). Aggregation decisions, however, are made with respect to every entity (physical or behavioral) and every process included in the model. Use of a 5-year time step to represent a process that occurs on a yearly (or shorter) basis constitutes temporal aggregation. Not including potentially salient personal attributes (say, for example, education level or occupation type) in decision-making models represents aggregation over attribute space. And so on.

5.7.4.2 Endogenous vs. Exogenous Factors

Any agent or process that is explicitly modeled within the model so that its attributes or behavior is determined within the model is said to be endogenous to the model. Conversely, factors that affect system performance but whose values are simply provided to the model as inputs are called exogenous factors. A fundamental step in any model design involves drawing the boundaries around the model, that is, determining what is to be included within the model vs. what will be excluded. As with the aggregation discussed above, these decisions will directly affect data and computing requirements, policy sensitivity, and process modeling options.

5.7.4.3 Process Type

Decisions must be made concerning how to model each endogenous process within the model. While a near-continuum of options exist, these can be broadly defined as falling into two categories: transition models and choice models (Wegener, 1995). Transition models use simple deterministic or probabilistic rules for determining changes in attributes, system states, etc., over time. Examples of transition models include most models for most demographic processes, such as deterministic population aging models (i.e., add 1 year to each person's age for each year being simulated) and fertility models, which express the probability of a woman giving birth to a child as a simple function of her age, marital status, etc. Choice models, on the other hand, attempt to model explicitly the choice process underlying a particular decision or action (random utility choice models and computational process models are both obvious examples of this class of model). Residential location choice, employment location choice, auto ownership, and activity–travel decisions are all examples of processes that one might typically model as choice processes within an integrated urban model.

While some processes may obviously fall into one category or the other (e.g., aging is a pure transition process), allocation of a given process to one type of modeling approach or the other is at least partially dependent on the application context, available data and modeling methods, computational resources, etc. For example, household formation and evolution in real life certainly are the results of complex interpersonal decision making. In most integrated urban models, however, such processes (if endogenously modeled at all) are represented using relatively simply transition models.

5.7.4.4 Model Specification

This includes both the selection of model functional form (logit model, etc.) and the explanatory variables to be included within the model. This issue is so integral to all model building that there is perhaps little that needs to be said with respect to it, except to point out the obvious facts that model specification determines theoretical soundness (and hence the fundamental credibility of the model), computational intensity, data requirements, and policy sensitivity (if a particular policy-relevant variable is not included in the model, then the model obviously will not be able to respond to the given policy).

5.7.5 Implementation Issues

All models require data, computational resources, and technical support to be developed, implemented, and maintained as an operational tool. Each of these issues is briefly discussed below.

5.7.5.1   Data Requirements

Historical data are required for both model estimation/calibration and validation. Estimation usually refers to the statistical estimation of model parameters that cause the model to best fit (in a statistically well-defined sense) observed, historical data (e.g., use of maximum likelihood estimation to estimate logit choice model parameters, or use of linear regression analysis to estimate trip generation model parameters). Calibration usually refers to postestimation parameter adjustments that force the model to better replicate observed data (e.g., use of K-factors in gravity trip distribution models to force the model to reproduce observed screen line or cordon counts). Given the complexity of most integrated urban models (typically involving many submodels, each one possessing its own level of complexity, often exercised within a simulation framework), a considerable amount of calibration, as opposed to estimation, is usually required in order to get these models working properly. This, in turn, implies the need for considerable experience and good professional judgment to be applied to the model development process.

Once a model has been estimated or calibrated, it should be validated as a forecasting tool by performing historical forecasts between two or more points in time in the past for which historical data are available. For example, a model may be calibrated using data from 1980 and 1990. Using 1990 as a base, it may then be used to "forecast" 2000 conditions. This 2000 "forecast" can then be compared with known data for 2000 in order to assess the ability of the model to predict beyond the time period covered by the calibration data.

The foregoing discussion indicates that integrated urban models typically require a considerable amount of historical data from multiple time periods in order to be calibrated and validated. The likely availability of historical data (what variables at what level of spatial detail for what years at what level of reliability, etc.) must be considered in the model design process, since there is no use in designing a modeling system that can not possibly be implemented due to data restrictions. Known, insurmountable data limitations will often drive the model design with respect to such important factors as time step, level of spatial aggregation, and choice of model specification.

Once a model is operational, it requires a new type of data to be used as a forecasting tool: estimated values of the exogenous inputs to the model for the future year(s) being simulated by the model. These estimates may come from policy scenarios, professional judgment, other models, etc., but, one way or another, they must be provided by the analyst to the model so that it can be run. These input data can be quite extensive, difficult to generate, and, of course, subject to error. In general, a classic trade-off exists in model design between specification error (which is built into the model due to model simplifications, abstractions, etc., which cause the model to fail to perfectly capture real-world behavior) and forecast error (error introduced during the forecasting process by inaccurate inputs). As with the model development data requirements, the forecast input data requirements must also be considered during the model design process and, again, may well impose significant practical constraints on model design with respect to the temporal, spatial, or behavioral representations that are feasible to achieve.

5.7.5.2   Computational Requirements

Integrated urban models by definition are computer based. The size of the computer (CPU, memory, disk space, etc.) required to house the model, the time required to execute a single run of the model (with obvious trade-offs between run time and computer size), and the software required to implement and support the model (i.e., the actual computer code within which the model is implemented, as well as the ancillary software — operating system, GIS, database management system (DBMS), statistical analysis systems, etc.) are all of critical concern within the model design process. Historically, the computing power cost-effectively available to researchers and planners has imposed significant limitations on the scale and scope of integrated urban models. However, past and continuing advances in computer technology are fast removing these barriers. The amazing power of desktop computers, the continuing emergence of parallel processing, the explosion of software, etc., are all extending the boundaries of what is feasible, to the point that computing power per se is probably no longer the primary constraint on practical modeling systems.

5.7.5.3 Technical Support Requirements

The discussion to this point has focussed on the model design and development process. Implementation of a model within a given planning agency, and then the ongoing maintenance and use of the model within this agency, requires significant technical support. In-house staff must be dedicated to the operation of the model; this staff must have appropriate professional backgrounds and must have been properly trained in the understanding and use of the model. An institutional, management-level commitment must exist within the planning agency to provide the time, money, and moral support required to get the model implemented and then to keep it operating effectively and efficiently. Adequate and ongoing support must also be available from the model developers (who usually will be external to the planning agency) with respect to training, troubleshooting, and ongoing system maintenance and upgrading. While largely implementation and operations, rather than design, oriented, the design implication of this issue is that an overly complex model design that is difficult to understand, operate, and maintain, or that is not robust with respect to its ease of use within an operational planning environment, will not be an attractive or even practical model for application within such contexts.

5.8 Concluding Remarks

Integrated land use–transportation models have the potential both to provide improved inputs into travel demand models and to permit a wider range of policy options affecting transportation system performance to be investigated. Such models have been evolving and improving in capabilities over the last four decades, to the point that many operational models exist and are in use worldwide. This trend toward greater reliance on integrated urban models is likely to be maintained over time as computing power, database quality, and modeling methods continue to improve, and as the needs for a more holistic analysis of urban land use and transportation policies continue to grow. This chapter has provided an overview of the basic structure of integrated models, the key modeling methods and assumptions employed in such models, and the major design and implementation issues involved in developing and using an integrated model in an operational planning setting.

References

Alonso, W., *Location and Land Use*, Harvard University Press, Cambridge, MA, 1964.

Anas, A., Discrete choice theory, information theory, and the multinomial logit and gravity models, *Transp. Res. B*, 17, 13–23, 1983.

Anas, A., *NYMTC Transportation Models and Data Initiative: The NYMTC Land Use Model*, Alex Anas & Associates, Williamsville, NY, 1998.

Ben-Akiva, M. and Lerman, S.R., *Discrete Choice Analysis: Theory and Application to Predict Travel Demand*, MIT Press, Cambridge, MA, 1985.

Brotchie, J.F., Dickey, J.W., and Sharpe, R., *TOPAZ Planning Techniques and Applications*, Lecture Notes in Economics and Mathematical Systems Series, Vol. 180, Springer-Verlag, Berlin, 1980.

Caindec, E.K. and Prastacos, P., A Description of POLIS: The Projective Optimization Land Use Information System, Working Paper 95-1, Association of Bay Area Governments, Oakland, CA, 1995.

de la Barra, T., *Integrated Land Use and Transport Modelling*, Cambridge University Press, U.K., 1989.

Echenique, M.H. et al., The MEPLAN models of Bilbao, Leeds and Dortmund, *Transp. Rev.*, 10, 309–322, 1990.

Ellickson, B., An alternative test of the hedonic theory of housing markets, *J. Urban Econ.*, 9, 56–79, 1981.

Goldner, W., The Lowry model heritage, *J. Am. Inst. Plann.*, 37, 100–110, 1971.

Hunt, J.D. and Simmonds, D.C., Theory and application of an integrated land-use and transport modelling framework, *Environ. Plann. B*, 20, 221–244, 1993.

Kim, T.J., *Integrated Urban Systems Modeling: Theory and Practice*, Martinus Nijhoff, Norwell, MA, 1989.

Landis, J.D., The California urban futures model: A new generation of metropolitan simulation models, *Environ. Plann. B*, 21, 399–422, 1994.

Lee, D.A., Requiem for large-scale models, *J. Am. Inst. Plann.*, 39, 163–178, 1973.

Lerman, S.R., Location, housing, auto ownership and mode to work: A joint choice model, *Transp. Res. Rec.*, 610, 6–11, 1976.

Lowry, I.S., *A Model of Metropolis*, RM-4035-RC, Rand Corp., Santa Monica, CA, 1964.

Martinez, F.J., The bid-choice land-use model: An integrated economic framework, *Environ. Plann. A*, 24, 871–875, 1992.

Martinez, F.J., MUSSA: Land use model for Santiago City, *Transp. Res. Rec.*, 1552, 126–134, 1996.

Miller, E.J., Kriger, D.S., and Hunt, J.D., Integrated Urban Models for Simulation of Transit and Land-Use Policies, Final Project Report to TCRP Project H-12, University of Toronto Joint Program in Transportation, Toronto, 1998 (published on-line by the Transportation Research Board, Washington, D.C., as Web Document 9 at www4.nas.edu/trb/crp.nsf).

Putman, S.H., Extending DRAM model: Theory-practice Nexus, *Transp. Res. Rec.*, 1552, 112–119, 1996.

Southworth, F., A Technical Review of Urban Land Use–Transportation Models as Tools for Evaluating Vehicle Travel Reduction Strategies, Report ORNL-6881, Oak Ridge National Laboratory, Oak Ridge, TN, 1995.

Waddell, P., An Urban Simulation Model for Integrated Policy Analysis and Planning: Residential Location and Housing Market Components of UrbanSim, paper presented at the 8th World Conference on Transport Research, Antwerp, Belgium, July 1998.

Webber, M., Pedagogy again: What is entropy? *Ann. Assoc. Am. Geogr.*, 67, 254–266, 1977.

Wegener, M., Operational urban models: State of the art, *J. Am. Plann. Assoc.*, 60, 17–29, 1994.

Wegener, M., Current and future land use models, in *Travel Model Improvement Program Land Use Modeling Conference Proceedings*, Shunk, G.A. et al., Eds., Travel Model Improvement Program, Washington, D.C., 1995, pp. 13–40.

Wilson, A.G., A statistical theory of spatial distribution models, *Transp. Res.*, 1, 253–269, 1967.

Yen, Y.-M. and Fricker, J.D., An Integrated Transportation Land Use Modeling System, paper presented at the 76th Annual Meeting of the Transportation Research Board, Washington, D.C., 1997.

Further Reading

An overview of the historical evolution of integrated land use–transportation modeling is provided in Chapter 6 of Meyer, M.D. and Miller, E.J., *Urban Transportation Planning: A Decision-Oriented Approach*, 2nd ed., McGraw-Hill, New York, 2001.

Much more extensive reviews of specific operational models are provided in Miller et al. (1998), Southworth (1995), and Wegener (1994), all cited above.

A major conference on land use modeling needs and methods was held in Fort Worth, Texas, February 19–21, 1995. The proceedings for this conference, *Travel Model Improvement Program Land Use Modeling Conference Proceedings* (edited by G.A. Shunk, P.L. Bass, C.A. Weatherby, and L.J. Engelke), are available through the U.S. Department of Transportation's Travel Model Improvement Program. They provide a very good discussion of planning needs and applications for integrated models, as well as a summary of the integrated modeling state of the art and practice that is still quite current.

Although now becoming somewhat dated, a very interesting comparison of a range of land use models was undertaken in the late 1980s, in which models from several cities around the world were cross-run against each other in each city included in the study. This study probably represents the most rigorous validation test of integrated models that has ever been attempted. The findings of the study are documented in the book *Urban Land-Use and Transport Interaction, Policies and Models, Report of the International Study Group on Land-Use/Transport Interaction (ISGLUTI)*, edited by F.V. Webster, P.H. Bly, and N.J. Paulley and published by Avebury, Avershot, in 1988.

Useful websites dealing with integrated models include:

DRAM/EMPAL: http://dolphin.upenn.edu/~yongmin/usl/intro.html
MEPLAN: http://www.meap.co.uk/meap/ME&P.htm

MUSSA: http://www.port.unican.es/50/
TRANUS: http://www.modelistica.com/
UrbanSim: http://urbansim.org/

6

Planning, Household Travel, and Household Lifestyles

CONTENTS

Kevin J. Krizek
University of Minnesota

6.1   Introduction

Concerns about urban sprawl, growth, and traffic are now among the most important issues facing the United States, edging out more traditional matters, such as crime and education. According to a series of new polls commissioned by the Pew Center for Civic Journalism (2000), respondents claim that spending "too much time in their car" was one of the main drawbacks to their quality of life. Embedded within concerns about "spending too much time in their car" lies at least three fundamental issues — automobile dependence, traffic congestion, and greenhouse gas emissions — each of which has risen to the point of needing urgent attention.

Consequently, transportation planners are looking to a variety of solutions. One prescription that has received increased attention as of late marries transportation planning with land use planning as a means to influence travel. Such ideas are not new; the question of how different forms of metropolitan development affect travel patterns has long been of concern to engineers and planners. One need only to examine the quotes below to understand past thought on this subject.

> If the problem of urban transportation is ever to be solved, it will be on the basis of bringing a larger number of institutions and facilities within walking distance of the home; since the efficiency of even the private motorcar varies inversely with the density of population and the amount of wheeled traffic it generates.
>
> **— Lewis Mumford**
> *The Urban Prospect, p. 70*

> In a nation that is both motorized and urbanized, there will have to be a closer relation between transportation and urban development. We will have to use transportation resources to achieve better communities and community planning techniques to achieve better transportation. The combination could launch a revolutionary attack on urban congestion that is long overdue.
>
> **— Wilfred Owen**
> *The Metropolitan Transportation Problem*

Historically, the bulk of the research exploring relationships between land use and transportation has centered on the effects of suburbanization in particular, and the degree to which compact vs. dispersed urban form affects household travel. At this level, the debate is a macroscale one, focusing on the overall structure of metropolitan regions. More recently, however, the spotlight has focused on the neighborhood, prompting a microscale debate. The fundamental question asks whether alternative types of urban, suburban, or ex-urban development engender different travel patterns. This line of inquiry focuses on the structure and travel patterns of a particular community or neighborhood within a metropolitan region. Such a discussion has prompted broader questions less concerned about documenting correlations between urban form and travel and more concerned about understanding the prospects of using land use planning to moderate travel given the myriad preferences, attitudes, and lifestyles among different households.

The land use planning initiatives being urged call for compact neighborhoods, a fine grain mix of land uses, neighborhood amenities, plus myriad improvements in urban design (e.g., sidewalks, street crossings, provisions for cyclists and transit users). This combination of features, it is presumed, will gel together at the neighborhood scale to provide residential and employment areas that make walking, cycling, and transit use more attractive. Increased development of such neighborhoods, it is hoped, will combat automobile dependence and its consequences (i.e., decreased social equity, increased pollution, increased fossil fuel consumption, loss of environmental lands).

These types of land use designs have been labeled neo-traditional development, transit-oriented development, traditional neighborhood design, or pedestrian pockets; such concepts have been recently rechristened new urbanism or smart growth. While different styles of development (new or old) may focus on different aspects (transit or pedestrian travel), each share a common underpinning from a transportation perspective. Each aims to provide increased levels of neighborhood accessibility (NA) that will allow residents to more easily drive fewer miles and more frequently use transit and walk (see Figure 6.1). Tables 6.1 and 6.2 contrast many of the characteristics for neighborhoods with high and low levels of NA. The proposed merits of high NA neighborhoods have been the focus of heated debate between academics, public officials, and policy decision makers over the past dozen or so years. In response, a considerable amount of research has been conducted examining relationships between urban form and travel.

This chapter is divided into seven parts. The aim is to provide the reader with a summary of past and current research, describe the relevance of this research to land use–transportation policy, and illuminate future thinking and research on this subject. The first section of the chapter describes the literature examining relationships between neighborhood-scale urban form and travel. This introduction sets the

| | Residential neighborhood exhibiting features characteristics of high NA: nearby commercial uses, narrow streets, on-street parking, sidewalks, and relatively higher density. |
| | Residential neighborhood exhibiting features characteristics of low levels of NA: curvilinear streets, driveways, lack of sidewalks, relatively low density. |

FIGURE 6.1 Photographic representations of neighborhoods with high and low levels of NA.

TABLE 6.1   Typology of Differences between High and Low Levels of NA

| Density | Levels of Neighborhood Accessibiliy | |
	High	Low (Post WWII Development)
Land use mix	Relatively higher residential densities Small home lots Mixed land uses and close proximity of land uses Convenient access to parks, recreation Distinct neighborhood centers	Relatively lower residential densities Large home lots Segregated, clustered land uses Access to a limited number of highly desirable land uses
Circulation framework and urban design	Interconnected, street patterns with small block size Separate paths for pedestrian and bicycles Narrow streets On-street parking Sidewalks, green spaces, and tree lining Variation in housing design and size Shallow setbacks Front porches and detached garages	Circuitous, meandering streets Strict attention to hierarchical street patterns (highways, arterials, collectors) Wide streets without on-street parking Missing or nonshaded sidewalks Homogeneous housing design Relatively large setbacks Dominating garages and driveways

stage for the remaining sections by identifying four primary gaps in previous research. The following four sections identify in detail the shortcomings of previous research and describe strategies to address such shortcomings. In each section, results from recent research by the author are used as examples to demonstrate the particular aspect being described. The setting for the results that are presented is the Seattle metropolitan area, and the data used in each piece of analysis come from the Puget Sound Transportation Panel (PSTP). The final section describes emerging thoughts on relationships between urban form and travel by suggesting a handful of future research needs and research topics.

6.2   Travel Behavior and Neighborhood Access: What Do We Know?

The potential of urban form in moderating travel has been the subject of almost 100 empirical studies. Any single review cannot do justice to the innumerable issues, approaches, findings, and shortcomings involved in the synthesizing of these studies. At least two bibliographies cover the literature in annotated form (Handy, 1992c; Ocken, 1993). A handful of literature reviews are also available (Handy, 1996a; Pickrell, 1996; Crane, 2000). As mentioned in Ewing and Cervero (2001), the reader may wonder whether another literature survey can add much value. For this reason, the review offered in this section does not examine in detail existing literature related to urban form and travel. The reader is urged to consult Handy (1996a) and Crane (2000) since both reviews focus on the different approaches used in past studies, explaining their techniques, strengths, and weaknesses. The focus of this section is twofold. The first is to provide the reader with a better understanding of both the complexity and disparity of existing research. The second is to clearly articulate four gaps of knowledge left open in previous research.

Given the complex array of issues at stake in such a research endeavor, any number of data, research approaches, and analysis strategies could be employed. Consequently, any review of such research could be organized in a variety of ways. For example, Boarnet and Crane (2001) list different strategies to organize studies (see Table 6.2). The first category relates to the travel (dependent) variables being analyzed. Depending on data availability, most studies separately examine one dimension of travel (e.g., trip generation for work vs. nonwork travel). Doing so reduces the extent to which different studies can be compared because they often analyze different phenomenon. A second strategy for organizing a review separates studies according to the independent variable. For example, Ewing and Cervero (2001) discuss different analyses according to their findings on at least four different dimensions of the built form: land use patterns, transportation network, urban design features, or composite

TABLE 6.2   Taxonomy of Ways to Classify Studies Related to Urban Form and Travel

Travel Outcome Measures	Urban Form and Land Use Measures	Methods of Analysis	Other Distinctions and Issues
Total miles traveled (e.g., VMT) Trip generation Vehicle trip generation Time spent traveling Car ownership Mode of travel Congestion Commute length Other commute measures (e.g., speed, time)	Density Land use pattern Land use mixing Traffic calming Circulation pattern Jobs and housing balance Pedestrian features (e.g., sidewalks, perceived safety, visual amenities) Composite indices	Simulation Description of observed travel behavior in different settings (e.g., commute length in large vs. small cities) Multivariate statistical analysis of observed behavior	Land use and urban design features as the trip origin vs. the destination vs. the entire route Composition of trip chains and tours (e.g., use of commute home to buy groceries) Use of aggregate vs. individual level traveler data and aggregate vs. site-specific urban form data

Source: From Boarnet, M.G. and Crane, R., *Travel by Design: The Influence of Urban Form on Travel*, Oxford University Press, New York, 2001. With permission.

indices.[1] The third category groups studies that use similar methods of analysis (e.g., simulation studies, aggregate analysis, disaggregate analysis, choice models, and activity-based analysis (Handy, 1996a)). But even within each grouping, there remains considerable variation.[2] Confounding issues stem from varying units of analysis (e.g., disaggregate vs. aggregate) or measuring only trips from certain origins or destinations.

Despite such disparities in methods, approaches, or data, it is helpful to shed light on some of the findings from an extremely rich and active line of research. Doing so provides a better appreciation for the range of issues discussed, the travel behavior variables used, the urban form measures employed, and the general pattern of results. Early work primarily used matched pair analysis and aggregate statistics to examine travel outcomes in neighborhoods with varying degrees of neighborhood access. Crudely simplifying this stream of research suggests the following:

- Fewer vehicle miles traveled (VMT) in neighborhoods with higher density and better transit access (Holtzclaw, 1994)
- Fewer VMT and more pedestrian and transit trips in neighborhoods that are more pedestrian friendly (1000 Friends of Oregon, 1993)
- Fewer total trips and slightly higher ratios of transit use and pedestrian activity in traditional neighborhoods vs. standard suburban neighborhoods (McNally and Kulkarni, 1997)
- Higher percentages of transit use for commuting in some transit neighborhoods relative to automobile neighborhoods (Cervero and Gorham, 1995)
- Two thirds more vehicle hours of travel per person for households in sprawling-type suburbs vs. comparable households in a traditional city (Ewing et al., 1994)
- More pedestrian activity in mixed-use centers with site design features that include sidewalks and street crossings (Hess et al., 1999)

In later work more disaggregate approaches analyze the travel behavior of individual households within neighborhoods to better understand travel choices and areas with high NA. These studies use analysis of variance, regression, or logit models to compare the relative influence of different urban form characteristics to sociodemographic characteristics. Again, simplifying the results suggests:

- More walking to shopping and potentially less driving to shopping for residents in some traditional neighborhoods (Handy, 1996b, c)
- Fewer vehicle hours of travel for residents in neighborhoods with higher accessibility (Ewing, 1995)
- Higher percentages for transit and nonmotorized trips for residents closer to the bus or rail and in higher density neighborhoods (Kitamura et al., 1997)
- Reduced trip rates and more nonauto travel for individuals living in neighborhoods with higher density, land use mix, and better pedestrian orientation (Cervero and Kockelman, 1997)
- Walking to transit stations more likely where retail uses predominate around stations (Loutzenheiser, 1997)

---

[1] While this strategy may help better understand the relative effect of each element, such an approach has at least two principal shortcomings. First, many studies examine more than one dimension of urban form in concert with other dimensions. Second, some studies use a single measure (e.g., street pattern) to represent the myriad dimensions of NA. Thus, assessing the independent effect of one variable without fully considering the range of other variables, often times does not do justice to the specific dimension under question. It speaks more to the limitations of singling out the individual effect of one element of the built environment as opposed to attempting to fully capture the myriad dimensions of NA.

[2] For example, studies with similar methods of analysis may still analyze different dependent variables, or they may employ different analysis techniques (e.g., regression models vs. discrete choice models).

- Higher trip frequency in areas of high accessibility to jobs or households (Sun et al., 1998)
- A reduced number of nonwork auto trips in zip code areas with higher retail employment densities (Boarnet and Greenwald, 2000)
- Higher transit passenger distance in areas with fewer jobs and grocery stores within 1 km (Pushkar et al., 2000)
- More walking and transit use, lower VMT, and less frequent auto trips in areas with higher composite indices (Lawton, 1997)
- Use of nonauto modes more likely in areas with greater mixing of commercial–residential uses (only in middle suburbs); auto use is less likely in areas (Pushkar et al., 2000)

Given such extensive research, it seems that we should be in a position to inform planning commissioners and decision makers about the capacity of land use policy in managing travel. Each of the above studies show that different dimensions of urban form appear to influence travel in hypothesized (and expected) directions. However, R^2 values rarely exceed 0.40 in such work, suggesting, in part, that there remain many unexplained factors that influence travel. Our knowledge of these issues is analogous to peeling an onion: as each layer is revealed, another layer is found. One study may find that NA is associated with shorter trip distance to conclude less travel; a different study may find that NA is associated with greater trip generation to conclude more overall travel. Still other approaches may look at mode split. Each study reveals new and different questions — questions that previous data, methodologies, or analysis leave unanswered. In general, at least four overarching issues confound past research endeavors; these four issues provide the framework for the remaining sections of this chapter.

- The first stems from concerns about existing data. The limited nature of its availability and the manner in which it is often operationalized to measure both travel and urban form is deficient for arriving at certain conclusions.
- The second confounding issue is that most studies fail to acknowledge the total demand for travel. Measuring associative relationships on a limited number of travel outcomes does not uncover the total travel of most households, and it is not able to capture trade-offs and interactions between trip frequency, trip distance, multipurpose trips, and mode split.
- Third, researchers are increasingly realizing that for their work to best address land use and transportation policy, they need to *better* disentangle the myriad factors that influence travel — the role that attitudes or preferences have vs. the role of urban form.
- Finally, past research also fails to recognize that relatively short-term decisions (e.g., where to travel and how) may not always be conditioned by relatively longer-term decisions (e.g., where to live and how many cars to own). These types of decisions serve to mutually inform one another and should be analyzed in tandem.

Echoing the sentiments expressed by both Handy (1996a) and Boarnet and Crane (2001), one can begin to see the difficulty involved in putting together pieces of a puzzle related to urban form, travel behavior, and residential location. Such complexities have even led some to contend that "not much can be said to policy makers as to whether the use of urban design and land use planning can help reduce automobile traffic" (Crane, 2000).

6.3   Understanding Data: Its Demands and Shortcomings

A considerable amount of discussion over past neighborhood-scale travel studies stems from issues related to data collection and processing. After briefly describing issues central to travel data, the bulk of this section focuses on issues central to the urban form data; this latter discussion is separated into three parts: availability, processing, and ability to capture multiple dimensions.

6.3.1 Travel Data

A thrust of the increased NA movement supposes that residents will shed their auto-using behavior in favor of walking, cycling, or using transit. To assess the merits of such claims obviously requires researchers to have adequate account of such travel. The problem lies, however, in that walking, cycling, and transit trips tend to be either: (1) underrepresented in typical travel surveys, (2) underreported using typical survey methods, or (3) a combination of both. Travel surveys are notorious for undersampling lower income populations who tend to rely on non-auto-based forms of transportation more frequently. Travel diaries often ask households to record only trips longer than 5 min in duration. The coding schemes for many surveys fail to consider the following activities as trips: a walk around the block, errands completed within the same block, a visit to a neighbor. But each of these types of trips is central to better understanding the difference in travel that neighborhood-scale design may have.

Data concerns transcend walking or transit travel, however. In travel surveys the distance of each trip is typically calculated for a given zonal origin–destination pair using the road network assignment procedure from the region-wide transportation model; all trips are assumed to start at the centroid of the traffic analysis zone (TAZ) and all trips are assumed to be loaded onto the network. The accuracy of this procedure tends to suffice for longer trips (i.e., over 5 mi), but does injustice to the accuracy of shorter trips (McCormack, 1999). For the same reason as walking trips, these shorter trips (e.g., trips that never leave a TAZ or those to neighboring zones) are of intense interest in this line of work and tend to be grossly misreported using data from typical travel surveys.

6.3.2 Urban Form Data

6.3.2.1 Data Availability

From an urban form standpoint, at least three issues stand out: data availability, the manner in which data are processed, and the need to capture multiple dimensions of urban form. An initial concern is that researchers aiming to understand the travel impacts of neighborhoods designed around the new urbanist paradigm have been somewhat stumped. Such neighborhoods are difficult to study because they are only slowly being developed and occupied; few have matured with full residential occupancy and well-established retail or schools. Researchers therefore rely on second-best strategies to examine the attributes in existing traditional neighborhoods thought to mirror many new urbanist characteristics (thus the term neo-traditional).

Using traditional neighborhoods as proxies for new urbanist neighborhoods draws attention to the ability to measure the attributes of such neighborhoods. Regional databases, while widely available, provide aggregate measures or coarse representations of the street network. Such data are hardly suitable to operationalize issues central to NA. Few municipalities maintain databases specifying detailed urban form features, such as the size and type of commercial activity centers, parking supplies, sidewalk and landscaping provisions, or the safety of street crossings. Density measures (available through the U.S. Census) provide block group data that are relatively disaggregate. Parcel-level GIS databases are becoming increasingly available in some metropolitan areas. But being inherently large and messy files, they are incomplete in many instances. Several research efforts have conducted extensive fieldwork to collect primary data, capturing many fine-grained measures of urban form (1000 Friends of Oregon, 1993; Cervero and Kockelman, 1997; Moudon et al., 1997; Bagley et al., 2000). Though comprehensive in their approach, these efforts usually prove prohibitively expensive to do over an entire metropolitan area.

6.3.2.2 Units of Analysis

Largely because of limited data, the majority of past research depicts the neighborhood unit by aggregating information to census tracts, zip code areas (TAZs). These units often do little justice to the central aim; they can be quite large, almost 2 mi wide, and contain over 1000 households. The problem is that an ecological fallacy arises because average demographic or urban form characteristics are assumed to apply to any given individual neighborhood resident.[3] Furthermore, census tracks or TAZs are often delineated by artificial boundaries (e.g., main arterial streets) that bear little resemblance to the neighborhood scale phenomenon being studied in terms of their size or shape. Consider a four-way intersection with retail activity on all four corners. TAZ geography may divide this retail center into different zones, thereby diluting the measure of commercial intensity for any single zone. In terms of affecting travel behavior, the commercial intensity of all four corners should be grouped together.

6.3.2.3 Capturing Multiple Dimensions

Any strategy to operationalize NA needs to be guided by the overall purpose of the study in combination with the nature of available data. Aggregate urban form measures suffice for uncovering general differences between two different neighborhoods (Friedman et al., 1994). Geographically detailed measures are usually preferred for more disaggregate modeling purposes (Cervero and Kockelman, 1997). In either case, however, the researcher needs to be able to sufficiently tease out and capture different dimensions of urban form.

A first distinction that needs to be made is that effects of the NA need to be differentiated from the urban form effects at the regional scale. Household travel may be influenced by both the immediate locale — the character of the particular neighborhood in which the household lives — and the position of the neighborhood[4] in the larger region. Using a single dimension of urban form, a given place may be very far from a few large activity centers or close to several small activity centers, yet the implications for travel behavior may be very different (Handy, 1993). The regional context of a neighborhood, too often neglected in research, may provide more opportunities that mean more travel. Or the regional structure may simply dwarf variation in NA.

A second issue relates to the way in which neighborhoods are measured — generally in one of three ways: binomial (matched pair), ordinal, or continuous. The first approach, binomial, is frequently used with quasi-experimental techniques, matching more compact and mixed-use neighborhoods with lower-density single-use neighborhoods (Handy, 1992a; Friedman et al., 1994; Cervero and Gorham, 1995; Cervero and Radisch, 1996; Dueker and Bianco, 1999; Hess et al., 1999). Two classifications, however, tend to define the extremes of development; many neighborhoods contain a mix of attributes. Several studies therefore use ordinal classifications to rank neighborhoods with similar characteristics (Ewing et al., 1994; Handy, 1996c; McNally and Kulkarni, 1997; Levine et al., 2000). While both binomial and ordinal approaches are easy to understand and straightforward to operationalize, they are limited in at least two respects. First, they tend to restrict the sample size because of the limited number of neighborhoods in which it is possible to control for other socioeconomic conditions. Second, individual urban form variables are used to group the neighborhoods. This often precludes the ability to assess the independent effect of different elements of urban form. A third strategy conceptualizes neighborhoods in a continuous manner and is relied on more recently as detailed urban form data become increasingly available (Hanson and Schwab, 1987; Frank and Pivo, 1994; Holtzclaw, 1994; Ewing, 1995; Cervero and Kockelman, 1997; Kitamura et al., 1997; Boarnet and Sarmiento, 1998;

---

[3] As an example, research in the Central Puget Sound identified almost one-hundred concentrations of multifamily housing within one mile of retail centers and/or schools (Moudon and Hess, 2000). By aggregating measures of commercial intensity, each zone reveals the same measure. However, each development pattern is likely to affect travel behavior differently. Because census tracks or TAZs average out these types of concentrations with adjacent lower-density development, it is difficult to associate many neighborhood-scale aspects with travel demand.

[4] Restricting attention to the physical-spatial dimensions, the neighborhood as first conceived by Perry (1929) was thought of as a geographic unit. He proposed that the neighborhood unit contain four basic elements: an elementary school, small parks, small stores, and buildings and streets all configured to allow all public facilities to be within safe pedestrian access. Many studies attempt to measure Perry's concept of neighborhood using a variety of units of analysis. Some efforts use relatively large districts of a metropolitan area (Cervero and Radisch, 1996). The other extreme does not describe any neighborhood boundaries; the term "neighborhood" assumes individual meanings for each respondent (Lansing et al., 1970; Lu, 1998). A middle ground defines neighborhood using a buffer distance around each household (Hanson and Schwab, 1987).

Crane and Crepeau, 1998; Frank et al., 2000). Continuous rankings of neighborhoods differ from matched pair or ordinal rankings because the individual urban form measures are often entered directly into the statistical analysis rather than used to classify neighborhood types. This allows at least two primary advantages. It typically allows a wider variation between neighborhoods and therefore larger sample sizes. Second, it allows the researcher a means to more easily assess the partial effect of urban form variables on either travel or residential location.

Finally, the researcher needs to ensure that different dimensions are sufficiently captured in any measure of NA. For example, density has long been used in land use–transportation research as a powerful predictor of travel behavior. In many contexts it is the only urban form variable used. Neighborhood attributes such as increased density, mixed land uses, and sidewalks usually coexist; such features represent a package of characteristics usually found together, particularly in areas more traditional in character. The predictive value of density is often relied on as a proxy measure for other difficult-to-measure variables that may more directly affect travel behavior (Steiner, 1994; Ewing, 1995).[5]

6.3.3   Recap and Policy Significance

Density (or any other single indicator of urban form) cannot always be relied on as a sole measure of NA. Imagine a tight cluster of residential-only apartments located in a suburban community away from other basic services. This cluster of buildings may be high density, but by itself does little in terms of decreasing travel distance to nonresidential uses. Residents would still need to travel considerable distances to buy a quart of milk. Even spreading basic services around this residential cluster would not guarantee the neighborhood to be well suited for walking or transit.[6] Would a neighborhood with high density and sidewalks but no diversity in land use lead to increased pedestrian activity and decreased driving? How about a neighborhood that is diverse in land use, but surrounded by fast-moving vehicles and eight-lane roadways?

The concept of NA embodies multiple, perhaps infinite dimensions. The conundrum from a research standpoint is uncovering the most effective strategy to capture these myriad dimensions. Measuring a single variable does not do justice to the multiple dimensions of NA. On the other hand, it is difficult to identify the partial effects of one characteristic over another; some contend that it may even be a futile endeavor to isolate the unique contribution of each and every aspect of the built environment (Cervero and Kockelman, 1997).

6.4   Understanding the Total Demand for Travel and Urban Form

A second important issue stems from the fact that travel behavior is often measured using a single dimension such as mode split, trip frequency, or travel distance. Simplifying the dependent variable in this way does not do justice to possible trade-offs between different dimensions of travel. The substance and nature of past research — primarily showing associative relationships — has only recently been brought into question.

For example, Handy's (1996b, c) work provides empirical evidence of Crane's (1996b) assertion that open and gridded circulation patterns make for shorter trip distances and may even stimulate trip taking. He argues that residents with higher neighborhood access may shop more often and drive more miles

---

[5]   In a study of transit-supportive designs across a number of U.S. cities, Cervero (1993) concluded that micro-design elements are often too 'micro' to exert any fundamental influence on travel behavior, more macro factors like density and the comparative cost of transit vs. automobile travel are the principal determinants of commuting choices.

[6]   The research by Moudon and Hess (2000), for example, identified several clusters of relatively high-density residential environments, all with nearby retail. Many of these clusters were found not to stimulate increased pedestrian activity, because they lacked, among other things, qualities such as good urban design and/or small block sizes. This finding prompts researchers to more fully consider the variety of characteristics that would promote areas with high levels of NA.

overall. Boarnet and Crane (2001) subsequently argue that basic relationships between urban form and travel have not been analyzed within a behavioral framework that considers basic tenets such as the cost (in terms of time or convenience) of each trip. This assertion echoes results found in recent work (Boarnet and Sarmiento, 1998; Crane and Crepeau, 1998) that remain skeptical about urban form's potential to moderate travel demand, especially with respect to vehicle trip generation.

If high NA prompts increased trip making, important policy questions lie in the degree to which additional trips (1) supplement trip making, (2) substitute for trip making (and if so, which types of trips), and (3) are made by environmentally benign modes. Only if additional trips are made by environmentally benign modes or substitute for other travel would there be advantages of NA from a travel behavior standpoint. Unfortunately, the question of substitution is an elusive and underresearched dimension of travel — one that can be best uncovered by combining quantitative and qualitative approaches.[7] Using regression or logit models on a limited number of dependent variables is able to shed light on only one piece of the puzzle.

An additional confounding issue stems from the fact that most studies analyze individual trips independently. This approach masks sequential and multipurpose travel because many trips are often a function of the preceding trip. The decision to drive to the dry cleaner may not be because a car was required for this trip; rather, it may be because the dry cleaner trip was done on the way to the grocer — a trip that required a car in the first place. Examining individual trips instead of the larger pattern of linked trips fails to work with the basic forces that generate and influence travel. It is also important to examine multiple trip purposes — both work and nonwork. Commute data are often analyzed because they are readily available and have long been considered the lion's share of metropolitan travel flow; nonwork trips are analyzed because they represent trip types most directly influenced by neighborhood access. Over two decades ago, Hanson (1980) stressed the importance of analyzing work and nonwork travel jointly, because separating trips by type fails to capture linked and multipurpose travel behavior that we know exists.

Unlike substitution travel, our understanding of linked travel has fortunately benefited from over two decades of research. A major shortcoming of such research, however, lies in the degree to which linked travel is married with NA. To develop a better understanding of how NA relates to household travel, the remaining part of this section is broken into four parts: (1) the typical range of services offered in areas with high NA, (2) the limitations of trip-based travel analysis, (3) travel tours (e.g., the sequence of trips that begin and end at home) and a typology of travel tours that consider different travel purposes, and (4) relationships between tour type and NA.

6.4.1   Understanding Accessible Neighborhoods and Travel Purpose

To the extent that travel is a derived demand (i.e., individuals travel to engage in activities in other places — work, recreation, shopping, health services), it is important to consider the types of activities that households engage in. The success of NA to influence travel behavior depends in large part on the opportunities that are provided for. It is axiomatic, yet worth repeating, that the variety, location, and type of destinations are critical.[8]

To date, this discussion is best addressed by Handy (1992b), who describes that commute patterns are relatively fixed; they are often constrained by larger forces such as time of day and route.[9] Therefore,

---

[7] To the extent substitution travel can be addressed, it is still likely to yield small travel savings. Even if the majority of residents in high NA neighborhoods substitute a walk to the corner store for driving, one attempt to quantify the savings in terms of vehicle miles is estimated on the order of 3.4 miles per month (Handy and Clifton, 2001).

[8] Crane (1996b) discusses in detail trip demand models that can be specified by type of urban design feature and trip purpose. The reader is urged to consult his application of the economic concepts of price and cost to issues of trip generation and accessibility. The discussion provides important, yet often overlooked, assumptions related to urban form and travel. He does not, however, speak to the different purposes of travel that may most likely be influenced by neighborhood access.

[9] This argument, however, realizes that the ubiquitous transportation network now found in most U.S. metropolitan areas considerably relaxes the assumption that households tend to choose residential locations primarily close to employment location. Generally speaking, the once prominent role of the work commute is diminishing in importance.

travel for purposes other than the commute (i.e., nonwork travel) remain more flexible and tend to be more likely influenced by different levels of NA. Subsequently, the simply disguised distinction between work and nonwork travel is one commonly considered in literature relating urban form and travel (Ewing et al., 1994; Ewing, 1995; Kockelman, 1996; Cervero and Kockelman, 1997; Boarnet and Sarmiento, 1998; Crane and Crepeau, 1998; Boarnet and Greenwald, 2000).

Such a simple classification, however, does not do justice to examining how travel behavior is affected by different patterns of urban form for at least two reasons. First, suggesting that nonwork travel is more influenced by levels of neighborhood access tends to oversimplify the range of services often included in such neighborhoods. Second, separating work trips from nonwork trips is unable to account for travel that links multiple purposes. Each shortcoming is addressed in more detail below.

The first question tackles how the ranges of activities in accessible neighborhoods compare with the types of activities for which households travel. As a starting point, Table 6.3 comments on the likelihood that eight different purposes of travel would be available in areas with high NA. At first glance, the potential to capture a variety of trip types appears to be relatively high. The right-hand column in Table 6.3 shows that travel types for five purposes (appointments, personal, college, school, and shopping) are likely to be contained in areas with high NA; in contrast, trip purposes related to work, free time, or visiting do not appear to have similar drawing power.

TABLE 6.3 Travel Purpose and the Likelihood That Purpose Will Be Available in Areas with High NA

Purpose of Travel (Definition)	Likelihood Travel Type Is Contained within an Area Considered to Have High Neighborhood Access
Work	When considering residential areas based on NA, major employment opportunities are not likely. Even a careful read of many designs for new urbanist villages reveals that employment is not a major feature of such designs. Furthermore, when employment opportunities would be available within the neighborhood, there is seldom a satisfactory match between the residents' skills or preferences and the jobs offered.
Personal (getting a service done or completing a transaction, e.g., banking, gas station, dry cleaning)	Advocates of NA would contend that most of these activities, if not all, would be available within the neighborhood domain.
Free time (non-task-oriented activities, e.g., entertainment, dining, theater, sports, church, clubs, library, exercise)	The relatively wide range of activities available in this category makes it difficult to posit which ones are likely to be within a community with high NA, though most would certainly be available.
Shopping (travel to buy concrete things); shopping services, as suggested by Handy (1992b), can be divided into three categories	Convenience shopping (e.g., bread, milk) is the activity most heralded by NA designs; every neighborhood based on principles of NA is urged to have a corner store.
	Comparison goods shopping (e.g., furniture, appliances, clothing) is increasingly being satisfied by big-box and superstores, which tend to locate on large tracts of land with ample parking. Such locations are typically the antithesis of areas with high NA.
	Specialty goods shopping (e.g., niche markets, boutiques) typically involves shopping that customers will put forth a special effort to visit. The size and nature of the shops meshes well with NA designs.
Appointment (activities to be done at a particular time, e.g., doctor's appointment, meeting)	One would expect a residential neighborhood to have standard appointment services (e.g., dentist, general physician), but not necessarily more specialized services.
Visiting	One would expect a close locale of people in highly accessible neighborhoods. However, personal, cultural, and sociodemographic preferences do not ensure that they will be nearby.
School	Schools are strongly urged, especially elementary schools. With each advance in education level, however, the likelihood of being within a residential neighborhood decreases rapidly.
College	Where colleges and universities are present in neighborhoods, they are most likely an intricate part of the community.

6.4.2   Introducing Tour-Based Analysis

Analysis that separates work from nonwork trips suffers from two related problems. First, it considers each type of trip in an isolated manner. Second, it does not allow a means of accounting that is able to capture travel that combines multiple purposes. Examining only individual trips instead of the larger pattern of linked trips fails to work with the basic forces that generate and influence travel. Furthermore, doing so provides an incomplete account of the travel behavior picture.

Two decades of research suggest strategies to circumvent what has been referred to as the isolated trip approach (Damm, 1982). A technique for taming the complexity of travel involves organizing travel into multistop trips, commonly known as tours or trip chains. Tours better recognize that travel is a function of the interaction between many factors, including types of destinations, previous destinations, subsequent destinations, travel mode, and household and individual characteristics. When multiple tours are jointly considered across a day, sequence of days, or even a week, they provide a means to more robustly track the schedule of activities in which individuals participate.

6.4.2.1   Approaches to Operationalizing Tours

While the idea of multistop journeys is straightforward, the concept is more difficult to operationalize. In this section, factors that influence the nature of tours, a task that has been partially completed in the literature reviewed thus far (see Thill and Thomas, 1987), are discussed. By convention, the literature most often defines tours in terms of the home-to-home loop (Bowman et al., 1998) to better understand how activities are spread throughout daily, bidaily, or weekly travel patterns. Tours are most commonly analyzed by the number of trips (i.e., stops). Simple tours contain two trips (e.g., home to work and then work to home); complex tours contain more than two trips. The complexity of tours has been measured in a variety of contexts. Adler and Ben-Akiva (1979) develop a theoretical model that explicitly accounts for the trade-offs involved in the choice of multiple-stop chains. Using a cross between qualitative and quantitative research, Clark et al. (1981) draw correlations between trip chain complexity, household characteristics, and life cycle. Recker and McNally's (1985) analysis shows that the likelihood of chaining trips is positively associated with the number of trips taken and negatively related to activity duration, employment status, and age. Williams (1988) considers household activity, trip frequency, and travel time in concert with accessibility indices to show that residents in less accessible areas have a higher likelihood to form trip chains and have higher trip frequencies. Strathman et al. (1994) analyze trip chaining differences among household types by developing models to estimate the propensity to link nonwork trips to the work commute and to estimate nonwork travel by three chain types: work commutes, multistop nonwork journeys, and unlinked trips. More recently, Wallace et al. (2000) estimate a model to predict the number of trips in a chain based on characteristics of the household, the traveler, trip type, and origin location.

Analyzing the nature and frequency of simple vs. complex tours, however, considers only one dimension of the tour: number of stops. It does not do justice to how a separate dimension of travel — purpose — influences the nature of tours. Travel purpose is important to consider because land use initiatives based on NA potentially capture different types of travel.

6.4.2.2   Accounting for Multipurpose Tours

Employing tours as a unit of analysis prompts the following challenge: how to assign a single purpose to what is often a multitrip or multipurpose tour. To better capture how different purposes of travel — a nominal variable — interact with trips, classification emerges as the preferred strategy. Although it is the lowest form of measurement, classification allows many variables to be considered simultaneously (e.g., the purpose and number of trips on a tour). Only a handful or so of studies present different ways to analyze travel behavior using tours (or chains) that specify different purposes of travel, presented in Table 6.4.[10]

Any classification scheme used depends on the particular purpose of the study or application. A detailed coding scheme (e.g., Golob, 1986) is advantageous because it provides a means to more precisely track the sequence of detailed travel purposes. While even 20 classifications of tour type do not capture all

TABLE 6.4 Different Strategies for Classifying Tours

Golob (1986)	Pas (1982)	Southworth (1985)
H-W-H	H-W-H	H-X-H
H-W-W-H	H-M-H	H-X- ... -X-H, where X is same purpose
H-W-S-H	H-D-H	H-X- ... -X-H, where X is any purpose
H-W-other than W/S-H	H-W-X-[X]-H	H-[X]-W-X-H
H-school-H	H-X- ... -X-H	W-X-W
H-school-X-H	**Bradley Research and**	**Strathman et al. (1994)**
H-P-H	**Consulting et al. (1998)**	H-W-[W]-H
H-SP-H	H-W-H	H-NW-[NW/W]-W-H
H-P/SP-other than D/S-H	H-M-H	H-W-[NW/W-]-NW-H
H-P/SP-S-H	H-D-H	H-NW-[-NW/W-]-W-[-NW/W-]-NW-H
H-P/SP-D-H	H-[X]-W-[X]-H	H-W-[NW/W-]-NW-[-NW/W-]-W-H
H-S-H	H-[NW]-M-[NW]-H	H-NW-H
H-S-S-H	H-D- ... -D-H	H-NW-[-NW-]-H
H-S-D-H	**McCormack (1999)**	**Hanson (1980), Ewing (1995)**
H-D-H	H-W W-NW	H-[X]-W-[X]-H
H-D-D-H	H-H NW-W	H-[NW]-NW-[NW]-H
H-D-S-H	H-NW NW-H	
H-other-H	W-H NW-NW	
H-other-other-H	W-W	
Anything else		

Note: H = home; M = maintenance; X = any purpose destination; W = work; D = discretionary; SR = social/recreational; S = shop; NW = nonwork; P = personal; SP = serve passenger.

trip–purpose combinations, the enormous number of tour combinations produced by matching merely eight trip purposes with number of trips would produce an overly complex and burdensome bookkeeping issue. On the other hand, simple coding schemes (e.g., Ewing et al., 1995; Hanson, 1980) are limited because they do not differentiate between various different types of nonwork activities — activities that may have very different travel characteristics.

Reichman (1976) first explained that while lifestyles and travel patterns may vary considerably between households, it is still possible to define three major classes of travel-related activities:

- *Subsistence* activities, to which members of the households supply their work and business services; commuting most commonly associated with this activity
- *Maintenance* activities, consisting of the purchase and consumption of convenience goods or personal services needed by the individual or household
- *Leisure or discretionary* activities, comprising multiple voluntary activities performed on free time, not allocated to work or maintenance activities

---

[10] Pas (1984) developed a similarity index of travel activity to identify single types of travel for a person over a day. Homogeneous types of travel were grouped together by a twelve cluster analysis and a five cluster application. A report from Bradley Research et al. (1998) used similar groups of activities, but allowed greater flexibility in how tours were coded. Golob (1986) developed an elaborate typology of tour-types analyzing the transitions between activities. Southworth (1985) used yet a different scheme in efforts to demonstrate a trip chaining simulation model. Ewing (1993) and Hanson (1980) used any work-related trip to binomially code tours as work/nonwork. McCormack coded tours by the origin-destination pair as defined by 90 minute cutoff. Similar efforts at classifying travel activity have been used by Recker and McNally (1985), Kansky (1967), and Oppenheim (1975).

Common themes emerge from these eight tour classification schemes. First, the predominant way of classifying a tour is the sequence of consecutive trip links that begin and end at home. Second, four of the studies use a simple binary system—work vs. nonwork—to differentiate between travel purpose within a tour. Other studies specify more detailed non-work trip purposes; Pas (1982) and Bradley Research (1998) categorize three types of activities whereas Golob uses six. All of the studies provide a separate category for simple tours, yet they all differ in terms of the combinations and permutations for more complex tours.

This typology of activities was employed by Pas (1982, 1984) to classify daily travel activity behavior. It has also been used more recently for daily activity modeling (Gould and Golob, 1997; Ma and Goulias, 1997; Bradley Research, 1998). Using this classification scheme, activities for work, school, or college trips are considered subsistence (or work). Maintenance activities include personal, appointment, and shopping. Discretionary activities would be visiting and free time.[11]

6.4.3   Research Results: Trips, Tours, and Urban Form

The final part of this section turns to analyzing the influence NA has on travel behavior using the above discussion as a foundation for analysis. As previously mentioned, the following analysis is based on travel data from the Puget Sound Transportation Panel (Murakami and Ulberg, 1997)[12] (see Figure 6.2) and urban form data described in Krizek (forthcoming).[13] The first look at the data tallies each trip by purpose (Table 6.5).[14] As shown, over 33% of the trips away from home are for work. This leaves over two thirds of the trips devoted to nonwork activities, leading many to assert that areas with higher NA could reduce travel. But again, this tells an incomplete story because of the interconnected way in which these trips are conducted.

Therefore, the individual trip data are classified into 10,569 tours, where each tour is classified according to one of nine different types based on a combination of the purpose of the trips and complexity

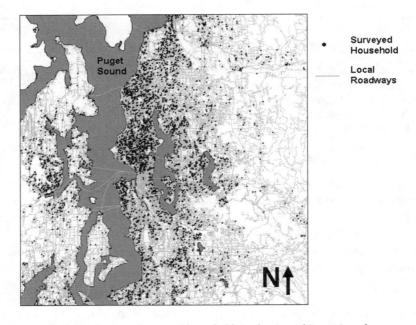

FIGURE 6.2   Geographical distribution of surveyed households in the Central Puget Sound.

---

[11] Aggregating the trip types in such manner provides a way to code and analyze different combinations of tours that is more parsimonious than using eight different activity types, but more detailed than the too simple work/nonwork dichotomy.

[12] The PSTP is the first general-purpose travel panel survey in the United States. It has been conducted annually for the past seven years by the Puget Sound Regional Council to track socio-demographic and travel behavior data of approximately the same 1700 households from King, Snohomish, Pierce, and Kitsap Counties. While the household is the unit of analysis for the panel data, travel behavior is recorded using a two-day trip diary completed by each household panel member at least 15 years of age. In addition to household and sociodemographic/economic characteristics, the travel diary data collected for each trip contains the purpose, mode, duration, and distance.

TABLE 6.5   Individual Trips by Purpose

Trip Type	# of Trips	% of Trips
Shopping	3210	14.4
Appointment	1145	5.1
Personal	5681	25.4
College	325	1.5
Subtotal	10,361	46.4
Free time	3306	14.8
Visiting	861	3.9
Work	7439	33.3
School	371	1.7
Subtotal	11,977	53.7

[13] The PSTP data include the composition of the households and the X-Y coordinate of both their residential and workplace locations. A household's urban form is measured for both its workplace and residential site because both are theorized to influence travel. The urban form around the work and home location is measured on two different scales: (a) the immediate locale—the character of a particular neighborhood, and (b) the position of the neighborhood in the larger region. The different scales are important to consider because it is theorized that each influences both residential location decisions and the nature of household travel. Attempting to understand the partial effect of each is important because issues of NA tend to be more central to current land use policy debates and new-urbanist initiatives. This analysis therefore focuses on the former (neighborhood accessibility) while attempting to control for the role of the later (regional accessibility).

The strategy used to measure the accessibility of a neighborhood within the larger region is computed using a standard gravity model. This approach is consistent with the aims of deriving a measure of activity concentrations that have drawing power from various centers of the Puget Sound region. Opportunities are measured using total retail employment. Of the many ways to account for travel impedance, the most common approach is employed which specifies an exponential function, f(impedance)=exp-b*tij. The result is a measure of regional accessibility that is similar to that specified by Shen (2000) and Handy (1993) and is specific for each TAZ as follows:

$$regional\_access_i = \sum_j \left[\frac{retail\_employment}{\exp(time_{ij} \times \beta)} \right]$$

where, $time_{ij}$ is the off-peak (free-flow) zone-to-zone travel times by automobile taken from the regional transportation model), and β is an empirically determined parameter (0.2) that best explains variations in distance for all trips.

A combination of three variables—density, land use mix, and street patterns—are used to measure levels of NA. Each variable is measured using units of analysis consisting of 150 meter grid cells; the attributes of each grid cell are not determined by the attributes of that cell alone, but rather influenced by adjoining cells. I therefore average the values for each grid cell over a walking distance of one-quarter mile. The substantive significance of these variables, their relation to neighborhood attributes, the new urbanism, and land use-transportation planning initiatives is documented elsewhere (Krizek, forthcoming) and is briefly described below.

Density, the most commonly used urban form variable, measures housing units per square mile at the individual block level according using U.S. Census data. Land use mix is captured by examining existing retail activity in each grid cell. For every business in the study area, detailed employment data from Washington State provides: (1) the two digit Standard Industrial Classification Code assigned to the business, (2) the number of employees, and (3) the X-Y coordinates. Rather than use employment for all sectors, I only use those business types considered to be representative of high NA. These business types include food stores, eating and drinking establishments, miscellaneous retail and general merchandise. To account for differences in drawing power of larger establishments, I sum the number of employees per grid cell (rather than number of businesses). Finally, the grain of the street pattern is used to proxy for the "traditionalness" of the neighborhood and other urban design amenities. Street pattern is operationalized by calculating the average block area per grid cell. Neighborhoods with higher intersection density—or lower average block area—more closely resemble the street patterns heralded by land use–transportation planners. A single measure of NA is therefore arrived at by combining the three measures into factor scores using principal component factor analysis.

[14] Because of the interest in the number of destinations to which residents travel, this tally counts only trips away from home (it does not count those trips returning to the residence).

TABLE 6.6   Tour Classification Scheme and Descriptive Statistics for PSTP Data

Type #	Tour Type	Coding	% of Tours	Mean Distance in Kilometers (Miles)
1	Simple work	H-W-H	23.9	35.6 (22.2)
2	Simple maintenance	H-M-H	20.4	18.1 (11.2)
3	Simple discretionary	H-D-H	12.2	23.9 (14.9)
4	Complex work only	H-W-W- ... -H	6.0	63.9 (39.7)
5	Complex maintenance only or	H-M-M- ... -H	9.9	32.5 (20.2)
	Complex discretionary only	H-D-D- ... -H		
6	Complex work + maintenance only	H-W-M- ... -H[a]	1.5	54.0 (33.6)
7	Complex work + discretionary only	H-W-D- ... -H[a]	12.8	53.6 (33.3)
8	Complex maintenance + discretionary only	H-M-D- ... -H[a]	4.2	53.0 (33.0)
9	Complex work + maintenance + discretionary	H-W-M-D-H[a]	9.1	50.3 (31.2)
				Mean distance = 36.6 (22.8)

[a] Trip making could take place in any order. H = home; W = work, subsistence,; M = maintenance; D = discretionary.

(number of stops) (Table 6.6). Identifying the nine different tour types sheds light on how households combine subsistence, maintenance, and discretionary purposes across tours. More importantly, deriving nine different tour classifications helps us to better understand tours, the purposes they contain, and the potential of NA to influence different tour types internal to the neighborhood.

To clearly articulate the expected relationships between travel tours and NA, three related sets of hypotheses are offered. These hypotheses are tested using a series of regression models for a sample of 1811 households. To evaluate the different hypotheses, regression models are used to predict the dependent (outcome) variable, which differs in each model, representing different tour characteristics. In efforts to ensure consistency between each of the models, the outcome variables are estimated as a function of the same set of independent variables, which include the household characteristics and measures of accessibility as previously described. The regression models generated are of the sort

$$T_{characteristic} = f(HC, CD, WA, RA)$$

where T is the household tour characteristic (number of tours by type, tour complexity, tour distance); HC is a vector of household characteristics (number of adults, number of employees, number of children, income, number of vehicles); CD is a household's commute distance; WA is a vector representing the accessibility of the workplace (regional and neighborhood); and RA is a vector representing the accessibility of the residence (regional and neighborhood).

6.4.3.1   Number of Tours and Tour Complexity

To establish relationships between accessibility and tour generation, I am guided by the threshold hypothesis (Adler and Ben-Akiva, 1979), which suggests that unfulfilled household activities accumulate until some critical threshold is reached. At this threshold, a tour is scheduled to complete some or all of the activities. More tours would therefore be expected in areas with higher NA because the cost (in terms of time and inconvenience) would be less for each. The corollary states that the complexity of each tour would then decrease. Consequently, let us hypothesize that (1) increases in NA would be directly related to increased tour generation, and (2) increases in accessibility would be directly related to a decreased propensity to link trips (decreased trips per tour).

Results from two models testing these hypotheses are shown, together with their estimated coefficients and other statistical indicators, in Table 6.7.[15] As expected, variables representing household

---

[15] Models predicting number of tours is estimated as a poisson regression because of the count nature of the data; models predicting number of trips per tour and tour distance are estimated using OLS regression because of the continuous nature of the distribution.

TABLE 6.7 Regression Results for Different Household Tour Characteristics

Explanatory Variables	Number of Tours (Poisson Regression)			Number of Trips per Tour (OLS Regression)		
	Coefficient	t-Statistic	Significance	Coefficient	t-Statistic	Significance
(Constant)	0.400305573	8.499	.000	3.411	32.484	.000
# of adults	0.406161374	23.034	.000	−0.137	−2.917	.004
# of employees	3.01E-02	1.792	.073	−8.571E-02	−2.064	.039
Household income	2.50E-06	4.787	.000	2.720E-06	2.208	.027
# of vehicles	1.57E-02	1.505	.132	−7.882E-03	32.484	.766
# of older children	0.205313869	18.766	.000	−2.511E-02	−.830	.406
Commute distance	−3.93E-03	−2.859	.004	−4.521E-03	−1.425	.154
Work neighborhood access	4.59E-02	2.532	.011	8.381E-02	2.044	.041
Work regional access	−4.00E-06	−1.368	.171	1.279E-05	1.914	.056
Residential neighborhood access	0.121298876	6.148	.000	−.129	−2.841	.005
Residential regional access	−7.56E-06	−1.377	.169	−1.126E-05	−.914	.361

Log-likelihood function at convergence: 3873.813
Initial: 4724.961
Pseudo δ^2 = 0.18

Adjusted R^2 = 0.028
F = 6.268, $p < 0.000$

Outcome Variables	Number of Simple Work Tours (Tour Type 1)			Number of Simple Maint. Tours (Tour Type 2)		
	Coefficient	t-Statistic	Significance	Coefficient	t-Statistic	Significance
(Constant)	0.232	3.672	.000	0.289	4.284	.000
# of adults	−9.214E-02	−3.242	.001	0.197	6.484	.000
# of employees	0.337	13.319	.000	−0.205	−7.580	.000
Household income	−2.732E-06	−3.653	.000	1.112E-07	0.139	.889
# of vehicles	2.416E-02	1.494	.135	−3.472E-02	−2.010	.045
# of older children	−5.495E-03	−0.297	.767	0.199	10.043	.000
Commute distance	3.267E-03	3.672	.089	−1.329E-03	−0.647	.518
Work neighborhood access	0.105	4.169	.000	−4.557E-02	−1.702	.089
Work regional access	−1.382E-05	−3.380	.001	3.509E-06	0.803	.422
Residential neighborhood access	6.973E-02	2.523	.012	0.120	4.077	.000
Residential regional access	8.934E-06	1.187	.235	−1.197E-05	−1.489	.137

Log-likelihood function at convergence: −2554.355
Initial: −3104.018
Pseudo δ^2 = 0.17

Log-likelihood function at convergence: −2540.955
Initial: −2864.248
Pseudo δ^2 = 0.10

Distance of Simple Maintenance Tours (ln) (OLS)

Explanatory Variables	Coefficient	t-Statistic	s Significance	Effect Analysis[a]
(Constant)	15.115	10.077	.000	
# of adults	.610	.936	.350	n.s.
# of employees	−.729	−1.325	.186	n.s.
Household income	1.709E-05	1.034	.301	n.s.
# of vehicles	.650	1.640	.101	n.s.
# of older children	−1.143	−2.935	.003	−21%
Commute distance	.192	4.238	.000	12%
Work neighborhood access	−1.150	−1.996	.046	−6%
Work regional access	−1.582E-05	−.162	.871	n.s.
Residential neighborhood access	−3.513	−5.461	.000	−15%
Residential regional access	−3.385E-04	−1.897	.058	−5%

[a]Shows the percentage change in the dependent variable that would result if the score for the independent variable increased from the median value to the value at the 75th percentile of all observations, substituting median values for other variables in the regression equation. (For example, change number of children from 0 to 2; change commute distances from 7.4 to 14.1.)

Adjusted R^2 = 0.162
F = 18.91, $p < 0.000$

Note: OLS =ordinary least squares; n.s. = not significant.

characteristics (number of adults, number of employees, number of children) are statistically significant and positive for tour generation. Conversely, the same variables for household characteristics are significant and inversely related to tour complexity; the greater the number of adults, employees, and children, the less likely the household linked trips. This is likely because there are more people within the household to spread the chores around, thereby reducing the trip chaining demands on any single individual. Commute distance was significant and negative for tour generation, showing that households with longer commute distances engage in fewer tours. The impact of commute distance on tour complexity, however, was not significant.

The impact of NA also shows to be statistically significant and in the expected direction for each model. Households with higher neighborhood accessibility make more tours. The model for number of trips per tour shows the average complexity of tours to be inversely related to levels of neighborhood accessibility; households that live in areas with higher NA are more likely to make tours with a fewer number of stops.

6.4.3.2 Tour Frequency by Purpose

Models of tour frequency and tour complexity do not shed light on the various types of trips contained within each tour. The trip purpose completed along each tour — particularly maintenance trips — is likely to vary based on differing levels of NA.[16] To test the hypothesis that different types of tours are likely to be generated by different levels of NA, regression models are used to predict the frequency in which households engage in each of the nine different tour types presented in Table 6.6. Of the nine different tour types modeled, the measure of residential NA proved significant and positive in only two of them.[17] Simple commute tours were significant, most likely because households living in highly accessible neighborhoods can more easily head out again in the evening. Therefore, they return home directly from work before doing so. The increased number of simple maintenance tours is entirely consistent with the arguments presented thus far.

In theory, we could expect households in high NA areas to more likely engage in at least two other types of tours. The first would be complex maintenance-only tours; these would be tours in which multiple maintenance errands (e.g., grocery, dry cleaner, bakery) would be satisfied within walking distance to one's home. The second type of tour would represent combined subsistence–maintenance tours; these would be the showcase tours that the new urbanists love to point to: commuting by transit and stopping at the neighborhood store on the way home to pick up groceries.

However, the findings from these models suggest that NA appears to have little influence on a household's propensity to engage in complex tours of any kind. This is likely because of two related reasons. First, consider the above-mentioned tour that combines subsistence and maintenance stops. It is conceivable that residents in high NA areas complete maintenance stops by foot on the way home from transit. However, households in areas with low NA may perform the same errands in the same order, but would do so driving from one neighborhood to another. Second, there remain a limited range of services that surround highly accessible neighborhoods; these services are more likely to satisfy maintenance type activities than subsistence and discretionary activities. Therefore, any tour containing these latter activities is *less* likely to be pursued locally. Satisfying these purposes is more likely to pull the traveler beyond the range local to one's neighborhood. Those households who live in high- or low-access neighborhoods have an equal propensity of leaving their neighborhood to complete trips for subsistence or discretionary purposes. Once they leave the neighborhood for these other types of services, there is similar likelihood of chaining trips.

---

[16] The previous discussion suggests that maintenance activities would be pursued as part of a tour closer to home since these types of trips could be more easily satisfied local to one's neighborhood.

[17] Not surprisingly, these models represent two types of simple tours: subsistence and maintenance. The models generated for the remaining seven tour types had exceptionally low explanatory power from a statistical standpoint or the measure of NA was not statistically significant at the 90% level.

6.4.3.3 Distance for Simple Maintenance Tours

Because of the theoretically important role of maintenance travel, the final part of this analysis focuses on simple maintenance tours to gain a better understanding of the extent to which access affects the nature of these tours. Refer to the descriptive statistics in the final column of Table 6.6, presenting the mean distance for each tour type. As expected, simple tours show shorter distances; specifically, maintenance-only tours are shortest. The mean values reported are from a univariate distribution of households spread throughout the region with varying degrees of neighborhood access.

Some households would be able to satisfy maintenance errands close to their residence; others must drive considerable distances for basic services. To better test the effect that NA has on travel across this univariate distribution, determine the extent to which the distance traveled for simple maintenance tours (tour type 2) is inversely related to levels of NA (results of regression model shown in Table 6.7). As expected, the coefficient for NA is significant and negative, indicating that higher levels are met with shorter tour distances. The relative importance of the factors influencing tour distance is summarized in the results of an effect analysis shown in the last column. Neighborhood access appears to impact tour distance more than any variable, other than the number of older children. However, increasing its value from the median value to the 75th percentile results in only a 10% reduction in tour distance. This suggests that NA does impact simple maintenance tour distance, but draws into question the influence of land use planning and, in particular, how often maintenance services are captured internal to the neighborhood. Therefore, Table 6.8 shows more detailed descriptive statistics for maintenance-only travel. Results are presented across the univariate distribution, as well as for bifurcated distributions of households that live in both the lower half and the upper decile (10%)[18] of accessible neighborhoods.

Examining median values, we see expected differences in travel distance. Households with high NA travel 3.2 km (2.0 mi) one way for maintenance activities vs. 8.1 km (5.0 mi) one way for households with lower NA. The differences in median distance — 3.2 vs. 8.1 km — support the expected hypothesis for high vs. low NA. But a distance of 3.2 km is hardly within the walking distance espoused by the new urbanists and other like-minded individuals. It is therefore helpful to know how

TABLE 6.8 Descriptive Statistics for Maintenance Trips or Tours for Neighborhoods with High and Low Neighborhood Accessibility

		Entire Sample	Households in Upper Decile (10%) of Neighborhood Access	Households in Lower Half (50%) of Neighborhood Access
Distance of simple maintenance tours	Median	11.9 km (6.0)*	6.44 (3.2)[a]	16.1 (8.1)[a]
	Mean	18.1 km (9.1)*	9.9 (5.0)[a]	22.2 (11.1)[a]
	Std. Dev.	20.8	10.8	21.3
		(n = 2150)	(n = 181)	(n = 1047)
Distance of maintenance trips	Median	6.3	3.9	7.9
	Mean	9.8	6.4	11.6
	Std. Dev.	11.2	8.7	11.4
		(n = 10,008)	(n = 931)	(n = 5030)
Percentile of simple, maintenance tours completed within …	3.2 km	4.5%	20%	1.7%
	4.8 km	13%	38%	5.4%
	6.4 km	22%	50%	12%

[a] Each way.

Note: 1 km = 0.62 mi. Maintenance trips include personal, appointment, and shopping.

---

[18] The upper decile was chosen because areas above this threshold were considered to contain representative characteristics of high access neighborhoods. Using other thresholds (e.g., quartiles) included many neighborhoods which, despite having relatively high neighborhood access scores, did not possess the urban form feel promoted by high levels of NA.

such trips are distributed for these two populations. Households in highly accessible neighborhoods complete 20% of their simple maintenance tours within 3.2 km (2.0 mi) of their home. This is compared to a mere 1.7% of simple maintenance tours for their low NA counterparts. While a distance of 3.2 km (2.0 mi) is still being beyond walking distance, it needs to be recognized that this represents a median value.

6.4.4　Recap and Policy Significance

Part four of this chapter helped answer two outstanding questions in urban form–travel research. First, how do neighborhood access and trip purpose relate? Second, how do neighborhood access and the manner in which such trips are combined — travel tours — relate? Most studies are unable to shed light on these questions because they employ a strictly trip-based approach to operationalize travel. Examining trips, instead of the larger pattern of linked travel, does not represent travel in a manner consistent with how travel decisions are made; examining only trips does not shed light on the relationships that may exist between trip frequency and trip chaining.

Several lessons are important to understand for land use and transportation planning or urban policy. The most specific evidence provided sheds light on an important land use–transportation issue: the extent to which maintenance travel (what has often been called nonwork travel) is captured internal to neighborhoods with high accessibility. Of the different tour types classified, households with higher levels of neighborhood access more frequently engage in specifically two types of tours: simple subsistence and simple maintenance. Such households make more simple maintenance tours, but they also pursue these simple maintenance trips closer to their home. Households that live in the top decile of accessible neighborhoods in the region visit maintenances services that are available within 3.2 km of their home for 20% of their simple maintenance tours (in contrast to a mere 1.7% for households in the lower half of neighborhood accessibility). On one hand, this lends comforting evidence for those who believe land use planning can be used to moderate travel — in particular, driving distance. On the other hand, however, it represents only a fraction of maintenance travel, much less all travel.

Such findings, however, need to be approached with caution for three reasons. What remains relatively unclear from this research is whether: (1) these maintenance tours substitute or complement other trips, (2) these maintenance tours tend to be pursued by nonmotorized mode, and (3) the majority of these tours are conducted local to one's neighborhood. For example, many households that live in areas replicating NA will continue to shop outside their immediate neighborhood. Maintenance-type errands are subject to a wide array of constraints related to consumer behavior — e.g., bargain hunting, comparison shopping, preference for variety, parking convenience — each of which prize destination and schedule flexibility (Nelson and Niles, 1998). A household's desired goods at a desired price are many times not located within walking distance to home. Basic preferences suggest that households will travel farther than their neighborhood center for many basic shopping needs. Each of these factors is likely to draw the shopper away from the neighborhood. Thus, while households with high neighborhood access may frequent the corner store periodically, it does not take but a few maintenance trips across town to increase mean values or to sway the median distance. A further stage of development would aim to uncover such relationships. A strict quantitative mode of inquiry is unable to shed light on the nuances of travel behavior decisions. A more qualitative mode of inquiry is likely necessary to better explain trade-offs related to substitution travel, mode choice, or local travel.

6.5　Understanding Causality Underlying Urban Form and Travel

A third issue important for land use policy stems from the myriad ways in which urban form influences travel behavior. To help clarify this issue, one can imagine three different scenarios by which urban form affects travel behavior. In the first case, urban form directly influences the *range of travel possibilities* that may exist for a particular household. Land use patterns help define the set of available travel choices. For example, transit service may not be available for a household that lives in a suburban setting. In addition to influencing the set of available travel choices, patterns of urban form can *influence the relative attractiveness* of each travel choice, illustrated by a second scenario. Transit service may be available but the suburban resident would still prefer to drive because of the constraints that a given urban form places on the transit system, among other things. In the first case, urban form helps define the choice set; in the second, it influences the relative attractiveness of each different travel choice.

A third scenario draws into question the direction of any causal relationship that may exist between urban form and travel. This scenario is concerned with the self-selection issue — of increasing interest to researchers struggling to untangle such relationships. Residents may locate in a residential neighborhood to realize their travel preferences. For example, residents that prefer to take transit may choose to reside where transit is available. The important point for land use policy is that differences in travel between households with different neighborhood designs should not be credited to urban form alone; such differences could be attributed to broader issues that triggered the choice to locate in a given neighborhood. Furthermore, it suggests that the relative magnitude of each (the influence of urban form vs. the influence of preferences) is a worthwhile question.

If there is a self-selection bias at work, policies designed to induce changes in household travel through altering land uses may not have the expected or desired effect — or their impact may be marginal. For example, using urban design tools to induce unwilling auto-oriented households to drive less may be futile for at least two reasons. First, their auto-using behavior may a function of larger issues such as their overall preference for auto-oriented behavior. Modifying an old phrase, You can take the family out of the suburban location but you can't take reliance on the Chevy Suburban out of the family. Second, it is unlikely that such auto-oriented households would locate in heavily transit-oriented neighborhoods in the first place. This in turn suggests that the success of the new urbanism may be based on the relatively small market of households that currently live in transit-oriented neighborhoods or those who will bring their non-auto-using behavior with them to newer neighborhoods.

Recent research has attempted to better understand the broader issues (e.g., preferences) related to urban form and at work in influencing household travel. Prevedouros (1992) measured personality characteristics and analyzed their association with choice of residential neighborhood type. Kitamura et al. (1997) used attitude surveys combined with travel diaries to conclude that general attitudes toward travel behavior better explain travel than urban form characteristics. Using the same data set within a system of structural equations, Bagley and Mokhtarian (2000) examined relationships between urban form and travel, incorporating attitudinal, lifestyle, and demographic variables. In terms of both direct and total effects, they concluded that attitudinal and lifestyle variables had the greatest impact on travel demand among all the explanatory variables. Finally, Boarnet and Sarmiento (1998) and Boarnet and Greenwald (2000) used instrumental variables representing residential location decisions to control for the possibility that households choose their residential locations based in part on their desired travel behavior.

6.5.1 Examining the Same Households in Different Neighborhoods

Each of the above-described efforts aim to disentangle different influences of household travel, in particular, residential location preferences or urban design features. Each approach, however, is limited by the ability to do so because they rely on cross-sectional data. Assuming time series data, an alternative strategy could examine the same household's revealed travel behavior in two different urban form settings. The longitudinal nature of the PSTP data, consisting of information gathered from the same units at different points in time, permits a research strategy ideal for analyzing changes in household travel. Doing so would go beyond cross-sectional exploration used to infer associative results and would use longitudinal data to shed light on causal relationships.

Over the seven waves of the PSTP, approximately 430 households changed their residential location from one year to the next. The travel data for these households provide before (pretest) and after (posttest) observations, where the changes that may affect travel behavior could be considered analogous to receiving a "treatment." For purposes of this research endeavor, the treatment of interest applies to the households that relocated between waves, and therefore changed the urban form surrounding their residence. Of course, such a treatment group is self-selected, and so this does not conform to the strict model of randomized experiments.

Between the two waves, one could assume that the regional stock in different types of neighborhoods does not substantially change from one year to the next. If movers were more or less in equilibrium with respect to neighborhood type before their move *and* there is no or very little excess demand for areas with high levels of NA, then they mainly moved for reasons *other* than to locate in neighborhoods more suited to their preferred travel environment. That is, they moved for reasons related to life cycle, income, employment, etc. Approaching the research design by analyzing travel of households between two consecutive years has two advantages. First, by assuming that households are in equilibrium with respect to neighborhood type over a 2-year period, this analysis controls for the case where preferences remain fixed from one year to the next. The same households between consecutive years and in two different urban form settings are compared. Second, the panel data allow changes to be measured directly on the respondents themselves, allowing one to infer that changes in one variable (e.g., a change in residential location) affect travel behavior. While statistical control is never fully introduced in any quasi-experimental research, the relatively short time between observations (1 year) helps us better isolate the role of urban form in influencing travel.[19]

Using this analysis framework, the factors that trigger changes in travel behavior are addressed on the basis of four regression models. The dependent variable differs between all four models and represents *change*[20] in travel behavior, defined by vehicle miles traveled, person miles traveled, number of trips, and number of trips per tour. These outcome variables are estimated as a function of base[21] values and change values between waves in the panel survey. The set of independent variables remain the same in each model (except for the base travel measure). Sociodemographic factors modeled include household income, number of vehicles, number of adults, number of children, and number of employees. Changes in residential location will most often change one's commute distance. It is important to identify the extent to which any changes in household travel are due to change in commute distance vs. changes in other travel. Consequently, the change in household commute distance is included as an additional control variable. Urban form (independent) variables relevant for land use planning and policy are the change in neighborhood-scale accessibility (both residential and workplace[22]) and the change in regional accessibility (both resi-

---

[19] Of the 430 households who moved, the bulk of them relocated close to their initial location, with approximately 20 percent of the households moving less than 2.5 miles. This finding confirms the prevailing notion in the residential location literature that households tend to relocate within corridors or within neighborhoods, most often for better and larger housing stock (Rossi, 1980). It stresses the importance of using detailed strategies to measure urban form to detect the extent to which urban form characteristics differ between two adjacent neighborhoods.

[20] The change value for each variable was calculated subtracting the pre-move value from the post-move value (POST − PRE = CHANGE). Only households with 2 consecutive years of data (between the panel waves) was used for analysis.

[21] Base values are included because the direction and magnitude of a change may vary based on starting levels of key factors, including travel characteristics in the base year.

[22] From the standpoint of urban form, one needs to decide from where to measure accessibility: home location, work location, or from each individual trip destination (Ewing, 1995). I assume that overall travel behavior is influenced primarily by the characteristics of one's residential location, but of course, not entirely. An individual's travel is likely to be influenced by the urban form characteristics surrounding one's workplace location or even by the characteristics along the corridor from home to work. For lack of a good way to capture the latter phenomena and a desire for a better strategy to capture the workplace site, I average workplace urban form variables within the same household.

dential and workplace). Households that do not relocate, by definition, have a zero value for changes in residential accessibilities.

6.5.2   Research Results: Examining Moved Households

6.5.2.1   Empirical Findings

Table 6.9 reports the regression models run for the entire sample of 6144 households, together with their estimated coefficients and other statistical indicators.[23] Of these models, none is preferred over another; they are presented and discussed jointly to provide a more comprehensive understanding of travel outcomes.

Initial observation shows that most base sociodemographic variables have a statistically significant effect on each of the change travel outcomes. That is, the propensity of a household to change their travel behavior between waves is reflected in their initial sociodemographic conditions (income, number of vehicles, number of adults, number of children, and number of employees). The largest influence in each of the models comes from the baseline travel behavior variable, showing that the greater the baseline value, the greater the propensity to decrease it (i.e., the greater distance a household travels, the more they want to decrease that amount). Neighborhood access in each of the models is also statistically significant. Households originally located in neighborhoods with higher access have a greater propensity to decrease their miles traveled and number of trips per tour while increasing number of tours. The other ten variables in the models operationalize change variables.[24]

The remaining discussion focuses on changes in travel behavior as triggered by changes in urban form variables. To more clearly articulate the relationships depicted by these models, Table 6.10 graphically depicts relationships for changes in residential accessibility. Model 1 shows that households that relocate to neighborhoods with both higher neighborhood and regional accessibility reduce their miles traveled by vehicle. To test how their overall travel (person miles, regardless of mode) is affected, model 2 is presented. Again, we see similar results, showing that increases in neighborhood and regional accessibility

---

[23] The models could be estimated for the sample of all households or on only those households who move. The former strategy reduces the risk of fitting models on a self-selected sample (i.e., movers); the latter strategy understands that the change variables for urban form are identified only from moved households. I chose to report the results from the sample of all 6144 households because using the full sample helps better identify the base variables. Doing so also increases the sample size, permits more degrees of freedom, and therefore produces a series of models that are more robust from the perspective of statistical significance (F-stat). Because so much interpretation rests on the behavior of the movers, however, it is important to understand: (a) if the population of movers differ significantly from the entire sample (and if so, in what ways), and (b) if the model results differ significantly when run on movers vs. the entire sample.

To address (a), a Chow test was conducted on the base variables to evaluate the stability of the regression coefficients across movers and nonmovers. The results indicate that there are no statistically significant differences in the independent variables across movers and nonmovers with one notable exception. The interaction term between the indicator variable for movers and regional accessibility was significant and had a negative coefficient suggesting that a relationship exists between the two. These findings from the Chow test are consistent with the results of the regression run using only the subsample of movers, which was done to address the concern stated in (b). In comparing the regression output for just the movers with that from the entire sample, the same policy relevant variables are significant and there are minor differences in parameter estimates. There is only one minor discrepancy, which may be related to the exception above: regional accessibility proved insignificant for Model 4 (which is not the case when run for the entire sample). The possible explanations for this relationship are many. However, these findings are not a major cause for alarm and therefore the results presented in Table 6.9 include the entire sample of movers and nonmovers.

For these regression models, a case from each wave pair (e.g., the difference between wave 5 and wave 4 values) was modeled independent of presence of that case in previous waves. Using only the first observation for multiple cases produced similar coefficients in the model.

[24] For example, change in commute distance is significant in models 1, 2, and 3. This finding shows that reducing commute distance results in decreasing miles traveled and increasing number of tours. This suggests that households who shorten their commute are more prone to participate in more tours through the course of the day.

TABLE 6.9   Regression Models of Variation in Change in Travel Behavior between Panel Waves

		Model 1 Vehicle Miles Traveled			Model 2 Person Miles Traveled			Model 3 Number of Tours			Model 4 Number of Trips/Tour		
		Coefficients	t-Statistic	Signif.	Coefficients	t-Statistic	Signif.	Coefficients	t-Statistic	Signif.	Coefficients	t-Statistic	Signif.
Sociodemographic variables (base)	Constant	37.387	16.326	.000	42.406	18.072	.000	1.309	19.757	.000	2.131	28.911	.000
	Income	1.384E-04	5.021	.000	1.202E-04	4.322	.000	1.735E-06	2.179	.029	1.799E-06	2.314	.021
	# of vehicles	3.919	6.892	.000	3.284	5.737	.000	5.247E-02	3.227	.001	1.072E-02	0.676	.499
	# adults	-7.146	-7.224	.000	-7.282	-7.289	.000	-9.819E-02	-3.467	.001	-0.109	-3.927	.000
	# of children	0.289	0.607	.544	0.240	0.498	.619	0.182	12.925	.000	-4.293E-02	-3.191	.001
	# of employees	4.389	5.204	.000	4.759	5.580	.000	2.848E-02	1.173	.241	3.911E-03	0.165	.869
Sociodemographic variables (change)	Δ income	1.372E-04	3.645	.000	1.416E-04	3.724	.000	2.394E-06	2.202	.028	9.751E-08	0.092	.927
	Δ # of vehicles	2.468	3.490	.000	1.905	2.667	.008	2.820E-02	1.386	.166	3.402E-02	1.712	.087
	Δ # adults	-9.143	-6.730	.000	-9.191	-6.698	.000	-0.233	-5.927	.000	-0.109	-2.840	.005
	Δ # children	-1.714	-1.279	.201	-2.499	-1.846	.065	-2.173E-02	-0.562	.574	-7.724E-02	-2.046	.041
	Δ # employees	3.887	4.057	.000	4.003	4.135	.000	6.263E-02	2.264	.024	1.814E-02	0.671	.502
Base travel measure		-0.540	-47.865	.000	-0.559	-49.047	.000	-0.640	-53.920	.000	-0.633	-52.250	.000
Change in commute distance		0.790	10.371	.000	0.832	10.795	.000	-7.709E-03	-3.516	.000	6.563E-04	0.306	.759
Urban form variables (base)	Neighborhood access	-5.857	-6.545	.000	-5.498	-6.095	.000	0.113	4.395	.000	-7.517E-02	-3.001	.003
	Regional access	-8.828E-04	-3.650	.000	-1.385E-03	-5.646	.000	1.235E-05	1.771	.077	3.071E-07	0.045	.964
	Work neighborhood access	-0.346	-0.423	.672	-1.272	-1.541	.123	2.980E-02	1.259	.208	2.754E-02	1.192	.233
	Work regional access	-1.169E-04	-0.912	.362	4.562E-04	3.515	.000	-3.947E-06	-1.065	.287	5.938E-06	1.641	.101
Urban form variables (change)	Δ neighborhood access	-5.766	-1.999	.046	-6.792	-2.331	.020	0.266	3.197	.001	-0.189	-2.329	.020
	Δ regional access	-2.265E-03	-2.940	.003	-2.309E-03	-2.966	.003	2.651E-05	1.194	.233	-4.999E-05	-2.304	.021
	Δ work neighborhood access	-0.697	-0.649	.516	-1.656	-1.526	.127	3.071E-02	0.989	.323	2.995E-02	0.987	.323
	Δ work regional access	-6.069E-04	-3.103	.002	-2.564E-05	-0.130	.897	-3.153E-06	-0.559	.576	1.247E-05	2.263	.024
N = 6144		Adjusted R^2 = 0.298 F = 131.31, $p < 0.000$			Adjusted R^2 = 0.310 F = 139.10, $p < 0.000$			Adjusted R^2 = 0.331 F = 151.47, $p < 0.000$			Adjusted R^2 = 0.313 F = 139.98, $p < 0.000$		

TABLE 6.10 Graphic Depiction of Regression Results

Change Variables ↓ Explanatory	Dependent →	Vehicle Miles Traveled	Person Miles Traveled	Number of Tours	Number of Trips/Tour
Policy relevant variables	Neighborhood access (residential)	−	−	+	−
	Regional access (residential)	−	−	n.s.	−

Note: All relationships are based on the regression models as presented in Table 6.2. The direction of the sign represents the effect of a positive change (an increase in accessibility) in the explanatory (policy relevant) variable. n.s. indicates that the relationship is not significant at the $p < 0.05$ level.

decrease the distance households travel. Households that may have changed their employment locations to one with higher regional access also decrease their vehicle miles of travel.

Model 3 predicts the average number of tours (or the number of forays away from home).[25] The positive coefficient indicates that increases in neighborhood accessibility result in an increase in number of tours. This finding is consistent with previous arguments (Crane, 1996b) claiming that because the cost of each trip (or tour) is lesser for households with higher accessibility, they may make more of them. Households with higher accessibility are more likely to go out and buy a quart of milk because the cost of each tour is lower. In contrast, households with poorer accessibility may likely accumulate many unfulfilled needs in order to form multidestination trips that satisfy many needs during a single tour. A change in number of tours, however, was not significant for changes in regional accessibility or either of the measures for workplace access. Whereas model 3 looks only at the number of tours, model 4 sheds light on the complexity of those tours by modeling the average number of trips within tour. By doing this, we can gain a better understanding of the trade-offs between changes in number of tours and trips per tour. As expected, we see that increases in neighborhood and regional accessibility result in decreases in number of trips per tour.

6.5.3 Recap and Policy Significance

This set of results focuses on four dimensions of travel behavior: vehicle miles traveled, person miles traveled, number of tours, and number of trips per tour. This prompts the question "What about the influence of urban form on other dimensions of travel behavior?" Additional analysis for mode split and trip generation (divorced from number of tours) revealed statistically significant models (F-test). However, these are not reported in this article because the effect of the policy significant variables was either not significant or suggested inconclusive results.[26] Such findings, however, do not preclude us from discussing their substantive significance.

First, it does not appear that changes in urban form settings trigger changes in overall mode split. Despite promises from the new urbanists, this further supports theories that household travel preferences remain fixed: transit-using households likely remain transit using, heavily auto-using households likely remain auto using. The effect of urban form on trip generation remains inconclusive. Upon further reflection, this should come as little surprise considering the aforementioned findings for number of tours and number of trips per tour. Any savings in trip number due to chaining may be compensated by an increase in number of tours.

---

[25] Models 3 and 4 are technically better estimated with discrete choice methods; when done so using a probit model they yield similar results in terms of significant variables. Because the practical effect in doing so rarely differs between the two estimation procedures, I use ordinary least squares modeling to provide consistency with models 1 and 2.

[26] The mode split model showed neither neighborhood or regional accessibility to be statistically significant. The trip generation models were relatively unstable and rarely showed neighborhood and regional accessibility to be significant at greater than the 92% level.

As a whole, this set of model results can be summarized by the following crude simplification.[27] When households move to traditional types of neighborhoods, they are more likely to go to the corner store to buy a pint of milk. They do so more often and are less likely to link this trip for milk with a trip to the dry cleaner or the day care. We do *not* know if such changes result in shifts in mode split. But despite the increased number of tours and because the corner store is closer to their home, they reduce their vehicle miles and overall miles of travel as well.

The magnitude of the changes in travel behavior can be gleaned from modeling a simulation using the above unstandardized coefficients. For the two residential urban form variables (neighborhood access and regional access), a typical relocation from a traditional single-family neighborhood in Seattle (Wallingford) to a suburban location in Bellevue is modeled. Assuming median values for all other variables, the model predicts that this relocation would result in an increase of over 5 vehicle miles of travel per day.[28] Thus, not only does the land use–transport relationship appear to make a difference at both the regional and neighborhood levels, but by most standards, the magnitude — reducing 5 mi of car driving — deserves attention from the planning community.

6.6 Understanding Household Lifestyles and Choices

The previously described research approach, combined with the aforementioned studies (Kitamura et al., 1997; Boarnet and Sarmiento, 1998; Bagley and Mokhtarian, 2000), offers strategies to understand the partial effect of urban form vs. preferences. In each application, however, the influence of longer-term decisions with respect to household composition, socioeconomic status, and stage of life cycle is largely considered exogenous to the choices related to travel behavior. It is important to acknowledge that longer-term choices (e.g., residential location) condition shorter-term decisions (e.g., daily household travel behavior); however, the possibility that short-term and long-term choices mutually inform one another is too often ignored.

The need to better understand the behavioral linkages between daily household activity and travel patterns, on the one hand, and long-term choices of housing and job locations and vehicle ownership, on the other, has been apparent for some time. Early pioneering efforts linked residential location, housing type, auto ownership, and travel mode to work in a multinomial logit model (Lerman, 1977). Related efforts aimed to extend this approach (Ben-Akiva et al., 1980; Weisbrod et al., 1980; Ben-Akiva and DePalma, 1986; Abraham and Hunt, 1997) and recent thinking links residential location to an activity-based model through a deeply nested logit model (Ben-Akiva and Bowman, 1998). But as a whole, there have been relatively few attempts to integrate and understand aspects of long-term household behavior with travel behavior.

6.6.1 A Hypothetical Example

To help describe how short- and long-term decisions may be interwoven, consider the following hypothetical scenario. A household consists of a young married couple, both spouses employed; they have no children and live in a condominium near downtown. They choose to live in an urban, dense environment because they devote substantial time to their early careers, working long hours. To accommodate their busy lifestyle, they take advantage of time-saving goods and services (e.g., take-out food, dry cleaning) and enjoy the walking and transit proximity from their condominium to entertainment opportunities in the city core. The wife uses transit to commute to work downtown, and the husband drives to work because of the relatively poor transit accessibility to his suburban work site. They own one car since parking is expensive and scarce.

---

[27] Of course, such a scenario is just one example that fits the results; the set of models don't actually show that this is what happens.

[28] The model predicts an increase of 10.07. But this represents the average household vehicle travel over 2 days; it was therefore halved it to provide a value over a single day.

Their lifestyle is a strategic combination of being married, postponing children, orienting time toward work, and living in an urban, transit-oriented manner. On a daily basis, their lifestyle may be manifested by working long hours, perhaps participating in fewer discretionary or maintenance activities, and frequently walking or using transit. It is easy enough to identify interdependencies between various aspects of this set of choices. The interdependencies represent a strategic combination of choices that are consistent with each other and with the stage of life cycle of the household.

Let us now examine the effect of a hypothetical change to this scenario. Consider an employment offer received by the wife from a firm at a suburban location. This employment offer prompts the household to make strategic choices that will substantially influence their lifestyle. By accepting the job, the couple will have to decide whether to continue living in town. Assuming the two workplaces are not in the same suburb, they may be forced to purchase a second car, and both may commute relatively longer distances. Alternatively, they could relocate to the suburbs. Now, they add to their strategic deliberations the question of having children. If they decide to have a child within the next few years, it would make the move to a suburban environment more attractive.

To what degree does a change in employment alter their lifestyle? If the job is not too far away and the couple will tolerate the dual commutes from their current residence, their lifestyle may adapt only marginally, shifting somewhat toward auto use and altering daily schedules to accommodate the new job. If the job is just far enough away that it causes their combined commutes from their present residence to seem untenable, then accepting the job would require complementary changes in residential location, auto ownership, and commute travel mode; it would trigger a shift to a substantially different lifestyle. They might be willing to make such a lifestyle change if they also decide that they will have children within a few years. In this case, a suburban house in a good-quality school district will seem much more appealing.

Knowing how each of the decisions relate to one another is central to understanding first the role of urban form, and second the value of alternative land use designs in mitigating travel. Both of these questions are essential to fully understand the merits of land use–transportation policies. Assessing the relative magnitude of different long-term choices does not do justice to the complex and mutually reinforcing phenomena at work. Each influence tends to reinforce and mutually inform one another, combining long-term decisions (e.g., where to live, vehicle ownership) with short-term decisions (e.g., daily travel, activity participation).

6.6.2 Introducing and Defining Lifestyles

These above-described related phenomena have been referred to as a "deftly spun, fragile, and tangled web of preference, constraint, and identity" (Jarvis, 2000). The particular knitting of the web is represented by what is referred to here as different lifestyles. While the term lifestyle is widely referred to in the literature on travel behavior, it is generally employed informally and with little analytical basis. The choice to use this term is guided in large part by Salomon (1983) and Ben-Akiva and Bowman (1998). Subsequently, Waddell (2000) uses the term as a framework for describing clusters of long-term household choices of residential location, labor force activity, and auto ownership that predispose or condition patterns of daily activity and travel behavior. Rather than unravel the web of household decision making, the approach of employing lifestyles provides a framework that treats household decisions as a single, integrated phenomenon. For this research, a household's lifestyle is defined herein by four different but related dimensions as follows:

$$L_{classification} = f\ (T,\ A,\ V,\ UF)$$

where

L = Lifestyle classification

T = Travel characteristics (measured by): Number of drive-along vehicle trips, carpool trips, transit trips, and walking trips; Number of tours; Average number of trips or tours; Travel distance; Household commute distance

A = Activity frequency[29] (measured by): Number of subsistence activities, maintenance activities, and discretionary activities

V = Auto ownership (measured by): Number of vehicles (0, 1, 2, or more)

UF = Urban form (measured by): Residential household density, street pattern, land use mix, and regional accessibility

6.6.3   Research Results: Analysis and Findings

6.6.3.1   Factor Analysis

To empirically uncover different lifestyles, two analytical strategies are employed. First, principal component analysis is employed to determine how the four described lifestyle dimensions relate to one another. Subsequently, cluster analysis is used to assign different households to lifestyle clusters.

Factor analysis (or principal component analysis) is a statistical technique to extract a small number of fundamental dimensions (factors) from a larger set of intercorrelated variables measuring various aspects of those dimensions. It has been used in previous land use–transportation applications to measure more narrowly defined or separate concepts. For example, Cervero and Kockelman (1997) used it to discern both the walking quality (a factor based on attributes such as sidewalk availability and block length) and intensity (a factor based on attributes such as population density and retail store availability) of neighborhoods. To et al. (1983) used factor analysis to define a housing quantity variable. Additionally, Bagley and Mokhtarian (2000) recently devised a multidimensional measure of neighborhood type.

Rather than restrict our attention to the variables contained within each independent lifestyle dimension, we combine each of the four dimensions to apply factor analysis on the 16 variables across the four dimensions. By doing so, we are able to better understand how specific elements within one dimension (e.g., number of maintenance activities) relate to outcomes in another dimension (e.g., number of carpool trips), thereby capturing possible interdependencies.

Factor analysis was performed using SPSS 8.0 on a disaggregate data set of 1907 households from the seventh wave (1997) of the PSTP. The results are shown in Table 6.11. For ease of interpretation, the variables are listed in order of the size of their factor loadings (i.e., coefficients) sequentially for each factor. A total of five factors was extracted, explaining almost 70% of the variation in the data. Each of the eigenvalues for the five factors is greater than 1, and no loading on any factor is lower than 0.57.

The factor loadings of each of the household measures onto each of the five factor components provides an initial understanding of the interdependencies between each of the variables. The first factor, which accounts for 23% of the total variation, clearly represents the dimension relating to *urban form* or overall accessibility. Given the high degree of association between density, land use mix, street patterns, and the computed measures of regional access, covariation between these measures is expected. The second factor (explaining 18% of the variation) represents travel dominated by *nonwork activities*, which tend to be done with someone else (carpool) and via many tours. Again, the loading of these particular variables makes sense since maintenance and discretionary activities are often completed with others (carpool) and many times as part of individual jettisons from home. The third factor (explaining 12% of the variation) clearly represents the market of households that travel in *environmentally benign* ways, picking up expected covariation between travel by transit and walking with lower rates of automobile ownership. The fourth factor (10%) represents a factor that is heavily influenced by the *work commute*, as seen by the loadings on the number of subsistence trips and commute distance. The final factor (7%) detects *auto-dependent* patterns, representative of complex tour making that often requires many vehicle trips.

---

[29] Reichman (1976) first explained how the basic travel of households falls into three general activity purposes: subsistence (work, school, college), maintenance (shopping, personal, appointment), and discretionary (visiting, free time). This scheme was employed by Pas (1982, 1984) to classify daily travel activity behavior and by Bowman et. al (1998) to forecast daily activities.

TABLE 6.11 Factor Analysis on Each of the Lifestyle Dimensions

| Household Measure | Factor Component | | | | |
	1 Urban Form	2 Nonwork Activities	3 Environmentally Benign Travelers	4 Work Commute	5 Auto- Dependent Travelers
Land use mix	**0.814**	6.243E-02	6.906E-02	−7.060E-02	−2.982E-02
Household density	**0.897**	2.227E-02	0.112	−9.715E-02	−5.879E-03
Block size	**−0.839**	−1.728E-02	−8.492E-02	7.192E-02	−2.594E-03
Regional accessibility	**0.870**	2.138E-02	7.390E-02	−0.127	8.167E-03
# of maintenance trips	−4.373E-02	**0.754**	2.031E-02	−0.188	0.374
# of discretionary trips	7.838E-02	**0.608**	−3.310E-02	−1.782E-02	0.212
# of carpool trips	−8.306E-02	**0.871**	3.205E-02	2.768E-03	−0.144
# of tours	0.271	**0.703**	−0.102	0.353	−4.237E-02
# of transit trips	6.637E-02	−4.592E-02	**0.811**	0.101	−6.747E-02
# of walk trips	8.146E-02	6.267E-02	**0.739**	0.112	8.128E-02
# of vehicles	−0.228	9.621E-02	**−0.566**	0.315	−8.244E-02
# of subsistence trips	0.114	−0.111	**0.255**	0.674	0.464
Household commute distance	−0.271	−0.100	−1.005E-02	**0.674**	−0.115
Travel distance	−0.0308	0.343	−3.928E-02	**0.754**	0.180
Average # of trips/tour	−0.175	0.231	0.284	**0.627**	**0.801**
# of vehicle trips	0.150	0.108	−0.405	−5.427E-02	**0.728**

Note: Extraction method: principal component analysis; rotation method: varimax with Kaiser normalization. Bold text indicates variables used to define each factor.

6.6.3.2 Cluster Analysis

Using the above factors as the foundation, the heart of this initial analysis aims to understand how each factor combines to represent different household lifestyle choices. Iterative cluster analysis[30] is employed to identify groupings of households with similar patterns of travel, activity, auto ownership, and neighborhood characteristics. These groupings are referred to as different lifestyles. Nine clusters best identified clearly distinguishable (and recognizable) lifestyles;[31] the values of the cluster centers for each lifestyle (A through I) are presented in Table 6.12. The length and direction of each bar represent the value of the cluster center for each of the five factors. For example, the dramatic spike for lifestyle E indicates a substantial (and positive) weighting of the factor representing transit–walk frequency.

6.6.3.3 Uncovering Different Lifestyle Classifications

The output displayed in Table 6.12 shows that interesting patterns of different lifestyles emerge. Of the five factors incorporated into the analysis, three of them were instrumental in defining their own lifestyle;

---

[30] The clustering uses the K-means statistical routine in the SPSS 8.0 statistical package and the analysis is based on the distance and similarity between the factor scores output for each of the 16 variables.

[31] An important issue to address up front is the most appropriate number of different types of lifestyles to accommodate the full range of housing and travel choices. The choice is ultimately guided by a combination of four factors: (a) statistical output, (b) the manner in which the output is transferable for land use-transportation policy, (c) lessons from past research efforts, and (d) common sense and intuition. For example, Wells and Tigert (1971) used factor analysis to reduce 300 or so statements about activities, interests, and opinions into 22 lifestyle dimensions. Pas (1982) who was interested in only examining different types of travel tours, found diminishing returns using more than five clusters. And Ma and Goulias (1997) used four clusters to represent combinations individuals' travel characteristics and four to represent their activity frequency.

A range of cluster values from 5 to 12 were tested. Specifying too few clusters (e.g., five) made it impossible to differentiate between important elements within each cluster and identified groups that were too broad. Specifying a dozen clusters too finely parsed the sample of households and provided diminishing returns in terms of variance explained. The results from the seven, eight, and nine cluster solutions produced stable and reasonably similar groups of lifestyles.

TABLE 6.12 Final Cluster Centers for Each of the Lifestyle Clusters

Factor	Lifestyle								
	A	B	C	D	E	F	G	H	I
High accessibility	0.698	0.608	−0.779	0.727	0.288	−0.883	0.245	−0.246	−1.258
High nonwork trip frequency, many tours, and car pool often	−0.325	−0.198	−0.646	−0.389	−0.100	0.314	1.790	−0.189	0.113
High transit/walk frequency, fewer vehicles owned	0.176	−0.281	−0.260	−0.396	3.332	−0.024	−0.235	−0.081	−0.155
High commute trip frequency, longer travel distance	−0.952	0.252	−0.618	0.324	0.316	−1.184	−0.020	1.575	0.757
High vehicle trip frequency, complex tours	−0.416	1.917	−0.398	−0.267	−0.067	1.324	−0.291	0.427	−0.522

most often, the other four factors did not appear to be instrumental in defining that factor. For example, lifestyle B is largely distinguished by the heavy weighting of the last factor (high vehicle trip frequency, complex tours). Lifestyle E is clearly dominated by transit users and walkers. Lifestyle G appears to be dominated by the second factor (high nonwork trip frequency, many tours, and frequent carpooling).

The largest group is lifestyle D, representing over one fifth of the sample. While these households do not appear to have travel or activity patterns that clearly distinguish themselves from others (i.e., they do not appear to have higher rates of transit ridership), the optimistic news for land use–transportation is that this cluster appears to value residential locations with higher levels of accessibility. The smallest clusters (in terms of number of households represented) are lifestyles E, F, and H — each constituting 6% or less of the sample.

The third observation notes how each of the clusters relates to the urban form (access) factor. In three of the nine clusters (lifestyles E, G, and H), the accessibility factor appears to play a very minimal role (as denoted by the relatively small magnitude of their cluster center). In 38% of the cases (lifestyles A, B, and D), higher levels of accessibility were important; in 29% of them (lifestyles C, F, and I), lower levels of access were prevalent. However, there appears to be no consistent pattern between each of the accessibility levels and the manner in which they relate to other dimensions of travel or activity frequency.

6.6.3.4 Covariation between Lifestyles and Household Sociodemographic Information

The first step in this part of this research used factor analysis to identify relevant lifestyle dimensions and the placement of objects (i.e., households) within those dimensions (plotting of factor scores). Cluster analysis then grouped households within those dimensions. The final step examines covariation that exists between each of the lifestyle clusters and four dimensions of the household's sociodemographic characteristics.

The first dimension relates to the type of household (stage of the life cycle across eight different classifications); the second relates to the number of children; the third is the number of employees in each household; and the fourth is income. Each of these characteristics is provided by data from the PSTP. The results of the cross-classification statistics for each lifestyle are provided in Tables 6.13 and 6.14, together with Pearson's chi-square statistic, degrees of freedom, and significance. In each case, we reject the null hypotheses stating that there is no association between lifestyle type and each sociodemographic characteristic.

The following discussion, combined with Tables 6.13 and 6.14, describes characteristics of each lifestyle to more fully understand the market of households that choose different lifestyles. The discussion below focuses on noticeable ways in which the actual number of households deviates from the expected number within each lifestyle (i.e., the residuals), to learn about patterns of covariation. The label for each is based on the combination of elements that help to define that particular lifestyle derived from cluster centers

TABLE 6.13 Cross-Tabulations of Lifestyle vs. Household Type

Lifestyle		Any child < 6	All children 6-17	1 adult, < 35	1 adult, 35-64	1 adult, 65+	2+ adults, <35	2+ adults, 35-64	2+ adults, 65+	Total
A Retirees	Counts	16	16	9	25	49	17	44	72	248
	% w/n lifestyle	6.50	6.50	3.60	10.10	19.80	6.90	17.70	29.00	100.0
	Residual	−14	−29	2	−3	29	5	−28	39	
B Single, busy urbanists	Counts	8	20	8	52	10	6	44	2	150
	% w/n lifestyle	5.30	13.30	5.30	34.70	6.70	4.00	29.30	1.30	100.00
	Residual	−10	−7	3	35	−2	2	0	−18	
C Homebodies	Counts	32	36	5	21	27	2	104	57	284
	% w/n lifestyle	11.30	12.70	1.80	7.40	9.50	0.70	36.60	20.10	100.0
	Residual	−2	−16	−4	−11	4	−12	21	20	
D Urbanists	Counts	44	76	17	50	12	34	150	23	406
	% w/n lifestyle	10.80	18.70	4.20	12.30	3.00	8.40	36.90	5.70	100.00
	Residual	−5	2	5	4	−20	14	32	−31	
E Transit users	Counts	6	11	10	38	11	3	7	15	101
	% w/n lifestyle	5.90	10.90	9.90	37.60	10.90	3.00	6.90	14.90	100.0
	Residual	−6	−7	7	27	3	−2	−23	2	
F Suburban errand runners	Counts	8	15	1	12	31	1	34	29	131
	% w/n lifestyle	6.10	11.50	0.80	9.20	23.70	0.80	26.00	22.10	100.00
	Residual	−8	−9	−3	−3	21	−6	−4	12	
G Activity participants	Counts	46	66	1	7	10	6	49	50	235
	% w/n lifestyle	19.60	28.10	0.40	3.00	4.30	2.60	20.90	21.30	100.0
	Residual	18	23	−6	−20	−9	−6	−20	19	
H Surburban workaholics	Counts	18	29	6	7	2	14	52	2	130
	% w/n lifestyle	13.80	22.30	4.60	5.40	1.50	10.80	40.00	1.50	100.0
	Residual	2	5	2	−8	−8	8	14	−15	
I Exurban, family commuters	Counts	52	78	1	4	0	12	72	2	221
	% w/n lifestyle	23.50	35.30	0.50	1.80	0.00	5.40	32.60	0.90	100.0
	Residual	25	38	−6	−21	−18	1	8	−27	
Total	Count	230	347	58	216	152	95	556	252	1906
	% w/n lifestyle	12.10	18.20	3.00	11.30	8.00	5.00	29.20	13.20	100.00

Pearson's Chi-Square = 715.66, degrees of freedom = 56, $p < 0.000$.

TABLE 6.14 Cross-Tabulations of Lifestyle vs. Presence/Absence of Children, Number of Employees, and Income

Children

	Lifestyle	No	Yes	Total
Count	A Retirees	217	33	250
& w/n lifestyle		86.80	13.20	1.0
Residual		43	-43	
Count	B Single, busy urbanists	121	28	149
& w/n lifestyle		81.20	18.80	100.0
Residual		17	-17	
Count	C Elderly homebodies	217	68	285
& w/n lifestyle		76.10	23.90	1.0
Residual		18	-18	
Count	D Urbanists w/ higher income	285	121	406
& w/n lifestyle		70.20	29.80	100.0
Residual		2	-2	
Count	E Transit users	84	17	101
& w/n lifestyle		83.20	16.80	1.0
Residual		14	-14	
Count	F Suburban errand runners	108	23	131
& w/n lifestyle		82.40	17.60	100.0
Residual		17	-17	
Count	G Activity oriented families	124	111	235
& w/n lifestyle		52.80	47.20	1.0
Residual		-40	40	
Count	H Suburbanites w/double income	84	46	130
& w/n lifestyle		64.60	35.40	100.0
Residual		-7	7	
Count	I Exurban, family commuters	90	130	220
& w/n lifestyle		40.90	59.10	100.0
Residual		-63	63	
Count	Total	1330	577	1907
& w/n lifestyle		69.70	30.30	100.0

Pearson's Chi-square = 188.35, degrees of freedom = 8, p < 0.000

Number of Employees

	Lifestyle	0	1	2 or more	Total
Count	A Retirees	134	78	38	250
& w/n lifestyle		53.60	31.20	15.20	100.0
Residual		74	-20	-54	
Count	B Single, busy urbanists	18	82	50	150
& w/n lifestyle		12.00	54.70	33.30	100.0
Residual		-18	23	-5	
Count	C Elderly homebodies	103	108	73	284
& w/n lifestyle		36.30	38.00	25.70	100.0
Residual		35	-3	-31	
Count	D Urbanists w/ higher income	22	186	197	405
& w/n lifestyle		5.40	45.90	48.60	100.0
Residual		-76	28	48	
Count	E Transit users	26	61	14	101
& w/n lifestyle		25.70	60.40	13.90	100.0
Residual		2	22	-23	
Count	F Suburban errand runners	78	32	21	131
& w/n lifestyle		59.50	24.40	16.00	100.0
Residual		46	-19	-27	
Count	G Activity-oriented families	74	81	80	235
& w/n lifestyle		31.50	34.50	34.00	100.0
Residual		17	-11	-6	
Count	H Suburbanites w/double income	2	34	94	130
& w/n lifestyle		1.50	26.20	72.30	100.0
Residual		-29	-17	46	
Count	I Exurban, family commuters	3	84	133	220
& w/n lifestyle		1.40	38.20	60.50	100.0
Residual		-50	-2	52	
Count	Total	460	746	700	1906
& w/n lifestyle		24.10	39.10	36.70	100.0

Pearson's Chi-square = 533.01, degrees of freedom = 16, p < 0.000

Income Level

	Lifestyle	$0-30k	$30-50k	$50k>	Total
Count	A Retirees	116	78	39	233
& w/n lifestyle		49.80	33.50	16.70	100.0
Residual		47	0	-47	
Count	B Single, busy urbanists	49	44	49	142
& w/n lifestyle		34.50	31.00	34.50	100.0
Residual		7	-4	-4	
Count	C Homebodies	100	78	91	269
& w/n lifestyle		37.20	29.00	33.80	100.0
Residual		21	-12	-9	
Count	D Urbanists	86	134	170	390
& w/n lifestyle		22.10	34.40	43.60	100.0
Residual		-29	4	25	
Count	E Transit users	44	28	26	98
& w/n lifestyle		44.90	28.60	26.50	100.0
Residual		15	-5	-10	
Count	F Suburban errand runners	47	41	35	123
& w/n lifestyle		38.20	33.30	28.50	100.0
Residual		11	0	-11	
Count	G Activity participants	53	78	88	219
& w/n lifestyle		24.20	35.60	40.20	100.0
Residual		-12	5	7	
Count	H Suburbanites workaholics	10	42	72	124
& w/n lifestyle		8.10	33.90	58.10	100.0
Residual		-27	1	26	
Count	I Exurban, family commuters	29	83	102	214
& w/n lifestyle		13.60	38.80	47.70	100.0
Residual		-34	11	23	
Count	Total	534	606	672	1812
& w/n lifestyle		29.50	33.40	37.10	100.0

Pearson's Chi-square = 160.822, degrees of freedom = 16, p < 0.000

as described, sociodemographic characteristics, anecdotal, and colloquial information (see Table 6.15). The label is not intended to specify every household within each cluster, but is used as a means to quickly identify the nature of the different clusters.

Retirees (A): The first lifestyle, representing 13% of the population, is labeled retirees. Their socio-demographic characteristics not only reveal high proportions of adults over the age of 65, but also the highest rate with no children (87%), the highest rate in the low-income category (50%), and close to the highest rate of households with no employees (53.6%). Accordingly, these households travel less, have very low commute trip frequency, travel shorter distances, and live in highly accessible locations (e.g., condominiums in relatively high-density areas).

Single, busy urbanists (B): Single, busy urbanists comprise the second lifestyle (7.8%) and appear to be heavily dominated by single, working types between the ages of 35 and 64. They too tend to live in areas with high accessibility and engage in highly complex tours, presumably participating in the many different types of activities that occupy their day.

Elderly homebodies (C): Homebodies represent the third lifestyle and have uncharacteristically low rates of activity frequency and travel distance. They appear to be equally distributed across each type of household, with slightly higher rates of households, with two or more adults (over age 35). Similar to retirees, they tend to be unemployed, have lower rates of income, and have slightly lower rates of children. However, their activity and travel characteristics differ from the retirees in that the homebodies appear to live in areas with lower levels of accessibility and have lower rates of walking and transit use.

Urbanists with higher income (D): The largest lifestyle group is the urbanists, comprising 21.3% of the sample. There are an equal number of households with and without children, but dramatically higher rates of employed persons. With the exception of slightly more households that have two or more adults, different household types are equally distributed across this lifestyle. As mentioned, the good news for land use–transportation initiatives is that these households appear to value residential locations with high accessibility.

Transit users (E): The fifth lifestyle represents the most distinctive lifestyle as output from the cluster analysis — those with high rates of walking and transit use. As expected, these households are represented by disproportionately high shares of lower-income households and single, working people with no children. Important news for transit advocates is that this lifestyle represents the lowest proportion of households, a mere 5.3% of the sample.[32]

Suburban errand runners (F): Suburban errand runners are households represented with higher rates of unemployed persons and lower rates of children. Again, there appears to be little disparity across household types, with slightly higher rates of single adults over the age of 65. Their label is derived primarily by the fact that these households live in areas with relatively lower rates of accessibility, and they complete many vehicle trips with complex tours.

Family and activity-oriented participants (G): This seventh lifestyle clearly represents families with children who are engaging in many activities throughout the day (e.g., soccer, chorus). Because they have a disproportionate share of children between the ages of 6 and 17 (ages at which children cannot drive), these households appear to engage in many non-work-related tours and carpool often.

Suburban workaholics (H): While the suburban workaholics appear to be proportionately represented across household types (with slightly higher rates for two or more adults, ages 35 to 64), they heavily loaded on households with two or more employees.

Exurban, family commuters (I): The final lifestyle, denoted as exurban, family commuters, also appears to be heavily work oriented, but with a relatively high proportion of children between the ages of 6 and 17. Their extremely low score on the accessibility factor suggests their propensity to find less expensive housing on the outskirts and a heavy commuting influence of the dual-income household — both in efforts to support the relatively larger family sizes.

---

[32] It is important to identify, however, that the "low" income breakdown used aggregates all households earning less than $30,000 into a single category. In reality only ten percent of the households in the entire sample have a household income less than $15,000.

TABLE 6.15 Labels and Descriptions for Different Lifestyles

Lifestyle	Lifestyle Label	Prevalent Sociodemographic/Economic Characteristics	Prevalent Lifestyle Characteristics	# of Households	% of Households
A	Retirees	Elderly (age 65+), without children, few working members, and of lower income	High accessibility, low commute trip frequency, and shorter travel distances	250	13.1
B	Single, busy urbanists	Many single households (age 35–64) who work	High accessibility and many vehicle trips with complex tours	149	7.8
C	Elderly homebodies	At least age 35, few working members, and of lower income	Low accessibility matched with fewer tours, shorter travel distances	284	14.9
D	Urbanists w/ higher income	Higher income and coupled workers	High accessibility and average other activity–travel dimensions	406	21.3
E	Transit users	Tend to be single and partial to lower income	Relatively high accessibility and high transit–walk frequency	101	5.3
F	Suburban errand runners	Fewer working members, older, and single	Low accessibility, low commute trip frequency, many vehicle trips, complex tours	131	6.9
G	Family and activity oriented	Many children	High nonwork trip frequency, many tours, and carpool frequently	235	12.3
H	Suburbanites w/double income	Higher income, more employees	Many commute trips and long travel distances	130	6.8
I	Exurban family commuters	Higher income, more employees with children	Low accessibility, many commute trips, and longer travel distances	221	11.6

6.6.4   Recap and Policy Significance

As opposed to previous efforts that attempt to pull apart the relative significance of different household decisions, the above framework approaches decisions as a *combined* phenomenon; doing so requires an analysis framework that recognizes that such decisions mutually inform one another. Using factor analysis and then cluster analysis, nine distinct lifestyles are uncovered. The characteristics of each of the nine lifestyle clusters have direct implications for land use–transportation planning and policy. They are representative of the variety of preferences and tastes that dictate where households live and how they travel. By understanding each lifestyle relative to its sociodemographic dimensions, researchers can gain a better understanding of: (1) how different phenomena interact, and (2) the potential market of households that may possibly respond to various land use–transportation planning initiatives.

Differing levels of accessibility (combined with other dimensions of activity or travel characteristics) appear to play a substantial role in defining household lifestyles. The good news for land use–transportation planners is that almost 60% of the households (five lifestyles) in the sample score positive on the accessibility factor. This finding shows that over half of the population live in neighborhoods with relatively higher levels of accessibility, representing a larger market of households than many would expect. It is important to recognize that high accessibility by itself may not be met with environmentally benign travel. One need only look at the lifestyle of the single, busy urbanists (B). While this population appears to live in accessible locations, they also have high vehicle trip frequency with complex tours.

Two of the lifestyles (18.4% of the sample) appear to replicate the classical pattern that new urbanists and other like-minded land use–transportation professionals espouse. Both the retirees (lifestyle A) and the transit users (lifestyle E) live in neighborhoods with relatively high access, own fewer vehicles, and have higher rates of transit use and walking. In addition, the other three factors (e.g., nonwork and vehicle trip frequency) of these households further support their relatively environmentally benign lifestyles. These households represent urbanites who may prefer neighborhoods that are more urban in character, with increased amenities and accessibility.

An important implication for policy is that both lifestyles represent populations that are increasing in size, as evidenced by prevailing sociodemographic trends (Myers and Gearin, 2001). The baby boom population certainly appears to represent a latent demand for the retiree lifestyle. The baby boomers represent a substantial size of the region's population who may eventually choose to escape their suburban homes and auto-reliant lifestyles to choose residential options that require less home maintenance and are more urban in character and less auto dependent. Secondly, as described above, transit users appear to be well represented by working, single adults between the ages of 35 and 64. One need only look to demographic projections showing modest increases in households of single, working types as a result of increased rates of separation and later marriages. It is important to point out, however, that both populations combined comprise less than 20% of the sample.

In contrast, lifestyles with negative scores on the accessibility factor comprise 40% of the sample. None of these lifestyles, as expected, score positive on the transit–walk factor. In fact, each lifestyle with a negative access score also demonstrates one or more cases of activity or travel characteristics (e.g., more vehicle trips, longer distances) that are counter to many land use–transportation initiatives.

The significance of this part of the research lies primarily with its approach and methods and secondarily with the results. An approach is presented that recognizes the integrated decision process of household decisions with respect to travel, activity, and neighborhood type. However, additional research is needed to more fully capture the underlying decision processes and, in particular, how households make complex trade-offs within and across these choice dimensions. These questions have direct policy significance, since the construction of neo-traditional neighborhoods, or of beltways or light-rail systems, may induce complex adjustment responses by households by influencing daily travel, activity, residential location, workplace, and vehicle ownership. The approach presented helps policy makers better understand how these phenomena interact within the context of both household lifestyle choices and land use–transportation policy.

6.7 Assessing the State of the Knowledge in Urban Form and Travel Research

6.7.1 Emerging Issues and Research

The past 15 or so years has witnessed an active line of research aiming to uncover relationships between urban form and travel behavior. Much progress has been made and our understanding of how these phenomena interact is undoubtedly richer. However, at least two meta-level research issues deserve additional attention. The first issue requires researchers to further the current stream of inquiry and understand how travel behavior relates to different urban forms. The second issue asks broader questions to learn the most effective strategies for land use–transportation policies and programs. To steer future research efforts in these endeavors, the following section describes six specific topics (not intended to be exhaustive) that encapsulate emerging issues and subjects for land use and transportation planning.

6.7.1.1 Understanding the Multidimensions of Urban Form

Some studies focus on identifying the relative contribution of different dimensions of urban form (e.g., the presence of sidewalks vs. density) (Ewing and Cervero, 2001). An alternative set of questions asks whether elements of urban form combine in ways to affect different travel behavior outcomes. That is, what proportions of mixed land use best complement design elements? Which aspects of travel are likely to be affected?

A corollary to the above questions asks the degree to which there may be significant thresholds along the continuum of accessibility. The task of identifying the extremes of high and low levels of NA is relatively straightforward; many researchers and planners understand the prevailing patterns of use in higher- and lower-density developments. The overwhelming amount of existing development, however, lies in what could be considered a gray area in between the high and the low: identifying the appropriate land use and urban design improvements to reduce auto dependence in this middle ground.

For example, once a neighborhood reaches a threshold of mixed land uses (all within attractive walking distance), the relative contribution of a few more shops becomes marginal in terms of advancing pedestrian use. There is likely a point of diminishing returns. In this case, the benefits gained from increasing accessibility may be asymptotic to a given measure of travel behavior (e.g., mode split). Frank and Pivo (1994) confirmed Pushkarev and Zupan's (1977) assertion that residential densities need to exceed eight housing units per acre before one can expect significant modal shifts from single-occupant vehicle to transit use. Furthermore, additional research is necessary to identify thresholds similar in nature using more precise measures of urban form. Such thresholds may exist for different dimensions of travel behavior (e.g., mode split vs. vehicle travel distance) or different ranges of neighborhood measurement (e.g., quarter mile vs. one-half mile).

6.7.1.2 Enhanced Data

To understand the detailed effects of urban form undoubtedly requires researchers to be able to measure different elements of urban form. Increased precision using Geographic Information Systems in concert with enhanced aerial photography and remote sensing will be paramount in the near future. Of course, such technical research capabilities require guidance about the specific types of elements that should be measured and at what scale.

Often considered the bane of travel behavior research, enhanced travel data are undoubtedly needed to soundly document hypothesized relationships. An ability to more precisely understand nonauto travel — in particular pedestrian behavior — tops the list of data needs. Better information on short-distance auto travel is also necessary. Significant improvements currently employ advanced monitoring systems and Global Positioning Systems to gather extremely detailed travel information. Continued deployment and refinement of these technologies will considerably aid travel behavior researchers.

6.7.1.3 Understanding the Role of Preferences vs. Urban Form

As highlighted in Section 6.5 above, it is important for researchers and policy officials to understand that differences in travel between households with different neighborhood designs should not be credited to urban form alone. The differences could be attributed to the broader *preferences* that triggered the choice to locate in a given neighborhood. Evidence is mounting that this second hypothesized mechanism is stronger; that is, neighborhood type may tend to act as a proxy for the true explanatory variables with which it is strongly associated. The important point is that the two effects — urban form vs. preferences — should be disentangled, and the relative magnitude of the independent effect of urban design on travel may become marginalized once preferences are accounted for.

The extent to which this assertion is true begs the question of knowing the relative magnitude of these two phenomena. Researchers have brought this issue to attention in the literature, and some studies have attempted to control for such preferences (Prevedouros, 1992; Kitamura et al., 1997; Boarnet and Sarmiento, 1998; Bagley and Mokhtarian, 1999), but continued work and refinement are required on this front. In particular, little work has explored how preferences are formed, what it means for preferences to determine travel behavior, and the degree to which preferences are appropriately measured and operationalized.

6.7.1.4 Matching Preferences, Maximizing Choice, and Understanding the Latent Demand

While some researchers aim to control for preferences, others claim that they are exactly the means through which land use and transportation planners and researchers should be orienting their work (Levine, 1999). The claim is that neighborhoods with high NA should not be based on their potential to reduce drive-alone travel, but rather their potential to expand households' choices in how to live and travel (Levine, 1999; Handy and Clifton, 2001). The choice set of available neighborhoods, many would argue, has been constrained by local land policies such as zoning that limits densities and mandates separation, transportation standards that call for wide streets and generous parking requirements, and fiscally motivated practices that restrict development of alternatives to the large-lot and single-family house. Increasing evidence suggests that there is a growing population of households currently frustrated by their residential location; they would prefer to live in areas with high NA if they were increasingly available at a reasonable price. Basing empirical work on only those households that currently live in high NA neighborhoods potentially misses a population segment who would prefer to locate in such neighborhoods given increased opportunity to do so. By this rationale, neighborhood self-selection, an expression of expanded choice, can actually work to reduce VMT.

If this is the case, a line of additional research asks what travel behavior (or residential location) changes will occur once barriers to land use and transportation choices are removed. Recent work has attempted to understand the disparity between where households prefer to live and where they actually live. One study has concluded that urban areas with a greater diversity of neighborhood types (e.g., Boston) allow residents to forge a closer connection with their preferences than does an urban area with relatively less diversity in neighborhood types (e.g., Atlanta) (Levine et al., 2000), suggesting a latent demand for highly accessible neighborhoods. But again, further work is needed to better understand this latent demand in variety of geographic settings using different kinds of research approaches that combine quantitative and qualitative approaches.

6.7.1.5 Understanding Household Decision Making

Research is also needed to better uncover the manner in which households combine what have previously been referred to as shorter- and longer-term decisions. The work presented in Section 6.6 sheds light on the correlations of choices that households make. It does not, however, illuminate the decision-making process itself, and it does not shed light on the trade-offs that households make. For example, do households first select how many autos to own and then a neighborhood in which to live, or vice versa? How do households select neighborhood location within the constraints of two-worker households and dual commutes? There are inevitably a number of trade-offs — higher-priced housing for better schools, poorer housing stock for increased levels of transit — embedded within each of these household decisions. Knowing what factors hinge on others is important for policy. Likewise, it is useful to know what households are willing to accept and at what cost. Further work is required to advance hierarchical modeling frameworks that best capture and represent these dynamic relationships. The researcher is currently guided by relatively little past work on this subject, providing an extremely fertile area for improvement.

6.7.1.6 How to Best Target Specific Households?

If understanding the behavior of household decision making is an initial step, the next step is to understand the ways in which households with differing composition or characteristics respond to land use–transportation policies and programs. The predominant thought in policy circles has been a one-size-fits-all approach, assuming that all households respond to programs in the same manner. However, several disaggregate studies show that subsectors of the population travel in different ways; the elderly, young, poor, race–ethnic minorities, immigrants, and first-generation Americans have all been tested for significance in econometric models. The findings from such models, however, fail to be integrated into the crafting of policies and programs. For example, how do different sociodemographic groups respond to urban configurations and how can land use–transportation programs be better tailored toward what we know of their household decision making?

As an example, a new practice for residential lending aims to encourage low- to moderate-income households to purchase homes in transit-rich neighborhoods. Because homes in such neighborhoods typically cost more than their suburban counterparts, they get passed up on account of affordability. But by wrapping costs saved from driving less into the mortgage, lenders can extend mortgages to a broader market of households. The increased availability of these properties is intended to influence residential location practices and, subsequently, land use–transportation patterns. Such a program represents just one innovative program that is available and targets specific populations based on their residential location and travel preferences.

6.7.2 Summary and Conclusions

Land use–transportation planning is a topic with a relatively long and beleaguered history. Interest in this topic has recently been sparked by planning movements christened as "new urbanist" or "smart growth." Consequently, the past 15 or so years has seen an extremely active line of research aiming to uncover how different elements of urban form, travel behavior, and residential location relate to one another. Significant progress has been made and the planning community undoubtedly has a better understanding of both the effect that urban form has on travel and the limitations of many land use policy initiatives.

The variety and complexity of past work has advanced knowledge of these relationships. But much like peeling an onion, previous work reveals greater complexity. Some studies may focus on pedestrian travel vis-à-vis specific urban design features; another study may examine household travel distance across a handful of communities with different landforms. Each of the sections in this chapter summarize the important points from past work — a task that is addressed in better detail in previous summaries (Handy, 1996a; Crane, 2000; Ewing and Cervero, 2001). But this chapter reviews the advancements, identifies knowledge gaps, and presents the results of recent research under a single, cohesive, and comprehensive cover. Specifically, Sections 6.3 to 6.6 identify and explain at least four overarching issues that confound past research endeavors.

The first issue stems from the deficiency of existing data. Many of the relationships researchers are trying to uncover are troubled by either inadequate or poorly operationalized data. The second confounding issue is that most studies fail to acknowledge the total demand for travel. The focus of most research to date — showing associative relationships — has recently been brought into question because it sheds light on only a limited number of travel outcomes. Most work neither uncovers the total travel of most households nor captures trade-offs and interactions between trip frequency, trip distance, multipurpose trips, and mode split. Third, researchers are increasingly realizing that for their work to be of most use for land use and transportation policy, they need to better disentangle the factors that trigger travel choices, in particular the role of urban form vs. the role of preferences. And finally, past research also fails to recognize that relatively short-term decisions (e.g., where to travel and how) may not always be conditioned by relatively longer-term decisions (e.g., where to live and how many cars to own); in fact, these types of decisions serve to mutually inform one another and should be analyzed in tandem.

In each section of this chapter, central issues are identified and the results of recent research from the Seattle area are used to demonstrate recent advancements that address some these shortcomings. Of course, there is substantial need for additional research. This section identifies a number of challenges that warrant further research and exploring.

Given increasing debate concerning the capacity of alternative land use planning as a means of travel demand, it is important for both planners and decision makers to more fully appreciate both the nature of household trips and tour making vis-à-vis the services usually contained within neighborhoods with high levels of access. Ultimately, the enormous complexity in the relationships among attitudes, household behavior, and social and economic constraints makes definitive progress on this front extremely difficult. In the end, however, answers to each of the questions posed, together with knowledge gained to date, will inevitably allow planners and modelers to better understand relationships between land use planning, travel behavior, and residential location. A more thorough understanding will ultimately assist policy makers in constructing more informed policies about our built environment.

References

Abraham, J.E. and Hunt, J.D., Specification and estimation of nested logit model of home, workplaces, and commuter mode choices by multiple-worker households, *Transp. Res. Rec.* 1606, 17–24, 1997.

Adler, T. and Ben-Akiva, M., A theoretical and empirical model of trip chaining behavior, *Transp. Res. B*, 13, 243–257, 1979.

Bagley, M.N. and Mokhtarian, P.L., The Role of Lifestyle and Attitudinal Characteristics in Residential Neighborhood Choice, paper presented at 14th International Symposium on Transportation and Traffic Theory, Oxford, 1999.

Bagley, M.N. and Mokhtarian, P.L., The impact of residential neighborhood type on travel behavior: A structural equations modeling approach, *Ann. Region. Sci.*, 6(2), 279–297, 2002.

Bagley, M.N. et al., A Methodology for the Disaggregate, Multidimensional Measurement of Residential Neighborhood Type, University of California, Davis, working paper, 2000.

Ben-Akiva, M. and Bowman, J.L., Integration of an activity-based model system and a residential location model, *Urban Stud.*, 35, 1131–1153, 1998.

Ben-Akiva, M. and DePalma, A., Analysis of dynamic residential location choice model with transaction costs, *J. Reg. Sci.*, 26, 321–341, 1986.

Ben-Akiva, M. et al., Understanding Prediction, and Evaluation of Transportation Related Consumer Behavior, MIT Center for Transportation Studies, 1980.

Boarnet, M.G. and Crane, R., *Travel by Design: The Influence of Urban Form on Travel*, Oxford University Press, New York, 2001.

Boarnet, M.G. and Greenwald, J., Land use, urban design, and non-work travel: reproducing other urban areas' empirical test results in Portland, Oregon, *Transp. Res. Rec.*, 1722, 27–37, 2000.

Boarnet, M.G. and Sarmiento, S., Can land-use policy really affect travel behavior? A study of the link between non-work travel and land-use characteristics, *Urban Stud.*, 35, 1155–1169, 1998.

Bowman, J.L. et al., Demonstration of an Activity-Based Model System for Portland, paper presented at 8th World Conference on Transport Research, Antwerp, Belgium, 1998.

Bradley Research and Consulting et al., A System of Activity-Based Models for Portland, U.S. DOT Report #DOT-T-99-02, Washington, D.C., 1998.

Cervero, R., Transit-Supportive Development in the United States: Experiences and Prospects, U.S. Department of Transportation, Federal Transit Administration, Washington, D.C., 1993.

Cervero, R. and Gorham, R., Commuting in transit versus automobile neighborhoods, *J. Am. Plann. Assoc.*, 61, 210–225, 1995.

Cervero, R. and Kockelman, K., Travel demand and the three Ds: density, diversity, and design, *Transp. Res. D*, 2, 199–219, 1997.

Cervero, R. and Radisch, C., Travel choice in pedestrian versus automobile oriented neighborhoods, *Transp. Policy*, 3, 127–141, 1996.

Clarke, M.I. et al., Some recent developments in activity–travel analysis and modeling, *Transp. Res. Rec.*, 1–8, 1981.

Crane, R., Cars and drivers in the new suburbs: linking access to travel in neotraditional planning, *J. Am. Plann. Assoc.*, 62, 51–65, 1996a.

Crane, R., On form versus function: will the new urbanism reduce traffic, or increase it? *J. Plann. Educ. Res.*, 15, 117–126, 1996b.

Crane, R., The influence of urban form on travel: an interpretative review, *J. Plann. Lit.*, 15, 3–23, 2000.

Crane, R. and Crepeau, R., Does neighborhood design influence travel? A behavioral analysis of travel diary and GIS data, *Transp. Res. D Transp. Environ.*, 3, 225–238, 1998.

Damm, D., Parameters of activity behavior for use in travel analysis, *Transp. Res. A*, 16, 135–148, 1982.

Dueker, K. and Bianco, M., Light-rail-transit impacts in Portland: the first ten years, *Transp. Res. Rec.*, 1685, 171–180, 1999.

Ewing, R., Beyond density, mode choice, and single purpose trips, *Transp. Q.*, 49, 15–24, 1995.

Ewing, R. and Cervero, R., Travel and the Built Environment: Synthesis, Transportation Research Board, Washington, D.C., 2001.

Ewing, R. et al., Getting around a traditional city, a suburban planned unit development, and everything in between, *Transp. Res. Rec.*, 1466, 53–62, 1994.

Frank, L.D. and Pivo, G., Impacts of mixed use and density on utilization of three modes of travel: single occupant vehicle, transit, and walking, *Transp. Res. Rec.*, 1466, 44–52, 1994.

Frank, L.D. et al., Linking land use with household vehicle emissions in the Central Puget Sound: methodological framework and findings, *Transp. Res. D*, 5, 173–196, 2000.

Friedman, B. et al., Effect of neotraditional neighborhood design on travel characteristics, *Transp. Res. Rec.*, 1466, 63–70, 1994.

Golob, T., A nonlinear canonical correlation analysis of weekly trip chaining behavior, *Transp. Res. A*, 20, 385–399, 1986.

Gould, J. and Golob, T., Shopping without travel or travel without shopping? An investigation of electronic home shopping, *Transp. Rev.*, 17, 355–376, 1997.

Handy, S.L., *How Land Use Patterns Affect Travel Patterns: A Bibliography*, CPL Bibliography, Chicago, 1992c.

Handy, S.L., Regional versus local accessibility: neotraditional development and its implications for non-work travel, *Built Environ.*, 18, 253–267, 1992a.

Handy, S.L., Regional versus local accessibility: variations in suburban form and the effects on non-work travel, in *City and Regional Planning*, Dissertation, University of California, Berkeley, 1992b.

Handy, S.L., Regional versus local accessibility: implications for nonwork travel, *Transp. Res. Rec.*, 1400, 58–66, 1993.

Handy, S.L., Methodologies for exploring the link between urban form and travel behavior, *Transp. Res. D*, 1, 151–165, 1996a.

Handy, S.L., Understanding the link between urban form and nonwork travel behavior, *J. Plann. Educ. Res.*, 15, 183–198, 1996b.

Handy, S.L., Urban form and pedestrian choices: study of Austin neighborhoods, *Transp. Res. Rec.*, 1552, 135–144, 1996c.

Handy, S.L. and Clifton, K.J., Local shopping as strategy for reducing automobile use, *Transp. Res. A*, 2001.

Hanson, S., The importance of the multi-purpose journey to work in urban travel behavior, *Transportation*, 9, 229–248, 1980.

Hanson, S. and Schwab, M., Accessibility and intraurban travel, *Environ. Plann. A*, 19, 735–748, 1987.

Hess, P.M. et al., Site design and pedestrian travel, *Transp. Res. Rec.*, 1674, 9–19, 1999.

Holtzclaw, J., Using Residential Patterns and Transit to Decrease Auto Dependence and Costs, Natural Resources Defense Council, San Francisco, 1994.

Jarvis, H., Understanding the Home-Work-Family Grid-Lock: The Case of Seattle, Working paper, University of Newcastle, Department of Geography, 2000.

Kansky, K., Travel patterns of urban residents, *Transp. Sci.*, 1, 261–285, 1967.

Kitamura, R. et al., A micro-analysis of land use and travel in five neighborhoods in the San Francisco Bay Area, *Transportation*, 24, 125–158, 1997.

Kockelman, K.M., Travel Behavior as a Function of Accessibility, Land Use Mixing, and Land Use Balance, in *City and Regional Planning*, Dissertation, University of California, Berkeley, 1996.

Krizek, K.J., Operationalizing neighborhood accessibility for land use–travel behavior research and modeling, *J. Plann. Educ. Res.*, in press, 2002.

Lansing, J.B. et al., *Planned Residential Environments*, Braun-Brumfield, Inc., Ann Arbor, MI, 1970.

Lawton, T.K., The Urban Environment Effects and a Discussion of Travel Time Budget, Portland Transportation Summit, Oregon, 1997.

Lerman, S., Location, housing, automobile ownership, and mode to work: a joint choice model, *Transp. Res. Rec.*, 610, 6–11, 1977.

Levine, J., Access to choice, *Access*, 14, 16–19, 1999.

Levine, J. et al., Innovation in Transportation and Land Use as Expansion of Household Choice, Association of Collegiate Schools of Planning, Atlanta, 2000.

Loutzenheiser, D., Pedestrian Access to Transit: Model of Walk Trips and Their Design and Urban Form Determinants around Bay Area Rapid Transit Stations, *Transp. Res. Rec.*, 1604, 40–49, 1997.

Lu, M., Analyzing migration decision making: relationships between residential satisfaction, mobility intentions, and moving behavior, *Environ. Plann. A*, 30, 1473–1495, 1998.

Ma, J. and Goulias, K.G., A dynamic analysis of person and household activity and travel patterns using data from the first two waves in the Puget Sound Transportation Panel, *Transportation* 24, 309–331, 1997.

McCormack, E.D., A chain-based exploration of work travel by residents of mixed land use neighborhoods, Dissertation, *Geography*, University of Washington, Seattle, 1997.

McCormack, E., Using a GIS to enhance the value of travel diaries, *ITE Journal*, January 1999, pp. 38–43.

McNally, M.G. and Kulkarni, A., Assessment of influence of land use–transportation system on travel behavior, *Transp. Res. Rec.*, 1607, 105–115, 1997.

Moudon, A.V. and Hess, P.M., Suburban clusters: the nucleation of multifamily housing in suburban areas of the Central Puget Sound, *J. Am. Plann. Assoc.*, 66, 243–264, 2000.

Moudon, A.V. et al., Effects of site design on pedestrian travel in mixed use, medium-density environments, *Transp. Res. Rec.*, 1578, 48–55, 1997.

Mumford, L., Neighborhood and neighborhood unit, in *The Urban Prospect*, Harcourt, Brace and World, New York, 1956, pp. 56–78.

Murakami, E. and Ulberg, C., *The Puget Sound Transportation Panel: Panels in Transportation Planning*, Kitamura, R., Ed., Kluwer Academic Publishers, Boston, 1997, pp. 159–192.

Myers, D. and Gearin, E., Current preferences and future demand for denser residential environments, *Housing Policy Debate*, 12, 633–659, 2001.

Nelson, D. and Niles, J.S., Market Dynamics and Nonwork Travel Patterns: Obstacles to Transit Oriented Development, paper presented at Proceedings of the Transportation Research Board, Washington, D.C., 1998.

Ocken, R., *Site Design and Travel Behavior: A Bibliography*, 1000 Friends of Oregon, Portland, 1993.

Oppenheim, N., A typological approach to individual urban travel behavior prediction, *Environ. Plann. A*, 7, 141–152, 1975.

Owen, W., *The Metropolitan Transportation Problem*, Brookings Institution, Washington, D.C., 1966.

Pas, E., Analytically derived classifications of daily travel–activity behavior: description, evaluation, and interpretation, *Transp. Res. Rec.*, 879, 9–15, 1982.

Pas, E., The effect of selected sociodemographic characteristics on daily travel–activity behavior, *Environ. Plann. A*, 16, 571–581, 1984.

Perry, C., The neighborhood unit, in *Neighborhood and Community Planning*, Regional Planning Association, New York, 1929.

Pew Center for Civic Journalism, http://www.percenter.org/doingcj/research/r_st2000nat1.html. Accessed on September 4, 2002.

Pickrell, D., Transportation and land use, in *Transportation and Cities*, Brookings Institution Press, Washington, D.C., 1996, pp. 403–435.

Prevedouros, P., Associations of personality characteristics with transport behavior and residence location decisions, *Transp. Res. A*, 26, 381–391, 1992.

Pushkar, A.O. et al., A Mulitvariate Regression Model for Estimated Greenhouse Gas Emissions from Alternative Neighborhood Designs, Transportation Research Board, Washington, D.C., 2000.

Pushkarev, B. and Zupan, J., *Public Transportation and Land Use Policy*, Indiana University Press, Bloomington, 1977.

Recker, W.W. and McNally, M.G., Travel/activity analysis: pattern recognition, classification and interpretation, *Transp. Res. A*, 19, 279–296, 1985.

Reichman, S., Travel adjustments and life styles: a behavioral approach, in *Behavioral Travel-Demand Models*, Stopher, P.R. and Meyburg, A.H., Eds., Lexington Books, Lexington, MA, 1976, pp. 143–152.

Rossi, P.H., *Why Families Move*, Sage, Beverly Hills, 1980.

Salomon, I., The use of the lifestyle concept in travel demand models, *Environ. Plann. A*, 15, 623–638, 1983.

Shen, Q., Spatial and social dimensions of commuting, *J. Am. Plann. Assoc.*, 66, 68–82, 2000.

Southworth, F., Multi-destination, multi-purpose trip chaining and its implications for locational accessibility: a simulation approach, *Pap. Reg. Sci. Assoc.*, 57: 108–123, 1985.

Steiner, R.L., Residential density and travel patterns: review of the literature, *Transp. Res. Rec.*, 1466, 37–43, 1994.

Strathman, J.G. et al., Effects of household structure and selected travel characteristics on trip chaining, *Transportation*, 21, 23–45, 1994.

Sun, A. et al., Household travel, household characteristics, and land use, *Transp. Res. Rec.*, 1617, 10–17, 1998.

Thill, J.-C. and Thomas, I., Toward conceptualizing trip-chaining behavior: a review, *Geogr. Anal.*, 19, 1–17, 1987.

1000 Friends of Oregon, *Making the Land Use Transportation Air Quality Connection: The Pedestrian Environment, Vol. 4A*, Portland, OR, LUTRAQ, with Cambridge Systematics, Inc., Calthorpe Associates, and Parsons Brinkerhoff Quade and Douglas, 1993.

To, M.C. et al., Externalities, preferences, and urban residential location: some empirical evidence, *J. Urban Econ.*, 14, 338–354, 1983.

Waddell, P., Towards a Behavioral Integration of Land Use and Transportation Modeling, paper presented at 9th International Association for Travel Behavior Research Conference, Queensland, Australia, 2000.

Wallace, B. et al., Evaluating the Effects of Traveler and Trip Characteristics on Trip Chaining, with Some Implications for TDM Strategies, Transportation Research Board, Washington, D.C., 2000.

Weisbrod, G.E. et al., Trade-offs in residential location decisions: transportation versus other factors, *Transp Policy Decision Making*, 1, 13–26, 1980.

Wells, W.D. and Tigert, D., Activities, interests, and opinions, *J. Advertising Res.*, 11, 27–35, 1971.

Williams, P., A recursive model of intraurban trip-making, *Environ. Plann. A*, 20, 535–546, 1988.

II

Data Collection and Analysis

7

Interactive Methods for Activity Scheduling Processes

CONTENTS

Sean T. Doherty
Wilfrid Laurier University

7.1 Introduction

In the field of transportation, a strong argument has been made for the use of an activity-based approach to improve the behavioral foundations of travel forecasting models (Axhausen and Gärling, 1992; Ettema and Timmermans, 1997). While this approach offers considerable theoretical appeal and potential, the data collection that it has inspired has been largely limited to a retooling of traditional diary-based survey methods from recording *trips* to recording *activities*. While activity diaries have several practical advantages, the implications for analysts is the more challenging task of trying to understand and model a more complex set of observed activities *and* travel patterns.

The main criticism of diary-based methods is that they focus on revealed outcomes, providing little, if any, information on the underlying behavioral process that led to the outcomes in the first place. To meet this need, a new class of survey methods has emerged that focuses on the activity scheduling decision process. Their main point of departure from traditional diary methods is an explicit focus on tracing the underlying process of *how* activity–travel decisions are planned, adapted, and executed over time, space, and across individuals — often termed an *activity scheduling process*. The results of this process are an observed pattern of activities and travel over time and space and across individuals — or an *activity schedule*.

It is only relatively recently that travel behavior researchers have begun to emphasize the need for in-depth research into the activity scheduling decision processes that underlie observed activity–travel patterns, as a means to both improve our understanding and provide a basis for new model development (e.g., Pas, 1985; Polak and Jones, 1994; Lee-Gosselin, 1996; Axhausen, 1998). Recent Transportation Research Board millennium papers also highlight the need for more judicious use of new technologies to augment existing survey techniques, the challenge of reducing the respondent burden in light of the demand for more detail (Goulias, 2000; Griffiths et al., 2000), and the need for realistic representation of decision-making behavior in travel forecasting models to improve their ability to forecast the more complex responses to travel demand management (TDM) strategies, such as telecommuting and congestion pricing (Bhat and Lawton, 2000). This latter point is particularly important, as emerging TDM and Intelligent Transportation Systems (ITS) solutions implicitly invoke a *rescheduling* response from individuals and households that is rarely confined to single trips, single people, or even single days, but rather has significant secondary effects across multiple activities, trips, days, and individuals — effects that may more or less contribute to the desired impacts of the policy.

In lieu of empirical insights into underlying behavioral processes, emerging activity scheduling models have had to make several types of assumptions that often limit their potential. Early scheduling models most often assumed either a simultaneous (all decisions made at once, and executed without revision) (e.g., Recker et al., 1986; Kawakami and Isobe, 1990) or a strict sequential decision process (decisions made in the same order as execution) (e.g., Kitamura and Kermanshah, 1983; van der Hoorn, 1983). More recent models attempt to replicate the process of schedule building by replicating the sequence of additions, modifications, and deletions to a schedule over time, based on a notion of the priority of activities (e.g., Ettema et al., 1993; Arentze and Timmermans, 2000). But even these most recent models continue to make stringent assumptions concerning activity priority (that it is fixed) and the sequence of scheduling decisions concerning the various attributes of activities (fixed sequences of choices for activity type, location, start and end times, duration, involved persons, and mode choice). Other models make similar assumptions concerning how tours are formed (e.g., Bowman and Ben-Akiva, 2001) or the sequence of decisions in the logit-based modeling structures (e.g., Bhat and Singh, 2000). Addressing the validity of these assumptions is a key step to future model development and their applicability to the forecasting of emerging policies that inherently invoke a rescheduling response.

7.2   Objectives

The focus of this chapter is on an emerging class of interactive survey methods that explicitly target activity scheduling decision processes. What is shared by these methods is a desire to interactively observe *how* decisions are made and their dynamics, not just the results of these decisions in the form of static observed activity–travel patterns. Given these dynamics, these methods tend to elicit and trace such behavior in a continuous and *interactive* way over a period of time — meaning that the sequence and types of questions asked depend on the particular responses and inputs of subjects.

An outline of the basic components of the activity scheduling decision process is first proposed. Each component of this framework is then discussed in depth in terms of specific survey method opportunities and challenges, including applied examples where applicable. The types of data that result and priority areas of analysis are then discussed. This chapter is based in part on cumulative experiences in developing, testing, and applying activity scheduling process surveys with collaborative research teams in Canada, the United States, and Europe. It includes discussion on the latest state-of-the-art techniques and technologies adopted in the field, many of which are still evolving in design. It is hoped that this chapter provides a framework that encourages further in-depth exploration of this exciting and emerging field of inquiry.

7.3   The Nature of the Activity Scheduling Decision Process

Figure 7.1 presents a simple schema of the major components of the activity scheduling process that are the focus of investigation in this chapter. The process takes as its starting point an *agenda* of household activities (similar words such as "listing" or "repertoire" could also be used). These activities are derived from the basic needs, desires, and goals of individuals and households, and embody a range of practical

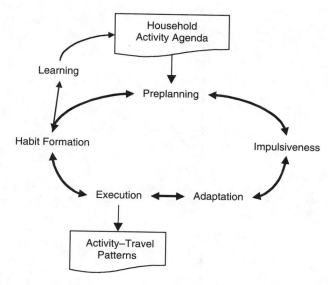

FIGURE 7.1 Simple schema of the main aspects of the activity scheduling decision process.

and physical constraints. An example agenda is shown in Table 7.1. Each activity in the agenda is defined as the act of satisfying a need that has unique attributes. These attributes influence how the activity is scheduled and eventually executed. On the agenda, these attributes are measured in terms of their relative degrees of *fixity* and *flexibility* and are meant to be "fuzzy" in nature — once executed, a final observed *static* choice of attributes is made.

Taking the activity agenda as given, the activity scheduling process depicted in Figure 7.1 is conceptualized as a dynamic and continuous process involving *preplanned*, *impulsive* (i.e., little or no preplanning involved), and *adaptive* decisions concerning the various components of activities: activity type, location, duration, start and end times, sequencing, involved persons, and mode and route choice. This process continues up to and during actual *execution* of activities, which leads to the formation of observed *activity–travel patterns* over time and space. As in all conceptualizations, an endless array of interdependencies (i.e., arrows) could be drawn in this diagram.

As a visual example of the scheduling process, consider Figure 7.2. The example is of a person who starts out with a preplanned schedule that includes empty time windows, but that goes through further

TABLE 7.1 A Simplified Household Weekly Agenda Example

Activity Label	Applicable Household Members	General Location	Attributes			
			Duration (mean)	Mean Frequency (per week)	# Perceived Locations	Etc.
Work	Male head	Home	2	2	1	
Telework	Male head	Out of home	8	5	1	
School	Child 1	Out of home	8	5	1	
Grocery shop	Female head	Out of home	1	3	12	
Grocery shop	Male and female	Out of home	2	1	12	
Active sport	Male	Out of home	1	1	2	
Active sport	Male and female	Out of home	2	1	1	
Chauffeuring	Male and female	Out of home	.5	5	1	
Socializing	Male and female	In or out of home	3	2	10	
Etc.						

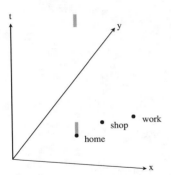

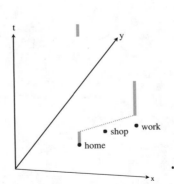

a) The most basic need, to sleep (or just be at home), is part of a long-standing routine, and forms a basic skeleton schedule. Note the unplanned time.

b) A work activity and associated travel are preplanned and added to the skeleton schedule.

Legend

┅┅┅ Travel between activity locations

─── Time spent conducting activity

● Activity location

t: Time of day

x, y: Location coordinates

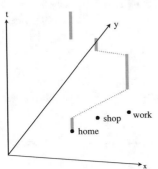

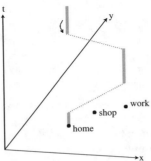

c) Upon execution of the preplanned schedule, unexpected congestion results in an impulsive increase to the travel time, and associate delay in work start.

d) A call from spouse during the day results in plan for dinner and movie together at home in the evening.

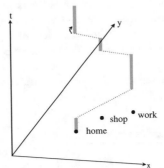

e) On drive home, impulsively decide to shop for a few grocery items for the evening.

f) Arrival home delayed slightly. Final outcome is the observed activity–travel pattern.

FIGURE 7.2 Step-by-step visualization of the activity scheduling decision process that underlies observed space–time paths.

planning, adaptation, and impulsive changes, leading to the final observed space–time pattern. Note that in reality, the planned activities at any stage in the process may only be partially elaborated — meaning that certain aspects of activities may be more or less planned, whereas the figure implies that all observed attributes are decided at once.

Also depicted in Figure 7.1 are two other important factors that influence scheduling in the longer term — *habits* and *learning*. Over time, habits in the form of set activity–travel decision routines may form, which are executed with very little thought during the process. These habits can be viewed as being realized through increased fixity of attributes of activities on the agenda over time, and as skeletal activities on a person's schedule (see Figure 7.2(a) and (b)). People may also seek out information during this process and learn of new aspects of activities, such as new locations, new involved persons to conduct them with, or even entirely new activities. Similar to habits, learning can be viewed as being realized through changes in the attributes of activities on the agenda (or new additions), but perhaps with more of a tendency toward increasing their flexibility (e.g., learn of more locations to conduct an activity).

7.3.1 Operational Definitions

The following operational definitions are adopted for this chapter:

Trip — Movement over space.

Activity — The act of satisfying a need that has unique attributes.

Activity attributes — A broad range of characteristics of activities that affect how they are planned and executed, generally measured in terms of their relative degrees of fixity and flexibility. See also, the listing in Section 7.4.2.1.

Activity agenda — A listing of activities and their attributes for an individual or household.

Activity schedule — A continuous pattern of activities and trips over time and space, including the observed choices of what activities to participate in, where, for how long, in what sequence, coupled with mode and route choices. (One way to visualize an activity schedule is as a time–space path, as shown in Figure 7.2(f)). Note how observed activity attributes (start time, end time, location, duration, etc.) differ from their associated fuzzy counterparts on the activity agenda (earliest start time, latest end time, perceived locations, duration distribution, etc.).

Activity scheduling process — The dynamic and continuous process of planning, adaptation, and execution of activities and their attributes over time and space and across individuals, leading to observed activity–travel patterns.

Decision rules — The behavioral mechanism applied to solve a choice problem. Could include traditional econometric random utility maximization, a range of other sub-optimal satisfying rules, or simple logical rules.

Habits or routines — Aspects of activity–travel patterns that are repeated on a regular basis and scheduled with very little contemplation, generally characterized with high levels of fixity.

Learning — The process of discovering new activity attribute information.

7.3.2 Specific Investigative Goals

The overriding goals of investigation are to improve our basic fundamental understanding of underlying decision processes, to assess and challenge the validity of existing scheduling process models and their assumptions, and to provide a new source of data for the estimation of new functions, algorithms, and choice models for scheduling and rescheduling processes. Given the conceptualization presented in the previous section, the specific questions of investigation concern the following:

- How the various decisions are organized and sequenced over time, including the "meta" style decisions *of when to* preplan, impulsively plan, adapt or reschedule, execute, and search for new information (i.e., learning)
- What components of activities (activity type, location, duration, start and end times, sequencing, involved persons, and mode and route choice) are decided upon at each point and in what sequence
- The sequence of rescheduling decisions in response to stimuli (what activities are chosen for change; what attributes are chosen for change, conflict resolutions, etc.)
- What rules are used to make choices at each stage in the process
- When and how are habits and routines formed over time
- The extent of learning that occurs in the short and long term

In addition, given the future modeling objectives, additional information should be sought on the activity attributes and situational factors that serve as potential explanatory factors in this process, including:

- Activity attributes (frequency, duration, spatial or temporal fixity, etc.)
- Travel characteristics (e.g., available modes)
- Personal and household characteristics (e.g., age, gender, personality, lifestyle, family life cycle)
- Structural characteristics (e.g., land uses and transport network, opening hours)
- Situational characteristics (e.g., time since last activity, time to next planned occurrence, available time windows, congestion)

To be realistic, it would be most useful to observe these processes and explanatory variables as they occur in real time. However, given our limited capacity to observe the workings of the human mind, we must rely on experiments and self-reports of such behavior as means for investigation. The next sections attempt to describe the various approaches and techniques for such investigations.

7.4 Investigating the Activity Scheduling Decision Process

Given the description above, it should come as no surprise that investigation of the scheduling process appears daunting. We are used to investigating observed and outward patterns of activity–travel behavior that can be recalled and recorded in sequence using simple diary techniques, which have become the primary focus of data collection and refinement over the past many decades. The scheduling decision process, on the other hand, is not outwardly viewable, and it involves a combination of a variety of scheduling decisions concerning *when* and *what* to schedule at any given moment, followed by the application of a variety of decision rules to make the choices.

However daunting this may seem, we must remember that the scheduling of our daily life is a problem that each and every one of us solves every day, and is thus a very familiar process. It is human nature to be aware of our needs and desires, and to consciously plan to meet these needs in some fashion. Asking people to self-report on their scheduling behaviors (What are you going to do today?) is perhaps just as familiar a task as asking them what they did (What did you do yesterday?). The key realization is that the answer to the former question will continue to change over time, whereas the latter is fixed. This implies a need for multiple observations over time to capture the true dynamics of the process.

As with any complex problem, it is convenient to separate out key concepts for separate investigation strategies — this is especially so when working with human subjects for which respondent burden is a key limiting factor. In this case, the most immediate and convenient separation would be between activity agenda formation and the activity scheduling process that follows. As shown in Figure 7.1, the key link between these two processes concerns the formation of habits (which may tend to make the attributes of activities on the agenda more fixed) and the learning of new opportunities and information (which may tend to make the attributes of activities more flexible). If one assumes that habit formation and learning processes are fixed in the short term (operationally meaning that the attributes of activities on the agenda are not updated in the short term), then one can conveniently consider that the scheduling process proceeds in a top-down fashion in the short term, taking activities from the agenda as a starting point and proceeding with preplanning, execution, etc., as shown in Figure 7.1. The implications of this approach for modeling are clear — two black boxes in sequence in the short term, with longer-term feedbacks and updating. The implications for investigation of each of these concepts are taken up in the next sections.

Thus, although interesting, the longer-term processes of habit formation and learning are conveniently separated out from the investigation of the scheduling process and left for investigation on their own (interested readers should see also Chapter 3). Most appropriately, it would seem that some form of regular updating of the agenda, perhaps with feedback from the scheduling, should be incorporated into longer-term forecasts of scheduling behavior, such as the case when a scheduling process model is integrated within a larger land use and transportation model.

7.4.1   Individuals vs. Households

All the methods described below could be applied to an individual or a household or family with due care. However, given that households will share an activity agenda and exhibit considerable interdependencies in activity scheduling, they are a much more logical choice as the unit of analysis. Whereas parents and spouses can often be assumed to act as good surrogates for recalling observed activity patterns of their children and partners, the validity of this assumption when it comes to the underlying decision process is more questionable. In fact, the differences and similarities in decision-making processes across household members should be embraced for investigation.

Thus, ideally, all adults and children of decision-bearing age (i.e., once they start generating and scheduling activities on their own) should be directly involved in any survey method in order to capture the true dynamics and interdependencies of decision-making processes. However, if only one partner or parents without children are chosen or available, then special efforts should be made to capture as much of the independencies as possible from single individuals. This can be done at the agenda investigation phase by quantifying the activities of other household members that have a bearing on the individual, either because they are joint or service activities or because they serve as important constraints on the individual. During investigation of activity scheduling decisions, special queries should then be adopted to trace not only joint activities, but also joint decisions and communication acts with other household members.

7.4.2   Investigating Activity Agendas

Although the conceptualization of a household activity agenda as a listing of activities and their attributes seems straightforward, operationalizing this list is a challenging and crucially important task for several reasons. Firstly, the flexibility or fixity of attributes of activities such as start and end times, frequency, duration, location, involved persons, and travel mode are obviously strong determinants of when and how an activity is subsequently planned and executed during the scheduling process. Thus, capturing the most salient attributes on the agenda is key to the success of the activity scheduling process to follow. Secondly, the attributes on the agenda are the means from which to assess the impacts of a variety of policy measures. For example, a program of telework inherently affects the spatial and temporal fixity of work activities via the modification of attributes such as location choices (e.g., not fixed to the workplace anymore) and the times at which the activity could be conducted (e.g., not fixed to office hours, 9 to 5). This in turn influences how the activity is scheduled, having a primary impact on work trips, but also secondary impacts on other activities and trips — thus providing a much more behaviorally realistic impact assessment.

7.4.2.1   Definition of Activities and Their Attributes

The first challenge faced in investigating activity agendas is deciding upon an operational definition of an activity. The traditional approach is to label a range of activities that involve travel with a set of generic labels such as work, shopping, recreation, etc. The trouble with this approach is that the set of activities defined is not universal in type or level of detail, hampering the transferability of the results. What is needed is an activity classification that focuses more on the fundamental attributes of activities that make them different from each other. From a scheduling perspective, these attributes may include their frequency, duration, involved persons, earliest start and latest end times, available locations, etc. The challenge is to narrow in on the key attributes, and then seek to define activities based on unique similarities and differences across these attributes.

One approach to meeting the activity definition challenge is to establish a set of rules to guide definition. For example (based in part on investigations of household agendas reported in Doherty and Miller (2000)):

1. Include all activities that involve travel or could potentially be replaced by travel.
2. Include activities that serve as important constraints upon other activities (e.g., attending to children at home), even if very short (e.g., dropping-off or picking-up activities).

3. Define separate activities of the same basic type when their attributes are significantly different.
4. Group multipurpose activities that always occur in sequence together (e.g., washing, dressing, and packing in the morning) or tend to consist of a variety of tasks (e.g., cleaning and maintenance around the house) to avoid unnecessary detail (this rule balanced against rule 3).

Rule 3 is particularly important to the eventual success of any scheduling model. For example, an employment activity on the agenda may be traditionally labeled as a "work" and have the following attributes: participated in an average of five times per week; normal duration of 6 to 10 h per day, earliest start at 8:00 A.M., latest end at 7:00 P.M.; located at the office *or at home*. However, given that conducting work at home implies a different set of attributes, rule 3 implies that it should be defined separately on the agenda, perhaps with the label "telework" and the following attributes: 2 days per week; duration of 6 to 10 h, earliest start at 6:00 A.M., latest end at 11:00 P.M.; located at home. The difference in attributes will have a strong effect on how this latter activity is scheduled. The challenge is to balance the level of detail in the agenda vs. the desired accuracy of scheduling results — if too general (e.g., consider only two activities: in home and out of home), the subsequent scheduling model will lack realism and forecasting power; if too specific, the model may break down and lack computational or operational realism (e.g., considering breathing, moving ones arm, etc., as activities).

Another challenge associated with activity agendas concerns the types of *attributes* for each activity that should be investigated. These may include:

- Frequency (usual, normal, minimum, maximum, distribution)
- Duration (usual, normal, minimum, maximum, distribution)
- Temporal flexibility (earliest start and latest end times, range of start and end times)
- Spatial flexibility (number of perceived locations, number of possible locations)
- Interpersonal dependency (household members involved/required/optional)
- Interactivity dependency (performance of one activity linked to another activity or longer-term project in time or space)
- Travel modes (available travel modes, most likely mode)
- Perceived travel times
- Costs/expenditures
- Etc.

It is important to draw a clear distinction here between the fuzzy attributes of activities on the agenda that indicate their relative degree of fixity, flexibility, or constraint (e.g., earliest start time) and their final "observed" static choice on a person's executed schedule (e.g., actual start time). These attributes also reflect the constraints imposed on activities. For example, household coupling constraints are realized through attributes such as required involved persons, whereas environmental constraints such as store opening hours are realized through earliest and latest start and end times. Other attributes on the agenda are meant to reflect peculiar aspects of scheduling behavior interdependencies, such as how activities may be linked together, sequenced, or assigned to people (e.g., an indicator that captures the tendency of one activity to follow another). Embedding these attributes and constraints within the agenda is perhaps a more natural way to capture their effects, as opposed to "hardwiring" them into an eventual model. For instance, a household constraint that parents be at home at a certain hour to care for their children would be represented as a preplanned skeletal activity with highly fixed time and location, as opposed to the inclusion of a variable reflecting the presence of children. For forecasting purposes, this is particularly valuable, as policy changes are often materialized in the form of modifications to the constraints imposed upon activities.

7.4.2.2 Quantifying Activity Agendas

In practice, in-depth investigation of household activity agendas can be a time-consuming and burdensome task for individuals and households. Three possible approaches to investigating activity agendas include:

1. Repeated observation of activities over a sufficiently long observation period to capture the variability in observed attributes that serve as an indicator of the relative flexibility or fixity
2. In-depth face-to-face interviews querying directly for stated attributes, using a computerized form to speed data entry
3. Similar to number 2, but conducting the interview using computer-automated prompts and dialogs

As an example of the first method, consider the results of a 6-week travel survey conducted recently in Germany (Axhausen et al., 2002). The observed variability in the duration of a range of activities could be taken as an indicator of their relative flexibility or fixity. This could also apply to durations, frequencies, locations, involved persons, and mode choices. The obvious challenge of this approach is deciding on the length of the survey, which must be a sufficient period long enough to capture the variability. Even then, assuming that observed variability is a good indicator of the actual flexibility may be questioned in the case of activity attributes that have become habitual. For instance, a grocery shopping activity may be observed to occur at the same location and time every week out of habit, but assuming it is relatively fixed in space to just one location based on this information alone may be wholly inaccurate.

A second alternative, or even supplemental approach, is to hold an in-depth interview in order to investigate a household's *stated* range of activities and their attributes. In order to structure such an interview, a set of preliminary activity types should be defined as a basis for initial discussions. A typical listing is provided in Table 7.2, although the exact number and types of activities should be tailored to

TABLE 7.2   Example Listing of Generic Activity Types That Could Be Used as a Starting Basis to Define a Household Activity Agenda

Basic Needs	Work or School	Household Obligations	
Night sleep	Work	Cleaning, maintenance	
Wash, dress, pack	School	Meal preparation	
Home-prepared meals	Day care	Chauffeuring	
Bagged lunches	Volunteer work	Chauffeuring and passively observing	
Restaurants	Special training	Attending to children	
Delivered or picked-up meal	Other work or school	Pick up involved person	
Coffee or snack shops		Other errands	
Other basic needs		Other obligations	

Pick Up or Drop Off	Services	Shopping	
People	Doctor	Minor groceries (<10 items)	
Food	Dentist	Major groceries (10+ items)	
Movie	Other professional	Housewares	
Miscellaneous items	Personal (salon, barber, laundry)	Clothing and personal items	
	Banking	Drug store	
	Video store	Mostly browsing	
	Library	Convenience store	
	Other service	Pick-up meal	
		Other shopping	

Just for Kids or Teens	Recreation or Entertainment	Social	Other
Tag along with parent	Exercise or active sports	Visiting	Tag along travel
Play, socializing	Movies, theatre	Hosting visitors	Pleasure driving
Homework	Other spectator events	Cultural events	
With babysitter	Playing with kids	Religious events	
Other just for kids	Parks, recreation areas	Planned social events	
	Regular TV programs	Bars, special clubs	
	Unspecific TV	Phone or e-mail >10 min.	
	Movie video	Helping others	
	Relaxing, pleasure reading, napping	Other social	
	Hobbies (crafts, gardening, etc.)		
	Other recreation, entertainment		

the unique sociodemographic nature of the population in question. Household members should be asked to describe, in their own words, the specific activities of each type that they perform, along with their attributes. In practice, some further structure is needed to guide households along in this procedure. An example protocol used in practice to define activity types and their typical locations is provided in Figure 7.3. Following this step, household members can be asked to describe the specific attributes of each of the activities mentioned. This can be done immediately or could be delayed until the end of an associated scheduling survey, which offers some advantage in terms of avoiding certain attribute questions for activities that were observed repeatedly enough.

The number, format, and precision of attribute responses garnished can vary considerably, as can the length of time needed for the interview. For example, the definition of an average of 42 activities per household, along with five main attributes (common locations; normal frequency; normal, minimum, and maximum duration; applicable days; earliest start and latest end times), required an average of 45 min to 1 h of interview time in two such studies in Canada (see also Doherty and Miller, 2000). Aside from these time constraints, the main difficulty that will arise is deciding what attributes to ask for, how to ask for them, and when (for example questions, see Figure 7.4). In particular, the attributes of activities often depended on who was doing it (one person or joint) and on what days (e.g., weekday vs. weekend). In these cases, a new activity could be defined separately for each person or day(s), but in practice, this is often not possible. Recording possible locations (especially for flexible activities such as shopping), possible time windows, and normal durations also poses unique challenges to respondents in terms of recall.

A third alternative for investigating household activity agendas is to automate all or a portion of the in-depth interview task using a sequence of computer-generated forms and dialogs. The face-to-face or telephone interview could be limited to asking households about the typical activities that they perform (using, for example, the protocol in Figure 7.3), entering them directly into a database in the household's own words, but coded at the same time with generic labels. The remainder of the attributes of activities could then be queried on computer (or the Internet) using a sequence of dialog boxes that the user completes independently. An example of queries for five possible attributes is shown in Figure 7.4. Using such an approach is likely to reduce the interview time, although the resulting implications for data quality are still relatively unknown.

---

1. Start by informing the subject that you will be asking them about the types of activities they do in a range of categories.
2. Then let them know the purpose of this exercise is to:
   a. Give them an idea of the level of detail that is sought during the week.
   b. Make it easier during the scheduling exercise to select activity types from a list.
3. Then let them know that they can always enter a "new" activity during the week if they miss it during the interview.
4. Finally, let them know that there are three types of activities that you will record for them:
   a. Generic activities: Activities that most everyone does, in which case you inform them only of the activity name.
   b. Catch-all activities: Activities similar to the above, except that the activity is a "catch all" for a variety of activities that are not of interest separately, and will make data entry easier (e.g., cleaning or chores).
   c. Personal activities: Other activities personal to them, in which case you ask them generally, "What type of <x> activities do you do," wherein the <x> values are the general category names: work, shopping, recreation, etc. Once they name a specific activity (e.g., grocery shopping), you should type a description that matches the subject's own words. Only if the subject does not come up with any ideas at first should a suggestion be made.
5. You should try to define as many activities as possible that are likely to be done in the coming week.

FIGURE 7.3 Interview protocol for defining the range of activity types performed by households.

a) Frequency

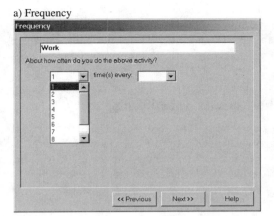

b) Duration

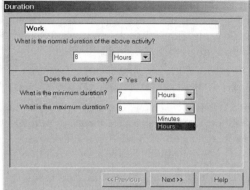

Note: if "No" selected, lower portion does not appear.

c) Locations

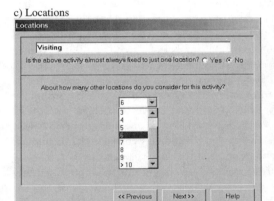

Note: if "No" selected, lower portion does not appear.

d) Timing

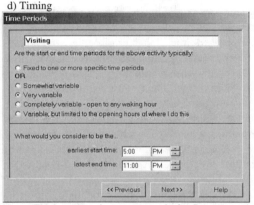

Note: The lower portion appears only for first two options.

e) Involved persons

Note: The lower portion appears only for the latter two options.

FIGURE 7.4 Example of automated computer dialog boxes for investigating activity attributes.

Owing to these challenges, it is too early to draw conclusions about the types and format of attributes that should be sought and the best methods. Which attributes are most influential in the scheduling process also remains an important analytical research question. Their choice should be balanced against our ability to eventually simulate them as part of a larger activity scheduling model. For instance, activity frequency and duration could be simulated using traditional diary data — if these two attributes were sufficient as predictors of the subsequent scheduling process, the task of agenda simulation would be eased. However, theoretically, it is likely that temporal, spatial, and interpersonal flexibility and fixity play a key role in the scheduling process, the effects of which require further rigorous testing before more definitive progress can be made.

7.4.3 Investigating the Dynamics of the Activity Scheduling Process

Taking activity agenda formation as given, and holding the process of habit formation and learning fixed in the short term, allows a clear focus on the dynamics of the scheduling process. As shown in Figure 7.1, the key components of this dynamic and continuous process include preplanning, impulsive and dynamic, and adaptive scheduling behaviors — each of which is explored in depth in the following sections.

7.4.3.1 Preplanning Decision Processes

In everyday life, it is very common for individuals and households to begin any given day or week with a set of routine, regular, or everyday activities that form a type of skeleton around which other scheduling decisions are made. In the least, this would include the act of sleeping, a necessity that bounds our daily behavior. For others, a range of other activities is included in their skeleton schedule, along with periods of unplanned time. Doherty and Miller (2000) have shown that the number of activities routinely planned in advance differs substantially between individuals, but averages about 40% of activities (~60% of the time).

Observing and eventually predicting what activities are preplanned on the skeleton schedule is a priority challenge. The traditional approach is to assume that activity types such as work and school are mandatory and thus constitute the primary pegs in a skeleton schedule. However, such an assumption may not necessarily hold for all people at all times — such as teleworkers or unemployed persons who have much more flexible schedules. A range of other activity types traditionally considered discretionary may also be included in the skeleton, especially if they share some of the same characteristics of the more mandatory activities (e.g., attending to children, sporting events). Addressing this assumption requires further investigation.

Predicting the activity types for inclusion on the skeleton is, however, only the start. The remaining specific attributes of the activities on the schedule — precise start and end times, location, involved persons, etc. — require further simulation, even if they are relatively fixed or highly constrained. The degree to which these interdependent decisions are decided in a fixed vs. variable sequence is an important area of investigation, especially in terms of eventual modeling assumptions. Is the timing of an activity or location decided first? What about mode, involved persons, etc.? What attributes may be left undecided? What alternative sequences are possible and under what conditions? Which attributes may later be modified, under what conditions, and to what degree? What agenda attributes and situational factors would serve as the best explanatory variables of whether the activity is preplanned on the skeleton? The answer to these questions is obviously important to the predictive and behavioral validity of models, since each decision is inherently constrained by earlier ones. Existing models most often assume a fixed or simultaneous decision sequence in lieu of any alternative information.

7.4.3.1.1 What to Ask About

In order to investigate these issues, individuals and households should be queried about what they have planned in advance for a given future day or week. In practice, the wording of the question requires considerable care. For example, people could be asked "What activities have you already thought about conducting for this week/day?" or "What have you planned for the coming day/week?" The difficulty for some people is deciding what "thought about" or "planned" really means. Do all attributes of an activity have to be "thought about" before it is considered planned? What about the fact that some attributes of activities may be preplanned, while others are not? Difficulty also arises with routine activities that people conduct with very little contemplation whatsoever.

Given these concerns, two types of queries could be asked of an individual or household concerning their preplanned activities:

1. *What activities have you planned for the future?* (including the naming of specific activity types, as well as the planning of unknown activity types)
2. *What attributes of these activities are planned/unplanned?*
   or
   To what degree of certainty is each attribute planned? (day, time, location, travel to, involved persons, etc.)

The first question may be structured by providing a list of activity groups or specific activity types for initial consideration (such as that provided in Table 7.2). Such a question may also follow a more in-depth investigation of an individual's or household's activity agenda, in which case the question becomes "What activities on your agenda have you already planned for the future?" People should also be given the opportunity to identify blocks of time (or locations) in which they are planning to conduct an unspecific activity (e.g., "I'm staying home Friday night — not sure what I will do, just that I want to stay home").

The second question will be fairly straightforward in the case where a given attribute has been planned (e.g., "I'm planning to shop at the mall") and is fairly certain (e.g., "I always shop at the mall and no where else"). However, in other cases a certain attribute may be planned (e.g., "I was thinking of shopping at the mall"), but still relatively uncertain (e.g., "But I still might decide to go downtown"). The second question provides considerable insight into the sequence of decision making in a sense that stated unplanned attributes can be assumed to have been decided after planned ones, and that attributes with a higher degree of stated certainty can be assumed to have been decided before those that are still relatively uncertain. In situations where all attributes are certain, or all are still relatively uncertain, some additional direct querying about the sequence of decisions may be needed ("Can you tell me which attributes you decided first?").

Recording and tracing people's responses to these questions can be done in an open-ended qualitative fashion or can involve some form of structured dialog in which responses are queried and recorded on paper or on computer. The key is not only to record responses, but also to trace the sequence of responses in some fashion for later analysis. Managing the supplemental prompts for information, such as that concerning attribute fixity or flexibility, is an additional concern in the design.

7.4.3.1.2 Open-Ended Interview Approach

An open-ended approach could involve the voice or video recording of an interview in which people are asked the two questions above for a series of planned activities for some future time period. This could include a selection of activities (e.g., one activity of every main category) for a selection of future time periods (e.g., tomorrow, a day next week), or could be more comprehensive in covering all planned activities for a longer future time period (e.g., a whole week, weeks). As activities are voiced, the interviewer's task would be to ensure that all attributes of interest are covered and that supplemental statements concerning attribute certainty are mentioned.

During such interviews, some structure in recording or displaying responses could be adopted, such as the use of display boards. For example, empty boxes for each attribute of an activity could be listed or displayed in circular format on a display board in order to serve as a basis for query. The circular design could be randomly rearranged for subsequent activities in order to reduce the tendency to voice attribute details in the order they are displayed. As attributes of activities are voiced, the interviewer's task would be to track the order in which they are mentioned — and, if ambiguous, to probe the subject concerning which attribute(s) were decided first. Additional structured questions or scale measures could be displayed to respondents for measuring relative degrees of certainty of each planned attribute (e.g., "By how much time could the start/end time (or duration) vary?" "If not fixed to one location, how many other possible locations could it be?").

7.4.3.1.3 Calendar Approach

An even more structured approach could involve presentation of a calendar or ordered listing of activities planned in the future. The calendar could appear similar to a typical day planner, as shown in Figure 7.5, in which the question becomes "looking at the calendar for <tomorrow/next week/other> and activities

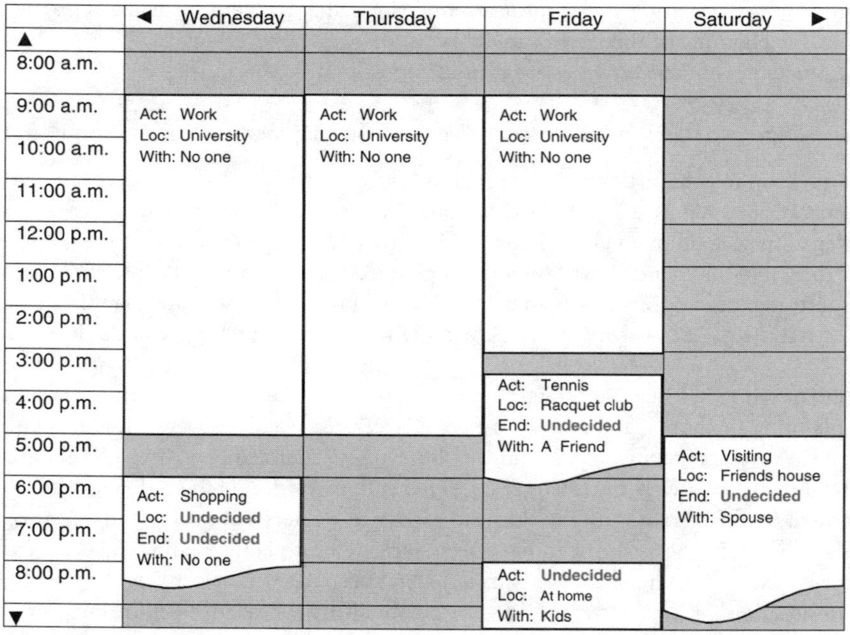

	◄ Wednesday	Thursday	Friday	Saturday ►
▲				
8:00 a.m.				
9:00 a.m.	Act: Work	Act: Work	Act: Work	
	Loc: University	Loc: University	Loc: University	
10:00 a.m.	With: No one	With: No one	With: No one	
11:00 a.m.				
12:00 p.m.				
1:00 p.m.				
2:00 p.m.				
3:00 p.m.				
4:00 p.m.			Act: Tennis / Loc: Racquet club / End: Undecided / With: A Friend	
5:00 p.m.				Act: Visiting / Loc: Friends house / End: Undecided / With: Spouse
6:00 p.m.	Act: Shopping			
7:00 p.m.	Loc: Undecided / End: Undecided			
8:00 p.m.	With: No one		Act: Undecided / Loc: At home / With: Kids	
▼				

FIGURE 7.5 Multiday calendar-style display for ongoing recording of scheduled activities.

that you have already planned." The display would include a number of days in the future listed across the top (at least 1 day displayed) and the time listed along the side, both with scroll bars for viewing more days or time periods. Basic menu commands should be provided in order to add, modify, or delete activities on screen. Planned activities would then appear as boxes on the screen displaying the planned or *undecided* attributes of activities. In the example in Figure 7.5, the day, time, activity type (act), location (loc), and involved persons (with) are shown as attributes. A range of other potential attributes of interest could also be included (e.g., who communicated with) or prompted for under a separate dialog.

In the context of Figure 7.5, travel could be treated as:

1. A separate activity all on its own
2. An attribute of an activity (travel to get to the activity)

If as a separate activity, then a box would be needed in Figure 7.5 to accommodate display of travel, including the mode(s), start and end times of travel, passengers, and other attributes of interest concerning the travel. A map could even be provided to trace routes. If travel is treated as an attribute of an activity, then each activity box would need space to specify the details of travel getting to the activity. Note that the travel that follows the activity would become the travel to get to the next activity, and does thus not need to be recorded twice. Treating travel as a separate activity will reduce the number of attributes for each activity, but would add additional boxes on the screen that may clutter the display or simply be too short to display adequately. Treating travel as an attribute of an activity may lead to a simpler display (i.e., only activities shown), but may be confusing in certain situations, such as when travel is not involved.

The preference would be to host this display on the computer in order to:

- Allow scrolling across different days and times of day (using ◄ ► buttons)
- Exactly and passively trace the sequence of entries
- Check for data inconsistencies or missing data automatically
- Create more consistent data entries across subjects
- Clearly display any planned or unplanned attributes

- Allow easy modification of previously entered activities to reflect modifications and rescheduling (discussed in more depth in the next section)
- Add other "bells and whistles" and colors, for enhancing the display and user-friendliness
- Automatically prompt for supplemental information following entry of an activity

This latter point is particularly valuable in terms of managing a series of additional prompts that could be custom tailored to particular entries, depending on the attributes entered or the time.

7.4.3.1.4 *Ordered List Approach*

The calendar approach to recording preplanned scheduling decisions has several disadvantages and potential biases. In practical terms, displaying short activity or travel segments on a time line could be difficult to manage on screen. Perhaps more importantly, displaying a person's schedule back to them may inadvertently encourage them to schedule more or less activities, especially as gaps and overlaps that a person may not be aware of are highlighted on the screen.

As an alternative, consider the multiday ordered list-style display for recording planned activities shown in Figure 7.6. The key difference is that the time line at the left is removed, and activities are displayed as equally sized boxes in order of planned occurrence. The advantage is that shorter activities and travel can be displayed just as easily as longer ones, and any unknown gaps or overlaps in a person's schedule are not overly highlighted on screen. The display is also much more compact (since longer activities do not take up large amounts of screen space), meaning that less scrolling is needed to view longer sequences of activities.

A second, even more compact list-style display is presented in Figure 7.7. In this case, only 1 day's preplanned activities are displayed on the screen as a series of rows, with each column representing a different attribute (timing, activity type or mode type, location, and involved persons). Again, no gaps or overlaps are displayed on the screen. Such a compact design would be more amenable to smaller displays available on hand-held computers or personal digital assistants (PDAs).

7.4.3.1.5 *Timing (or When and How Often to Ask)*

It is important to recognize that regardless of when you initiate observation, the scheduling process observed will always be a work-in-progress, including some established routines for which decisions may not be readily recalled, and future unobserved decisions yet to be made that will affect those made in the present. Thus, the question of *when* and how often to query an individual about his or her preplanned

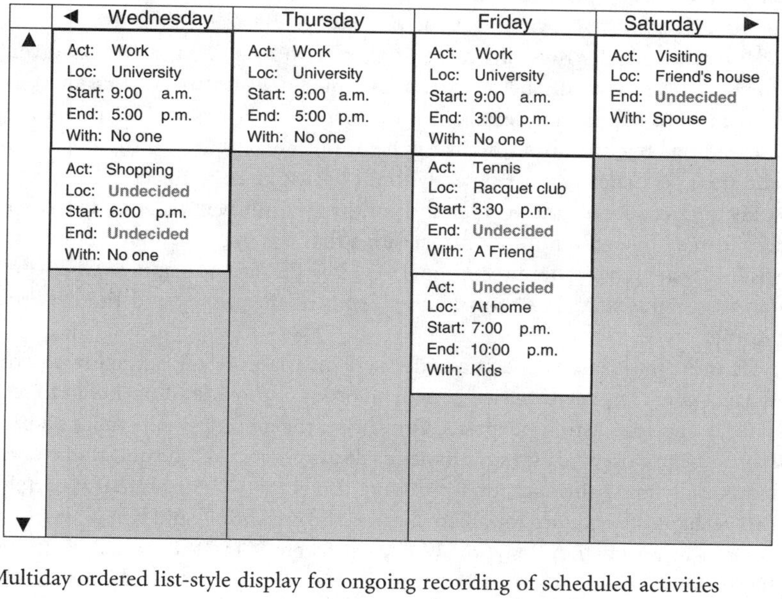

◄ Wednesday	Thursday	Friday	Saturday ►
Act: Work Loc: University Start: 9:00 a.m. End: 5:00 p.m. With: No one	Act: Work Loc: University Start: 9:00 a.m. End: 5:00 p.m. With: No one	Act: Work Loc: University Start: 9:00 a.m. End: 3:00 p.m. With: No one	Act: Visiting Loc: Friend's house End: Undecided With: Spouse
Act: Shopping Loc: Undecided Start: 6:00 p.m. End: Undecided With: No one		Act: Tennis Loc: Racquet club Start: 3:30 p.m. End: Undecided With: A Friend	
		Act: Undecided Loc: At home Start: 7:00 p.m. End: 10:00 p.m. With: Kids	

FIGURE 7.6 Multiday ordered list-style display for ongoing recording of scheduled activities

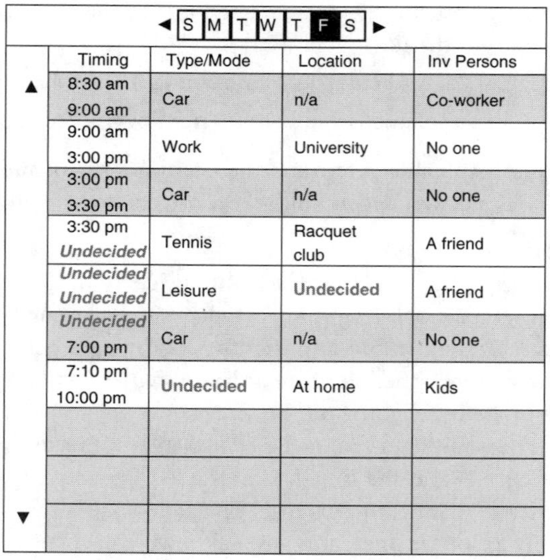

◄ S M T W T F S ►			
Timing	Type/Mode	Location	Inv Persons
8:30 am 9:00 am	Car	n/a	Co-worker
9:00 am 3:00 pm	Work	University	No one
3:00 pm 3:30 pm	Car	n/a	No one
3:30 pm *Undecided*	Tennis	Racquet club	A friend
Undecided *Undecided*	Leisure	Undecided	A friend
Undecided 7:00 pm	Car	n/a	No one
7:10 pm 10:00 pm	Undecided	At home	Kids

FIGURE 7.7 Compact single-day ordered list display for ongoing recording of scheduled activities.

activities is just as important as *what* is asked. No matter when (day before execution, week before, month before), the actual tracing of decision sequences will be left-censored, in a sense that a set of scheduling decisions will already have been made up to that point. For this set of decisions, the best we can do is to *ask* people to recall when the various decision was actually made, which may be more difficult for a respondent to recall accurately, and hence more difficult to reconstruct for the purposes of analysis and prediction. For decisions to follow, we have a much greater opportunity to actually *trace* when the decisions were made as they occur, and the all-important circumstances of these decisions.

Thus, the choice of when to initiate querying of preplanned activities is important. As the timing of the initial query moves closer to execution, the actual number of left-censored decisions will increase and the opportunities for tracing decrease. For instance, the day before execution, a person may have made decisions concerning various component attributes of three activities. The person could be queried the day before about what components were decided and when ("When did you originally make the decision to conduct this activity?" "When was the preplanned start/end time decided?" "When was the location decided?" etc.). Consider how difficult this may be to recall accurately. However, if the person was queried a week, or even a month, before execution, much fewer of these decisions would have been made already, and thus an opportunity exists to trace them as they occur. This implies repeated observations over time leading up to execution, allowing the timing and circumstances of decisions to be *traced* as they occur and their sequence reconstructed without having to *ask* subjects.

Thus, the goal is to query subjects at regular and continuous intervals leading up to execution of their schedule in order to maximize the number of decisions that are *traced* as they occur and minimize the number of decisions that have to be directly queried after the fact to capture when the decision was made. This requires careful balancing between the length of the survey and the desired accuracy and validity of the results.

In any event, for those left-censored decisions that will invariably occur, some form of query is needed to assess after the fact when the decisions were made in order to provide some ability to reconstruct their sequence for analysis and forecasting purposes. Otherwise, the assumption is that as a group, they were made as a result of some unobserved and simultaneous decision process. The question could be generically "When did you originally make the decision to conduct this activity?" However, if it pertains to a specific activity attribute (perhaps in sequence for all attributes), the question could be "When did you originally decide upon the <location/start/end/day/involved persons/etc.> for this activity?" In either case, the possible after-the-fact responses could include:

1. Just prior to the activity
2. Prior to the activity on the same day
3. Before the day of the activity
4. Did not really give it much thought — it happened as part of a regular routine
5. Cannot recall

If response 1 is chosen, then the activity or specific attribute decision can be assumed to have been made impulsively, close to the actual start time of the activity. Responses 2 and 3 indicate that relatively more preplanning was involved. In these cases, even more detail could be sought in the form of a follow-up query asking for approximate time on the same day (for response 2) or on what previous day or even month (for response 3) that the decision was made. The researcher must decide on how much detail in terms of being able to place the decision in sequence of previous decisions, balanced against issues of respondent burden and fatigue effects associated with multiple prompts.

Response 4 is meant to be associated with routine or habitual activities. In this case, it can be assumed that the decision was made long before any other decisions and has been repeated without any further contemplation since. Alternatively, it could be argued that the implications of this routine decision were not realized until they were consciously considered again at some point before execution, even if just before. To investigate this latter effect, respondents could be further asked "Can you recall when you last thought about the planning of this activity (attribute)?" followed by responses 1 to 3 and 5 above. If they cannot recall this, they could further be asked "For how long have you been doing this activity in this way?" (weeks, months, years).

The other timing-related issue concerns not only when, but *how often* to ask the above after-the-fact questions vs. when to rely simply on tracing the decisions as they occur before the fact. One approach would be to start the survey on one day (e.g., Wednesday in Figure 7.5 or 7.6), but make a selection of future days (e.g., Friday and Saturday) for which all the detailed prompts concerning planning are queried. In this way, more of the decisions leading up to the selected days will likely be traced, rather than have to be asked about after the fact. The longer the duration between survey start day and the selected days, the more tracing that will occur, thereby increasing the validity and accuracy of the results for a fewer number of selected days and minimizing respondent burden. This approach has the added advantage of avoiding too many repeated observations that may eventually confound statistical analysis.

7.4.3.2 Impulsive and Dynamic Scheduling Processes

As depicted in Figure 7.1, the preplanning of activities is followed by a more dynamic and continuous series of shorter-term preplanning, impulsive and adaptive decision making leading up to the actual execution of activities (Hayes-Roth and Hayes-Roth, 1979; Doherty and Axhausen, 1999; Doherty and Miller, 2000). This includes a large portion of activities planned impulsively (~30%) or close to execution (planned day of or day before, ~30%), with continuous modification during execution, often jumping out several days to make subplans (Doherty and Miller, 2000).

Observing and replicating the true dynamics of this process is a challenging task. Given that these decisions are made continuously over time, a continuous tracing method would be most appropriate. However, given our limited ability to passively trace a human's decisions, at best we can attempt to query individuals concerning their scheduling decisions at regular intervals. Too often, and the instrument will run into obvious observation biases; too rarely, and it will become difficult for respondents to recall and report on decision sequences accurately and completely.

At a bare minimum, people could be queried at one point about their preplanned activities, then once again following execution of the planned activities. The same forms depicted in Figures 7.5 to 7.7 and the after-the-fact prompts could be used to accomplish this. Differences from the preplan and the final observed activities could be used as a basis to distinguish preplanned from unplanned activities (at a binary level) and (final) modifications to preplanned activities. However, this approach is rather limiting in a sense that the actual sequence of decisions and modifications between preplanning and execution will be difficult or impossible to reconstruct accurately. The choice of when to query for preplanned activities (day before, week before) will also highly influence the results, as discussed in the previous section. Additionally, trying to explain the array of decisions that are taken will be a quite difficult task, especially as decisions during this phase of scheduling will be highly sensitive not only to the attributes of activities, but to the specific circumstances of the situation.

Capturing the true dynamics and the key situational factors requires more continuous observation of the scheduling process as it unfolds in as realistic circumstances as possible. Respondents should be asked to self-report their scheduling decisions on a more regular basis, such as once or more daily over a multiday survey period (e.g., 1 week). The focus should be on recording the accumulated scheduling decisions since last logging them, along with completing any time periods that are in the past. Under the calendar or list-style approaches depicted in Figures 7.5 to 7.7, a respondent could be instructed to examine past *and* future time periods in order to add new activities and attributes, modify activity attributes, and delete activities as they have changed since the last time the respondent logged the survey. In cases where past activities or activity attributes are added or modified, an after-the-fact query could be used to determine more precisely when the decision was made. If the list-style display of Figure 7.6 or 7.7 is adopted, it could be designed to revert back to a calendar-style display for days that are in the past (or after the fact) in order to highlight gaps and overlaps that need completion or resolution. The potential fill-up bias at this stage is presumed negligible, since activities and travel decisions have already been made.

The key is to record the sequence in which activity decisions are made, along with observed activity patterns on display. A computerized display is a necessity in this case, not only to allow respondents to interactively change activities and display the results on screen, but to automatically trace the timing and sequencing of data entry, and seamlessly prompt for additional information when and where needed (e.g., after-the-fact prompt) — tasks that are quite manageable via computer programming, but horrendously complex via paper-and-pencil surveys. This allows respondents to focus on the singular task of continuously updating their schedules (something we all think about every day), leaving the tracing and prompting up to the computer. A computer-based survey will also enable a vast array of additional queries of information prompts that can be set to appear only in certain circumstances as the survey proceeds over the multiday period, including:

- Closed-end prompt for why a certain decision was made
- Open-end prompts to query for explanations, or to get respondents to record a verbal protocol concerning a specific decision they just made (using a microphone or even telephone)
- Data checking functionality (e.g., checking for completed days, missing travel)
- On-line help

Additionally, observing scheduling over several days also allows the sequence and, more importantly, the situational factors of each decision in context of past and future plans to be examined in much more detail as they actually evolve. Most importantly, this includes:

- State of the person's schedule at the moment of the decision, including past and future planned events and available unplanned time
- Scheduling state of other household members and individuals with which they interact (e.g., how much free time they have, what they have already planned)
- Personal or household characteristics and resources or constraints (e.g., available modes)
- Environmental characteristics, including land use and transport network characteristics

Many of these characteristics will remain static during scheduling (e.g., person characteristics, road network), whereas many others will change dynamically as each decision is made and the schedule updated over time. For example, the time since (past dependence) and time to the next occurrence of (future dependence) activities on the agenda will likely be highly influential in the choice of how many activities may be added to the schedule at each step in the process; the length of available time windows will be influential in determining observed durations and start and end times.

7.4.3.2.1 *The Use of Emerging Technologies for Passive and Selective or Random Tracing*

Throughout the design process, a substantial challenge is to maximize the validity and accuracy of scheduling decisions while minimizing respondent burden. The use of computers described in the previous sections is meant to do just that — *passively* trace decision sequences and automatically manage additional prompts and data checking with a minimum of *active* intervention from subjects. Doherty and Miller (2000) have demonstrated that the calendar-style activity scheduling survey can be implemented with a reasonably low respondent burden — an average of about 16 min per adult per day. However, this and subsequent applications revealed several key future challenges related to increasing the depth at which scheduling decisions are traced (especially in terms of tracing how individual attributes of activities are differentially planned) and minimizing display biases, while needing to reduce survey costs and respondent burden.

One practical way to reduce respondent burden is to consider shortening components of the survey. Although shortening the total number of survey days would hamper the ability to trace the true dynamics of the scheduling process, the amount of information queried for on a daily basis, or the number of days for which detailed scheduling decisions are sought, could be reduced to minimize burden. In fact, observing a single individual or household over multiple days introduces a high level of repeated observations that may in many cases confound eventual statistical analysis anyway. However, what is still needed is a multiday decision tracing period prior to a shorter number of future target days to which all decisions pertain. Shortening of the survey could also imply a reduction in the number of decisions that are queried in more depth (e.g., for one particular day, for a systematic selection of activities, or for a random selection of decisions), preferably when the subject has the time to do so.

A range of emerging technologies will offer other opportunities for meeting these challenges of increasing the amount of information on scheduling decisions that can be passively traced, and for enhancing opportunities for more selective prompting for active information.

One way to increase the accuracy of spatial dimension in activity scheduling surveys is through the use of Geographic Information Systems (GIS). Kreitz (2001) used a GIS to display a map and allow subjects to zoom in and interactively click on activity locations (thereby storing the geocoordinates) as an alternative to specifying locations by street address or nearest intersection. Kreitz further utilized the GIS to allow selection and highlighting of travel routes.

While the GIS makes observed location and route choices easier to specify and more accurate, it may not make tracing decision processes any easier. Global Positioning Systems (GPS), on the other hand, offer the opportunity to passively trace activity–trip start and end times, activity locations, and travel routes. Murakami and Wagner (1999) demonstrated how passively collected GPS data could be linked to a hand-held computer used to prompt individuals for trip purposes and other information immediately prior to an actual trip, thereby obtaining information analogous to a traditional trip diary, but with much more accurate trip start and end times, locations, and route information. Transferring such data to a temporal-based scheduling interface, such as those described in Figures 7.5 to 7.7, would allow the information to be displayed as series of activity boxes on the main screen, and subsequently updated and merged within the scheduling process. The main advantages would be that people would be free to update and enhance the GPS data on a home-based computer (using Figure 7.6) or person-based PDA (using Figure 7.7) at a time that is convenient for them, which allows more in-depth probing of decision process information. The GPS information could even be used to assist with determining where and when a subject would be more amenable to a PDA reminder prompt to complete the survey. Comparison of preplanned activity attributes to those passively detected would also allow automated detection of schedule modifications. Overall, it is expected that such an approach would lead to a substantial increase in accuracy and detail, but with an overall reduction in respondent burden (Doherty et al., 2001).

As PDA, GPS, and wireless phone technologies continue to merge over the coming years, even more opportunities will exist to develop wearable survey devices that bring us closer to the practically unobtainable — being able to "plug into" a person's psyche. They will allow further regular, selective, or random querying with respect to ongoing scheduling decisions and the state of respondents' schedules at a given moment, along with continued passive tracing, and live transfer of data to serve as a basis for reminders and data checking. With careful design, such technologies could be used to increase accuracy and validity while at the same time reducing respondent burden.

7.4.3.2.2 Behavioral Rules

The emerging technologies described above could also bring us closer to the actual decision-making process, allowing more in-depth qualifying of the reasons for certain decisions or the behavioral "rules" adopted throughout the scheduling process — a key component of the investigation. A debate currently exists over two broadly defined decision-making frameworks for activity schedule modeling: random utility maximization based in microeconomic theory (e.g., Train, 1986; Ortuzar and Willumsen, 1990) vs. more rule-based approaches rooted in cognitive psychology (e.g., Newell and Simon, 1972; Svenson, 1979; Payne et al., 1992). Rule-based approaches have emerged more recently in response to the behavioral shortcomings of random utility theory. A long list of decision rules have been identified based on verbal reporting methods in which people are asked to think aloud during a problem-solving or decision-making exercise (see also Ericsson and Simon, 1993). Such methods work best when they are applied as close to the timing of the actual decision as possible.

Although the application of the rule-based approach to travel behavior modeling has been taken up in the literature (e.g., Gärling et al., 1986; Lundberg, 1988), very little direct empirical evidence has been published on the nature of the decision rules utilized during the activity scheduling process. This includes both "meta" scheduling rules and activity–travel specific "choice" rules. Meta rules cover the basic mechanics of scheduling — for example, when to start and stop scheduling, what attributes of activities to schedule at any given time and their sequence, when modifications or cancellations may be needed, what information to seek, and how to resolve conflicts. Activity–travel specific rules would cover the actual observed choice of activity attributes governed by the meta-decision rules — for example, exact activity types, start and end times, locations, involved persons, modes, etc., as well as the extent of activity attribute modifications. It is suspected that these rules will exhibit more stability across individuals than the vast array of observed activity patterns that result. Our search should focus on establishing these basic rules, analogous to the use of a car-following rule in traffic flow simulation models that leads to the replication of quite complex queuing patterns on highways.

The use of emerging technologies could assist in recording information on behavioral rules. A PDA scheduling process device equipped with a wireless telephone (now available on the market) could be used to first query for scheduling decisions, and then to direct subjects to discuss certain decisions (when triggered) with a live interviewer via the telephone, analogous to a think-aloud protocol. For example, when attempting to resolve a conflict in their schedule, such as the preplanning of an activity under time pressure, respondents could be automatically directed to talk on the phone concerning how they solved this problem. Additional and wireless transfer of GPS and planned activity data could be used to identify the most opportune time to query an individual for a verbal protocol.

7.4.3.3 Adaptive and Rescheduling Decision Processes

Thus far, the methods described focus on capturing scheduling processes used in everyday life. While many examples of everyday adaptive and rescheduling behaviors will invariably be captured, more direct means of capturing these processes would be valuable, especially in the context of specific policy changes or future scenarios. Stated adaptation (SA) survey methods appear amenable to this task. SA methods are a class of stated response methods in which *hypothetical behaviors* are elicited under modified constraints. Unlike in stated preference surveys, in SA surveys the possible elicited behaviors are left completely undefined (Lee-Gosselin, 1996). While SA methods may take many forms, they generally involve so-called reflexive methods, in which respondents are participant observers of themselves in a novel situation (Turrentine and Kurani, 1998). The basic components of an SA interview include:

- Creating and displaying a base of revealed data to personalize the game
- Framing the hypothetical situation to which participants will react
- Providing ground rules for the game
- Taking note of when and how decisions are made and the circumstances of choice

A computerized calendar and list-type scheduling survey method such as that described in the previous exercise would serve as a highly effective means to display the revealed bases of household activity and travel, analogous to previous SA designs, such as HATS (Jones et al., 1989) and CUPIG (Lee-Gosselin, 1989). Subjects could essentially be directed to "Look at your schedule on screen and tell us what you would have done/changed if ... " The interviewer's primary task thereafter would be to confirm that the consequences of decisions have been adequately considered, without unintentionally introducing strategies or solutions unknown to the household. The key differences from these past techniques would be the ability to automatically trace the underlying decision processes that lead to the chosen alternative, in addition to observing the final choice. Being able to distinguish planned from unplanned activities is also a key advance that, according to Gärling et al. (1998a), is important to our understanding of future reductions in auto travel, since habitual behaviors are more difficult to change and implement than impulsive ones. In addition to tracing the rescheduling decisions on computer, participants could be asked to verbalize their thought process during the game, or to think aloud, to reveal further insights into rescheduling decision rules.

7.5　Applications: Separate and Combined Investigations

This chapter has laid out the major components of the activity scheduling process in need of further investigation, along with proposing emerging techniques for doing so. These include activity agendas and their salient attributes, preplanning behavior, impulsive and adaptive decision processes, decision rules, and rescheduling in response to future scenarios (and, to a lesser extent, agenda formation, habit formation, and learning processes). These investigations could be further applied to individuals or households, can cover a range of multiday periods, and can focus generally on more "meta" activity scheduling decision processes or, more specifically, on how each attribute of an activity is differentially planned over time and space. A range of computerized, GPS, GIS, and other information technologies can also be utilized to improve accuracy and validity and reduce respondent burden. Obviously, a very multidimensional problem with a variety of observation methodologies is possible.

At this early stage in our understanding, a strong argument can be made for a variety of focused, small-sample, but thorough investigations of activity scheduling in order to examine the basic fundamentals of the process, demonstrate their complexity, test methodologies, and provide grounds for future projects.

Of course, the temptation is always to first attempt to develop as comprehensive a data collection as possible, hitting upon as many of the aspects above as possible, as accurately and behaviorally sound as possible, with a minimum of respondent burden and costs. This could take the form of an in-depth up-front interview to investigate activity agendas in the household with all household members present, followed by an interactive multiday (or even multiweek) activity scheduling exercise on the computer using forms similar to those in Figure 7.5 or 7.6, combined with GPS tracing and verbal protocols concerning decision rules, followed by an SA interview focusing on rescheduling in response to emerging policies and scenarios, perhaps followed up by waves of subsequent surveying at a future date to examine actual changes in scheduling behavior in response to real-world changes.

In reality, it is simply impractical and too biasing to put individuals or households through such a rigorous and time-consuming survey. We are forced to divide the problem into more management components for investigation with appropriately sized samples. The most amenable combinations of component surveys include combinations of up-front agenda investigation and scheduling behavior, scheduling behavior and follow-up SA experiments, and scheduling behavior and live in-depth verbal reports. A variety of other singular possibilities exist in order to focus on specific behaviors: how individual attributes of activities are planned and sequenced, household dynamics, and behavioral rules under specific circumstances. Regardless of the method, respondent time commitments should be estimated, and appropriate incentives offered to encourage accurate and timely completion of surveys. Upper time limits on up-front or follow-up interviews (e.g., 1 h) and daily commitments on multiday surveys (e.g., less than 20 min) should also be established.

7.6   Data Types and Analysis

The new data collection instruments described in this chapter invariably lead to fundamentally new forms of quantitative and qualitative data for analysis. Traditional diary techniques typically lead to a static listing of observed activities and trip *records*, along with their observed attribute *fields*. These new methods typically lead to a linked sequence of scheduling decision records ending in the observed activity, along with links to attributes of activities on the household agenda (such as spatial and temporal fixity) and supplemental information on decision timing, reasoning, and decision rules (which may consist of verbal records). The sequence of decision records will include the original addition, followed by a sequence of one or more update decisions (e.g., adding a location to a previously planned activity) and modification decisions (e.g., changing the end time of an extended activity), or possible deletion. Using the new information of the sequence of the decision in the scheduling process, further data processing can be used to cycle through the records to determine scheduling state variables unique to this type of data, such as the amount of free time available at the time of the decision, number of future activities of the same type planned, time since last activity, time to next planned occurrence, nature of activities before and after a time window, etc.

These new agenda and scheduling state variables, along with traditional sociodemographic and household indicators, will go a long way toward explaining how and why certain decisions are made. Utilizing such new data sources for modeling purposes is an explicit future goal, especially for scheduling process models. Instead of making static assumptions about decision sequences and the priority of scheduling different activities and their attributes, explicit models can be developed that are sensitive to the different attributes of activities on a person's or household's agenda (not just the type of activity) and to the situation at hand, thereby making them more realistic and dynamic in nature.

At present, modeling forms that explicitly replicate scheduling decision processes based on empirical data of this sort are currently being developed by a select few research teams around the world, such as in Canada (Miller and Salvini, 2000) and in The Netherlands (Arentze and Timmermans, 2000). As these developments proceed, amendable analysis techniques and procedures are expected to develop, further utilizing and informing these new data sources. In the short term, however, data of this form can also be used in a more limited way to enhance existing models that have attempted to replicate sequential decision processes (e.g., Kitamura et al., 1997; Vaughn et al., 1997), challenge the assumptions of other modeling approaches that assume static or simultaneous decision structures (e.g., Recker et al., 1986; Kawakami and Isobe, 1990), and support new dynamic scheduling models that are beginning to emerge (e.g., Gärling et al., 1998b; Arentze and Timmermans, 2000). Priority areas for analysis include:

- Regrouping activities into a new set of categories that more accurately reflect their more salient attributes (e.g., groups of similar activities sharing similar frequency, duration, relative spatial or temporal fixity). Currently, generic activity type definitions, such as mandatory vs. discretionary, are largely inadequate to capture the subtleties in activities that affect their scheduling.

- Development of algorithms, rules, or models to predict what activities and their attributes are likely to be of highest priority for preplanning in a given tour, day, or week, using explanatory variables that focus on key activity agenda attributes such as spatial and temporal fixity, frequency, and duration (not just activity type).

- Development of algorithms, rules, or models to predict the priority and sequence of further preplanned, impulsive, and adaptive decisions in event-oriented simulation or sequentially based models (including tour-based, trip chaining, and nested logit models) based on activity agenda attributes *and* key situational variables that make such decisions more dynamic in nature (e.g., history and future dependence, time windows, spouses' schedules).

- Development of algorithms, rules, or models to resolve conflicts (e.g., insertion of activities under time pressure) and mimic rescheduling in response to scenarios of change.

- Identification of efficient scheduling practices and rules that could be used to generally inform policy development and provide new consumer messages.

The types of models developed could at first focus on traditional logit, regression, and other common multivariate techniques that in the first instance could be used to differentiate the most important explanatory variables. For eventual model application, more behaviorally rich techniques could be explored, such as machine learning, artificial intelligence, sequence alignment, and genetic algorithms.

7.7 Discussion and Conclusions

Despite the growing need for more in-depth investigation of the activity scheduling decision process, very little empirical investigation has been conducted. Part of the problem is deciding where to start, since traditional data collection methods provide little, if any, initial insight into the problem. This chapter has attempted to break the problem down into several major components that can be investigated together or separately, but under the same conceptual framework. This includes investigation of activity agendas and their salient attributes, preplanning behavior, impulsive and adaptive decision processes, decision rules, and rescheduling in response to future scenarios. Other longer-term processes of interest include activity agenda formation, habit formation, and learning processes.

The key challenge for activity agenda investigation is deciding exactly how to define an activity and how much detail to seek on activity attributes and their relative flexibility or fixity. Key opportunities for investigation include the use of stated or observed data (or a combination of the two) to derive the agenda, and identification of those attributes that serve as the most important determinants of the scheduling process to follow. In terms of scheduling processes, the key challenge is predicting what activities and their attributes are preplanned and the sequencing of decisions concerning their specific observed attributes (start and end times, location, involved persons, mode, route choices, etc.). Prospective surveys using computerized calendar and list-based displays are suggested. With reference to the dynamic scheduling process that follows preplanning, the key challenge is attempting to observe the variety of decisions made at different timescales, along with key situational factors, and linking these to observed activity–travel outcomes. Doing so with a minimum of respondent burden is also a key challenge. A variety of opportunities for investigation were suggested and discussed in detail in this chapter, including the use of emerging technologies such as GPS and verbal protocol analysis.

Of course, the ideal data collection instrument would encompass all components of the scheduling process with a high amount of detail and realism and a minimum of respondent burden. While such an approach is feasible, perhaps assisted by the judicious use of computer, GIS, and GPS technologies, it would involve a very high respondent burden, and thus be limited to small samples. A more practical and feasible alternative would be to start targeting the specific components or limited combinations of the activity scheduling problem for more in-depth investigation. Assuming that the underlying decision processes and rules are generic to all human beings, they may be stable enough to warrant collection of small samples of individuals or households in a given locale. Such a survey could serve as a complement to the larger-sample, more traditional diary surveys of observed activity–travel patterns, which tend to vary much more between households and locales.

Reduction in survey costs and continued application, even of modest-sized data samples, will no doubt bring about new insights and directions for model development. The question remains as to whether this will lead to further tweaking of existing approaches, policies, and models or to fundamental shifts in model form. Both are real possibilities, as models based on traditional methods are being expanded in an effort to become less static in the way observed schedules are, and new dynamic scheduling models are beginning to emerge. The current priority should be placed on continued analysis and understanding of scheduling fundamentals, with an eye toward developing specific components of the scheduling process in isolation or within the most promising existing modeling frameworks.

Acknowledgments

Financial support for this chapter was provided from the Social Sciences and Humanities Research Council of Canada. The author acknowledges the cooperation and support from colleagues involved in collaborative research projects, especially Martin Lee-Gosselin, Eric Miller, Guido Rindsfüser, John Polak, Kay Axhausen, Jean Andrey, and Marion Kreitz.

References

Arentze, T.A. and Timmermans, H.J.P., *Albatross: A Learning Based Transportation Oriented Simulation System*, The European Institute of Retailing and Services Studies, Eindhoven, The Netherlands, 2000.

Axhausen, K.W., Can we ever obtain the data we would like to have? in *Theoretical Foundations of Travel Choice Modeling*, Gärling, T., Laitila, T., and Westin, K., Eds., Elsevier Science Ltd., Oxford, 1998, pp. 305–323.

Axhausen, K.W. and Gärling, T., Activity-based approaches to travel analysis: Conceptual frameworks, models, and research problems, *Transp. Rev.*, 12, 323–341, 1992.

Axhausen, K.W. et al., Observing the rhythms of daily life: A six-week travel diary, *Transportation*, 29(2), 95–124, 2002.

Bhat, C.R. and Lawton, K.T., Passenger Travel Demand Forecasting, Millennium Paper, Transportation Research Board, National Research Council, Washington, D.C., 2000.

Bhat, C.R. and Singh, S.K., A comprehensive daily activity–travel generation model system for workers, *Transp. Res. A Policy Pract.*, 34, 1–22, 2000.

Bowman, J.L. and Ben-Akiva, M.E., Activity-based disaggregate travel demand model system with activity schedules, *Transp. Res. A Policy Pract.*, 35, 1–28, 2001.

Doherty, S.T. and Axhausen, K.W., The development of a unified modelling framework for the household activity–travel scheduling process, in *Traffic and Mobility: Simulation-Economics-Environment*, Brilon, W. et al., Eds., Springer, Berlin, 1999, pp. 35–56.

Doherty, S.T. and Miller, E.J., A computerized household activity scheduling survey, *Transportation*, 27, 75–97, 2000.

Doherty, S.T. et al., Moving beyond observed outcomes: Integrating global positioning systems and interactive computer-based travel behaviour surveys, in *Personal Travel: The Long and Short of It*, Transportation Research Board, National Research Council, Washington, D.C., 2001, E-C206, pp. 449–466.

Ericsson, K.A. and Simon, H.A., *Protocol Analysis: Verbal Reports as Data*, MIT Press, Cambridge, MA, 1993.

Ettema, D., Borgers, A., and Timmermans, H., Simulation model of activity scheduling behavior, *Transp. Res. Rec.*, 1413, 1–11, 1993.

Ettema, D. and Timmermans, H., Theories and models of activity patterns, in *Activity-Based Approaches to Travel Analysis*, Ettema, D. and Timmermans, H., Eds., Pergamon, Oxford, 1997, pp. 1–36.

Gärling, T., Gillholm, R., and Gärling, A., Reintroducing attitude theory in travel behavior research: The validity of an interactive interview procedure to predict car use, *Transportation*, 25, 129–146, 1998a.

Gärling, T. et al., Computer simulation of household activity scheduling, *Environ. Plann. A*, 30, 665–679, 1998b.

Gärling, T. et al., The spatiotemporal sequencing of everyday activities in the large-scale environment, *J. Environ. Psychol.*, 6, 261–280, 1986.

Goulias, K., Traveler Behavior and Values Research for Human-Centered Transportation Systems, Millennium Paper, Transportation Research Board, National Research Council, Washington, D.C., 2000.

Griffiths, R., Richardson, A.J., and Lee-Gosselin, M.E.H., Travel Surveys, Millennium Paper, Transportation Research Board, National Research Council, Washington, D.C., 2000.

Hayes-Roth, B. and Hayes-Roth, F., A cognitive model of planning, *Cognit. Sci.*, 3, 275–310, 1979.

Jones, P., Bradley, M., and Ampt, E., Forecasting household response to policy measures using computerised, activity-based stated preference techniques, in *Travel Behaviour Research*, The International Association for Travel Behaviour, Ed., Avebury, Aldershot, U.K., 1989, pp. 41–63.

Kawakami, S. and Isobe, T., Development of a one-day travel–activity scheduling model for workers, in *Developments in Dynamic and Activity-Based Approaches to Travel Analysis*, Jones, P., Ed., Avebury, Aldershot, U.K., 1990, pp. 184–205.

Kitamura, R., Chen, C., and Pendyala, R., Generation of synthetic daily activity–travel patterns, *Transp. Res. Rec.*, 1607, 154–162, 1997.

Kitamura, R. and Kermanshah, M., Identifying time and history dependencies of activity choice, *Transp. Res. Rec.*, 944, 22–30, 1983.

Kreitz, M., Methods for Collecting Spatial Data in Household Travel Surveys, paper presented at International Conference on Transport Survey Quality and Innovation, Kruger Park, South Africa, August 2001.

Lee-Gosselin, M., In-depth research on lifestyle and household car use under future conditions in Canada, in *Travel Behaviour Research*, The International Association for Travel Behaviour, Ed., Avebury, Aldershot, U.K., 1989, pp. 102–118.

Lee-Gosselin, M., Scope and potential of interactive stated response data collection methods, in *Household Travel Surveys: New Concepts and Research Needs*, Transportation Research Board Conference Proceedings 10, National Academy Press, Irvine, CA, 1996, pp. 115–133.

Lundberg, C.G., On the structuration of multiactivity task-environments, *Environ. Plann. A*, 20, 1603–1621, 1988.

Miller, E.J. and Salvini, P.A., The Integrated Land Use, Transportation, Environment (ILUTE) Microsimulation Modelling System: Description and Current Status, paper presented at International Association of Travel Behaviour Research (IATBR), Gold Coast, Queensland, Australia, 2000.

Murakami, E. and Wagner, D.P., Can using Global Positioning System (GPS) improve trip reporting? *Transp. Res. C*, 7, 149–165, 1999.

Newell, A. and Simon, H., *Human Problem Solving*, Prentice-Hall, Englewood Cliffs, 1972.

Ortuzar, J.D. and Willumsen, L.G., *Modeling Transport*, John Wiley and Sons, New York, 1990.

Pas, E.I., State of the art and research opportunities in travel demand: Another perspective, *Transp. Res. A*, 19, 460–464, 1985.

Payne, J.W. et al., A constructive process view of decision making: Multiple strategies in judgment and choice, *Acta Psychol.*, 80, 107–141, 1992.

Polak, J. and Jones, P., A Tour-Based Model of Journey Scheduling under Road Pricing, paper presented at 73rd Annual Meeting of the Transportation Research Board, Washington, D.C., 1994.

Recker, W.W., McNally, M.G., and Root, G.S., A model of complex travel behavior. Part II: An operational model, *Transp. Res. A*, 20, 319–330, 1986.

Svenson, O., Process descriptions of decision making, *Organ. Behav. Hum. Perform.*, 23, 86–112, 1979.

Train, K., *Qualitative Choice Analysis*, MIT Press, Cambridge, MA, 1986.

Turrentine, T. and Kurani, K., Adapting interactive stated response techniques to a self-completion survey, *Transportation*, 25, 121–129, 1998.

van der Hoorn, T., Development of an activity model using a one-week activity-diary data base, in *Recent Advances in Travel Demand Analysis*, Carpenter, S. and Jones, P., Eds., Gower, Aldershot, U.K., 1983, pp. 335–349.

Vaughn, K.M., Speckman, P., and Pas, E.I., Generating Household Activity–Travel Patterns (HATPS) for Synthetic Populations, paper presented at 76th Annual Meeting of the Transportation Research Board, Washington, D.C., January 12–16, 1997.

8

Statistical and Econometric Data Analysis

CONTENTS

Konstadinos G. Goulias
Pennsylvania State University

8.1 Introduction

One can look at statistical analysis of data as a medium "for extracting information from observed data and dealing with uncertainty" (Rao, 1989, p. 98). Another way of saying the same thing is to consider *statistics* as a group of methods that are used to collect, analyze, present, and interpret data. From the myriad of methods available to us for data analysis (Snedecor and Cochran, 1980; Spanos, 1999), regression methods are one family of data analysis that are comprehensive in their ability to address data issues, efficient in their ability to extract large amounts of information in a concise way, and widely available, because even spreadsheet software provides for facilities to estimate simple regression models. Regression methods, particularly when one considers generalized regression models, can also be considered as the general family of models that contains analysis of variance and a variety of other methods for the analysis of experiments as special cases. Regression methods and models are also the techniques dominating *econometrics* — the art and science of analyzing economic data, which, when considering the leading textbooks on the subject, is nothing but the study of regression models (see Amemiya, 1985; Greene, 2000; Johnston and DiNardo, 1997; Pindyck and Rubinfeld, 1998; among many others).

In the previous chapters of this book different authors pointed out the richness and variety of data available for understanding and predicting travel behavior. In this chapter we set out to accomplish a few humble goals:

1. Give a short introduction on statistical and econometric methods and introduce a road map of transportation data analysis, mentioning a few major milestones
2. Provide an introduction to the next steps in data analysis methods and introduce the next three chapters
3. Provide selective references to transportation planning books and articles where these methods were used successfully and where new information may be found in the future

8.2   Models

Statistical methods are used in a wide variety of instances in transportation planning to help us identify, study, and solve many complex practical problems. For example, in the public involvement arena these methods enable decision makers, planners, and managers to make informed decisions, consistent with legislation, about the elements of their policies, plans, and programs. This is accomplished by collecting data from a variety of persons and groups using a wide variety of techniques to collect qualitative and quantitative data that are combined to yield answers to specific policy and planning questions. In another area, regional travel demand forecasting, regional models, and statistics are developed to build large-scale simulation models of a region or even an entire state to help identify alternate urban and regional designs, economic activity locations, and new or improved infrastructure system components. After data are collected from individuals and their households, statistical models are estimated and their equations are used in a spreadsheet-like format or embedded into a computer program code. Then other input data are provided to create predictions for each individual, household, or even geographical area, using these statistical models. In this way statistical models are at the heart of these simulation systems and any errors, omissions, misrepresentations, or other approximations may be amplified and provide the wrong indications. This is the key reason we continuously look for better and more precise and accurate model-building methods. Figure 8.1 provides a pictorial representation of a linear sequential version of this process. The feedback in the figure can be used to improve the models within a given project or to provide recommendations for improvements in a sequence of projects (lagged feedback).

As expected, data analysis is needed to support actions and project development in these contexts. Before moving into the details of data analysis, a digression to define a few terms is required. Decision makers take action based on *knowledge* about an issue. In the path from data to knowledge one can envision a sequence of transformations that lead to increase in power and confidence for the decisions to be made using the data. The sequence starts from *data* to *information*, which is the transformation of data to something that is relevant to a specific decision problem. Then the information becomes a group of *facts*, when statements can be supported by the data at hand. Facts, in turn, become knowledge when they are used to complete the decision process. Finally, knowledge aids actions when there is an implementation plan. A statistical or econometric analysis and estimated models are the enabling devices (or vehicles) to move from data to knowledge. Therefore the statistical or econometric models play a very important role in this example too because they summarize in a concise way the myriad pieces of information in a database. In a way similar to that for prediction and simulation, any errors, omissions, misrepresentations, or other approximations may be amplified by the decision process and lead to the wrong actions, which in turn may cause dramatic damages to humans and their environment. For this reason data analysis methods, together with operations research methods, have been considered of paramount importance in the decision sciences. Both data analysis and operations research are quantitative methods, and they have a long history of development. There are, however, other data collection and transformation devices and tools that have received very little systematic attention in transportation planning (Goulias, 2001), but they are beyond the scope of this chapter.

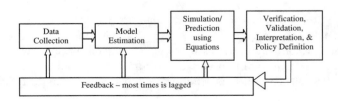

FIGURE 8.1 A sequential version of a system in which regression models are estimated and used.

8.3   Type of Data and Levels of Measurement

Classifications and taxonomies of data analysis methods abound (e.g., Judge et al., 1985; Jobson, 1991; Gelman et al., 1995), and they depend on the purpose of the reviews. Since we focus on regression methods and models, we will use one classification that is consistent with most textbooks and research in travel behavior.

In typical transportation surveys information is collected using qualitative or quantitative data. Qualitative data, such as the color of your car, is not computable by arithmetic operations. The color is a label that informs us about a category, a group, a region, or any other classification in which a person or artifact falls. These are named the categorical variables. On the other side of this classification we find data that are measured on the real line and take any value on it (e.g., a ratio or proportion). There are very few examples in transportation where the data can be considered (completely) continuous because we consider either finite countable and integer quantities, such as trips, cars, sites to visit, and so forth, or variables that may be characterized by a limited range of variation (e.g., a proportion can take values between 0 and 1).

The presentation of the models available and resources to study and apply them is divided into two major groups: categorical (discrete) data and continuous data. Each of these groups contains a variety of other models, depending on the more specific nature of the variable, the variation of which we are trying to explain (dependent or to be explained). The variation of this variable is explained by explanatory variables and parameters that we need to estimate (the combination of which is named systematic variation) and a random variation that we cannot explain. It is also important to stress that the classification we use here is not based on the variables we use as explanatory (predictor) variables. They can also be of any type, and there are ways to incorporate almost any type of explanatory variable in a regression model by converting it into some sort of numerical coding that can be handled by the software (Greene (2000) provides a discussion on this; Kennedy (1998) and Pindyck and Rubinfeld (1998) also provide a good discussion and examples).

Emphasis in this chapter is placed on cross-sectional data (data collected for individuals and household at one time point). Longitudinal data analysis methods are also starting to emerge in transportation planning, and within each section a short mention is made to this type of data analysis, emphasizing panel surveys (time series are excluded from this presentation entirely). In addition, emphasis is given to the single equation because the issues are similar when one considers each equation of a system of equations (Pendyala in Chapter 2 of this book provides examples of multiple equations issues). In addition, Chapter 11 provides an overview of structural equations and models, which are the premier methods when one wants to consider multiple dependent variables jointly. Goulias also addresses the multiple equations issue, incorporating time and social levels, in Chapter 9, on multilevel models.

8.4   Categorical (Discrete) Data

For the sake of convenience, these can be further divided into choice models, nonchoice models, and models for counting processes. The models for counting processes can be further divided into event count and duration models. This is described in additional detail below, and key references are provided.

8.4.1   Choice

In transportation planning a typical example is mode choice, when for a trip a person decides which mode to choose from among a finite set of modes. A typical model in discrete choice will be a model of the probability of choosing a mode as a nonlinear function of mode attributes, trip characteristics, and traveler demographics. The usual formulation shows the (indirect) utility of each mode as the function from which we depart, and then making certain assumptions about its stochastic nature, we derive the probability shape. This enables us to use specialized algorithms for estimation of the parameters driving the function. Most transportation planning and modeling textbooks contain the basic theory and examples of mode choice (Ortuzar and Willumsen, 2001; Meyer and Miller, 2001). There is also a monograph dedicated to the theory, data collection, and experience with choice problems, edited by Gärling et al. (1998). Many milestones in the past literature and important developments are reviewed in these three books.

However, *classic* references to discrete choice models are: for the logit model and initial formulations, Domencich and McFadden (1975); for the probit model and a very good detail on estimation, Daganzo (1979); and for a comprehensive review with clear examples, Ben-Akiva and Lerman (1985). Many subsequent developments have improved the original algorithms in these books, and widely available software exists for model estimation (see the websites at the end of this chapter).

There are, however, many very important and more recent developments in model formulation and estimation that are expanding the scope of discrete choice models, making them by far more flexible and usable than in the past. In a recent handbook, Bhat (2000a) and Koppelman and Sethi (2000) review some of these developments. Bhat in Chapter 10 of this book provides the latest review of developments and identifies many important issues that have been resolved. The analysis of repeated choices, however, has not received wide attention in discrete choice. It is expected that with the increasing use of stated choice and preference data (Louviere et al., 2000), we may see new developments in the field.

8.4.2   Nonchoice

This area is also known as *contingency table analysis* and *cross-classification categorical data analysis* and contributes one of the richest groups of models that have immense flexibility and potential applications in transportation. Fienberg's (1977) book is still one of the best presentations of the original methods. Agresti's (1990) textbook is one of the most comprehensive surveys linking contingency table analysis methods to logit for binary and multicategory data. Two somewhat newer expositions are Powers and Xie (1999) and Le (1998).

One early application due to Goodman's way of looking at contingency data analysis is reported in Kitamura et al. (1990); extensive use of this method was done for the design of a microsimulator in Goulias (1991). More recently, repeated observations of the same individuals have been analyzed with contingency table methods that contain latent classes (Goulias, 1999), and a connection between latent class and discrete choice has inspired some very interesting model-building work that has become available only recently (see the discussion by Golledge and Gärling in Chapter 3 in this book). Kitamura (2000) also reviews some of these models from a model formulation viewpoint.

8.4.3   Counting Processes

This group of models targets event counts — the number of times an event occurs. In probability, this is the realization of a nonnegative integer random variable. Counts and durations seem to be the two sides of the same coin:

> An event may be thought of as the realization of a point process governed by some specified rate of occurrence of the event. The number of events may be characterized as the total number of such realizations over some unit of time. The dual of the event count is the interarrival time, defined as the length of the period between events. (Cameron and Trivedi, 1998, p. 4)

In travel behavior there are many examples for both the count and duration regression models. In the case of duration models the study of activity episode durations (see Pendyala in Chapter 2) has received considerable attention in the past few years. The earliest examples are cited in Kim and Mannering (1997), and a review can be found in Bhat (2000b). Counts using Poisson and negative binomial models also abound for the number of trips, activities per day, number of departures, and so forth. Two of the earlier examples are Mannering (1989) and Monzon et al. (1989). Arentze and Timmermans (2000) and Ma and Goulias (1999) provide updates on count data models and more recent examples. A comprehensive review can also be found in Andersen et al. (1992).

One particular class of models emerges when the count is ordered, for example, the number of vehicles a household owns. In this case having three cars is more than having two cars, and having two cars is more that having one, etc. These models are particularly attractive because they allow use of certain estimation tricks. Greene (2000) provides an extensive discussion on *ordered models*. An earlier example of ordered regression is reported in Kitamura and Bunch (1990), in which repeated observations are also used. These models can also be used in attitudinal responses and judgments (Kim et al., 2001).

8.5 Continuous Data

Every introduction to regression and econometrics departs from a model with a continuously varying dependent variable. The usual treatment follows the same sequence with a discussion about the simple linear model and then removing each of a number of assumptions (very often referred to as the Gauss–Markov theorem assumptions). In this way, more and more complex and flexible models are built. Because the majority of time and effort in introductory econometrics courses and texts is dedicated to the linear regression model, and because its assumptions are consistently violated by transportation data, the references about this model are limited here to a few key texts and emphasize transportation applications of the limited dependent variable variety. For linear regression models, Greene's textbook is one of the best and most comprehensive references. The textbook also contains a very nice section on nonlinear regression models (Greene, 2000, Chap. 10, pp. 416–453).

When the dependent variable is limited (e.g., cannot take values below or above a value), special attention needs to be paid in computing its mean, but also in estimating the regression coefficients. Again, the standard textbook is Greene (2000), but a very good reference is also Maddala (1983). The typical example of a limited dependent variable is the Tobit model (see Monzon et al., 1989). There are also simpler methods, as illustrated in the practical application in Goulias and Kitamura (1993).

8.5.1 Other References and Journals

Consistently through the past 20 years transportation researchers have utilized many of the new regression methods almost immediately after they have been developed, and very often transportation problems have offered motivation for statisticians and econometricians to develop new methods. A notable example is D. McFadden, who won the Nobel Prize in 2000. The methods of the other person who won the Nobel Prize for econometric contributions in 2000, J. Heckman, are also used very often in transportation data analysis. Transportation journals and conference proceedings always contain papers and chapters that will either provide a review of new methods or apply a new method to a transportation problem. When seeking these new developments, one should examine the following:

Transportation Research Record — A journal of the Transportation Research Board
Transportation Research — A Pergamon international journal that is divided into parts dedicated to a specific focus
Transportation — A Kluwer international journal

The proceedings of the conferences mentioned in Chapter 1 of this book are also very good sources. There are also many websites with extensive treatment on statistical and econometric models. The two sites with the best and most up-to-date links for statistical and econometric software are:

http://www.feweb.vu.nl/econometriclinks/software.html
http://www.fas.harvard.edu/~stats/survey-soft/survey-soft.html

8.5.2 What Is Next?

Pendyala in Chapter 2 provided a state-of-the-art presentation of more sophisticated and informative models in travel behavior with the stochastic frontier models, mixtures of discrete and continuous dependent variable models, and the duration models. In discrete choice, Bhat, in Chapter 10, discusses other directions focusing on microeconometric data. Goulias, in Chapter 9, illustrates extensions of the linear regression that incorporate multiple hierarchies in the data, multiple equations, and multiple ways to incorporate randomness. Golob's review (Chapter 11) also provides another set of directions, along which we will see new advances. Finally, another direction of data analysis that we are starting to see develop is in the nonparametric data analysis methods, such as the example in Kharoufeh and Goulias (2002).

References

Agresti, A., *Categorical Data Analysis*, Wiley, New York, 1990.

Amemiya, T., *Advanced Econometrics*, Harvard University Press, Cambridge, MA, 1985.

Andersen, P.K. et al., *Statistical Models Based on Counting Processes*, Springer-Verlag, New York, 1992.

Arentze, T. and Timmermans, H., *Albatross: A Learning Based Transportation Oriented Simulation System*, European Institute of Retailing and Service Studies, Technical University of Eindhoven, Netherlands, 2000.

Ben-Akiva, M. and Lerman, S.R., *Discrete Choice Analysis*, MIT Press, Cambridge, MA, 1985.

Bhat, C.R., Flexible model structures for discrete choice analysis, in *Handbook of Transport Modelling*, Hensher, D.A. and Button, K.J., Eds., Pergamon, Amsterdam, 2000a, pp. 71–89.

Bhat, C.R., Duration Modeling, in *Handbook of Transport Modelling*, Hensher, D.A. and Button, K.J., Eds., Pergamon. Amsterdam, 2000b, pp. 91–110.

Cameron, A.C. and Trivedi, P.K., *Regression Analysis of Count Data*, Cambridge University Press, U.K., 1998.

Daganzo, C., *Multinomial Probit: The Theory and Its Application to Demand Forecasting*, Academic Press, New York, 1979.

Domencich, T. and McFadden, D., *Urban Travel Demand: A Behavioral Analysis*, Elsevier/North Holland, Amsterdam, 1975.

Fienberg, S.E., *The Analysis of Cross-Classified Categorical Data*, MIT Press, Cambridge, MA, 1977.

Gärling, T., Laitila, T., and Westin, K., *Theoretical Foundations of Travel Choice Modeling*, Elsevier, Amsterdam, 1998.

Gelman, A. et al., *Bayesian Data Analysis*, Chapman & Hall/CRC Press, Boca Raton, FL, 1995.

Goulias, K.G., Long-Term Forecasting with Dynamic Microsimulation, unpublished Ph.D. dissertation, University of California, Davis, 1991.

Goulias, K.G., Longitudinal analysis of activity and travel pattern dynamics using generalized mixed Markov latent class models, *Transp. Res. B*, 33, 535–557, 1999.

Goulias, K.G., On the role of qualitative methods in travel surveys, workshop report on qualitative methods Q-5, International Conference in Transport Survey Quality and Innovation, Kruger National Park, South Africa, August 5–10, CD-ROM, 2001.

Goulias, K.G. and Kitamura, R., Analysis of binary choice frequencies with limit cases: Comparison of alternative estimation methods and application to weekly household mode choice, *Transp. Res. B Methodol.*, 27, 65–78, 1993.

Greene, W.H., *Econometric Analysis*, 4th ed., Prentice Hall, Upper Saddle River, NJ, 2000.

Jobson, J.D., *Applied Multivariate Analysis*, Vols. 1 and 2, Springer, New York, 1991.

Johnston, J. and DiNardo, J., *Econometric Methods*, 4th ed., McGraw-Hill, New York, 1997.

Judge, G.G. et al., *The Theory and Practice of Econometrics*, 2nd ed., Wiley, New York, 1985.

Kennedy, P., *A Guide to Econometrics*, 4th ed., MIT Press, Cambridge, MA, 1998.

Kharoufeh, J.P. and Goulias, K.G., Nonparametric identification of daily activity durations using Kernel density estimators, *Transp. Res. B Methodol.*, 36, 59–82, 2002.

Kim, T., Koza, S.A., and Goulias, K.G., Analysis of the resident component in PennPlan's public involvement survey: Survey overview and item nonresponse selectivity issues, paper preprint 01-2772, *Transp. Res. Rec.*, 1780, 145–154, 2001.

Kim, S. and Mannering, F., Panel data and activity duration models: Econometric alternatives and applications, in *Panels for Transportation Planning: Methods and Applications*, Golob, T., Kitamura, R., and Long, L., Eds., Kluwer, Boston, 1997, pp. 349–373.

Kitamura, R., Longitudinal methods, in *Handbook of Transport Modelling*, Hensher, D.A. and Button, K.J., Eds., Pergamon, Amsterdam, 2000, pp. 113–128.

Kitamura, R. and Bunch, D.S., Heterogeneity and state dependence in household car-ownership: A panel analysis using ordered-response Probit models with error components, in *Transportation and Traffic Theory*, Koshi, M., Ed., Elsevier/North Holland, Amsterdam, 1990, pp. 477–496.

Kitamura, R., Nishii, K., and Goulias, K.G., Trip chaining behavior by central city commuters: A causal analysis of time–space constraints, in *Developments in Dynamic and Activity-Based Approaches to Travel Analysis*, Jones, P., Ed., Avebury, Aldershot, U.K., 1990, pp. 145–170.

Koppelman, F.S. and Sethi, V., Closed-form discrete-choice models, in *Handbook of Transport Modelling*, Hensher, D.A. and Button, K.J., Eds., Pergamon, Amsterdam, 2000, pp. 211–225.

Le, C.T., *Applied Categorical Data Analysis*, Wiley, New York, 1998.

Louviere, J.J., Hensher, D.A., and Swait, J.D., *Stated Choice Methods: Analysis and Applications*, Cambridge University Press, Cambridge, U.K., 2000.

Ma, J. and Goulias, K.G., Application of Poisson regression models to activity frequency analysis and prediction, *Transp. Res. Rec.*, 1676, 86–94, 1999.

Maddala, G.S., *Limited Dependent and Qualitative Variables in Econometrics*, Cambridge University Press, U.K., 1983.

Mannering, F., Poisson analysis of commuter flexibility in changing route and departure times, *Transp. Res. B*, 23, 53–60, 1989.

Meyer, M.D. and Miller, E.J., *Urban Transportation Planning*, 2nd ed., McGraw-Hill, Boston, 2001.

Monzon, J., Goulias, K.G., and Kitamura, R., Trip generation models for infrequent trips, *Transp. Res. Rec.*, 1220, 40–46, 1989.

Ortuzar, J. de D. and Willumsen, L.G., *Modelling Transport*, 3rd ed., Wiley, Chichester, U.K., 2001.

Pindyck, R.S. and Rubinfeld, D.L., *Econometric Models and Economic Forecasts*, 4th ed., McGraw-Hill, Boston, 1998.

Powers, D.A. and Xie, Y., *Statistical Methods for Categorical Data Analysis*, Academic Press, New York, 1999.

Rao, C.R., *Statistics and Truth: Putting Chance to Work*, International Co-Operative Publishing House, Fairland, MD, 1989.

Snedecor, G.W. and Cochran, W.G., *Statistical Methods*, 7th ed., Iowa State University Press, Ames, 1980.

Spanos, A., *Probability Theory and Statistical Inference: Econometric Modeling with Observational Data*, Cambridge University Press, U.K., 1999.

9

Multilevel
Statistical Models

CONTENTS

Konstadinos G. Goulias
Pennsylvania State University

9.1   Introduction

Recent travel demand forecasting systems and related data analyses target individual and household variations in behavior not only as functions of individual and household characteristics, but also as functions of other variables to capture the effect of social and geographical context on individual and household behavior. As discussed in previous chapters new conceptual and theoretical ideas in travel behavior are increasingly offered, providing analytical frameworks for human behavior in geographic space, social space, and time. To test hypotheses within these frameworks and to capture the relationships within and across different dimensions (levels), suitable data analytic techniques are needed.

Assuming we are able to identify and clearly define levels of social groupings, such as the family, the neighborhood, or the professional group, our interest centers on explaining individual behavior not only as a function of personal motivational factors, but also as a function of group influence, such as task allocation(s) and role assignments within a group (e.g., the household). At a more macro (aggregate) level we are also interested in the role personal factors play in shaping group behavior(s). Techniques to accomplish this must support behavioral theories that aim to explain behavior using factors that influence behavior at the same level of the behavioral unit of analysis (named micro-to-micro relationships), at one level higher (more aggregated) from processes taking place (named macro-to-micro effects), and at one level lower from processes taking place (named micro-to-macro effects).

Data analyses that include variables from different levels (e.g., a person, household, neighborhood, city, state) are inherently operating at multiple levels. These are called multilevel analyses because they examine the relationships among variables that are defined at different and multiple levels. Figure 9.1 provides an example of a hierarchy of this type. Each observation is a time point at which a person's behavior has been recorded or reported (within the time dimension we can have another hierarchy of

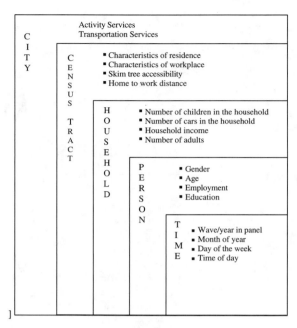

FIGURE 9.1 Pictorial representation of one possible data hierarchy.

different temporal entities, indicated in Figure 9.1 as different bullet items within the box labeled "time"). All these observations are about each person, and each person belongs to a household for which we have recorded many characteristics. Then each household resides in a neighborhood (indicated as a census tract for convenience) for which we have also measured characteristics, such as number of locations where activities can be pursued, accessibility indicators for each mode among the centers of these tracts, and so forth. Studies that do not account for the simultaneous influence of variables at multiple levels may lead to ecological fallacy or atomistic fallacy (for a more extensive discussion on this, see the review in Hox (1995), who also provides the earliest known scientific papers on multilevel theories and related conceptual risks and fallacies). The consequence of these fallacies is wrong inference about the effects of policies and the relationships among observed behavioral variables. Multilevel analyses, however, require data that are informative enough to unravel some of these relationships.

Multilevel analysis in travel behavior and transportation planning research is greatly facilitated by the availability of a specific type of travel behavior data. Data widely available to transportation planners are household survey data that contain a convenient natural hierarchy, allowing the study of context on behavior. Some household surveys contain travel or activity diaries of all the members in a household. Other surveys contain only a subset of the household members that satisfy some sample selection condition (e.g., in an American survey all adults and persons older than 15 years are included in the survey diary because 16 is the age at which individuals are allowed to start driving in the United States). We may also have repeated observations of each person's behavior. For example, when the survey contains a travel diary for a single day, the repetition is on behavioral indicators by different times in a day (i.e., each activity episode and trip made by the observed person). When the survey contains diaries from multiple days, the repetitions are the survey days and within each day the multiple trips made by the individual. When the survey is a panel, the hierarchy becomes even more interesting because the temporal repetition contains different years at which persons are interviewed. Within each year it also contains different days, and within each day different episodes (the trips and activities with clear start and end times). In this way time may be considered to contain three subdimensions (minutes and hours within a day, each individual day, and each individual year) along which change (variation from minute to minute, day to day, year to year) takes place, allowing the study of the dynamics of behavior in more detail. In addition, if the households

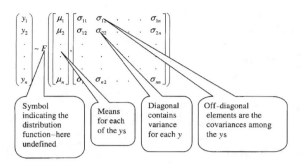

FIGURE 9.2 Joint distribution of y values.

can be grouped into other categories (e.g., based on the sampling criteria or area of residence), we may have yet another group of hierarchical dimensions based on spatial organization. One of the objectives in analyzing data of this sort is to decompose variation of a given indicator of interest (behavioral variable) into multiple dimensions to study human behavior (as shown in matrix form in Figure 9.2). For example, we would like to know if the bulk of variation is due to reasons within a person that change over time (e.g., taste, mood, and so forth), personal characteristics that we can observe (e.g., age, gender, employment), or even more stable factors such as personality. We would also like to know if a portion of this variation is due to household influences (e.g., task allocation within a household) and area of residence characteristics (e.g., density of locations at which leisure activities can be pursued).

Multilevel regression models are statistical techniques that (1) account for the data hierarchy, and (2) allow us to develop functions explaining the relationships among the different variables from the different levels in the hierarchy. These regression models consider one or more variables as the *dependent* variables, the variation of which we are trying to explain using independent (*explanatory*) variables. When dependent variables are depicting the behavior for each individual (the person) in a given sample, key explanatory variables are each individual's known characteristics (e.g., age, gender, education, employment, race). The relationships, usually represented by regression coefficients, between the explanatory variables and the dependent variables may depict all three types of relationships discussed above (micro to micro, macro to micro, and micro to macro). In most regression equations we find dependent and independent variables defined at the same level, depicting micro-to-micro relationships. To reflect and capture the effect of a higher-level social unit, e.g., the household, on the individual's behavior, we can also include explanatory variables that describe the household itself, such as number of children by age group, number of vehicles owned and available, and number of employed persons among other variables. This relationship represents the macro-to-micro relationships (the effect of the household as a unit on each household member's behavior). Information from units that are below (within) the individual could also be included using explanatory variables. One example is when we have data from the repetition of observation of this same individual (e.g., behavior at different days of the week) or the activity and travel episodes of the same person within a day. When behavior is explained by variables depicting these "within a person" behavior and we formulate models at the person level, we have an example of a model capturing a micro-to-macro relationship. These techniques provide a tool to quantify social context effects while at the same time capturing the relationships among factors within the same level. For this reason, the terms *multilevel statistical model* and *multilevel regression model* are used to label the techniques.

Multilevel regression techniques are superior to single-level regression models in four distinct ways. First, behavioral models can be improved if proper consideration of the contexts in which people act is reflected in the models. This is the regression analog (but not derived from it directly) to the activity theory approach in Chapter 1. For example, person-based models need to consider observed and unobserved within-household interactions. In fact, travel behavior researchers are developing theories and testing hypotheses about the interaction of persons within households. An early example and a comprehensive literature review about persons and their role in a household can be found in Townsend (1987).

As argued and demonstrated by van Wissen (1989) and much later by Golob and McNally (1997) using structural equations models, the interaction in time use decisions within a household between two persons is of paramount importance in modeling travel behavior. Similarly, behavioral understanding can be improved when joint participation in activities is studied in more detail as Gliebe and Koppelman (2002), focusing on a two-person time allocation example, demonstrate. In addition, as shown in Chandrasekharan (1999) and Chandrasekharan and Goulias (1999), consideration of joint activity participation and travel not only improves understanding of behavior, but also yields better estimates of some quantitative indicators (e.g., vehicle occupancy) that are used in the most popular regional forecasting models worldwide.

Second, model misspecification (the bias introduced by excluding important explanatory factors of behavior) of these models can be attenuated when we incorporate observed and unobserved heterogeneity using models with more informative random structures. For example, in models of the number of trips a person makes in a day, including variables that describe the person's household may capture the effects of the role each person plays within a household (for a comprehensive definition of roles, see Townsend (1987)), diminishing the negative effects of excluding significant explanatory factors that may have not been measured during the survey process.

Third, for forecasting model systems that use models in which behavioral dynamics are explicitly modeled, observed and unobserved longitudinal variation should be accounted for and explicitly represented because persons with the same characteristics may follow different paths of behavioral change (for an example using latent class models in transportation, see Goulias (1999a)). As demonstrated in another paper (Goulias, 2002) multilevel models applied to the repeated observation of the same persons over time (panel survey data) allow the building of trajectories of change that, in turn, can be used as building blocks of a forecasting model system.

Fourth, the usual single-level regression model assumption of independent random error terms implies that the observations used to estimate the model parameters are independent, given the explanatory variables in the regression model. When groups of observations are from the same household, and when we do not have access to all the variables that explain the behavior of each person, it is likely that the error terms in this model are correlated. This is similar to serial correlation (i.e., data points are correlated over subsequent time points) and spatial correlation (i.e., data points are correlated because they are from neighboring points). Neglecting this social correlation in regression estimation may lead to larger standard errors of the coefficient estimates (Kennedy, 1995), increasing the risk to exclude significant explanatory variables from our model. Intuitively, this inefficiency is due to our mistake not to consider the additional information contained in the data, which are the relationships within groups of observations.

In the remainder of the chapter the basic regression model and its variants are described. Then the basic multilevel model is provided with a numerical example. This is followed by a section presenting a multiequation (multivariate) multilevel model and another numerical example to illustrate interpretation and use of this approach. The chapter ends with a brief summary and a section on further reading material.

9.2 The Basic Model

Suppose we have set out to study the amount of time (y) a person j allocates in a day to some particular type of activity (e.g., leisure) as a function of a person's characteristic (x). Also assume that we have observed each person at multiple time points and have stored this information in our database.

Before proceeding with a more detailed presentation of the multilevel models, it is worth pointing out a key idea that underlies research and empirical data analysis work using regression models. This is the idea of independent and identically distributed random variables in the context of linear regression. Let us focus on the random variables $y_1, y_2, \ldots, y_n$ with a joint distribution $f(y_1, y_2, \ldots, y_n; \theta)$. θ contains all the usually unknown parameters in a regression model (μ and σ values). Let us name the joint distribution above F (if we would assume that it is normal, we would write N). In matrix format we can write

$$
\begin{pmatrix} y_1 \\ y_2 \\ \cdot \\ \cdot \\ \cdot \\ y_n \end{pmatrix} \sim F \left(\begin{bmatrix} \mu_1 \\ \mu_2 \\ \cdot \\ \cdot \\ \cdot \\ \mu_n \end{bmatrix}, \begin{bmatrix} \sigma_{11} & \sigma_{12} & \cdot & \cdot & \cdot & \sigma_{1n} \\ \sigma_{12} & \sigma_{22} & \cdot & \cdot & \cdot & \sigma_{2n} \\ \cdot & \cdot & \cdot & & & \cdot \\ \cdot & \cdot & & \cdot & & \cdot \\ \cdot & \cdot & & & \cdot & \cdot \\ \sigma_{n1} & \sigma_{n2} & \cdot & \cdot & \cdot & \sigma_{nn} \end{bmatrix} \right)
$$

(9.1)

Equation (9.1) contains $n + 1/2 \, (n(n + 1))$ unknown parameters, and usually we have only n observations from which to estimate these parameters. When we add the assumption that all the n observations are independent (they do not vary jointly, but vary independently) we obtain

$$
\begin{pmatrix} y_1 \\ y_2 \\ \cdot \\ \cdot \\ \cdot \\ y_n \end{pmatrix} \sim F \left(\begin{bmatrix} \mu_1 \\ \mu_2 \\ \cdot \\ \cdot \\ \cdot \\ \mu_n \end{bmatrix}, \begin{bmatrix} \sigma_{11} & 0 & \cdot & \cdot & \cdot & 0 \\ 0 & \sigma_{22} & \cdot & \cdot & \cdot & 0 \\ \cdot & \cdot & \cdot & & & \cdot \\ \cdot & \cdot & & \cdot & & \cdot \\ \cdot & \cdot & & & \cdot & \cdot \\ 0 & 0 & \cdot & \cdot & \cdot & \sigma_{nn} \end{bmatrix} \right)
$$

(9.2)

Equation (9.2) requires us to estimate the n μ values and the σ_{11} to σ_{nn} variances. If the n observations are persons from a random sample and they do not coordinate their activities in a day (or at least the day of the interview), the assumption of independent observations is reasonable; otherwise, we are neglecting a relationship by imposing zero covariances.

We can simplify Equation (9.2) even further if all y values are also identically distributed with mean μ and variance σ^2:

$$
\begin{pmatrix} y_1 \\ y_2 \\ \cdot \\ \cdot \\ \cdot \\ y_n \end{pmatrix} \sim F \left(\begin{bmatrix} \mu \\ \mu \\ \cdot \\ \cdot \\ \cdot \\ \mu \end{bmatrix}, \begin{bmatrix} \sigma^2 & 0 & \cdot & \cdot & \cdot & 0 \\ 0 & \sigma^2 & \cdot & \cdot & \cdot & 0 \\ \cdot & \cdot & \cdot & & & \cdot \\ \cdot & \cdot & & \cdot & & \cdot \\ \cdot & \cdot & & & \cdot & \cdot \\ 0 & 0 & \cdot & \cdot & \cdot & \sigma^2 \end{bmatrix} \right)
$$

(9.3)

This time all we need to estimate is one μ and one σ. This spectacular reduction in unknown parameters to be estimated (moving from Equation (9.1) to Equation (9.3)) is also one of the practical advantages of the usual simple unilevel linear regression model. Equation (9.3), however, is too restrictive and does not contain the relationship we are interested in, which is the link between X and Y (capital letters are used here to indicate vectors and matrices).

The relationship between Y and X using linear regression can be written as:

$$
y_j = \beta_0 + \beta_1 x_j + \varepsilon_j
$$

(9.4)

For example, the variable x_j represents the age of person j and the variable ε represents a random fluctuation with mean zero and a given amount of variance (σ^2). When a person's age is zero, the intercept

β_0 represents the amount of time allocated to leisure. When the person is 20 years old, his or her expected value of the amount of time allocated to leisure in a day will be $\beta_0 + 20\beta_1$. This can also be written in the following format:

$$
\begin{pmatrix} y_1 \\ y_2 \\ \cdot \\ \cdot \\ \cdot \\ y_n \end{pmatrix} \sim F \left(\begin{bmatrix} \beta_0 + \beta_1 x_1 \\ \beta_0 + \beta_1 x_2 \\ \cdot \\ \cdot \\ \cdot \\ \beta_0 + \beta_1 x_n \end{bmatrix}, \begin{bmatrix} \sigma^2 & 0 & \cdot & \cdot & \cdot & 0 \\ 0 & \sigma^2 & \cdot & \cdot & \cdot & 0 \\ \cdot & & \cdot & & & \cdot \\ \cdot & & & \cdot & & \cdot \\ \cdot & & & & \cdot & \cdot \\ 0 & 0 & \cdot & \cdot & \cdot & \sigma^2 \end{bmatrix} \right) \tag{9.5}
$$

The model in Equation (9.5) is the same as the simple linear regression model based on which we built a series of other regression models. When one compares Equation (9.4) with Equation (9.5), the increase in the number of additional parameters to estimate is only one (the β_0, and β_1, instead of just μ). This, however, can make Equation (9.5) very flexible when additional x values are added. In fact, most linear regression models we encounter in travel behavior analysis contain many more x values as explanatory variables, and each additional x increases the number of parameters to estimate by one unit, while at the same time it captures another piece of the variation in y.

A small digression is needed here to discuss centering because it is used in many multilevel models. We can also rewrite the linear regression model by transforming x as a deviation from the mean:

$$
y_j = \beta_0 + \beta_1(x_j - \bar{x}) + \varepsilon_j \tag{9.6}
$$

Interpretation of the β coefficients is somewhat different in Equation (9.6). If this person has an age equal to the mean, indicated by $\bar{x}$, then β_0 is the expected amount of time this person allocates to leisure in a day. β_1 represents the effect of a unit increase in age on leisure allocation (i.e., if age is measured in years, it represents the difference in time allocation between two persons of a year difference in age; this may not be the same as the effect of aging by 1 year). Note that Equations (9.4) and (9.6) are regression equations capturing the microlevel effects of age on the time allocated to leisure by a person, which is a microlevel dependent variable. According to these two equations, the effect of age on leisure is the same among persons because it does not change with a person's index. Another variant often used in multilevel model building is one that allows regression coefficients to change among the observations at hand.

In fact, one can increase the flexibility of this model by allowing the base time allocation to be different among persons. This can be written as:

$$
y_j = \beta_{0j} + \beta_1(x_j - \bar{x}) + \varepsilon_j \tag{9.7}
$$

This model is able to capture the differences among persons as differences among the β_{0j} values that in essence shift the regression line up and down with each individual observation. Equation (9.3) is not very different from the classic linear regression model in econometrics. When data are available, consistent and efficient estimates of the regression coefficients in this equation can be obtained using ordinary least squares. However, a problem may arise in interpreting the intercepts as representations of the population when we do not include *all* the population units, as is the usual practice in travel behavior. In addition, we need to estimate as many coefficients as the individuals in the study, which means that we need to have more observations than the j = 1, ..., n persons (the usual rule of thumb in regression models is that we should have at least ten observations per coefficient estimated). One way to resolve this is by assuming that the intercept is a randomly varying effect among the n observations, resulting in the random effects model. The usual added assumption is for this random effect to have a variance that is the same among observations (in this way, both the random intercept and the random residual are assumed to have a variance that does not change with each observation — called homoskedastic random error term). Multilevel models are able to release this homoskedasticity assumption to yield richer and more informative specifications; random error terms that are not homoskedastic are called heteroskedastic.

Further, we can imagine the effect of age on time allocation to also vary with each individual. If we have no information about systematic ways in which this effect may vary, we can assume that the β_1 values are randomly varying. A typical way of expressing this variation is the following:

$$\beta_{1j} = \gamma_1 + u_j \tag{9.8}$$

$$\beta_{0j} = \gamma_0 + v_j \tag{9.9}$$

The γ values in the above equations represent the mean effects around which each individual's behavior differs according to a randomly distributed variable (v for the intercept and u for the slope). The time allocation equation can then be written as:

$$y_j = \gamma_0 + \gamma_1(x_j - \bar{x}) + [u_j(x_j - \bar{x}) + \varepsilon_j + v_j] \tag{9.10}$$

Equation (9.10) shows the fixed and random parts of the model. The first two terms containing the coefficients γ are the intercept and slope of the fixed part. The last three terms within the brackets contain the three random components of the random part. If we were to neglect the complex nature of the random part, assume that it was made of independent identically distributed random variables, and apply ordinary least squares to estimate the γ values, we would obtain consistent parameter estimates but inconsistent standard errors of coefficient estimates, and most likely inefficient estimates.

In econometrics, the study of this type of models has focused on the issues raised by Balestra and Nerlove (1966) in their demand for energy study among the American states, providing a first formulation of a model with random effects. In terms of the levels we discuss here, each state is observed at different time points (years) leading to a two-level data hierarchy. The number of observations in this case is the number of states in their study, N, times the calendar time points, T. The Balestra and Nerlove study also introduced a plethora of other models that go beyond the focus of this chapter. A key contribution, however, to the analysis of data with hierarchies was the demonstration that observations of this type contain information that may not be captured by the observed explanatory variables in a regression model, and for this, requiring the use of information in their heterogeneous random error terms. In this way unobserved heterogeneity, in its heteroskedasticity form, is viewed as a source of additional information instead of a problem to eliminate.

Subsequently, in another fundamental contribution, that Swamy offered 30 years ago, emphasis was given to random coefficients, as in Equation (9.4) (creating the random coefficient regression model). This type of model is discussed extensively with other random coefficient models in Swamy (1974). In addition, different versions of Equation (9.4) that are based on repeated observations of the same groups of persons (known as panel data) led to a populous group of methods known in econometrics as models of panel data (Greene, 1997), econometric analysis of panel data (Baltagi, 1995), and analysis of panel data (Hsiao, 1986). The emphasis in this type of analysis is given to the individual (a person, firm, or state) and discrete time points at which the behavioral unit is measured or surveyed. A review book on models and methods for panels with many transportation examples from around the world is the edited volume by Golob et al. (1997). In an earlier experiment using a database similar to the one used in this chapter, Liao (1994) identified, discussed, and illustrated some estimation issues for the random coefficient model and the need for data variation within groups (e.g., for each person across time) when estimating models of this type. Similar issues are key to the multilevel models as well, and we will discuss them later in the chapter. It should be noted, however, that instead of using the typical econometric approach to model building, the following section describes multilevel models using conventions and an exposition that has been used in applied statistics.

9.3   The Basic Multilevel Model

The multilevel models described here are more general than panel data models because they allow many more dimensions than the two dimensions, individual unit and time, of the panels. Unlike more traditional multilevel presentations, we will start with panel data models and then move to more complex multilevel models, but first let us define a few terms that are specific to multilevel models.

The models and the type of regression analysis used here are known by different names in different fields of research for different reasons. For example, they have been named random coefficient models (Longford, 1993; Greene, 1997, p. 669) because emphasis is given to the varying nature of the regression coefficients and their specific pattern of variation, as shown in Equations (9.8) and (9.9). They have also been named multilevel models (Goldstein, 1995) to emphasize the measurement of the dependent variable at different levels (e.g., income can be measured for each person, but also as a household or neighborhood average or median value). Another group of researchers name these models *mixed models* (Searle et al., 1992) to emphasize the presence of fixed and random coefficients in the same regression model. Bryk and Raudenbush (1992) use the name *hierarchical models* to indicate that the data structures are from hierarchies. Some of the labels in this family of models indicate subtle but important differences revealing the researchers' modeling emphasis. All models share one element — *the arrangement of data into groups and the exploitation of group membership to unveil hidden aspects of data variation*. However, some of these labels are also confusing because some adjectives in the labels have also been used to indicate different classes of models or their properties. For example, Searle et al. (1992) use the term hierarchical model to indicate a model that is specified in a sequence of hierarchical stages. In addition, the term mixed model can be easily confused with the term *mixture* in statistics, indicating a different family of statistical models.

To avoid confusion and to be consistent with a few of the key references used here and the software employed to estimate the examples in this chapter, the term *hierarchical data* is used to indicate the nested nature of the data at hand and *multilevel models* to indicate:

1. Models containing an explicit recognition in their formulation of the hierarchical, multiple-level, and nested structure of the data to analyze
2. Model specification that uses three groups of regression components in the same regression model (fixed coefficients, random components of coefficients, and random error term residual)

The first group, fixed coefficients, assumes constant sensitivity to explanatory variables among the units of analysis, representing the mean effect of an explanatory variable on the dependent variable (we use the Greek letter γ for these coefficients). The second group, random coefficients, assumes a random deviation around this mean as in Equations (9.4) and (9.5) (we use u, v, and w to indicate these components). The third group is the usual random error term(s) of the regression equation (we use the Greek letter ε for this component). If we want to examine many dependent variables in a system of equations, we will have as many random errors (ε values) as the dependent variables.

To demonstrate the differences with other regression models, we rewrite the regression equation in a somewhat different way by introducing a second index and eliminating the centering (deviation from the mean) of the explanatory variable. Assume we have two levels: persons for whom we use the index j. Each person was observed at a few time points, and for the time points we use the index i.

$$y_{ij} = \beta_{0ij}x_{0ij} + \beta_{1j}x_{1ij} + \gamma_2 x_{2ij} + \gamma_3 x_{3ij} + \gamma_4 x_{4ij} \tag{9.11}$$

Equation (9.11) indicates that we have five explanatory variables. The variable x_{0ij} is the equivalent of the intercept (constant) in regression models that takes the value of 1 for all observations when we consider the person level alone. As we will see below, it is its random coefficient that contains some interesting components. A second explanatory variable (x_{1ij}) also has a random coefficient that changes with the person index (randomly varying across persons). The other three explanatory variables have coefficients γ that are neither functions of other variables nor randomly varying (i.e., they take one single unknown value for each observation). In addition, the two random coefficients can be written as

$$\beta_{0ij} = \gamma_0 + v_j + \varepsilon_{ij} \tag{9.12}$$

$$\beta_{1j} = \gamma_1 + u_j \tag{9.13}$$

Equation (9.12) indicates that all observations have one common fixed intercept γ_0, a randomly varying intercept among persons (that we also assume has $E(v_j) = 0$ and $Var(v_j) = \sigma^2_v$) and a randomly varying component with time and with persons (that we also assume has $E(\varepsilon_{ij}) = 0$ and $Var(\varepsilon_{ij}) = \sigma^2_\varepsilon$), which is the usual regression residual. Therefore, $E(\beta_{0ij}) = \gamma_0$.

Equation (9.13) contains two components, the fixed slope γ_1, indicating that all observations have one common slope (multiplier) for variable x_1, but that they differ in their behavior according to a random u (with $E(u_j) = 0$ and $Var(u_j) = \sigma^2_u$). In addition, the random part of this slope and the random part of the intercept are assumed to be correlated with $Cov(v_j\ u_j) = \sigma_{vu}$. Note that in Equations (9.11) to (9.13) we have modeled the variation in behavior among persons, and the only entities varying with time (and within persons) are the x values and the residual ε.

In the example here the model defined by Equations (9.11) to (9.13) is called model C (for reasons that will become clear later). In Equation (9.13), we can define the random slope as fixed ($\beta_{1j} = \gamma_1$), eliminating its randomly varying part with persons and the correlation with the random component of the intercept (u). This is called model B. If we eliminate all explanatory variables (x values), we obtain a third model (model A) that contains only an intercept defined by Equation (9.12). The parameters to be estimated for each model are:

Model A: $\quad\quad\quad\quad \gamma_0, \sigma^2_v, \sigma^2_\varepsilon$

Model B: $\quad\quad\quad\quad \gamma_0, \gamma_1, \gamma_2, \gamma_3, \sigma^2_v, \sigma^2_\varepsilon$

Model C: $\quad\quad\quad\quad \gamma_0, \gamma_1, \gamma_2, \gamma_3, \sigma^2_v, \sigma^2_\varepsilon, \sigma^2_u, \sigma_{vu}$

The estimates from model A can be used to compute a useful quantity called the intraclass correlation, ρ, using the following (Hox, 1995):

$$\rho = \frac{\sigma^2_v}{\sigma^2_v + \sigma^2_\varepsilon} \tag{9.14}$$

Estimation of all the fixed (γ values) and random (σ values) parameters can be accomplished by a few different methods. The estimation of one set of these parameters depends on the other. The key idea here is that the covariance components are not known, and for this reason, they need to be estimated with the fixed parameters. In general, most estimation techniques are based on maximizing a likelihood function.

In fact, full information maximum likelihood (FIML), which is applied to Y directly, and restricted maximum likelihood (REML), applied to the least squares residuals, which can be used in tandem with a generalized least squares approach, have been used in the past. Longford (1993), Bryk and Raudenbush (1992), and Goldstein (1995) provide a comprehensive review of estimation techniques, their relative performance, and details about implementation and algorithms. Kreft and De Leeuw (1998) provide an overview and a discussion about software and Internet websites with additional information (see also, the end of this chapter). van der Leeden (1998) also mentions the use of Bayesian techniques and one application of a data augmentation technique (see also Schafer, 1999) to the estimation of multilevel models.

In this chapter, Goldstein's (1995) iterative generalized least squares (IGLS) approach is used; it separates estimation of the fixed from the random parameters at different steps in sequence repeatedly until no change is observed in the estimates in subsequent steps. Goldstein (1995) has also improved the IGLS algorithm when based on FIML using a modified IGLS called RIGLS. In fact, this method provides standard errors of coefficient estimates that are conservative (larger), and for this, leading to more parsimonious models. In a series of experiments performed in a few studies using this same data set and reported elsewhere (Goulias, 2002), IGLS and RIGLS gave similar results and identical conclusions about the significance of variables.

For each estimate standard errors can also be computed (e.g., as an output of a maximum likelihood estimation) and hypotheses tests about their significance performed. A general agreement seems to exist in the multilevel literature that we can test for significance of the fixed coefficients using a test that is based on the ratio between a coefficient estimate and its estimate of its standard error (also known as the Wald test in honor of the first developer in the 1940s). Bryk and Raudenbush (1992) suggest the use of a t-test instead of a z-test. In practice, however, and because in the travel behavior examples we have a large number of observations, the two tests would yield very similar indications about significance.

In contrast, testing for significance of the random parameters (variances) is not as straightforward and simple, particularly for variances that are very small. As explained by Bryk and Raudenbush (1992) and Hox (1995), a solution to hypothesis testing for the significance of these variances is to use a test based on the likelihood ratio (the same ratio used in many other models such as the discrete choice models in travel behavior when models can be considered to have a nested specification structure).

Maximum likelihood estimation is the derivation of parameter estimates by finding the maximum of the function called likelihood using an iterative method. Most maximum likelihood algorithms produce a series of iterations that are stopped based on a rule of convergence to a solution, which is the maximum of the likelihood, beyond which no improvement in the parameters and value of the maximum are observed (e.g., computing numerically the first derivatives and finding them to be very close to a computable zero). At the end of the iterations that find the maximum of the likelihood function, the deviance is computed and defined as -2 logarithm of the likelihood evaluated at the maximum. If we estimate two models that have the same specification in terms of explanatory variables, but differ in the number of variances (let us assume that one model has k variances to be estimated and the other model has k–q), then each model will yield a deviance that we will indicate as D_k and D_{k-q}, respectively. The difference of these two quantities is χ^2 distributed with degrees of freedom equal to q. If the inclusion of the q parameters leads to a significantly better goodness of fit (a deviance that is much smaller in a statistical sense), then we should prefer the model with the q additional parameters; otherwise, we should prefer its competitor with k–q parameters.

9.3.1   Data Example 1: Time Allocation to Leisure Activities

In this chapter data from the one and only current (general-purpose) panel survey specifically designed for transportation planning in the United States are used. This survey, called the Puget Sound Transportation Panel (PSTP) and described in Murakami and Watterson (1990), Goulias and Ma (1996), and Murakami and Ulberg (1997), is a unique source of data for regional travel demand forecasting. Unfortunately, its potential has not been put to good use in practical applications yet. The Puget Sound Regional Council has plans, however, to use models derived from this data set in its regional forecasting model system. In addition, the recent addition of questions about information technology and traveler information use leads to unprecedented possibilities for studying traffic management strategies in Seattle and the surrounding region, as illustrated in this chapter in a later example.

A panel is a survey administered repeatedly on the same observations over time. Each survey, conducted at each point in time (in PSTP a year of interview), is called a *wave*. PSTP contains three groups of data:

TABLE 9.1 Average Sample Characteristics of the Data Used Here (Standard Deviation in Parentheses)

Variable	1989	1990	1992	1993	1994
Leisure (minutes/day) by a person	120.0 (159.2)	105.8 (155.9)	103.7 155.5)	109.7 (158.0)	99.5 (157.8)
Age	46.7 (13.3)	47.9 (13.7)	50.0 (13.7)	51.0 (13.7)	52.1 (13.7)
# of children ages 1 to 5 in household	0.213 (0.53)	0.200 (0.51)	0.158 (0.48)	0.147 (0.46)	0.133 (0.48)
# of children ages 6 to 17 in household	0.437 (0.80)	0.440 (0.80)	0.450 (0.82)	0.438 (0.80)	0.433 (0.80)
Numbers of cars in the household	2.34 (1.10)	2.34 (1.10)	2.36 (1.10)	2.27 (0.97)	2.26 (0.98)
Percent employed in household	69.8	69.4	73.9	65.8	64.5

household demographics, people's social and economic information, and reported travel behavior in a 2-day travel diary (additional details are available in Goulias and Ma (1996) for the first four waves of PSTP). The data used in this paper are from the first five waves of PSTP conducted in 1989, 1990, 1992, 1993, and 1994. These travel diaries cover a period of 48 h. Each person was interviewed on the same 2 days in all waves, and the travel diary includes every trip a person made during these 2 days. For each trip reported we have the trip purpose, mode used, departure time, arrival time, travel duration minutes and miles, origin, and destination. Activity participation information can be derived for all out-of-home activity engagement events using the trip purposes and for a portion of the in-home activities pursued between the first departure from home (e.g., in the morning) and the last arrival at home (e.g., in the evening). The duration of each activity episode (d) is computed by the difference between the start time of the next trip (t + d, departure from a given location) and the end time of the current trip (arrival at a given location, t), giving the sojourn time at an activity location (d).

In the first few waves of the PSTP database, trip purposes are classified into nine different types: work, school, college, shopping, personal business, appointments, visiting (other persons), free time, and home during the day. In past analyses by Ma (1997) using this same data set, activities were grouped in subsistence (work, school, college), maintenance (shopping, personal business, appointments), leisure (visiting, free time, home during the day), and travel. In this example we use data from five time points (first day of each wave) for 1201 persons in 758 households whose characteristics are provided in Table 9.1. For simplicity, only the stayers (persons who participated in all five waves) are used for model estimation in this example. However, the models presented in this chapter do not require an equal number of observations for each person.

A first group of three two-level models (models A, B, and C) are estimated using the data above to illustrate a few aspects of multilevel modeling. Table 9.2 shows the estimates (fixed and random) for these three models. At each level, time, and person, we have level-specific variance–covariance terms (the σ values for ε, u, and v in model A). The significance of the elements in each of the three matrices can be tested using goodness-of-fit measures based on the deviance, which is the difference in the −2 log-likelihood at convergence between two nested (in terms of specification) models. In addition, the γ values

TABLE 9.2 Leisure Time in a Day: Models A, B, and C

Model Component Fixed effect	Model A		Model B		Model C	
	Coefficient	SE	Coefficient	SE	Coefficient	SE
Fixed intercept (γ_0)	107.5		96.5 (1.4)		103.2 (1.1)	
Employed (=1, 0 otherwise) (γ_1)			−58.9	4.84	−59.5	6.47
Male (=1, 0 otherwise) (γ_2)			−8.3 (1.4)	4.87	−10.9 (1.1)	4.80
Driver (=1, 0 otherwise) (γ_3)			57.6 (1.1)	11.55 (1.1)	52.5 (1.1)	12.2 (1.1)
Random effects	(σ^2)		(σ^2)		(σ^2)	
Temporal variation within persons (ε)	20365.75	403.6	20080.60	397.8	19013.39	386.6
Variation between persons (v)	4764.07	385.0	4059.68	349.2	10235.76	1037.4
Between persons for employment (u)					10450.59	1540.6
Covariance (u with v)					−8999.4	1144.7
−2 log-likelihood (deviance)	77440.96		77277.56		77122.19	

Note: SE = standard error.

can also be tested if they are significantly different from zero using a z-test. This is applied in the same fashion as for (unilevel) linear regression.

Model A in Table 9.2 contains no explanatory variables. It is called the null model or fully unconditional model, and it is used as a benchmark to assess other model specifications that include explanatory variables and regression coefficients (fixed or random) at each level. As expected, the lion's share in the proportion of variance is within persons and across time points. The intraclass correlation is 0.19, which is the estimated percent of variation explained by the hierarchy assumed in this data set. Model B contains three additional coefficients for employment, gender, and driver's license. Employment and driver's license are significantly different from zero. The gender, however, is not significantly different from zero when we use a cutoff value of 2. A comparison between models A and B can also be done using the difference in the deviance, which is 163.4, indicating that model B is a significantly improved model over model A. Model C is a model with the coefficient for employment randomly varying among persons. This time the gender coefficient is significantly different from zero and the variance components are also fairly large. The covariance between u and v is negative. Applying the χ^2 test to compare models B and C, we obtain 155.46 with two degrees of freedom, which also indicates that model C is a significant improvement over model B.

From a travel behavior viewpoint, model C shows that on average an employed person is likely to have 59.5 min less leisure than an unemployed person. Similarly, males seem to spend on average 10.9 min less than females, but drivers tend to spend 52.5 min more than nondrivers in leisure activities. The large variance exhibited by the random component of the employment may be a signal of wide variation in the allocation of time to leisure among persons that may depend on other factors, including higher levels of aggregation of these persons, such as household characteristics. This has been analyzed in a much more detailed fashion using the same database as in this chapter in Goulias (2002).

9.4   Multivariate Multilevel Model

One of the advantages in analyzing data using the somewhat newer and more sophisticated techniques such as structural equations and multilevel models is our ability to study relationships among indicators from a more comprehensive viewpoint, allowing multiple relationships to be modeled simultaneously.

Single-equation regression models do not explain the interdependencies among explanatory variables. Some of these interdependencies may be very important because of potential trade-offs, feedback, and chicken-and-egg causalities. In fact, travel behavior research contains many examples (e.g., automobile ownership and use (Train, 1986)). The key advantage of estimating simultaneous equation models is the ability to represent more complex correlation patterns in the data and to obtain a clearer picture about the influence of one variable on another.

This capability is of paramount importance in the more recent activity-based approaches to travel demand because when we study time allocation, simultaneity of relationships and trade-offs is more likely than in other travel behavior aspects that can be divided into epochs of occurrence (e.g., residence location decisions may be easier to separate from leisure activity participation because these two blocks of decisions require different planning and execution time frames and horizons). The second example in this chapter is a typical case study of simultaneity in the relationship between activity and travel. Using four equations we study temporal causation among the dependent variables, and we can study the effect of information and telecommunication technology ownership and use on activity participation and travel in a more comprehensive way.

Telecommunications has been consistently looked at as a possible solution to urban transportation problems (Salomon, 2000). Transportation and telecommunications interaction, however, is a complex two-way relationship (for an overview, see Mokhtarian and Salomon, 2002). From the many aspects in this complex system we chose the relationship between telecommunications ownership and allocation of time to travel and to activities outside one's home that are greatly influenced by a variety of contextual factors within a household and outside (e.g., facilities at the workplace and school). In addition, the more recent mobile communication technology has opened possibilities of work and play that are unprecedented (e.g., browsing the Web from our wireless phone and receipt of tailored information on a personal digital assistant (PDA) that can communicate directly to our office computer, updating a dynamic to-do list) and very interesting. To assess the potential impact of these technologies we would like to know if persons that own and use mobile communication technologies travel more than others who do not use these technologies. In addition, we would like to know if the effect of these technologies is the same across different days and across different persons. We would also like to know the role played by households (e.g., presence of children, employment mix, residence location and accessibility) in determining the relationship between telecommunications and travel.

To do this type of analysis we can first write a system of equations representing the relationships described above as follows:

$$y_{jk}^{T1} = \beta_{jk}^{T1} + \gamma_1^{T1} x_{1jk}^{T1} + \dots + \gamma_{m_{T1}}^{T1} x_{m_{T1}jk}^{T1} \tag{9.15}$$

$$y_{jk}^{A1} = \beta_{jk}^{A1} + \gamma_1^{A1} x_{1jk}^{A1} + \dots + \gamma_{m_{A1}}^{A1} x_{m_{A1}jk}^{A1} \tag{9.16}$$

$$y_{jk}^{T2} = \beta_{jk}^{T2} + \gamma_1^{T2} x_{1jk}^{T2} + \dots + \gamma_{m_{T2}}^{T2} x_{m_{T2}jk}^{T2} \tag{9.17}$$

$$y_{jk}^{A2} = \beta_{jk}^{A2} + \gamma_1^{A2} x_{1jk}^{A2} + \dots + \gamma_{m_{A2}}^{A2} x_{m_{A2}jk}^{A2} \tag{9.18}$$

$$\beta_{jk}^{q} = \gamma_0^{q} + v_k^{q} + u_{jk}^{q}, \text{ where } q = \text{T1, A1, T2, and A2} \tag{9.19}$$

y_{jk}^{T1} is the total amount of time traveled in day 1 by a person j within his or her household k (with j = 1, 2, …, number of people in household k; k = 1, 2, …, number of households in the sample). Similarly, we define the other three dependent variables as total amount of time allocated to all other activities except travel in day 1 as y_{jk}^{A1}, total amount of time traveled in day 2 as y_{jk}^{T2}, and total amount of time allocated to activities in day 2 as y_{jk}^{A2}.

The first term on the right-hand side of each equation in this multilevel model system is a random intercept. This component has a specific meaning. For example, β_{jk}^{T1} is the travel expenditure of person j in household k for day 1 when all other explanatory variables are zero, which is similar to the definition in the previous section. The term u_{jk}^{T1} is a random person-to-person variation (also called within-household variation), and it is a deviation of travel expenditure around γ_0^{T1}. The term v_j^{T1} is a random household-to-household variation, and it is also a deviation of travel expenditure around γ_0^{T1}.

These are also called random error components and are assumed to be normally distributed with $E(u^q) = E(v^q) = 0$, and $Var(u^q) = \sigma_{u^q}^2$ and $Var(v^q) = \sigma_{v^q}^2$ (with q indicating each of the four variables as defined in Equation (9.19)). The random components (u^q and v^q) and their variance represent unobserved heterogeneity at the person and household levels, respectively. As in the single equation model, the γ coefficients are the fixed parameters (similar to the coefficients in a typical regression model). Although all the coefficients of explanatory variables are defined as fixed in the model specification above, the coefficients (β values) can be defined as random with a mean and a variation around their mean γ values, as illustrated in the single-equation model for leisure. In this way we could define a more general model at each of these levels to represent heterogeneous behavior due to either personal or household variation. With this multilevel model system approach we can assess the effects of each telecommunication technology on activity and travel behavior, while at the same time controlling for complex correlations within a person's behavior (one day to the next), within a household (one person to the next), and among households.

9.4.1 Data Example 2: Time Allocation to Activities and Travel on Different Days

In 1997 the PSTP (wave 7) asked the panel participants to report their personal use and attitudes toward existing and potential new (travel) information sources (in addition to the travel diary infor-

TABLE 9.3   Summary of Socioeconomic Characteristics of the Wave 7 Sample

	Number of Households	1910
	Number of Persons	3450
	Characteristics	Percent (N = 3450)
Gender	Male	47.9
	Female	52.1
Age	15–24	8.6
	25–44	34.2
	45–64	39.2
	65 and above	18.1
Occupation	Professional	23.6
	Managerial	9.8
	Secretary	8.5
	Sales	4.4
	Other	14.1
	Unemployed	39.7
Number of Vehicles in Household	No vehicles	1.4
	1 vehicle	17.6
	2 vehicles	46.1
	3 or more vehicles	34.9
Household Income	Less than $35,000	23.0
	$35,000 to $74,999	47.8
	$75,000 or more	22.3
	No answer	6.9

mation). These respondents were also asked about their use of electronic equipment and information services. For example, respondents provided information regarding their use of a desktop computer at home or at work, with access to the Internet at least once a week on average. Other questions asked if the respondents carried a personal cellular phone, pager, laptop computer (with modem), or PDA at least ten times a month. In this chapter we use data from 3450 persons (from 1910 households), who provided valid information to both travel daily and their personal daily information and communication choices survey.

Table 9.3 summarizes the social, demographic, and economic characteristics of the sample used in this section. The majority of the respondents in the sample are between 25 and 64 years old. In terms of employment characteristics, 40% of the sample is unemployed. Among the employed, professionals occupy the largest portion. In terms of income, 70% of the sample belongs to middle and upper-middle income categories ($35,000 to $75,000), and as a result, there is a very small fraction of the sample without cars. Given the emphasis on presenting multilevel models in this chapter, no additional comparisons are made among the residents of the four-county area in the Puget Sound region. Table 9.4 presents the technology use characteristics in the sample. There are about 50% of the respondents who use computers in their daily lives, and males seem to use computers more than females. For about 30% of the sample (27.5 to 33.7%), computers are not part of their daily lives. In terms of mobile technology, the use of mobile devices has not yet reached the level of market penetration of desktop computers. In fact, more than 60% of the survey participants do not use any of the mobile technologies. Men seem to use mobile technologies more than women, with the exception of cellular phones, which are used more by women (30.5%) than men (27.1%).

The bottom portion of Table 9.4 is key to the analysis here because it reports the values of the variables that are used as dependent variables in the analysis. Total out-of-home activity time includes the entire time each person spends in activities outside of the home in a day. Total travel time includes the sum of travel time durations for all trips made by a person in a day. In terms of activity–travel, the sample spends an average of about 400 min participating in various activities outside of the home and an average of about 80 min traveling per day. These are very similar to the time allocations from past waves, an example of which can be found in Goulias (2002). Table 9.5 provides a list of the variables and their symbols used in the estimation tables.

TABLE 9.4 Summary of Technology Use in the Wave 7 Sample

Technology	Male ($N_1 = 1654$) (%)	Female ($N_2 = 1796$) (%)
Use desktop computer at work/school	54.2	44.9
Use desktop computer at home	56.4	48.8
Use Internet at work/school	36.1	25.2
Use Internet at home	36.6	26.6
None of these	27.5	33.7
Carry a portable cellular phone	27.1	30.5
Carry a personal pager	16.1	8.3
Carry a portable computer	7.8	3.1
Carry a personal digital assistant (PDA)	1.0	0.2
None of these	62.4	65.0

Average Total Out-of-Home Activity and Travel Durations (Standard Deviation in Parentheses)

	Day 1	Day 2
Total out-of-home activity (minutes/day)	405.14 (266.12)	402.08 (273.48)
Total travel (minutes/day)	83.03 (63.25)	80.01 (60.36)

TABLE 9.5 List of Variables Used in the Multivariate Multilevel Models

	Dependent Variables
T1	Total travel duration in day 1 [min] (min: 0; max: 780)
A1	Total out-of-home activity duration in day 1 [min] (min: 0; max: 1440)
T2	Total travel duration in day 2 [min] (min: 0; max: 598)
A2	Total out-of-home activity duration in day 2 [min] (min: 0; max: 1440)

Explanatory Variables

Household Level

HHSIZE	Number of people in the household
TOT1_5	Number of children who are younger than 5
TOT6_17	Number of children whose age is between 6 and 17
NUMVEH	Number of vehicles in household
MIDINC	Indicator, 1 if \$35,000 ≤ annual household income < \$75,000; 0 otherwise
HIGHINC	Indicator, 1 if annual household income ≥ \$75,000; 0 otherwise

Person Level

GENDER	Indicator, 1 = male; 0 = female
AGE2544	Indicator, 1 if 25 ≤ age ≤ 44; 0 otherwise
AGE4564	Indicator, 1 if 45 ≤ age ≤ 64; 0 otherwise
AGE65_	Indicator, 1 if 65 ≤ age; 0 otherwise
STUDENT	Indicator, 1 if a student; 0 otherwise
SECRET	Indicator, 1 if in a secretary position; 0 otherwise
SALES	Indicator, 1 if in a sales position; 0 otherwise
UNEMP	Indicator, 1 if unemployed; 0 otherwise
WK5	Indicator, 1 if work 5 times or more per week; 0 otherwise
LICENSE	Indicator, 1 if have a driver's license; 0 otherwise
BUSPASS	Indicator, 1 if have a bus pass; 0 otherwise

Technology Usage or Ownership

COMWORK	Indicator, 1 if use computer at work/school; 0 otherwise
COMHOME	Indicator, 1 if use computer at home; 0 otherwise
WEBWORK	Indicator, 1 if use Internet at work/school; 0 otherwise
WEBHOME	Indicator, 1 if use Internet at home; 0 otherwise
WEB	Indicator, 1 if use Internet at work/school and home; 0 otherwise
CELL	Indicator, 1 if carry a cellular phone; 0 otherwise
PAGER	Indicator, 1 if carry a pager; 0 otherwise
LAPTOP	Indicator, 1 if carry a laptop; 0 otherwise

TABLE 9.6   Multivariate Multilevel Error Component Model for Wave 7 Data

Model Component Fixed Effect	T1		A1		T2		A2	
	Coefficient	SE	Coefficient	SE	Coefficient	SE	Coefficient	SE
Grand mean (γ_0)	82.44	1.19	401.5	5.08	79.61	1.13	399.1	5.09
Random effects	σ^2	%	σ^2	%	σ^2	%	σ^2	%
Person variation within households (u_{ij})	3073.1	77.1	52983.7	74.5	2903.5	79.6	60619.0	80.8
Between households variation (v_j)	912.6	22.9	18157.3	25.5	745.1	20.4	14404.6	19.2
Total	3985.7	100.0	71141.0	100.0	3648.6	100.0	75023.6	100.0
−2 log-likelihood				169604.4				

Variance–Covariance Matrices (Upper Triangle Correlations)

	Between Persons				Between Households			
	T1	A1	T2	A2	T1	A1	T2	A2
T1	3073.1	0.208	0.468	0.159	912.6	0.331	0.298	0.143
A1	2654.4	52983.7	0.122	0.659	1347.8	18157.3	0.353	0.825
T2	1397.0	1517.3	2903.5	0.236	245.8	1299.3	745.1	0.406
A2	2173.9	37333.6	3127.8	60619.0	517.9	13359.0	1329.7	14404.6

Note: SE = standard error.

There are two levels in this model representing hierarchical entities: the household and within each household the persons that responded to the survey. Variance decomposition will be examined in these two levels. The multivariate model contains one additional dummy level in the implementation of multilevel model estimation in Rasbach et al. (2001). This level allows the assembly of four equations, two for activity time and two for travel time, and the estimation of the cross-equation correlations among their random error terms.

In a way similar to that for the single-equation multilevel model, one can estimate an error components model (model A above) that provides an idea of within-class correlation and that is used as the baseline model. Table 9.6 shows the estimation results of this error component model, which contains no explanatory variables. This model is used as a benchmark to assess other model specifications that include explanatory variables. As shown in the random effects portion of Table 9.6, the proportion of variance of the household level variance is about one fourth to one third of the person level variance, depending on the variable in Table 9.6, which indicates that it should not be neglected in model specifications, and thus multilevel specification appears to be justified and desirable. In addition, it confirms that it is necessary to specify models using explanatory variables depicting not only person characteristics but also household characteristics to reduce the unexplained variation. Unlike simultaneous equations systems in econometrics (Greene, 1997) and most structural equation implementations (see Chapters 2 and 11 in this handbook), multilevel models estimate a variance–covariance matrix and associated correlation coefficients for each of the levels, decomposing the variance and covariance parameters into multiple levels. A comparison between simultaneous equations in econometrics and multilevel models shows similar estimates between the two methods (Goulias and Kim, forthcoming). The information in multilevel models, however, provides deeper insights about unobserved heterogeneity and complex correlations at each level. The bottom half of Table 9.6 contains the estimated variance–covariance matrix and the estimated correlation coefficients at the two levels (person level and household level) for the combination of the four dependent variables in this example.

The estimates for Equations (9.16) to (9.19) are provided in Table 9.7. The models have estimated mean baseline values between 44.74 and 49.07 min for traveling and between 438.90 and 460.80 min for out-of-home activity participation per day. The presence of children ages 0 to 5 negatively affects the amount of traveling and has no significant effect on the total amount of activity participation. Children ages 6 to 17, however, have a significant and positive effect on out-of-home activity participation, with each child contributing an additional 21 to 32 min per day. High-income groups (annual household

TABLE 9.7 Multivariate Multilevel Model Fixed Effect Estimates

	T1			A1			T2			A2		
	Coefficient	SE	t-Statistic	Coefficient	SE	t-Statistic	Coefficient	SE	t-Statistic	Coefficient	SE	t-Statistic
Constant	44.74	6.36	7.04	438.90	25.30	17.35	49.07	6.07	8.08	460.80	26.79	17.20
HHSIZE	4.11	1.14	3.61	-21.05	5.03	-4.18	0.74	1.06	0.70	-27.71	5.15	-5.38
TOT1_5	-3.94	2.72	-1.45				-3.14	2.54	-1.23			
TOT6_17												
NUMVEH	-0.12	1.16	-0.10	20.79	6.69	3.11	2.53	1.08	2.33	32.35	6.84	4.73
MIDINC	-7.73	2.34	-3.31	0.49	3.90	0.13	-0.42	2.18	-0.19	11.00	4.00	2.75
HIGHINC				17.69	9.24	1.91				-4.37	9.52	-0.46
GENDER	6.00	1.95	3.07	63.00	11.39	5.53	8.94	1.90	4.70	28.21	11.72	2.41
AGE2544	8.40	3.55	2.37	28.62	6.92	4.14	10.72	3.39	3.16	37.43	7.49	5.00
AGE4564	11.33	3.16	3.59	-43.65	17.66	-2.47	10.33	3.02	3.42	-71.02	18.79	-3.78
AGE65_				-43.35	17.83	-2.43				-63.85	19.03	-3.36
STUDENT	10.24	4.44	2.30	-103.50	20.48	-5.05	12.98	4.28	3.03	-126.20	21.74	-5.80
SECRET				188.50	18.75	10.05				171.20	19.99	8.56
SALES				-28.16	12.92	-2.18				-16.11	13.74	-1.17
UNEMP	-12.45	2.88	-4.32	-71.45	16.95	-4.22	-15.51	2.77	-5.60	-72.59	18.02	-4.03
WK5				-175.40	13.29	-13.20				-165.80	14.15	-11.72
LICENSE	14.99	4.95	3.03	119.10	11.74	10.14	7.82	4.74	1.65	123.20	12.48	9.87
BUSPASS	24.23	3.18	7.61				17.42	3.05	5.72			
COMWORK	6.56	2.90	2.26	18.31	10.74	1.70	2.49	2.78	0.90	-2.20	11.37	-0.19
COMHOME				54.04	8.87	6.09				55.71	9.45	5.90
WEBWORK	-0.38	3.61	-0.10	-3.54	9.07	-0.39	7.35	3.44	2.13	-19.02	9.56	-1.99
WEBHOME	4.19	3.12	1.34				4.18	2.97	1.41			
WEB	-11.86	4.74	-2.50	-14.97	9.62	-1.56	-16.55	4.54	-3.65	-12.04	10.10	-1.19
CELL	11.56	2.44	4.74				13.75	2.33	5.91			
PAGER	11.35	3.27	3.47	9.07	8.18	1.11	3.83	3.15	1.22	18.19	8.62	2.11
LAPTOP	13.10	4.69	2.79	26.35	11.04	2.39	13.68	4.49	3.05	25.72	11.75	2.19
-2 log-likelihood				167126.9								

Note: Deviance from error component model (Table 9.6) = 2477.5, with 76 degrees of freedom. SE = standard error.

income = $75,000) spend much more time for out-of-home activities than other groups. However, this is accompanied by large differences among days. Men tend to travel from 6 to 9 min per day more than women and spend an average of 29 to 37 min per day more than women on activities. In terms of age, all age groups spend more time traveling than the senior group. The two groups with the highest level of mobility in terms of travel time are the age groups 25 to 44 and 45 to 64. Presumably, older individuals do not travel as long because their total amount of activity participation in out-of-home locations is also the lowest, as shown by the large negative coefficients in the two activity equations.

Employment is consistently a key factor in determining travel and activity behavior. As expected, there are large differences in the daily travel and out-of-home activity expenditure between employed and unemployed persons. Interestingly, however, the unemployed spend on average 12 to 16 min less time for traveling than their employed counterparts. This is an additional indication of the decreasing role of commuting on traffic. The type of occupation does not seem to have an effect on average travel time, but persons involved in specific professions (secretarial and sales) tend to spend less time in out-of-home activities. To the contrary, however, workers who work during all weekdays (five times or more per week) spend approximately 2 h more time for out-of-home activities per day. As expected, drivers tend to travel more that nondrivers, and persons who have a bus pass travel on average between 17 and 24 min more in a day, but only half as much as drivers.

In this system of equations each of the telecommunication and information technologies appears to impact travel and activity participation in different ways. Persons with access to computers at work or school travel an average of approximately 2 to 7 min more (than persons not having access to computers at work), and they spend an average of 54 to 56 min more on activities than the nonusers. In contrast, persons with access to computers at home seem to travel less and spend less time in out-of-home activities. Having access to the Internet (World Wide Web) at work does not seem to influence activity participation, and it has an extremely variable effect on travel. Having access to the Internet at home has a positive effect on travel and a negative effect on out-of-home activity participation. This is particularly interesting because it may be pointing out the tendency to make short trips for persons that have access to information at home. Interestingly, persons that use the Internet both at work and at home tend to travel between 12 and 17 min less than persons having no access at all. When we consider all these indications together, we see that there is a systematic difference in the daily traveling behavior between regular Internet users and nonusers, and these differences are complex, presumably depending on the way information is used. It may also be an indication of the large differences in the activity scheduling of all these groups and the need to examine their choices and lifestyles (see also Chapter 6 in this handbook) in more detail, looking at the different population segments separately (the propensity to own these technologies was analyzed and reported earlier in Viswanathan et al. (2001)).

A much clearer picture, however, is offered by the mobile technology. Users of mobile technologies such as cellular phones and pagers are usually involved in more traveling and longer activity times. The use of laptop computers does not significantly affect the amount of time spent in out-of-home activities, but it does show a consistent positive effect (of approximately 13 min/day) on travel time. Another technology, the PDA, does not seem to have a significant influence on travel and activity behavior (very few persons used this technology in 1997). Wireless telephone users ("cell" in Table 9.7), however, spend more time on the road either pursuing activities or traveling.

The strengths of the multivariate multilevel models here and their key advantage over other simultaneous equations models are the additional insights about unobserved heterogeneity, at both the person and household levels. The variance–covariance matrix in Table 9.8 shows that there is a significant portion of unexplained variance for all four dependent variables, in days and at both the person and household levels, even after we added approximately 25 explanatory variables. Compared to Table 9.6, the amount of unexplained variation in Table 9.8 is much lower, because the mix of explanatory variables captured a portion of the variance in activity and travel expenditures. As expected and also seen elsewhere (Goulias, 2002), there is greater variation in activity and travel expenditure between persons within a household than between households. This is a clear indication that in travel behavior we will capture variation in

TABLE 9.8   Multivariate Multilevel Model Variance–Covariance Matrices (Upper Triangle Correlations) [Standard Error]

	Between Persons				Between Households			
	T1	A1	T2	A2	T1	A1	T2	A2
T1	2795.9 [97.0]	0.095	0.422	0.043	864.7 [93.6]	0.353	0.245	0.085
A1	913.0 [236.9]	33208.9 [1146.0]	−0.011	0.472	930.2 [223.7]	8041.9 [1039.6]	0.161	0.612
T2	1159.4 [72.8]	−104.2 [230.9]	2698.7 [93.0]	0.146	178.8 [66.6]	359.0 [208.2]	616.3 [83.3]	0.247
A2	459.4 [258.1]	17223.9 [980.7]	1522.6 [255.2]	40097.8 [1369.1]	190.5 [229.6]	4186.2 [854.7]	467.3 [219.4]	5814.5 [1125.4]

a more efficient way by formulating person-based models instead of household models. Since we also see that a large portion of the variation is attributable to households, we are by far better off in formulating models that contain and model both sources of variation (person and household). This is the most important advantage of multilevel models.

The cross-equation covariance estimates indicate that there are strong positive correlations for the time allocated to travel between the two days (T1 and T2). The correlation between T1 and T2 is stronger at the person level (0.422) and decreases to 0.245 at the household level. For out-of home activity times across two days (A1 and A2), the person level correlation is 0.472 and increases at the household level to 0.612. This may be an indication that at the household level, where we sum the activities of persons, we tend to look at a stronger consistency in activity time over the different days of the week and this is accomplished by employing different traveling options among the days. In contrast, for travel time individuals have a stronger consistency than their household sums. This is particularly interesting because most of the past research shows travel to be more restricted than activity participation and, for this reason, less variable. It is also important to note that the other correlations (T1 with A2, and A1 with T2) are small and not significantly different from zero as one would expect. The likelihood ratio (LR) statistic, which is $-2(L(c) - L(\beta)) = -2(-84802.2 - (-83563.45)) = 2477.5$, with 76 degrees of freedom, suggests that the choice of explanatory variables is satisfactory and that the model fits the data better than the naïve error components model of Table 9.6.

9.5   Summary

In this chapter two examples of multilevel models are offered to illustrate their versatility and potential uses in transportation planning and, more specifically, travel behavior analysis. The key advantage of these models over their unilevel counterparts is the possibility to estimate correlations within and among units of measurement and to pull the units into different groups while at the same time studying contributions to variation at the group level. This is particularly important when surveys do not (and sometimes cannot) ask explicit questions about allocations of tasks, scheduling, learning, experimentation, and a variety of other processes taking place in parallel with the activity participation by the survey respondents. Multivariate multilevel models have not been used in transport analysis very often. One application was the introduction of multilevel and contextual philosophy in Ma and Goulias (1997) that did not use the multiple variance components, but it sets the stage for the models here. Later, Goulias (1999b) used an error components four-level binary analysis for mode choice constraints to analyze a large database from Germany. More recently, Goulias and Kim (2001) estimated a multinomial multilevel model for activity and travel patterns, and they compared it to the more traditional multinomial logit model. In the telecommunications and travel analysis, Viswanathan et al. (2001) have also employed the models to assess the correlation of telecommunications and travel.

FIGURE 9.3 MLWIN interface in equations.

FIGURE 9.4 MLWIN interface at the end of iterations.

9.5.1 Further Reading

Textbooks providing introductions and more in-depth presentation for multilevel models abound. There are, however, a few books that are easier to follow and have better coverage of the key principles underlying these methods and the key elements of data analysis and interpretation. The most interesting of these books is by Hox (1995), *Applied Multilevel Analysis*, which contains a very good discussion about multilevel approaches in social sciences and some of the early rationale for considering context in our analysis. Another textbook type of presentation is the book by Bryk and Raudenbush (1992), *Hierarchical Linear Models*. The emphasis in this book is on formulation and estimation, and there is a very good discussion of the basic principles underlying the many models in this area.

The most useful website, with a plethora of information and easy-to-use software, can be found at http://multilevel.ioe.ac.uk/index.html (accessed in March 2002). The site contains extensive information about the software used in this chapter, MLWIN. One of the key advantages of the software is a graphics dynamic interface that allows one to specify the models using equations. Figure 9.3 provides an example of this interface. As estimation iterations are completed, the coefficient estimates are updated in this interface and the output looks like that in Figure 9.4. The software also contains a variety of diagnostic tests, switches between estimation methods, and other data manipulation options that are needed when building models.

Acknowledgments

Krishnan Viswanathan provided expert support in data management during earlier stages of the estimation here, and Tae-Gyu Kim provided help with estimation of the models here using MLWIN and other related software. Both are greatly acknowledged for help with portions of this chapter. Credit for errors or omissions remains with the author.

References

Balestra, P. and Nerlove, M., Pooling cross section and time series data in the estimation of a dynamic model: the demand for natural gas, *Econometrica*, 34(3), 585–612, 1996.

Baltagi, B.H., *Econometric Analysis of Panel Data*, Wiley, Chichester, U.K., 1995.

Bryk, A.S. and Raudenbush, S.W., *Hierarchical Linear Models*, Sage, Newberry Park, CA, 1992.

Chandrasekharan, B., Cross Sectional, Longitudinal, and Spatial Analysis of Joint and Solo Travel Patterns, M.S. thesis, Department of Civil and Environmental Engineering, College of Engineering, Pennsylvania State University, University Park, 1999.

Chandrasekharan, B. and Goulias, K.G., Exploratory longitudinal analysis of solo and joint trip making in the Puget Sound transportation panel, *Transp. Res. Rec.*, 1676, 77–85, 1999.

Gliebe, J.P. and Koppelman, F.S., A model of joint activity participation between household members, *Transportation*, 29, 49–72, 2002.

Goldstein, H., *Multilevel Statistical Models*, Edward Arnold, New York, 1995.

Golob, T.F., Kitamura, R., and Long, L., Eds., *Panels for Transportation Planning: Methods and Applications*, Kluwer, Boston, 1997.

Golob, T.F. and McNally, M.G, A model of activity participation and travel interactions between household heads, *Transp. Res. B*, 31, 177–194, 1997.

Goulias, K.G., Longitudinal analysis of activity and travel pattern dynamics using generalized mixed Markov latent class models, *Transp. Res. B*, 33, 535–557, 1999a.

Goulias, K.G., Multilevel random effects analysis of modal use constraints and perceptions on public transportation using data from Germany, in *Urban Transport V: Urban Transport and the Environment for the 21st Century*, Sacharov, L.J., Ed., WIT Press, Southampton, U.K., 1999b, pp. 181–190.

Goulias, K.G., Multilevel analysis of daily time use and time allocation to activity types accounting for complex covariance structures using correlated random effects, *Transportation*, 29, 31–48, 2002.

Goulias, K.G. and Kim, T., Multilevel analysis of activity and travel patterns accounting for person- and household-specific observed and unobserved effects simultaneously, *Transp. Res. Rec.*, 1752, 23–31, 2001.

Goulias, K.G. and Ma, J., Analysis of Longitudinal Data from the Puget Sound Transportation Panel: Task B: Integration of PSTP Databases and PSTP Codebook, Final Report 9619, Pennsylvania Transportation Institute, Pennsylvania State University, University Park, 1996.

Greene, W.H., *Econometric Analysis*, 3rd ed., Prentice Hall, Englewood Cliffs, NJ, 1997.

Hox, J.J., *Applied Multilevel Analysis*, TT Publications, Amsterdam, 1995.

Hsiao, C., *Analysis of Panel Data*, Cambridge University Press, Cambridge, U.K., 1986.

Kennedy, P., *A Guide to Econometrics*, MIT Press, Cambridge, MA, 1995.

Kreft, I. and deLeeuw, J., *Introducing Multilevel Modeling*, Sage Publications, London, 1998.

Liao, C.-Y., An Exploratory Analysis of Random Coefficient Regression Models for Transportation Demand, M.E. thesis, The Pennsylvania State University, University Park, 1994.

Longford, N.T., *Random Coefficient Models*, Clarendon Press, Oxford, 1993.

Ma, J., An Activity-Based and Micro-Simulated Travel Forecasting System: A Pragmatic Synthetic Scheduling Approach, unpublished Ph.D. dissertation, Department of Civil and Environmental Engineering, Pennsylvania State University, University Park, 1997.

Ma, J. and Goulias, K.G., An analysis of activity and travel patterns in the Puget Sound Transportation Panel, in *Activity-Based Approaches to Travel Analysis*, Ettema, D.F. and Timmermans, H.J.P., Eds., Pergamon, Amsterdam, 1997, pp. 189–207.

Mokhtarian, P.L. and Salomon, I., Emerging travel patterns: Do telecommunications make a difference? in *In Perpetual Motion: Travel Behavior Research Opportunities and Application Challenges*, Mahmassani, H., Ed., Elsevier, Amsterdam, 2002.

Murakami, E. and Ulberg, C., The Puget Sound transportation panel, in *Panels for Transportation Planning Methods and Applications*, Golob, T.F., Kitamura, R., and Long, L., Eds., Kluwer, Boston, 1997, pp. 159–192.

Murakami, E. and Watterson, W.T., Developing a household travel panel survey for the Puget Sound region, *Transp. Res. Rec.*, 1285, 40–48, 1990.

Rasbach, J. et al., *A User's Guide to M1wiN*, University of London, London, 2001.

Salomon, I., Can telecommunications help solve transportation problems? in *Handbook of Transport Modelling*, Hensher, D.A. and Button, K.J., Eds., Pergamon, Amsterdam, 2000, chap. 27.

Schafer, J.L., *Analysis of Incomplete Multivariate Data*, Chapman & Hall/CRC, Boca Raton, FL, 1999.

Searle, S.R., Casella, G., and McCulloch, C.E., *Variance Components*, Wiley, New York, 1992.

Swamy, P.A.V.B., Linear models with random coefficients, in Zarembka, P., Ed., *Frontiers in Econometrics*, Academic Press, New York, 1974, pp. 143–168.

Townsend, T.A., The Effects of Household Characteristics on the Multi-day Time Allocations and Travel/Activity Patterns of Households and Their Members, Ph.D. dissertation, Northwestern University, Evanston, IL, 1987 (available via UMI 8723720).

Train, K., *Qualitative Choice Analysis: Theory, Econometrics, and an Application to Automobile Demand*, MIT Press, Cambridge, MA, 1986.

van der Leeden, R., Multilevel analysis of longitudinal data, in Bijleveld, C.J.H. et al., Eds., *Longitudinal Data Analysis: Designs, Models, and Methods*, Sage Publications, London, 1998, pp. 269–317.

van Wissen, L., A Model of Household Interactions in Activity Patterns, paper presented at the International Conference on Dynamic Travel Behavior Analysis, Kyoto University, Japan, July 16–17, 1989.

Viswanathan, K., Goulias, K.G., and Jovanis, P.P., Use of traveler information in the Puget Sound region: preliminary multivariate analysis, *Transp. Res. Rec.*, 1719, 94–102, 2000.

Viswanathan, K., Goulias, K.G., and Kim, T., On the relationship between travel behavior and information and communications technology (ICT): what do travel diaries show? in Sacharov, L.J. and Brebbia, C.A., Eds., *Urban Transport VII, Urban Transport and the Environment for the 21st Century*, WIT Press, Southampton, U.K., 2001, pp. 213–222.

10

Random Utility-Based Discrete Choice Models for Travel Demand Analysis

CONTENTS

Chandra R. Bhat
University of Texas

10.1 Introduction

This chapter is an overview of the motivation for, and structure of, advanced discrete choice models derived from random utility maximization. The discussion is intended to familiarize readers with structural alternatives to the multinomial logit. Before proceeding to review advanced discrete choice models, we first summarize the assumptions of the multinomial logit (MNL) formulation. This is useful since all other random utility maximizing discrete choice models focus on relaxing one or more of these assumptions.

There are three basic assumptions that underlie the MNL formulation. The first assumption is that the random components of the utilities of the different alternatives are independent and identically distributed (IID) with a type I extreme value (or Gumbel) distribution. The assumption of *independence* implies that there are no common unobserved factors affecting the utilities of the various alternatives. This assumption is violated, for example, if a decision maker assigns a higher utility to all transit modes (bus, train, etc.) because of the opportunity to socialize or if the decision maker assigns a lower utility to all the transit modes because of the lack of privacy. In such situations, the same underlying unobserved factor (opportunity to socialize or lack of privacy) impacts the utilities of all transit modes. As indicated by Koppelman and Sethi (2000), presence of such common underlying factors across modal utilities has implications for competitive structure. The assumption of *identically distributed* (across alternatives) random utility terms implies that the extent of variation in unobserved factors affecting modal utility is the same across all modes. In general, there is no theoretical reason to believe that this will be the case. For example, if comfort is an unobserved variable whose values vary considerably for the train mode (based on, say, the degree of crowding on different train routes) but little for the automobile mode, then the random components for the automobile and train modes will have different variances. Unequal error variances have significant implications for competitive structure.

The second assumption of the MNL model is that it maintains homogeneity in responsiveness to attributes of alternatives across individuals (i.e., an assumption of response homogeneity). More specifically, the MNL model does not allow sensitivity (or taste) variations to an attribute (for example, travel cost or travel time in a mode choice model) due to unobserved individual characteristics. However, unobserved individual characteristics can and generally will affect responsiveness. For example, some individuals by their intrinsic nature may be extremely time-conscious, while other individuals may be laid back and less time-conscious. Ignoring the effect of unobserved individual attributes can lead to biased and inconsistent parameter and choice probability estimates (see Chamberlain, 1980).

The third assumption of the MNL model is that the error variance–covariance structure of the alternatives is identical across individuals (i.e., an assumption of error variance–covariance homogeneity). The assumption of identical variance across individuals can be violated if, for example, the transit system offers different levels of comfort (an unobserved variable) on different routes (that is, some routes may be served by transit vehicles with more comfortable seating and temperature control than others). Then, the transit error variance across individuals along the two routes may differ. The assumption of identical error covariance of alternatives across individuals may not be appropriate if the extent of substitutability among alternatives differs across individuals. To summarize, error variance–covariance homogeneity implies the same competitive structure among alternatives for all individuals, an assumption that is generally difficult to justify.

The three assumptions discussed above together lead to the simple and elegant closed-form mathematical structure of the MNL. However, these assumptions also leave the MNL model saddled with the independence of irrelevant alternatives (IIA) property at the individual level (Luce and Suppes (1965); see also Ben-Akiva and Lerman (1985) for a detailed discussion of this property). Thus, relaxing the three assumptions may be important in many choice contexts.

In this chapter, we focus on three classes of discrete choice models that relax one or more of the assumptions discussed above *and* nest the multinomial logit model. The first class of models, which we will label as heteroskedastic models, relax the identically distributed (across alternatives) error term assumption, but do not relax the independence assumption (part of the first assumption above) or the assumption of response homogeneity (second assumption above). The second class of models, which we will refer to as generalized extreme value (GEV) models, relax the independently distributed (across alternatives) assumptions, but do not relax the identically distributed assumption (part of the first assumption above) or the assumptions of response homogeneity (second assumption). The third class of models, which we will label as flexible structure models, are very general; models in this class are flexible enough to relax the independence and identically distributed (across alternatives) error structure of the MNL as well as the assumption of response homogeneity. We do not focus on the third assumption implicit in the MNL model since it can be relaxed within the context of any given discrete choice model by parameterizing appropriate error structure variances and covariances as a function of individual attributes (see Bhat (1997) for a detailed discussion of these procedures).

The rest of this paper is structured in three sections: Section 10.2 discusses heteroskedastic models, Section 10.3 focuses on GEV models, and Section 10.4 presents flexible structure models. The final section concludes the paper. Within each of Sections 10.2 to 10.4, the material is organized as follows. First, possible model formulations within that class are presented and a preferred model formulation is selected for further discussion. Next, the structure of the preferred model structure is provided, followed by the estimation of the structure, a brief discussion of transport applications of the structure, and a detailed presentation of results from a particular application of the structure in the travel behavior field.

10.2 Heteroskedastic Models

10.2.1 Model Formulations

Three models have been proposed that allow nonidentical random components. The first is the negative exponential model of Daganzo (1979), the second is the oddball alternative model of Recker (1995), and the third is the heteroskedastic extreme value (HEV) model of Bhat (1995).

Daganzo (1979) used independent negative exponential distributions with different variances for the random error components to develop a closed-form discrete choice model that does not have the IIA property. His model has not seen much application since it requires that the perceived utility of any alternative not to exceed an upper bound (this arises because the negative exponential distribution does not have a full range). Daganzo's model does not nest the multinomial logit model.

Recker (1995) proposed the oddball alternative model, which permits the random utility variance of one "oddball" alternative to be larger than the random utility variances of other alternatives. This situation might occur because of attributes that define the utility of the oddball alternative, but are undefined for other alternatives. Then random variation in the attributes that are defined only for the oddball alternative will generate increased variance in the overall random component of the oddball alternative relative to others. For example, operating schedule and fare structure define the utility of the transit alternative, but are not defined for other modal alternatives in a mode choice model. Consequently, measurement error in schedule and fare structure will contribute to the increased variance of transit relative to other alternatives. Recker's model has a closed-form structure for the choice probabilities. However, it is restrictive in requiring that all alternatives except one have identical variance.

Bhat (1995) formulated the heteroskedastic extreme value (HEV) model, which assumes that the alternative error terms are distributed with a type I extreme value distribution. The variance of the alternative error terms is allowed to be different across all alternatives (with the normalization that the error terms of one of the alternatives has a scale parameter of 1 for identification). Consequently, the HEV model can be viewed as a generalization of Recker's oddball alternative model. The HEV model does not have a closed-form solution for the choice probabilities, but involves only a one-dimensional integration regardless of the number of alternatives in the choice set. It also nests the multinomial logit model and is flexible enough to allow differential cross-elasticities among all pairs of alternatives. In the rest of our discussion of heteroskedastic models, we will focus on the HEV model.

10.2.2 HEV Model Structure

The random utility of alternative i, U_i, for an individual in random utility models takes the form (we suppress the index for individuals in the following presentation)

$$U_i = V_i + \varepsilon_i \tag{10.1}$$

where V_i is the systematic component of the utility of alternative i (which is a function of observed attributes of alternative i and observed characteristics of the individual) and ε_i is the random component of the utility function. Let C be the set of alternatives available to the individual. Let the random components in the utilities of the different alternatives have a type I extreme value distribution with a location parameter equal to zero and a scale parameter equal to θ_i for the ith alternative. The random components are assumed to be independent, but nonidentically distributed. Thus, the probability density function and the cumulative distribution function of the random error term for the ith alternative are

$$f(\varepsilon_i) = \frac{1}{\theta_i} e^{-\frac{\varepsilon_i}{\theta_i}} e^{-e^{-\frac{\varepsilon_i}{\theta_i}}} \quad \text{and} \quad F_i(z) \int_{\varepsilon_i=-\infty}^{\varepsilon_i=z} f(\varepsilon_i) d\varepsilon_i = e^{-e^{-\frac{z}{\theta_i}}} \tag{10.2}$$

The random utility formulation of Equation (10.1), combined with the assumed probability distribution for the random components in Equation (10.2) and the assumed independence among the random components of the different alternatives, enables us to develop the probability that an individual will choose alternative i (P_i) from set C of available alternatives:

$$P_i = \text{Prob}(U_i > U_j), \text{ for all } j \neq i, j \in C$$

$$= \text{Prob}(\varepsilon_j \leq V_i - V_j + \varepsilon_i), \text{ for all } j \neq i, j \in C \tag{10.3}$$

$$= \int_{\varepsilon_i=-\infty}^{\varepsilon_i=+\infty} \prod_{j\in C, \, j\neq i} \Lambda\left[\frac{V_i - V_j + \varepsilon_i}{\theta_j}\right] \frac{1}{\theta_i} \lambda\left(\frac{\varepsilon_i}{\theta_i}\right) d\varepsilon_i$$

where $\lambda(.)$ and $\Lambda(.)$ are the probability density function and cumulative distribution function, respectively, of the standard type I extreme value distribution and are given by (see Johnson and Kotz, 1970):

$$\lambda(t) = e^{-t} e^{-e^{-t}} \quad \text{and} \quad \Lambda(t) = e^{-e^{-t}} \tag{10.4}$$

Substituting $w = \varepsilon_i/\theta_i$ in Equation (10.3), the probability of choosing alternative i can be rewritten as follows:

$$P_i = \int_{w=-\infty}^{w=+\infty} \prod_{j\in C, \, j\neq i} \Lambda\left[\frac{V_i - V_j + \theta_i w}{\theta_j}\right] \lambda(w) dw \tag{10.5}$$

If the scale parameters of the random components of all alternatives are equal, then the probability expression in Equation (10.5) collapses to that of the multinomial logit (the reader will note that the variance of the random error term ε_i of alternative i is equal to $U_i = V_i + \varepsilon_i$, where θ_i is the scale parameter).

The HEV model discussed above avoids the pitfalls of the IIA property of the multinomial logit model by allowing different scale parameters across alternatives. Intuitively, we can explain this by realizing that the error term represents unobserved characteristics of an alternative; that is, it represents uncertainty associated with the expected utility (or the systematic part of utility) of an alternative. The scale parameter of the error term, therefore, represents the level of uncertainty. It sets the relative weights of the systematic and uncertain components in estimating the choice probability. When the systematic utility of some alternative l changes, this affects the systematic utility differential between another alternative i and the alternative l. However, this change in the systematic utility differential is tempered by the unobserved random component of alternative i. The larger the scale parameter (or equivalently, the variance) of the random error component for alternative i, the more tempered the effect of the change in the systematic utility differential (see the numerator of the cumulative distribution function term in Equation (10.5)) and the smaller the elasticity effect on the probability of choosing alternative i. In particular, two alternatives will have the same elasticity effect due to a change in the systematic utility of another alternative only if they have the same scale parameter on the random components. This property is a logical and intuitive extension of the case of the multinomial logit, in which all scale parameters are constrained to be equal and, therefore, all cross-elasticities are equal.

Assuming a linear-in-parameters functional form for the systematic component of utility for all alternatives, the relative magnitudes of the cross-elasticities of the choice probabilities of any two alternatives i and j with respect to a change in the kth level-of-service variable of another alternative l (say, x_{kl}) are characterized by the scale parameter of the random components of alternatives i and j:

$$\eta_{x_{kl}}^{P_i} > \eta_{x_{kl}}^{P_j} \text{ if } \theta_i < \theta_j; \; \eta_{x_{kl}}^{P_i} = \eta_{x_{kl}}^{P_j} \text{ if } \theta_i = \theta_j; \; \eta_{x_{kl}}^{P_i} < \eta_{x_{kl}}^{P_j} \text{ if } \theta_i > \theta_j \qquad (10.6)$$

10.2.3 HEV Model Estimation

The HEV model can be estimated using the maximum likelihood technique. Assume a linear-in-parameters specification for the systematic utility of each alternative given by $V_{qi} = \beta X_{qi}$ for the qth individual and i^{th} alternative (we introduce the index for individuals in the following presentation since the purpose of the estimation is to obtain the model parameters by maximizing the likelihood function over all individuals in the sample). The parameters to be estimated are the parameter vector β and the scale parameters of the random component of each of the alternatives (one of the scale parameters is normalized to 1 for identifiability). The log-likelihood function to be maximized can be written as

$$\mathcal{L} = \sum_{q=1}^{q=Q} \sum_{i \in C_q} y_{qi} \log \left\{ \int_{w=-\infty}^{w=+\infty} \prod_{j \in C_q, j \neq i} \Lambda \left[\frac{V_{qi} - V_{qj} + \theta_i w}{\theta_j} \right] \lambda(w) dw \right\} \qquad (10.7)$$

where C_q is the choice set of alternatives available to the qth individual and y_{qi} is defined as follows:

$$y_{qi} = \begin{cases} 1 \text{ if the qth individual chooses alternative i} \\ 0 \text{ otherwise,} \quad (q = 1, 2, \ldots, Q, \quad i = 1, 2, \ldots I) \end{cases} \qquad (10.8)$$

The log-likelihood function in Equation (10.7) has no closed-form expression, but can be estimated in a straightforward manner using Gaussian quadrature. To do so, define a variable $u = e^{-w}$. Then, $\lambda(w)dw = -e^{-u}du$ and $w = -\ln u$. Also define a function G_{qi} as:

$$G_{qi}(u) = \prod_{j \in C_q, j \neq i} \Lambda \left[\frac{V_{qi} - V_{qj} - \theta_i \ln u}{\theta_j} \right] \qquad (10.9)$$

Then we can rewrite Equation (10.7) as

$$\mathcal{L} = \sum_q \sum_{i \in C_q} y_{qi} \log \left\{ \int_{u=0}^{u=\infty} G_{qi}(u) e^{-u} du \right\} \qquad (10.10)$$

The expression within braces in the above equation can be estimated using the Laguerre Gaussian quadrature formula, which replaces the integral by a summation of terms over a certain number (say K) of support points, each term comprising the evaluation of the function $G_{qi}(.)$ at the support point k multiplied by a probability mass or weight associated with the support point (the support points are the roots of the Laguerre polynomial of order K, and the weights are computed based on a set of theorems provided by Press et al. (1992, p. 124).

10.2.4 Transport Applications

The HEV model has been applied to estimate discrete choice models based on revealed choice (RC) data as well as stated choice (SC) data.

The multinomial logit, alternative nested logit structures, and the heteroskedastic model are estimated using RC data in Bhat (1995) to examine the impact of improved rail service on intercity business travel in the Toronto–Montreal corridor. The nested logit structures are either inconsistent with utility maximization principles or not significantly better than the multinomial logit model.

The heteroskedastic extreme value model, however, is found to be superior to the multinomial logit model. The heteroskedastic model predicts smaller increases in rail shares and smaller decreases in nonrail shares than the multinomial logit in response to rail service improvements. It also suggests a larger percentage decrease in air share and a smaller percentage decrease in auto share than the multinomial logit.

Hensher et al. (1999) applied the HEV model to estimate an intercity travel mode choice model from a combination of RC and SC choice data (they also discuss a latent-class HEV model in their paper that allows taste heterogeneity in a HEV model). The objective of this study was to identify the market for a proposed high-speed rail service in the Sydney–Canberra corridor. The revealed choice set includes four travel modes: air, car, bus or coach, and conventional rail. The stated choice set includes the four RC alternatives and the proposed high-speed rail alternative. Hensher et al. (1999) estimate a pooled RC–SC model that accommodates scale differences between RC and SC data as well as scale differences among alternatives. The scale for each mode turns out to be about the same across the RC and SC data sets, possibly reflecting a well-designed stated choice task that captures variability levels comparable to actual revealed choices. Very interestingly, however, the scales for all noncar modes are about equal or substantially less than that of the car mode. This indicates much more uncertainty in the evaluation of noncar modes than of the car mode.

Hensher (1997) has applied the HEV model in a related stated choice study to evaluate the choice of fare type for intercity travel in the Sydney–Canberra corridor conditional on the current mode used by each traveler. The current modes in the analysis include conventional train, charter coach, scheduled coach, air, and car. The projected patronage on a proposed high-speed rail mode is determined based on the current travel profile and alternative fare regimes.

Hensher (1998), in another effort, has applied the HEV model to the valuation of attributes (such as the value of travel time savings) from discrete choice models. Attribute valuation is generally based on the ratio of two or more attributes within utility expressions. However, using a common scale across alternatives can distort the relative valuation of attributes across alternatives. In Hensher's empirical analysis, the mean value of travel time savings for public transport modes is much lower when a HEV model is used than a MNL model, because of confounding of scale effects with attribute parameter magnitudes. In a related and more recent study, Hensher (1999) applied the HEV model (along with other advanced models of discrete choice, such as the multinomial probit and mixed logit models, which we discuss later) to examine valuation of attributes for urban car drivers.

Munizaga et al. (2000) evaluated the performance of several different model structures (including the HEV and the multinomial logit model) in their ability to replicate heteroskedastic patterns across alternatives. They generated data with known heteroskedastic patterns for the analysis. Their results show that the multinomial logit model does not perform well and does not provide accurate policy predictions in the presence of heteroskedasticity across alternatives, while the HEV model accurately recovers the target values of the underlying model parameters.

10.2.5 Detailed Results from an Example Application

Bhat (1995) estimated the HEV model using data from a 1989 Rail Passenger Review conducted by VIA Rail (the Canadian national rail carrier). The purpose of the review was to develop travel demand models to forecast future intercity travel and estimate shifts in mode split in response to a variety of potential rail service improvements (including high-speed rail) in the Toronto–Montreal corridor (see KPMG Peat Marwick and Koppelman (1990) for a detailed description of this data). Travel surveys were conducted in the corridor to collect data on intercity travel by four modes (car, air, train, and bus). This data included sociodemographic and general trip-making characteristics of the traveler, and detailed information on the current trip (purpose, party size, origin and destination cities, etc.). The set of modes available to travelers for their intercity travel was determined based on the geographic location of the trip. Level-of-service data were generated for each available mode and each trip based on the origin–destination information of the trip.

Bhat focused on intercity mode choice for paid business travel in the corridor. The study is confined to a mode choice examination among air, train, and car due to the very small number of individuals choosing the bus mode in the sample, and also because of the poor quality of the bus data (see Forinash and Koppelman, 1993).

Five different models were estimated in the study: a multinomial logit model, three possible nested logit models, and the heteroskedastic extreme value model. The three nested logit models were: (1) car and train (slow modes) grouped together in a nest that competes against air, (2) train and air (common carriers) grouped together in a nest that competes against car, and (3) air and car grouped together in a nest that competes against train. Of these three structures, the first two seem intuitively plausible, while the third does not.

The final estimation results are shown in Table 10.1 for the multinomial logit model, the nested logit model with car and train grouped as ground modes, and the heteroskedastic model. The estimation results for the other two nested logit models are not shown because the log-sum parameter exceeded 1 in these specifications. This is not globally consistent with stochastic utility maximization (McFadden, 1978; Daly and Zachary, 1978).

A comparison of the nested logit model with the multinomial logit model using the likelihood ratio test indicates that the nested logit model fails to reject the multinomial logit model (equivalently, notice

TABLE 10.1 Intercity Mode Choice Estimation Results

Variable	Multinomial Logit		Nested Logit with Car and Train Grouped		Heteroskedastic Extreme Value Model	
	Parameter	t-Statistic	Parameter	t-Statistic	Parameter	t-Statistic
Mode Constants (Car is Base)						
Train	−0.5396	−1.55	−0.6703	−2.14	−0.1763	−0.42
Air	−0.6495	−1.23	−0.5135	−1.31	−0.4883	−0.88
Large City Indicator (Car is Base)						
Train	1.4825	7.98	1.3250	6.13	1.9066	6.45
Air	0.9349	5.33	0.8874	5.00	0.7877	4.96
Household Income (Car is Base)						
Train	−0.0108	−3.33	−0.0101	−3.30	−0.0167	−3.57
Air	0.0261	7.02	0.0262	7.42	0.0223	6.02
Frequency of service	0.0846	17.18	0.0846	17.67	0.0741	10.56
Travel cost	−0.0429	−10.51	−0.0414	−11.03	−0.0318	−5.93
Travel Time						
In-vehicle	−0.0105	−13.57	−0.0102	−12.64	−0.0110	−9.78
Out-of-vehicle	−0.0359	−12.18	−0.0353	−13.86	−0.0362	−8.64
Log-sum parameter[a]	1.0000	—	0.9032	1.14	1.0000	—
Scale Parameters (Car Parameter = 1)[b]						
Train	1.0000	—	1.0000	—	1.3689	2.60
Air	1.0000	—	1.0000	—	0.6958	2.41
Log-likelihood at convergence[c]	−1828.89		-1828.35		−1820.60	
Adjusted log-likelihood ratio index	0.3525		0.3524		0.3548	

[a] The logsum parameter is implicity constrained to one in the multinomial logit and heteroskedastic model specifications. The t-statistic for the log-sum parameter in the nested logit is with respect to a value of one.

[b] The scale parameters are implicity constrained to one in the multinomial logit and nested logit models and explicitly constrained to one in the constrained "heteroskedastic" model. The t-statistics for the scale parameters in the heteroskedastic model are with respect to a value of one.

[c] The log likelihood value at zero is −3042.06 and the log likelihood value with only alternative specific constants and an IID error covariance matrix is −2837.12.

Source: From Bhat, C.R., *Transp. Res. B*, 29, 471, 1995. With permission.

the statistically insignificance of the log-sum parameter relative to a value of 1). However, a likelihood ratio test between the heteroskedastic extreme value model and the multinomial logit strongly rejects the multinomial logit in favor of the heteroskedastic specification (the test statistic is 16.56, which is significant at any reasonable level of significance when compared to a chi-squared statistic with two degrees of freedom). Table 10.1 also evaluates the models in terms of the adjusted likelihood ratio index ($\bar{\rho}^2$).[1] These values again indicate that the heteroskedastic model offers the best fit in the current empirical analysis (note that the nested logit and heteroskedastic models can be directly compared to each other using the nonnested adjusted likelihood ratio index test proposed by Ben-Akiva and Lerman (1985); in the current case, the heteroskedastic model specification rejected the nested specification using this nonnested hypothesis test).

In the subsequent discussion on interpretation of model parameters, the focus will be on the multinomial logit and heteroskedastic extreme value models. The signs of all the parameters in the two models are consistent with a priori expectations (the car mode is used as the base for the alternative specific constants and alternative specific variables). The parameter estimates from the multinomial logit and heteroskedastic models are also close to each other. However, there are some significant differences. The heteroskedastic model suggests a higher positive probability of choice of the train mode for trips that originate, end, or originate and end at a large city. It also indicates a lower sensitivity of travelers to frequency of service and travel cost; i.e., the heteroskedastic model suggests that travelers place substantially more importance on travel time than on travel cost or frequency of service. Thus, according to the heteroskedastic model, reductions in travel time (even with a concomitant increase in fares) may be a very effective way of increasing the mode share of a travel alternative. The implied cost of in-vehicle travel time is $14.70 per hour in the multinomial logit and $20.80 per hour in the heteroskedastic model. The corresponding figures for out-of-vehicle travel time are $50.20 and $68.30 per hour, respectively.

The heteroskedastic model indicates that the scale parameter of the random error component associated with the train (air) utility is significantly greater (smaller) than that associated with the car utility (the scale parameter of the random component of car utility is normalized to 1; the t-statistics for the train and scale parameters are computed with respect to a value of 1). Therefore, the heteroskedastic model suggests unequal cross-elasticities among the modes.

Table 10.2 shows the elasticity matrix with respect to changes in rail level-of-service characteristics (computed for a representative intercity business traveler in the corridor) for the multinomial logit and

TABLE 10.2 Elasticity Matrix in Response to Change in Rail Service for Multinomial Logit and Heteroskedastic Models

Rail Level-of-Service Attribute	Multinomial Logit Model			Heteroskedastic Extreme Value Model		
	Train	Air	Car	Train	Air	Car
Frequency	0.303	−0.068	−0.068	0.205	−0.053	−0.040
Cost	−1.951	0.436	0.436	−1.121	0.290	0.220
In-vehicle travel time	−1.915	0.428	0.428	−1.562	0.404	0.307
Out-of-vehicle travel time	−2.501	0.559	0.559	−1.952	0.504	0.384

Note: The elasticities are computed for a representative intercity business traveler in the corridor.
Source: From Bhat, C.R., *Transp. Res. B*, 29, 471, 1995. With permission.

---

[1] The adjusted likelihood ratio index is defined as follows;

$$\bar{\rho}^2 = 1 - \frac{L(M) - K}{L(C)}$$

where L(M) is the model log-likelihood value, L(C) is the log-likelihood value with only alternative specific constants and an IID error covariance matrix, and K is the number of parameters (besides the alternative specific constants) in the model.

heteroskedastic extreme value models.[2] Two important observations can be made from this table. First, the multinomial logit model predicts higher percentage decreases in air and car choice probabilities and a higher percentage increase in rail choice probability in response to an improvement in train level of service than the heteroskedastic model. Second, the multinomial logit elasticity matrix exhibits the IIA property because the elements in the second and third columns are identical in each row. The hetero-skedastic model does not exhibit the IIA property; a 1% change in the level of service of the rail mode results in a larger percentage change in the probability of choosing air than auto. This is a reflection of the lower variance of the random component of the utility of air relative to the random component of the utility of car. We discuss the policy implications of these observations in the next section.

The observations made above have important policy implications at the aggregate level (these policy implications are specific to the Canadian context; caution must be exercised in generalizing the behavioral implications based on this single application). First, the results indicate that the increase in rail mode share in response to improvements in the rail mode is likely to be substantially lower than what might be expected based on the multinomial logit formulation. Thus, the multinomial logit model overestimates the potential ridership on a new (or improved) rail service and, therefore, overestimates revenue projections. Second, the results indicate that the potential of an improved rail service to alleviate auto traffic congestion on intercity highways and air traffic congestion at airports is likely to be less than that suggested by the multinomial logit model. This finding has a direct bearing on the evaluation of alternative strategies to alleviate intercity travel congestion. Third, the differential cross-elasticities of air and auto modes in the heteroskedastic logit model suggest that an improvement in the current rail service will alleviate air traffic congestion at airports more than auto congestion on roadways. Thus, the potential benefit from improving the rail service will depend on the situational context, that is, whether the thrust of the congestion alleviation effort is to reduce roadway congestion or air traffic congestion. These findings point to the deficiency of the multinomial logit model as a tool to making informed policy decisions to alleviate intercity travel congestion in the specific context of Bhat's application.

10.3 The GEV Class of Models

The GEV class of models relaxes the IID assumption of the MNL by allowing the random components of alternatives to be correlated, while maintaining the assumption that they are identically distributed (i.e., identical, nonindependent random components). This class of models assumes a type I extreme value (or Gumbel) distribution for the error terms. All the models belonging to this class nest the multinomial logit and result in closed-form expressions for the choice probabilities. In fact, the MNL is also a member of the GEV class, though we will reserve the use of the term "GEV class" to those models that constitute generalizations of the MNL.

The general structure of the GEV class of models was derived by McFadden (1978) from the random utility maximization hypothesis, and generalized by Ben-Akiva and Francois (1983). Several specific GEV structures have been formulated and applied within the GEV class, including the nested logit (NL) model (Williams, 1977; McFadden, 1978; Daly and Zachary, 1978); the paired combinatorial logit (PCL) model (Chu, 1989; Koppelman and Wen, 2000); the cross-nested logit (CNL) model (Vovsha, 1997); the ordered GEV (OGEV) model (Small, 1987); the multinomial logit–ordered GEV (MNL-OGEV) model (Bhat, 1998c); the product differentiation logit (PDL) model (Bresnahan et al., 1997); and the generalized nested logit model (Wen and Koppelman, 2001).

The nested logit model permits covariance in random components among subsets (or nests) of alternatives (each alternative can be assigned to one and only one nest). Alternatives in a nest exhibit an

---

[2]Since the objective of the original study for which the data was collected was to examine the effect of alternative improvements in rail level of service characteristics, we focus on the elasticity matrix corresponding to changes in rail level of service here.

identical degree of increased sensitivity relative to alternatives not in the nest (Williams, 1977; McFadden, 1978; Daly and Zachary, 1978). Each nest in the NL structure has associated with it a dissimilarity (or log-sum) parameter that determines the correlation in unobserved components among alternatives in that nest (see Daganzo and Kusnic, 1993). The range of this dissimilarity parameter should be between 0 and 1 for all nests if the NL model is to remain globally consistent with the random utility maximizing principle. A problem with the NL model is that it requires a priori specification of the nesting structure. This requirement has at least two drawbacks. First, the number of different structures to estimate in a search for the best structure increases rapidly as the number of alternatives increases. Second, the actual competition structure among alternatives may be a continuum that cannot be accurately represented by partitioning the alternatives into mutually exclusive subsets.

The paired combinatorial logit model initially proposed by Chu (1989) and recently examined in detail by Koppelman and Wen (2000) generalizes, in concept, the nested logit model by allowing differential correlation between each pair of alternatives (the nested logit model, however, is not nested within the PCL structure). Each pair of alternatives in the PCL model has associated with it a dissimilarity parameter (subject to certain identification considerations that Koppelman and Wen are currently studying) that is inversely related to the correlation between the pair of alternatives. All dissimilarity parameters have to lie in the range of 0 to 1 for global consistency with random utility maximization.

Another generalization of the nested logit model is the cross-nested logit model of Vovsha (1997). In this model, an alternative need not be exclusively assigned to one nest as in the nested logit structure. Instead, an alternative can appear in different nests with different probabilities based on what Vovsha refers to as allocation parameters. A single dissimilarity parameter is estimated across all nests in the CNL structure. Unlike in the PCL model, the nested logit model can be obtained as a special case of the CNL model when each alternative is unambiguously allocated to one particular nest. Vovsha proposes a heuristic procedure for estimation of the CNL model. This procedure appears to be rather cumbersome and its heuristic nature makes it difficult to establish the statistical properties of the resulting estimates.

The ordered GEV model was developed by Small (1987) to accommodate correlation among the unobserved random utility components of alternatives close together along a natural ordering implied by the choice variable (examples of such ordered choice variables might include car ownership, departure time of trips, etc.). The simplest version of the OGEV model (which Small refers to as the standard OGEV model) accommodates correlation in unobserved components between the utilities of each pair of adjacent alternatives on the natural ordering; that is, each alternative is correlated with the alternatives on either side of it along the natural ordering.[3] The standard OGEV model has a dissimilarity parameter that is inversely related to the correlation between adjacent alternatives (this relationship does not have a closed form, but the correlation implied by the dissimilarity parameter can be obtained numerically). The dissimilarity parameter has to lie in the range of 0 to 1 for consistency with random utility maximization.

The MNL-OGEV model formulated by Bhat (1998c) generalizes the nested logit model by allowing adjacent alternatives within a nest to be correlated in their unobserved components. This structure is best illustrated with an example. Consider the case of a multidimensional model of travel mode and departure time for nonwork trips. Let the departure time choice alternatives be represented by several temporally contiguous discrete time periods in a day, such as A.M. peak (6 to 9 A.M.), A.M. midday (9 A.M. to noon), P.M. midday (noon to 3 P.M.), P.M. peak (3 to 6 P.M.), and other (6 P.M. to 6 A.M.). An appropriate nested logit structure for the joint mode–departure time choice model may allow the joint choice alternatives to share unobserved attributes in the mode choice dimension, resulting in an increased sensitivity among time-of-day alternatives of the same mode relative to the time-of-day alternatives across modes. However, in addition to the uniform correlation in departure time alternatives sharing the same mode, there is likely to be increased correlation in the unobserved random utility components of each pair of adjacent departure time alternatives due to the natural ordering among the departure time

---

[3]The reader will note that the nested logit model cannot accommodate such a correlation structure because it requires alternatives to be grouped into mutually exclusive nests.

alternatives along the time dimension. Accommodating such a correlation generates an increased degree of sensitivity between adjacent departure time alternatives (over and above the sensitivity among non-adjacent alternatives) sharing the same mode. A structure that accommodates the correlation patterns just discussed can be formulated by using the multinomial logit formulation for the higher-level mode choice decision and the standard ordered generalized extreme value formulation (see Small, 1987) for the lower-level departure time choice decision (i.e., the MNL-OGEV model).

More recently, Wen and Koppelman (2001) proposed a general GEV model structure, which they refer to as the general nested logit model. Swait (2001), independently, proposed a similar structure, which he refers to as the choice set generation logit (GenL) model; Swait's derivation of the GenL model is motivated from the concept of latent choice sets of individuals, while Wen and Koppelman's derivation of the generalized nested logit (GNL) model is motivated from the perspective of flexible substitution patterns across alternatives. Wen and Koppelman (2001) illustrate the general nature of the GNL model formulation by deriving the other GEV model structures mentioned earlier as special restrictive cases of the GNL model or as approximations to restricted versions of the GNL model.

The GNL model is conceptually appealing because it is a very general structure and allows substantial flexibility. However, in practice, the flexibility of the GNL model can be realized only if one is able and willing to estimate a large number of dissimilarity and allocation parameters. The net result is that the analyst will have to impose informed restrictions on the general GNL model formulation that are customized to the application context under investigation.

The advantage of all the GEV models discussed above is that they allow relaxations of the independence assumption among alternative error terms while maintaining closed-form expressions for the choice probabilities. The problem with these models is that they are consistent with utility maximization only under rather strict (and often empirically violated) restrictions on the dissimilarity parameters. The origin of these restrictions can be traced back to the requirement that the variance of the joint alternatives be identical in the GEV models. In addition, the GEV models do not relax the response homogeneity assumption discussed in the previous section.

In the rest of the discussion on GEV models, we will focus on the GNL model since it subsumes other GEV models proposed to date as special cases.

10.3.1   GNL Model Structure

The GNL model can be derived from the GEV postulate using the following function:

$$G = (y_1, y_2, \ldots, y_1) = \sum_m \left[\sum_{i' \in N_m} (\alpha_{i'm} y_{i'})^{1/\rho_m} \right]^{\rho_m} \tag{10.11}$$

where N_m is the set of alternatives belonging to nest m, α_m represents an allocation parameter characterizing the portion of alternative i assigned to nest m ($0 < \alpha_{im} < 1$; $\sum_m \alpha_{im} = 1 \; \forall \; i$), and ρ_m is a dissimilarity parameter for nest m ($0 < \rho_m \leq 1$). Then it is easy to verify that G is nonnegative, homogenous of degree 1, tending toward $+\infty$ when any argument y_i tends toward $+\infty$, and whose n^{th} nonpartial derivatives are nonnegative for odd n and nonpositive for even n because $0 < \rho_m < 1$. Thus the following function represents a cumulative extreme value distribution:

$$F = (\varepsilon_1, \varepsilon_2, \ldots, \varepsilon_1) = \exp \left\{ -\sum_m \left[\sum_{i' \in N_m} (\alpha_{i'm} \rho^{-\varepsilon_{i'}})^{1/\rho_m} \right]^{\rho_m} \right\} \tag{10.12}$$

To obtain the probability of choice for each alternative i in the GNL model, consider a utility maximizing decision process where the utility of each alternative i (U_i) is written in the usual form as the sum of a deterministic component (V_i) and a random component E_i. If the random components follow the cumulative distribution function (CDF) in Equation (10.12), then, by the GEV postulate, the probability of choosing the ith alternative is:

$$P_i = \frac{\sum_m \left[(\alpha_{im} e^{V_i})^{1/\rho_m} \left(\sum_{i' \varepsilon N_m} (\alpha_{i'm} e^{V_{i'}})^{1/\rho_m} \right)^{\rho_m - 1} \right]}{\sum_m \left(\sum_{i' \varepsilon N_m} (\alpha_{i'm} e^{V_{i'}})^{1/\rho_m} \right)^{\rho_m}} \tag{10.13}$$

The cross-elasticity of a pair of alternatives i and j, which appear in one or more common nests, is:

$$\eta_{x_i}^{P_j} = - \left[P_i + \frac{\sum_m \left(\frac{1}{\rho_m} - 1 \right) P_m P_{i/m} P_{j/m}}{P_j} \right] \beta' x_i \tag{10.14}$$

If the two alternatives i and j do not appear in any common nest, the cross-elasticity reduces to zero. Wen and Koppelman (2001) also demonstrate that the correlation between two alternatives i and j is a function of both the allocation parameters and the dissimilarity parameters.

10.3.2   GNL Model Estimation

The GNL model may be estimated using the commonly used maximum likelihood method. The parameters to be estimated in the GNL structure include variable coefficients, the dissimilarity parameters ρ_m (m = 1, 2, ..., M), and the allocation parameters α_{im} (i = 1, 2, ..., I; m = 1, 2, ..., M). All the dissimilarity and allocation parameters need to be between 0 and 1, and the allocation parameters for each alternative should sum to 1. Wen and Koppelman (2001) used a constrained maximum likelihood procedure to estimate the model. It should be noted that the maximum number of dissimilarity parameters that can be estimated is one less than the number of pairs of alternatives.

10.3.3   GNL Model Applications

The GNL model was proposed recently by Wen and Koppelman (2001). The results of their application are discussed in detail in the next section. In most practical situations, the analyst will have to impose informed restrictions on the GNL formulation. Such restrictions might lead to models such as the PCL, OGEV, MNL-OGEV, and CNL models. In addition, the NL model can also be shown to be essentially the same as a restricted version of the GNL. Since there have been several applications of the NL model, and we have reviewed studies that have used the other GEV structures, we proceed to a detailed presentation of the GNL model by Wen and Koppelman.

10.3.4   Detailed Results from an Application of the GNL Model

Wen and Koppelman (2001) use the same Canadian rail data set used by Bhat (1995) and discussed in Section 10.2.5. They examined intercity mode choice in the Toronto–Montreal corridor. The universal choice set includes air, train, bus, and car.

TABLE 10.3 Comparison between MNL and GNL Model Estimates

Variable	MNL Model		GNL Model	
	Parameter	Standard Error	Parameter	Standard Error
Mode Constants				
Air	8.2380	0.429	6.2640	0.321
Train	5.4120	0.267	4.9810	0.285
Car	4.4210	0.301	5.1330	0.253
Bus (Base)				
Frequency	0.0850	0.004	0.0288	0.002
Travel cost	−0.0508	0.003	−0.0173	0.002
In-vehicle time	−0.0088	0.001	−0.0031	0.0002
Out-of-vehicle time	−0.0354	0.002	−0.0110	0.001
Log-Sum Parameters				
Train–car			0.0146	0.002
Air–car			0.2819	0.032
Train–car–air			0.01	—
Allocation Parameters				
Train–car nest				
Train			0.2717	0.033
Car			0.1057	0.012
Air–car nest				
Air			0.6061	0.040
Car			0.4179	0.046
Train–Car–Air Nest				
Train			0.5286	0.031
Car			0.2741	0.029
Air			0.3939	0.041
Train nest			0.1998	0.025
Car nest			0.2024	0.032
Bus nest			1.0000	
Log-likelihood at convergence	−2784.6		−2711.3	
Likelihood Ratio Index				
vs. zero	0.4896		0.5031	
vs. market share	0.3205		0.3382	
Value of Time (Per Hour)				
In-vehicle time	$10		$11	
Out-of-vehicle time	$42		$38	
Significance test rejecting MNL model (χ^2, degrees of freedom, significance)	—		146.6, 11, < 0.0001	

Source: From Wen, C.-H. and Koppelman, F.S., *Transp. Res. B*, 35, 627–641, 2001. With permission.

Table 10.3 shows the results that Wen and Koppelman obtained from the GNL and MNL models. Wen and Koppelman also estimated several NL structures, a PCL model, and CNL models. However, the GNL model provided a better data fit in their application context.

Table 10.3 provides the expected impacts of the level-of-service variables. The table also indicates that the model parameters tend to be smaller in magnitude in the GNL model than in the MNL model. However, the values of time are about the same for the two models. Most importantly, the differences in coefficient between the two models, combined with the correlation patterns generated by the GNL model, are likely to produce different mode share forecasts in response to policy actions or investment decisions.

10.4 Flexible Structure Models

The HEV and GEV class of models have the advantage that they are easy to estimate; the likelihood function for these models either includes a one-dimensional integral (in the HEV model) or is in closed form (in the GEV model). However, these models are restrictive since they only partially relax the IID error assumption across alternatives. In this section, we discuss model structures that are flexible enough to completely relax the independence and identically distributed error structure of the MNL *as well as* to relax the assumption of response homogeneity. This section focuses on model structures that explicitly nest the MNL model.

10.4.1 Model Formulations

Two closely related model formulations may be used to relax the IID (across alternatives) error structure or the assumption of response homogeneity: the mixed multinomial logit (MMNL) model and the mixed GEV (MGEV) model.

The mixed multinomial logit model is a generalization of the well-known multinomial logit model. It involves the integration of the multinomial logit formula over the distribution of unobserved random parameters. It takes the structure shown below:

$$P_{qi}(\theta) = \int_{-\infty}^{+\infty} L_{qi}(\beta)f(\beta|\theta)d(\beta), \ L_{qi}(\beta) = \frac{e^{\beta'x_{qi}}}{\sum_{j} e^{\beta'x_{qj}}} \tag{10.15}$$

where P_{qi} is the probability that individual q chooses alternative i, x_{qi} is a vector of observed variables specific to individual q and alternative i, β represents parameters that are random realizations from a density function $f(.)$, and θ is a vector of underlying moment parameters characterizing $f(.)$.

The MMNL model structure of Equation (10.15) can be motivated from two very different (but formally equivalent) perspectives. Specifically, an MMNL structure may be generated from an intrinsic motivation to allow flexible substitution patterns across alternatives (error components structure) or from a need to accommodate unobserved heterogeneity across individuals in their sensitivity to observed exogenous variables (random coefficients structure), as discussed later in Section 10.4.2 (of course, the MMNL structure can also accommodate both a non-IID error structure across alternatives and response heterogeneity).

The MGEV class of models use a GEV model as a core, and superimposes a mixing distribution on the GEV core to accommodate response heterogeneity or additional heteroskedasticity or correlation across alternative error terms. A question that arises here is "Why would one want to consider an MGEV model structure when an MMNL model can already capture response heterogeneity and any identifiable pattern of heteroskedasticity or correlation across alternative error terms?" That is, why would one want to consider a GEV core to generate a certain interalternative error correlation pattern when such a correlation pattern can be generated as part of an MMNL model structure? Bhat and Guo (2002) provide situations where an MGEV model may be preferred to an equivalent MMNL model. Consider, for instance, a model for household residential location choice. It is possible, if not very likely, that the utility of spatial units that are close to each other will be correlated due to common unobserved spatial elements.

A common specification in the spatial analysis literature for capturing such spatial correlation is to allow alternatives that are contiguous to be correlated. In the MMNL structure, such a correlation structure will require the specification of as many error components as the number of pairs of spatially contiguous alternatives. In a residential choice context, the number of error components to be specified will therefore be very large (in the 100s or 1000s). This will require the computation of very high dimensional integrals (in the order of 100s of 1000s) in the MMNL structure. On the other hand, a carefully specified GEV model can accommodate the spatial correlation structure within a closed-form formulation. However, the GEV model structure cannot accommodate unobserved random heterogeneity across individuals. One could superimpose a mixing distribution over the GEV model structure to accommodate such heterogeneity, leading to a parsimonious and powerful MGEV structure.

In the rest of this section, we will focus on the MMNL model structure, since all the concepts and techniques for the MMNL model are readily transferable to the MGEV model structure.

10.4.2 MMNL Model Structure

In this section, we discuss the MMNL structure from an error components viewpoint as well as from a random coefficient viewpoint.

10.4.2.1 Error Components Structure

The error components structure partitions the overall random term associated with each alternative's utility into two components: one component that allows the unobserved error terms to be nonidentical and nonindependent across alternatives, and one component that is specified to be independent and identically (type I extreme value) distributed across alternatives. Specifically, consider the following utility function for individual q and alternative i:

$$
\begin{aligned}
U_{qi} &= \gamma' y_{qi} + \zeta_{qi} \\
&= \gamma' y_{qi} + \mu' z_{qi} + \varepsilon_{qi}
\end{aligned}
\tag{10.16}
$$

where $\gamma' y_{qi}$ and ζ_{qi} are the systematic and random components of utility and ζ_i is further partitioned into two components: $\mu' z_{qi}$ and ε_{qi}. z_{qi} is a vector of observed data associated with alternative i, some of whose elements might also appear in the vector y_{qi}. μ is a random vector with zero mean. The component $\mu' z_{qi}$ induces heteroskedasticity and correlation across unobserved utility components of the alternatives. Defining $\beta = (\gamma', \mu')'$ and $x_{qi} = (y'_{qi}, z'_{qi})'$, we obtain the MMNL model structure for the choice probability of alternative i for individual q.

The emphasis in the error components structure is to allow a flexible substitution pattern among alternatives in a parsimonious fashion. This is achieved by the clever specification of the variable vector z_{qi}, combined with (usually) the specification of independent normally distributed random elements in the vector μ. For example, z_i may be specified to be a row vector of dimension M, with each row representing a group m (m = 1, 2, ..., M) of alternatives sharing common unobserved components. The row(s) corresponding to the group(s) of which i is a member take(s) a value of 1 and other rows take a value of 0. The vector μ (of dimension M) may be specified to have independent elements, each element having a variance component σ_m^2. The result of this specification is a covariance of σ_m^2 among alternatives in group m and heteroskedasticity across the groups of alternatives. This structure is less restrictive than the nested logit structure in that an alternative can belong to more than one group. Also, by structure, the variance of the alternatives is different. More general structures for $\mu' z_i$ in Equation (10.16) are presented by Ben-Akiva and Bolduc (1996) and Brownstone and Train (1999).

10.4.2.2 Random Coefficients Structure

The random coefficients structure allows heterogeneity in the sensitivity of individuals to exogenous attributes. The utility that an individual q associates with alternative i is written as

$$U_{qi} = \beta_q' x_{qi} + \varepsilon_{qi} \tag{10.17}$$

where x_{qi} is a vector of exogenous attributes, β_q is a vector of coefficients that varies across individuals with density $f(\beta)$, and ε_{qi} is assumed to be an independently and identically distributed (across alternatives) type I extreme value error term. With this specification, the unconditional choice probability of alternative i for individual q is given by the mixed logit formula of Equation (10.15). While several density functions may be used for $f(\beta)$, the most commonly used is the normal distribution. A lognormal distribution may also be used if, from a theoretical perspective, an element of β has to take the same sign for every individual (such as a negative coefficient for the travel time parameter in a travel mode choice model).

The reader will note that the error components specification in Equation (10.16) and the random coefficients specification in Equation (10.17) are structurally equivalent. Specifically, if β_q is distributed with a mean of γ and deviation μ, then Equation (10.17) is identical to Equation (10.16), with $x_{qi} = y_{qi} = z_{qi}$. However, this apparent restriction for equality of Equations (10.16) and (10.17) is purely notational. Elements of x_{qi} that do not appear in z_{qi} can be viewed as variables whose coefficients are deterministic in the population, while elements of x_{qi} that do not enter in y_{qi} may be viewed as variables whose coefficients are randomly distributed in the population with mean zero (with cross-sectional data, the coefficients on the alternative specific constants have to be considered deterministic).

Due to the equivalence between the random coefficients and error components formulations, and the more compact notation of the random coefficients formulation, we will use the latter formulation in the next section in the discussion of the estimation methodology for the mixed logit model.

10.4.3 MMNL Estimation Methodology

This section discusses the details of the estimation procedure for the random coefficients mixed logit model using each of three methods: the cubature method, the pseudo-Monte Carlo (PMC) method, and the quasi-Monte Carlo (QMC) method.

Consider Equation (10.17) and separate out the effect of variables with fixed coefficients (including the alternative specific constant) from the effect of variables with random coefficients:

$$U_{qi} = \alpha_{qi} + \sum_{k=1}^{K} \beta_{qk} x_{qik} + \varepsilon_{qi} \tag{10.18}$$

where α_{qi} is the effect of variables with fixed coefficients. Let $\beta_{qk} \sim N(\mu_k, \sigma_k)$, so that $\beta_{qk} = \mu_k + \sigma_k s_{qk}$ (q = 1, 2, ..., Q; k = 1, 2, ..., K). In this notation, we are implicitly assuming that the β_{qk} values are independent of one another. Even if they are not, a simple Choleski decomposition can be undertaken so that the resulting integration involves independent normal variates (see Revelt and Train, 1998). s_{qk} (q = 1, 2, ..., Q; k = 1, 2, ..., K) is a standard normal variate. Further, let $V_{qi} = \alpha_{qi} + \sum_k \mu_k x_{qik}$.

The log-likelihood function for the random coefficients logit model may be written as:

$$\mathcal{L} = \sum_q \sum_i y_{qi} \log P_{qi}$$

$$= \sum_q \sum_i y_{qi} \log \left\{ \int_{s_{q1}=-\infty}^{s_{q1}=+\infty} \int_{s_{q2}=-\infty}^{s_{q2}=+\infty} \cdots \int_{s_{qK}=-\infty}^{s_{qK}=+\infty} \frac{e^{V_{qi} + \sum_k \sigma_k s_{qk} x_{qik}}}{\sum_j e^{V_{qj} + \sum_k \sigma_k s_{qk} x_{qik}}} d\Phi(s_{q1}) d\Phi(s_{q2}) \dots d\Phi(s_{qK}) \right\} \tag{10.19}$$

where $\Phi(.)$ represents the standard normal cumulative distribution function and

$$y_{qi} = \begin{cases} 1 \text{ if the qth individual chooses alternative i} \\ 0 \text{ otherwise,} \qquad (q = 1, 2, \ldots, Q,\ i = 1, 2, \ldots, I) \end{cases} \qquad (10.20)$$

The cubature, PMC, and QMC methods represent three different ways of evaluating the multidimensional integral involved in the log-likelihood function.

10.4.3.1 Polynomial-Based Cubature Method

To apply the cubature method, define $\varpi_k = s_{qk}/\sqrt{2}$ for all q. Then the log-likelihood function in Equation (10.19) takes the following form:

$$\mathcal{L} = \sum_q \sum_i y_{qi}$$

$$\log\left\{ \left(\frac{1}{\sqrt{\pi}}\right)^K \int_{\varpi_1=-\infty}^{\varpi_1=+\infty} \int_{\varpi_2=-\infty}^{\varpi_2=+\infty} \cdots \int_{\varpi_K=-\infty}^{\varpi_K=+\infty} \frac{e^{V_{qi}+\sqrt{2}\sum_k \sigma_k \varpi_k x_{qik}}}{\sum_j e^{V_{qj}+\sqrt{2}\sum_k \sigma_k \varpi_k x_{qik}}} e^{-\varpi_1^2} e^{-\varpi_2^2} \cdots e^{-\varpi_K^2} d\varpi_1 d\varpi_2 \cdots d\varpi_K \right\} \qquad (10.21)$$

The above integration is now in an appropriate form for application of a multidimensional product formula of the Gauss–Hermite quadrature (see Stroud, 1971).

10.4.3.2 Pseudo-Random Monte Carlo (PMC) Method

This technique approximates the choice probabilities by computing the integrand in Equation (10.19) at randomly chosen values for each s_{qk}. Since the s_{qk} terms are independent across individuals and variables, and are distributed standard normal, we generate a matrix s of standard normal random numbers with Q*K elements (one element for each variable and individual combination) and compute the corresponding individual choice probabilities for a given value of the parameter vector to be estimated. This process is repeated R times for the given value of the parameter vector, and the integrand is approximated by averaging over the computed choice probabilities in the different draws. This results in an unbiased estimator of the actual individual choice probabilities. Its variance decreases as R increases. It also has the appealing properties of being smooth (i.e., twice differentiable) and strictly positive for any realization of the finite R draws. The parameter vector is estimated as the vector value that maximizes the simulated log-likelihood function. Under rather weak regularity conditions, the PMC estimator is consistent, asymptotically efficient, and asymptotically normal. However, the estimator will generally be a biased simulation of the maximum likelihood (ML) estimator because of the logarithmic transformation of the choice probabilities in the log-likelihood function. The bias decreases with the variance of the probability simulator; that is, it decreases as the number of repetitions increase.

10.4.3.3 Quasi-Random Monte Carlo (QMC) Method

The quasi-random Halton sequence is designed to span the domain of the S-dimensional unit cube uniformly and efficiently (the interval of each dimension of the unit cube is between 0 and 1). In one dimension, the Halton sequence is generated by choosing a prime number r ($r \geq 2$) and expanding the sequence of integers 0, 1, 2, ..., g, ..., G in terms of the base r:

$$g = \sum_{l=0}^{L} b_l r^l, \text{ where } 0 \leq b_l \leq r - 1 \text{ and } r^L \leq g < r^{L+1} \qquad (10.22)$$

Thus, g (g = 1, 2, ..., G) can be represented by the r-*adic* integer string $b_1 \ldots b_1 b_0$. The Halton sequence in the prime base r is obtained by taking the radical inverse of g (g = 1, 2, ..., G) to the base r by reflecting through the radical point:

$$\varphi_r(g) = 0.b_0 b_1 \ldots b_L (\text{base } r) = \sum_{l=0}^{L} b_l r^{-l-1} \tag{10.23}$$

The sequence above is very uniformly distributed in the interval (0, 1) for each prime r. The Halton sequence in K dimensions is obtained by pairing K one-dimensional sequences based on K pairwise relatively prime integers (usually the first K primes):

$$\psi_g = (\varphi_{r_1}(g), \varphi_{r_2}(g), \ldots, \varphi_{r_S}(g)) \tag{10.24}$$

The Halton sequence is generated number theoretically rather than randomly, and so successive points at any stage "know" how to fill in the gaps left by earlier points, leading to a uniform distribution within the domain of integration.

The simulation technique to evaluate the integral in the log-likelihood function of Equation (10.19) involves generating the K-dimensional Halton sequence for a specified number of draws R for each individual. To avoid correlation in simulation errors across individuals, separate independent draws of R Halton numbers in K dimensions are taken for each individual. This is achieved by generating a Halton matrix Y of size G × K, where G = R*Q + 10 (Q is the total number of individuals in the sample). The first ten terms in each dimension are then discarded because the integrand may be sensitive to the starting point of the Halton sequence. This leaves a (R*Q) × K Halton matrix, which is partitioned into Q submatrices of size R × K, each submatrix representing the R Halton draws in K dimensions for each individual (thus, the first R rows of the Halton matrix Y are assigned to the first individual, the second R rows to the second individual, and so on).

The Halton sequence is uniformly distributed over the multidimensional cube. To obtain the corresponding multivariate normal points over the multidimensional domain of the real line, the inverse standard normal distribution transformation of Y is taken. By the integral transform result, X = Φ^{-1}(Y) provides the Halton points for the multivariate normal distribution (see Fang and Wang, 1994, Chap. 4). The integrand in Equation (10.19) is computed at the resulting points in the columns of the matrix X for each of the R draws for each individual, and then the simulated likelihood function is developed in the usual manner as the average of the values of the integrand across the R draws.

Bhat (2001) proposed and introduced the use of the Halton sequence for estimating the mixed logit model and conducted Monte Carlo simulation experiments to study the performance of this quasi-Monte Carlo simulation method vis-à-vis the cubature and pseudo-Monte Carlo simulation methods (this study, to the author's knowledge, is the first attempt at employing the QMC simulation method in discrete choice literature). Bhat's results indicate that the QMC method outperforms the polynomial cubature and PMC methods for mixed logit model estimation. Bhat notes that this substantial reduction in computational cost has the potential to dramatically influence the use of the mixed logit model in practice. Specifically, given the flexibility of the mixed logit model to accommodate very general patterns of competition among alternatives or random coefficients, the use of the QMC simulation method of estimation should facilitate the application of behaviorally rich structures for discrete choice modeling. Another subsequent study by Train (1999) confirms the substantial reduction in computational time for mixed logit estimation using the QMC method. Hensher (1999) has also investigated Halton sequences and compared the findings with random draws for mixed logit model estimation. He notes that the data fit and parameter values of the mixed logit model remain almost the same beyond 50 Halton draws. He concludes that the quasi-Monte Carlo method "is a phenomenal development in the estimation of complex choice models" (Hensher, 1999).

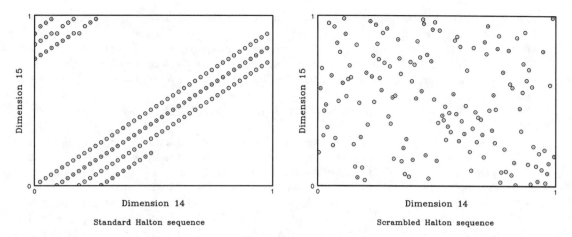

FIGURE 10.1 A total of 150 draws of standard and scrambled Halton sequences. (From Bhat, C.R., *Transp. Res. B*, forthcoming. With permission.)

10.4.3.4 Scrambled and Randomized QMC Method

Bhat (2002) notes that a problem with the Halton sequence is that there is strong correlation between higher coordinates of the sequence. This is because of the cycles of length r for the prime r. Thus, when two large prime-based sequences, associated with two high dimensions, are paired, the corresponding unit square face of the S-dimensional cube is sampled by points that lie on parallel lines. For example, the fourteenth dimension (corresponding to the prime number 43) and the fifteenth dimension (corresponding to the prime number 47) consist of 43 and 47 increasing numbers, respectively. This generates a correlation between the fourteenth and fifteenth coordinates of the sequence. This is illustrated diagrammatically in the first plot of Figure 10.1. The consequence is a rapid deterioration in the uniformity of the Halton sequence in high dimensions (the deterioration becomes clearly noticeable beyond five dimensions).

Number theorists have proposed an approach to improve the uniformity of the Halton sequence in high dimensions. The basic method is to break the correlations between the coordinates of the standard Halton sequence by scrambling the cycles of length r for the prime r. This is accomplished by permutations of the coefficients b_l in the radical inverse function of Equation (10.23). The resulting scrambled Halton sequence for the prime r is written as

$$\varphi_r(g) = \sum_{l=0}^{L} \sigma_r(b_l(g)) r^{-l-1} \qquad (10.25)$$

where σ_r is the operator of permutations on the digits of the expansion $b_l(g)$ (the standard Halton sequence is the special case of the scrambled Halton sequence, with no scrambling of the digits $b_l(g)$). Different researchers (see Braaten and Weller, 1979; Hellekalek, 1984; Kocis and Whiten, 1997) have suggested different algorithms for arriving at the permutations of the coefficients b_l in Equation (10.25). The permutations used by Braaten and Weller are presented in the appendix for the first ten prime numbers. Braaten and Weller have also proved that their scrambled sequence retains the theoretically appealing N^{-1} order of integration error of the standard Halton sequence.

An example would be helpful in illustrating the scrambling procedure of Braaten and Weller. These researchers suggest the following permutation of (0, 1, 2) for the prime 3: (0, 2, 1). As indicated earlier, the fifth number in base 3 of the Halton sequence in digitized form is 0.21. When the permutation above is applied, the fifth number in the corresponding scrambled Halton sequence in digitized form is 0.21, which when expanded in base 3 translates to $1 \times 3^{-1} + 2 \times 3^{-2} = 5/9$. The first eight numbers in the scrambled sequence corresponding to base 3 are 2/3, 1/3, 2/9, 8/9, 5/9, 1/9, 7/9, and 4/9.

The Braaten and Weller method involves different permutations for different prime numbers. As a result of this scrambling, the resulting sequence does not display strong correlation across dimensions, as does the standard Halton sequence. This is illustrated in the second plot of Figure 10.1, which plots 150 scrambled Halton points in the fourteenth and fifteenth dimensions. A comparison of the two plots in Figure 10.1 clearly indicates the more uniform coverage of the scrambled Halton sequence relative to the standard Halton sequence.

In addition to the scrambling of the standard Halton sequence, Bhat (2002) also suggests a randomization procedure for the Halton sequence based on a procedure developed by Tuffin (1996). The randomization is useful because all QMC sequences (including the standard Halton and scrambled Halton sequences discussed above) are fundamentally deterministic. This deterministic nature of the sequences does not permit the practical estimation of the integration error. Theoretical results exist for estimating the integration error, but these are difficult to compute and can be very conservative.

The essential concept of randomizing QMC sequences is to introduce randomness into a deterministic QMC sequence that preserves the uniformly distributed and equidistribution properties of the underlying QMC sequence (see Shaw, 1988; Tuffin, 1996). One simple way to introduce randomness is based on the following idea. Let $\psi^{(N)}$ be a QMC sequence of length N over the S-dimensional cube $\{0, 1\}^S$ and consider any S-dimensional uniformly distributed vector in the S-dimensional cube ($u \in \{0, 1\}^S$). $\psi^{(N)}$ is a matrix of dimension $N \times S$, and u is a vector of dimension $1 \times S$. Construct a new sequence $\chi^{(N)} = \{\psi^{(N)} + u \otimes 1^{(N)}\}$, where $\{.\}$ denotes the fractional part of the matrix within parenthesis, $\otimes$ represents the kronecker or tensor product, and $1^{(N)}$ is a unit column vector of size N (the kronecker product multiplies each element of u with the vector $1^{(N)}$). The net result is a sequence $\chi^{(N)}$ whose elements χ_{ns} are obtained as $\psi_{ns} + u_s$ if $\psi_{ns} + u_s = 1$, and $\psi_{ns} + u_s - 1$ if $\psi_{ns} + u_s > 1$. It can be shown that the sequence $\chi^{(N)}$ so formed is also a QMC sequence of length N over the S-dimensional cube $\{0, 1\}^S$. Tuffin provides a formal proof for this result, which is rather straightforward but tedious. Intuitively, the vector u simply shifts the points of each coordinate of the original QMC sequence $\psi^{(N)}$ by a certain value. Since all the points within each coordinate are shifted by the same amount, the new sequence will preserve the equidistribution property of the original sequence. This is illustrated in Figure 10.2 in two dimensions. The first diagram in Figure 10.2 plots 100 points of the standard Halton sequence in the first two dimensions. The second diagram plots 100 points of the standard Halton sequence shifted by 0.5 in the first dimension and 0 in the second dimension. The result of the shifting is as follows. For any point below 0.5 in the first dimension in the first diagram (for example, the point marked 1), the point gets moved by 0.5 toward the right in the second diagram. For any point above 0.5 in the first dimension in the first diagram (such as the point marked 2), the point gets moved to the right, hits the right edge, bounces off this edge to the left edge, and is carried forward so that the total distance of the shift is 0.5 (another way to visualize this shift is to transform the unit square into a cylinder with the left and right edges "sewn" together; then the shifting entails moving points along the surface of the cylinder and perpendicular to the cylinder axis). Clearly, the two-dimensional plot in the second diagram of Figure 10.2 is also well distributed because the relative positions of the points do not change from that in Figure 10.1; there is simply a shift of the overall pattern of points. The last diagram in Figure 10.2 plots the case where there is a shift in both dimensions: 0.5 in the first and 0.25 in the second. For the same reasons discussed in the context of the shift in one dimension, the sequence obtained by shifting in both dimensions is also well distributed.

It should be clear from above that any vector $u \in \{0, 1\}^S$ can be used to generate a new QMC sequence from an underlying QMC sequence. An obvious way of introducing randomness is then to randomly draw u from a multidimensional uniform distribution.

An important point to note here is that randomizing the standard Halton sequence as discussed earlier does not break the correlations in high dimensions because the randomization simply shifts all points in the same dimension by the same amount. Thus, randomized versions of the standard Halton sequence will suffer from the same problems of nonuniform coverage in high dimensions as the standard Halton sequence. To resolve the problem of nonuniform coverage in high dimensions, the scrambled Halton sequence needs to be used.

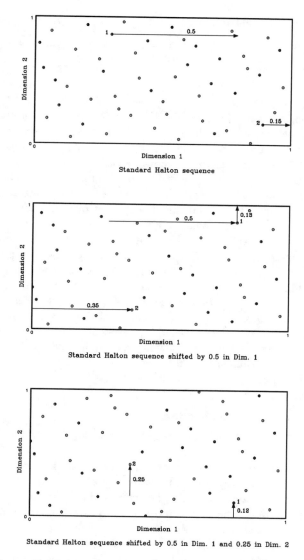

FIGURE 10.2 Shifting the standard Halton sequence. (From Bhat, C.R., *Trans. Res. B*, forthcoming. With permission.)

Once a scrambled and randomized QMC sequence is generated, Bhat (2002) proposes a simulation approach for estimation of the mixed logit model that is similar to the standard Halton procedure discussed in the previous section.

10.4.3.5 Bayesian Estimation of MNL

Some recent papers (Brownstone, 2000; Train, 2001) have considered a Bayesian estimation approach for MMNL model estimation, as opposed to the classical estimation approaches discussed above. The general results from these studies appear to suggest that the classical approach is faster when mixing distributions with bounded support, such as triangulars are considered, or when there is a mix of fixed and random coefficients in the model. On the other hand, the Bayesian estimation appears to be faster when considering the normal distribution and its transformations, and when all coefficients are random and are correlated with one another. However, overall, the results suggest that the choice between the two estimation approaches should depend more on interpretational ease in the empirical context under study than on computational efficiency considerations.

10.4.4   Transport Applications

The transport applications of the mixed multinomial logit model are discussed under two headings: error components applications and random coefficients applications.

10.4.4.1   Error Components Applications

Brownstone and Train (1999) applied an error components mixed multinomial logit structure to model households' choices among gas, methanol, compressed natural gas (CNG), and electric vehicles, using stated choice data collected in 1993 from a sample of households in California. Brownstone and Train allow nonelectric vehicles to share an unobserved random component, thereby increasing the sensitivity of non-electric vehicles to one another, compared to an electric vehicle. Similarly, a non-CNG error component is introduced. Two additional error components related to the size of the vehicle are also introduced: one is a normal deviate multiplied by the size of the vehicle, and the second is a normal deviate multiplied by the luggage space. All these error components are statistically significant, indicating non-IIA competitive patterns.

Brownstone et al. (2000) extended the analysis of Brownstone and Train (1999) to estimate a model of choice among alternative-fuel vehicles using both stated choice and revealed choice data. The RC data were collected about 15 months after the SC data, and recorded actual vehicle purchase behavior since the collection of the SC data. Brownstone et al. (2000) maintain the error components structure developed in their earlier study, and also accommodate scale differences between RC and SC choices.

Bhat (1998a) applied the mixed multinomial logit model to a multidimensional choice situation. Specifically, his application accommodates unobserved correlation across both dimensions in a two-dimensional choice context. The model is applied to an analysis of travel mode and departure time choice for home-based social–recreational trips using data drawn from the 1990 San Francisco Bay Area household survey. The empirical results underscore the need to capture unobserved attributes along both the mode and departure time dimensions, both for improved data fit and more realistic policy evaluations of transportation control measures.

10.4.4.2   Random Coefficients Applications

There have been several applications of the mixed multinomial logit model motivated from a random coefficients perspective.

Bhat (1998b) estimated a model of intercity travel mode choice that accommodates variations in responsiveness to level-of-service measures due to both observed and unobserved individual characteristics. The model is applied to examine the impact of improved rail service on weekday business travel in the Toronto–Montreal corridor. The empirical results show that not accounting adequately for variations in responsiveness across individuals leads to a statistically inferior data fit and also to inappropriate evaluations of policy actions aimed at improving intercity transportation services.

Bhat (2000) formulated a mixed multinomial logit model of multiday urban travel mode choice that accommodates variations in mode preferences and responsiveness to level of service. The model is applied to examine the travel mode choice of workers in the San Francisco Bay Area. Bhat's empirical results indicate significant unobserved variation (across individuals) in intrinsic mode preferences and level-of-service responsiveness. A comparison of the average response coefficients (across individuals in the sample) among the fixed coefficient and random coefficient models shows that the random coefficients model implies substantially higher monetary values of time than the fixed coefficient model. Overall, the empirical results emphasize the need to accommodate observed and unobserved heterogeneity across individuals in urban mode choice modeling.

Train (1998) used a random coefficients specification to examine the factors influencing anglers' choice of fishing sites. Explanatory variables in the model include fish stock (measured in fish per 1000 ft of river), aesthetics rating of the fishing site, size of each site, number of campgrounds and recreation access at the site, number of restricted species at the site, and the travel cost to the site (including the money value of travel time). The empirical results indicate highly significant taste variation across anglers in sensitivity to almost all the factors listed above. In this study as well as Bhat's (2000) study, there is a very dramatic increase in data fit after including random variation in coefficients.

Mehndiratta (1997) proposed and formulated a theory to accommodate variations in the resource value of time in time-of-day choice for intercity travel. Mehndiratta then proceeded to implement his theoretical model using a random coefficients specification for the resource value of disruption of leisure and sleep. He used a stated choice sample in his analysis.

Hensher (2000) undertakes a stated choice analysis of the valuation of nonbusiness travel time savings for car drivers undertaking long-distance trips (up to 3 h) between major urban areas in New Zealand. Hensher disaggregates overall travel time into several different components, including free flow travel time, slowed-down time, and stop time. The coefficients of each of these attributes are allowed to vary randomly across individuals in the population. The study finds significant taste heterogeneity to the various components of travel time, and adds to the accumulating evidence that the restrictive travel time response homogeneity assumption undervalues the mean value of travel time savings.

In addition to the studies identified in Section 10.4.4.1 and this section, some recent studies have included both interalternative error correlations (in the spirit of an error components structure) and unobserved heterogeneity among decision-making agents (in the spirit of the random coefficients structure). Such studies include Hensher and Greene (2000), Bhat and Castelar (2002), and Han and Algers (2001).

10.4.5 Detailed Results from an Example Application

Bhat (1998a) uses an error components motivation for the analysis of mode and departure time choice for social–recreational trips in the San Francisco Bay Area. Bhat suggests the use of an MMNL model to accommodate unobserved correlation in error terms across both the model and temporal dimension simultaneously. The data for this study are drawn from the San Francisco Bay Area Household Travel Survey conducted by the Metropolitan Transportation Commission (MTC) in the spring and fall of 1990. The modal alternatives include drive alone, shared ride, and transit. The departure time choice is represented by six time periods: early morning (12:01 to 7 A.M.), A.M. peak (7:01 to 9 A.M.), A.M. off-peak (9:01 A.M. to noon), P.M. off-peak (12:01 to 3 P.M.), P.M. peak (3:01 to 6 P.M.), and evening (6:01 P.M. to midnight). For some individual trips, modal availability is a function of time of day (for example, transit mode may be available only during the A.M. and P.M. peak periods). Such temporal variations in modal availability are accommodated by defining the feasible set of joint choice alternatives for each individual trip. Level-of-service data were generated for each zonal pair in the study area and by five time periods: early morning, A.M. peak, midday, P.M. peak, and evening. The sample used in Bhat's paper comprises 3000 home-based social–recreational person trips obtained from the overall single-day travel diary sample. The mode choice shares in the sample are as follows: drive alone (45.7%), shared ride (51.9%), and transit (2.4%). The departure time distribution of home-based social–recreational trips in the sample is as follows: early morning (4.6%), A.M. peak (5.5%), A.M. off-peak (10.3%), P.M. off-peak (17.2%), P.M. peak (16.1%), and evening (46.3%).

Bhat estimated four different models of mode departure time choice: (1) the multinomial logit model, (2) the mixed multinomial logit model that accommodates shared unobserved random utility attributes along the departure time dimension only (the MMNL-T model), (3) the mixed multinomial logit model that accommodates shared unobserved random utility attributes along the mode dimension only (the MMNL-M model), and (4) the proposed mixed multinomial logit model that accommodates shared unobserved attributes along both the mode and departure time dimensions (the MMNL-MT model). In the MMNL models, the sensitivity among joint choice alternatives sharing the same mode (departure time) was allowed to vary across modes (departure times). It is useful to note that such a specification generates heteroskedasticity in the random error terms across the joint choice alternatives. In the MMNL-T and MMNL-MT models, the shared unobserved components specific to the morning departure times (i.e., early morning, A.M. peak, and A.M. off-peak periods) were statistically insignificant. Consequently, the MMNL-T and MMNL-MT model results restricted these components to zero.

The level-of-service parameter estimates, implied money values of travel time, data fit measures, and the variance parameters in $[\Sigma]$ and $[\Omega]$ from the different models, are presented in Table 10.4. The signs of the level-of-service parameters are consistent with a priori expectations in all the models. Also, as

TABLE 10.4 Level-of-Service Parameters, Implied Money Values of Travel Time, Data Fit Measures, and Error Variance Parameters

Attributes/Data Fit Measures	MNL Model	MMNL-T Model	MMNL-M Model	MMNL-MT Model
Level of Service[a]				
Travel cost (in cents)	−0.0031 (−3.13)	−0.0036 (−3.02)	−0.0044 (−2.88)	−0.0045 (−2.83)
Total travel time (in minutes)	−0.0319 (−3.15)	−0.0336 (−2.87)	−0.0382 (−3.22)	−0.0408 (−3.33)
Out-of-vehicle time/distance	−0.2363 (−3.42)	−0.2429 (−4.82)	−0.2508 (-4.19)	−0.2589 (−4.26)
Implied Money Values of Time ($ per Hour)				
In-vehicle travel time	6.17	5.60	5.21	5.44
Out-of-vehicle travel time[b]	13.66	12.23	10.80	11.09
Log-likelihood at convergence[c]	−6393.6	−6382.9	−6387.7	−6375.8
Error Variance Parameters				
$\delta_{\text{pm off-peak}}$	—	0.8911 (2.76)	—	0.9715 (2.96)
$\delta_{\text{pm peak}}$	—	0.7418 (2.83)	—	0.3944 (1.88)
δ_{Evening}	—	1.9771 (2.70)	—	1.6421 (3.02)
$\sigma_{\text{Drive alone}}$	—	—	0.6352 (1.91)	0.5891 (1.98)
$\sigma_{\text{Shared ride}}$	—	—	1.9464 (3.06)	1.9581 (3.20)
σ_{Transit}	—	—	0.7657 (1.73)	0.7926 (2.07)

[a] The entries in the different columns correspond to the parameter values and their t-statistics (in parentheses).
[b] Money value of out-of-vehicle time is computed at the mean travel distance of 6.11 miles.
[c] The LL (log-likelihood) at equal shares is ~8601.24 and the LL with only alternative specific constants and an IID error covariance matrix is −6812.07.
Source: From Bhat, C.R., *Trans. Res. B*, 32, 455, 1998. With permission.

expected, travelers are more sensitive to out-of-vehicle travel time than in-vehicle travel time. A comparison of the magnitudes of the level-of-service parameter estimates across the four specifications reveals a progressively increasing magnitude as we move from the MNL model to the MMNL-MT model (this is an expected result since the variance before scaling is larger in the MNL model than in the mixture models, and in the MMNL-M and MMNL-T models than in the MMNL-MT model; see Revelt and Train (1998) for a similar result). The implied money values of in-vehicle and out-of-vehicle travel times are less in the mixed multinomial logit models than in the MNL model.

The four alternative models in Table 10.4 can be evaluated formally using conventional likelihood ratio tests. A statistical comparison of the multinomial logit model with any of the mixture models leads to the rejection of the multinomial logit. Further likelihood ratio tests among the MMNL-M, MMNL-T, and MMNL-MT models result in the clear rejection of the hypothesis that there are shared unobserved attributes along only one dimension; that is, the tests indicate the presence of statistically significant shared unobserved components along both the mode and departure time dimensions (the likelihood ratio test statistic in the comparison of the MMNL-T model with the MMNL-MT model is 14.2; the corresponding value in the comparison of the MMNL-M model with the MMNL-MT model is 23.8; both these values are larger than the chi-squared distribution, with three degrees of freedom at any reasonable level of significance). Thus, the MNL, MMNL-T, and MMNL-M models are misspecified.

The variance parameters provide important insights regarding the sensitivity of joint choice alternatives sharing the same mode and departure time. The variance parameters specific to departure times (in the MMNL-T and MMNL-MT models) show statistically significant shared unobserved attributes associated with the afternoon–evening departure periods. However, as indicated earlier, there were no statistically significant shared unobserved components specific to the morning departure times (i.e., early morning, A.M. peak, and A.M. off-peak periods). The implication is that home-based social–recreational trips pursued in the morning are more flexible and more easily moved to other times of the day than trips pursued later in the day. Social–recreational activities pursued later in the day may be more rigid because of scheduling considerations among household members or because of the inherent temporal fixity of late evening activities (such as attending a concert or a social dinner). The magnitude of the departure time variance parameters reveal that late evening activities are most rigid, followed by activities pursued during the P.M. off-peak hours. The P.M. peak social–recreational activities are more flexible than the P.M. off-peak and late evening activities. The variance parameters specific to the travel modes (in the MMNL-M and MMNL-MT models) confirm the presence of common unobserved attributes among joint choice alternatives that share the same mode; thus, individuals tend to maintain their current travel mode when confronted with transportation control measures such as ridesharing incentives and auto use disincentives. This is particularly so for individuals who rideshare, as can be observed from the higher variance associated with the shared-ride mode than with the other two modes. In the context of home-based social–recreational trips, most ridesharing arrangements correspond to travel with children or other family members; it is unlikely that these ridesharing arrangements will be terminated after implementation of transportation control measures such as transit use incentives.

The different variance structures among the four models imply different patterns of interalternative competition. To demonstrate the differences, Table 10.5 presents the disaggregate self- and cross-elasticities (for a person trip in the sample with close-to-average modal level-of-service values) in response to peak period pricing implemented in the P.M. peak (i.e., a cost increase in the drive-alone P.M. peak alternative). All morning time periods are grouped together in the table since the cross-elasticities for these time periods are the same for each mode (due to the absence of shared unobserved attributes specific to the morning time periods).

The MNL model exhibits the familiar independence from irrelevant alternatives property (that is, all cross-elasticities are equal). The MMNL-T model shows equal cross-elasticities for each time period across modes, a reflection of not allowing shared unobserved attributes along the modal dimension. However, there are differences across time periods for each mode. First, the shift to the shared-ride P.M. peak and transit P.M. peak is more than that to the other non-P.M. peak joint choice alternatives. This is, of course, because of the increased sensitivity among P.M. peak joint choice alternatives generated by the error variance term specific to the P.M. peak period. Second, the shift to the evening period alternatives are lower than the shift to the P.M. off-peak period alternatives for each mode. This result is related to the heteroskedasticity in the shared unobserved random components across time periods. The variance parameter in Table 10.4 associated with the evening period is higher than that associated with the P.M. off-peak period; consequently, there is less shift to the evening alternatives (see Bhat (1995) for a detailed

TABLE 10.5 Disaggregate Travel Cost Elasticities in Response to a Cost Increase in the Drive-Alone (DA) Mode during P.M. Peak

Effect on Joint Choice Alternative	MNL Model	MMNL-T Model	MMNL-M Model	MMNL-MT Model
DA morning periods[a]	0.0072	0.0085	0.0141	0.0165
DA P.M. off-peak	0.0072	0.0060	0.0141	0.0131
DA P.M. peak	−0.1112	−0.0993	−0.1555	−0.1423
DA evening	0.0072	0.0042	0.0141	0.0099
SR morning periods[a]	0.0072	0.0085	0.0059	0.0072
SR P.M. off-peak	0.0072	0.0060	0.0059	0.0055
SR P.M. peak	0.0072	0.0120	0.0059	0.0079
SR evening	0.0072	0.0042	0.0059	0.0045
TR morning periods[a]	0.0072	0.0085	0.0119	0.0131
TR P.M. off-peak	0.0072	0.0060	0.0119	0.0106
TR P.M. peak	0.0072	0.0120	0.0119	0.0150
TR evening	0.0072	0.0042	0.0119	0.0082

[a] The morning periods include early morning, A.M. peak, and A.M. off-peak. The cross-elasticities for the morning periods within each mode with respect to a P.M. peak cost increase in the drive alone mode are the same in the mixture logit models because of the absence of shared unobserved attributes specific to the morning time periods.

Note: SR = shared ride; TR = transit.

Source: From Bhat, C.R., *Trans. Res. B*, 32, 455, 1998. With permission.

discussion of the inverse relationship between cross-elasticities and the variance of alternatives). The MMNL-M model shows, as expected, a heightened sensitivity of drive-alone alternatives (relative to the shared-ride and transit alternatives) in response to a cost increase in the drive-alone P.M. peak alternative. The higher variance of the unobserved attributes specific to the shared-ride alternative (relative to transit; see Table 10.4) results in the lower cross-elasticity of the shared-ride alternatives than of the transit alternatives. The MMNL-MT model shows higher cross-elasticities for the drive-alone alternatives as well as for the non-drive-alone P.M. peak period alternatives since it allows shared unobserved attributes along both the mode and time dimensions.

The drive-alone P.M. peak period self-elasticities in Table 10.5 are also quite different across the models. The self-elasticity is lower in the MMNL-T model than in the MNL mode. The MMNL-T model recognizes the presence of temporal rigidity in social–recreational activities pursued in the P.M. peak. This is reflected in the lower self-elasticity effect of the MMNL-T model. The self-elasticity value from the MMNL-M model is larger than that from the MMNL-T model. This is because individuals are likely to maintain their current travel mode (even if it means shifting departure times) in the face of transportation control measures. But the MMNL-T model accommodates only the rigidity effect in departure time, not in travel mode. As a consequence, the rigidity in mode choice is manifested (inappropriately) in the MMNL-T model as a low drive-alone P.M. peak self-elasticity effect. Finally, the self-elasticity value from the MMNL-MT model is lower than the value from the MMNL-M models. The MMNL-M model ignores the rigidity in departure time; when this effect is included in the MMNL-MT model, the result is a depressed self-elasticity effect.

The substitution structures among the four models imply different patterns of competition among the joint mode–departure time alternatives. We now turn to the aggregate self- and cross-elasticities to examine the substantive implications of the different competition structures for the level-of-service variables.

Table 10.6 provides the cost elasticities obtained for the drive-alone and transit joint choices in response to a congestion pricing policy implemented in the P.M. peak. The aggregate cost elasticities reflect the same general pattern as the disaggregate elasticities discussed earlier. Some important policy relevant observations that can be made from the aggregate elasticities are as follows. The drive-alone P.M. peak self-elasticities show that the MNL and MMNL-T models underestimate the decrease in peak period congestion due to peak period pricing, while the MMNL-M model overestimates the decrease. Thus, using the drive-alone P.M. peak cost self-elasticities from the MNL and MNL-T models will make a policy analyst much more conservative than he or she should be in pursuing peak period pricing strategies. On the other hand, using the drive-alone P.M. peak cost self-elasticity from the MMNL-M model provides an overly optimistic projection of the congestion alleviation due to peak period pricing. From a transit standpoint, the MNL and MMNL-T models underestimate the increase in transit share across all time periods due to P.M. peak period pricing. Thus, using these models will result in lower projections of the increase in transit ridership and transit revenue due to a peak period pricing policy. The MMNL-M model underestimates the projected increase in transit share in all the nonevening time periods, and overestimates the increase in transit share for the evening time period. Thus, the MNL, MMNL-T, and MMNL-M models are likely to lead to inappropriate conclusions regarding the necessary changes in transit provision to complement peak period pricing strategies.

10.5 Conclusions

This chapter presents the structure, estimation techniques, and transport applications of three classes of discrete choice models: heteroskedastic models, GEV models, and flexible structure models. Within each class, alternative formulations are discussed. The formulations presented are quite flexible (this is especially the case with the flexible structure models), though estimation using the maximum likelihood technique requires the evaluation of one-dimensional integrals (in the heteroskedastic extreme value model) or multidimensional integrals (in the flexible model structures). However, these integrals can be approximated using Gaussian quadrature techniques or simulation techniques. In this regard, the recent use of quasi-Monte Carlo simulation techniques seems to be particularly effective.

TABLE 10.6　Aggregate Travel Cost Elasticities in Response to a Cost Increase in the Drive-Alone Mode during P.M. Peak

Effect on Joint Choice Alternative	MNL Model	MMNL-T Model	MMNL-M Model	MMNL-MT Model
		Drive-Alone Alternatives		
Early morning	0.0146	0.0202	0.0290	0.0392
A.M. peak	0.0125	0.0166	0.0259	0.0334
A.M. off-peak	0.0121	0.0155	0.0250	0.0317
P.M. off-peak	0.0123	0.0136	0.0254	0.0265
P.M. peak	−0.1733	−0.1536	−0.2355	−0.2192
Evening	0.0146	0.0088	0.0293	0.0204
		Transit Alternatives		
Early morning	0.0197	0.0260	0.0280	0.0371
A.M. peak	0.0188	0.0237	0.0283	0.0358
A.M. off-peak	0.0163	0.0195	0.0236	0.0291
P.M. off-peak	0.0168	0.0175	0.0246	0.0251
P.M. peak	0.0218	0.0393	0.0333	0.0485
Evening	0.0205	0.0120	0.0299	0.0203

Source: From Bhat, C.R., *Transp. Res. B*, 32, 455, 1998. With permission.

The advanced model structures presented in this chapter should not be viewed as substitutes for careful identification of systematic variations in the population. The analyst must always explore alternative and improved ways to incorporate systematic effects in a model. The flexible structures can then be superimposed on *models that have attributed as much heterogeneity to systematic variations as possible*. Another important issue in using flexible structure models is that the specification adopted should be easy to interpret; the analyst would do well to retain as simple a specification as possible while attempting to capture the salient interaction patterns in the empirical context under study.

The confluence of continued careful structural specification with the ability to accommodate very flexible substitution patterns and unobserved heterogeneity should facilitate the application of behaviorally rich structures in transportation-related discrete choice modeling in the years to come.

References

Ben-Akiva, M. and Bolduc, D., Multinomial Probit with a Logit Kernel and a General Parametric Specification of the Covariance Structure, working paper, Department of Civil and Environmental Engineering, Massachusetts Institute of Technology, Cambridge, MA, and Département d'économique, Université Laval, Sainte-Foy, Quebec, Canada, 1996.

Ben-Akiva, M. and Francois, B., Homogenous Generalized Extreme Value Model, working paper, Department of Civil Engineering, MIT, Cambridge, MA, 1983.

Ben-Akiva, M. and Lerman, S.R., *Discrete Choice Analysis: Theory and Application to Travel Demand*, MIT Press, Cambridge, MA, 1985.

Bhat, C.R., A heteroskedastic extreme-value model of intercity mode choice, *Transp. Res. B*, 29, 471–483, 1995.

Bhat, C.R., Recent Methodological Advances Relevant to Activity and Travel Behavior Analysis, invitational resource paper presented at International Association of Travel Behavior Research Conference, Austin, Texas, September 1997.

Bhat, C.R., Accommodating flexible substitution patterns in multidimensional choice modeling: Formulation and application to travel mode and departure time choice, *Transp. Res. B*, 32, 425–440, 1998a.

Bhat, C.R., Accommodating variations in responsiveness to level-of-service measures in travel mode choice modeling, *Transp. Res. A*, 32, 495–507, 1998b.

Bhat, C.R., An analysis of travel mode and departure time choice for urban shopping trips, *Transp. Res. B*, 32, 361–371, 1998c.

Bhat, C.R., Incorporating observed and unobserved heterogeneity in urban work travel mode choice modeling, *Transp. Sci.*, 34, 228–238, 2000.

Bhat, C.R., Quasi-random maximum simulated likelihood estimation of the mixed multinomial logit model, *Transp. Res. B*, 35, 677–693, 2001.

Bhat, C.R., Simulation estimation of mixed discrete choice models using randomized and scrambled Halton sequences, *Transp. Res.*, 2002, forthcoming.

Bhat, C.R. and Castelar, S., A unified mixed logit framework for modeling revealed and stated preferences: Formulation and application to congestion pricing analysis in the San Francisco Bay Area, *Transp. Res. B*, 36, 593–616, 2002.

Bhat, C.R. and Guo, J., A Spatially Correlated Logit Model: Formulation and Application to Residential Choice Modeling, draft working paper, Department of Civil Engineering, University of Texas at Austin, 2002.

Braaten, E. and Weller, G., An improved low-discrepancy sequence for multidimensional quasi-Monte Carlo integration, *J. Comput. Phys.*, 33, 249–258, 1979.

Bresnahan, T.F., Stern, S., and Trajtenberg, M., Market segmentation and the sources of rents from innovation: Personal computers in the late 1980s, *RAND J. Econ.*, 28, 17–44, 1997.

Brownstone, D., Discrete Choice Modeling for Transportation, resource paper presented at 2000 IATBR Conference, Gold Coast, Australia, July 2000.

Brownstone, D., Bunch, D.S., and Train, K., Joint mixed logit models of stated and revealed preferences for alternative-fuel vehicles, *Transp. Res. B*, 34, 315–338, 2000.

Brownstone, D. and Train, K., Forecasting new product penetration with flexible substitution patterns, *J. Econometrics*, 89, 109–129, 1999.

Chamberlain, G., Analysis of covariance with qualitative data, *Rev. Econ. Stud.*, 47, 225–238, 1980.

Chu, C., A paired combinatorial logit model for travel demand analysis, paper presented at Proceedings of the Fifth World Conference on Transportation Research, Ventura, CA, 1989, pp. 295–309.

Daganzo, C., *Multinomial Probit: The Theory and Its Application to Demand Forecasting*, Academic Press, New York, 1979.

Daganzo, C.F. and Kusnic, M., Two properties of the nested logit model, *Transp. Sci.*, 27, 395–400, 1993.

Daly, A.J. and Zachary, S., Improved multiple choice models, in *Determinants of Travel Choice*, Hensher, D.A. and Dalvi, M.Q., Eds., Saxon House, Westmead, U.K., 1978.

Fang, K.-T. and Wang, Y., *Number-Theoretic Methods in Statistics*, Chapman & Hall, London, 1994.

Forinash, C.V. and Koppelman, F.S., Application and interpretation of nested logit models of intercity mode choice, *Transp. Res. Rec.*, 1413, 98–106, 1993.

Han, A. and Algers, S., A Mixed Multinomial Logit Model for Route Choice Behavior, paper presented at 9th WCTR Conference, Seoul, Korea, July 2001.

Hellekalek, P., Regularities in the distribution of special sequences, *J. Number Theory*, 18, 41–55, 1984.

Hensher, D.A., A practical approach to identifying the market for high speed rail in the Sydney–Canberra corridor, *Transp. Res. A*, 31, 431–446, 1997.

Hensher, D.A., Extending valuation to controlled value functions and non-uniform scaling with generalized unobserved variances, in *Theoretical Foundations of Travel Choice Modeling*, Gärling, T., Laitila, T., and Westin, K., Pergamon, Oxford, 1998, pp. 75–102.

Hensher, D.A., The Valuation of Travel Time Savings for Urban Car Drivers: Evaluating Alternative Model Specifications, technical paper, Institute of Transport Studies, University of Sydney, Australia, 1999.

Hensher, D.A., Measurement of the valuation of travel time savings, in special issue of *Transportation Economics and Policy*, in honor of Michael E. Beesley, 2000, forthcoming.

Hensher, D.A. and Greene, W., Choosing between Conventional, Electric, and UPG/LNG Vehicle in Single Vehicle Households, technical paper, Institute of Transport Studies, University of Sydney, Australia, 2000.

Hensher, D.A., Louviere, J., and Swait, J., Combining sources of preference data, *J. Econometrics*, 89, 197–221, 1999.

Johnson, N. and Kotz, S., *Distributions in Statistics: Continuous Univariate Distributions*, John Wiley, New York, 1970, chap. 21.

Kocis, L. and Whiten, W.J., Computational investigations of low-discrepancy sequences, *ACM Trans. Math. Software*, 23, 266–294, 1997.

Koppelman, F.S. and Sethi, V., Closed-form discrete-choice models, in *Handbook of Transport Modelling*, Hensher, D. and Button, K., Eds., Pergamon, Oxford, 2000, 211–225.

Koppelman, F.S. and Wen, C.-H., The paired combinatorial logit model: Properties, estimation and application, *Transp. Res. B*, 34, 75–89, 2000.

KPMG Peat Marwick and Koppelman, F.S., Analysis of the Market Demand for High Speed Rail in the Quebec–Ontario Corridor, report produced for Ontario/Quebec Rapid Train Task Force, KPMG Peat Marwick, Vienna, VA, 1990.

Luce, R. and Suppes, P., Preference, utility and subjective probability, in *Handbook of Mathematical Psychology*, Vol. 3, Luce, R., Bush, R., and Galanter, E., Eds., Wiley, New York, 1965.

McFadden, D., Modeling the choice of residential location, *Transp. Res. Rec.*, 672, 72–77, 1978.

Mehndiratta, S., Time-of-Day Effects in Intercity Business Travel, Ph.D. thesis, Department of Civil Engineering, University of California, Berkeley, 1997.

Munizaga, M.A., Heydecker, B.G., and Ortuzar, J., Representation of heteroscedasticity in discrete choice models, *Transp. Res. B*, 34, 219–240, 2000.

Press, W.H., Teukolsky, S.A., and Nerlove, M., Numerical Recipes in C: The Art of Scientific Computing, Cambridge University Press, Massachusetts, 1992.

Recker, W.W., Discrete choice with an oddball alternative, *Transp. Res. B*, 29, 201–212, 1995.

Revelt, D. and Train, K., Mixed logit with repeated choices: Households' choices of appliance efficiency level, *Rev. Econ. Stat.*, 80, 647–657, 1998.

Shaw, J.E.H., A quasi-random approach to integration in Bayesian statistics, *Ann. Stat.*, 16, 895–914, 1988.

Small, K.A., A discrete choice model for ordered alternatives, *Econometrica*, 55, 409–424, 1987.

Stroud, A.H., *Approximate Calculation of Multiple Integrals*, Prentice Hall, Englewood Cliffs, NJ, 1971.

Swait, J., Choice set generation within the generalized extreme value family of discrete choice models, *Transp. Res. B*, 35, 643–666, 2001.

Train, K., Recreation demand models with taste differences over people, *Land Economics*, 74, 230–239, 1998.

Train, K., Halton Sequences for Mixed Logit, technical paper, Department of Economics, University of California, Berkeley, 1999.

Train, K., A Comparison of Hierarchical Bayes and Maximum Simulated Likelihood for Mixed Logit, technical paper, Department of Economics, University of California, Berkeley, 2001.

Tuffin, B., On the use of low discrepancy sequences in Monte Carlo methods, *Monte Carlo Methods Appl.*, 2, 295–320, 1996.

Vovsha, P., The Cross-Nested Logit Model: Application to Mode Choice in the Tel-Aviv Metropolitan Area, paper presented at the 1997 Annual Transportation Research Board Meeting, Washington, D.C., 1997.

Wen, C.-H. and Koppelman, F.S., The generalized nested logit model, *Transp. Res. B*, 35, 627–641, 2001.

Williams, H.C.W.L., On the formation of travel demand models and economic evaluation measures of user benefit, *Environ. Plann. A*, 9, 285–344, 1977.

Appendix

Permutations for Scrambled Halton Sequences

Prime r	Permutation of (0, 1, 2, ..., r − 1)
2	(0 1)
3	(0 2 1)
5	(0 3 1 4 2)
7	(0 4 2 6 1 5 3)
11	(0 5 8 2 10 3 6 1 9 7 4)
13	(0 6 10 2 8 4 12 1 9 5 11 3 7)
17	(0 8 13 3 11 5 16 1 10 7 14 4 12 2 15 6 9)
19	(0 9 14 3 17 6 11 1 15 7 12 4 18 8 2 16 10 5 13)
23	(0 11 17 4 20 7 13 2 22 9 15 5 18 1 14 10 21 6 16 3 19 8 12)
29	(0 15 7 24 11 20 2 27 9 18 4 22 13 26 5 16 10 23 1 19 28 6 14 17 3 25 12 8)

Source: From Braaten, E. and Weller, G., *J. Comput. Phys.*, 33, 249–258, 1979. With permission.

11

Structural Equation Modeling[1]

CONTENTS

Thomas F. Golob
University of California

11.1 Introduction

Structural equation modeling (SEM) is an extremely flexible linear-in-parameters multivariate statistical modeling technique. It has been used in transportation research since about 1980, and its use is rapidly accelerating, partially due to the availability of improved software. The number of published studies, now known to be more than 50, has approximately doubled in the past 3 years. This review of SEM is intended to provide an introduction to the field for those who have not used the method, and a compendium of applications for those who wish to compare experiences and avoid the pitfall of reinventing previously published research.

Structural equation modeling is a modeling technique that can handle a large number of endogenous and exogenous variables, as well as latent (unobserved) variables specified as linear combinations (weighted averages) of the observed variables. Regression, simultaneous equations (with and without error term correlations), path analysis, and variations of factor analysis and canonical correlation analysis are all special cases of SEM. It is a confirmatory, rather than exploratory, method because the modeler is required to construct a model in terms of a system of unidirectional effects of one variable on another. Each direct effect corresponds to an arrow in a path (flow) diagram. In SEM one can also separate errors in measurement from errors in equations, and one can correlate error terms within all types of errors.

---

[1]Based in large part on the article "Structural Equation Modeling for Travel Behavior Research," to appear in *Transportation Research Part B* (in press, 2002). With permission.

Estimation of SEM is performed using the covariance analysis method (method of moments). There are covariance analysis methods that can provide accurate estimates for limited endogenous variables, such as dichotomous, ordinal, censored, and truncated variables. Goodness-of-fit tests are used to determine if a model specified by the researcher is consistent with the pattern of variances–covariances in the data. Alternative SEM specifications are typically tested against one another, and several criteria are available that allow the modeler to determine an optimal model out of a set of competing models.

SEM is a relatively new method, having its roots in the 1970s. Most applications have been in psychology, sociology, the biological sciences, educational research, political science, and market research. Applications in travel behavior research date from 1980. Use of SEM is now rapidly expanding as user-friendly software becomes available and researchers become comfortable with SEM and regard it as another tool in their arsenal.

The remainder of this chapter is divided into two main parts: an introduction to SEM, and a review of applications of SEM in travel behavior research. Citations in the applications section are limited to models of travel demand, behavior, and values. Applications involving transportation from the perspectives of urban modeling, land use, regional science, geography, or urban economics are generally not included.

11.2 Methodology

11.2.1 SEM Resources

SEM is firmly established as an analytical tool, leading to hundreds of published applications per year. Textbooks on SEM include Bollen (1989), Byrne (2001), Hayduk (1987), Hoyle (1995), Kaplan (2000), Kline (1996), Loehlin (1998), Maruyama (1998), Mueller (1996), Schoenberg (1989), and Shipley (2000). Overviews of the state of the method can be found in Cudeck et al. (2001), Jöreskog (1990), Mueller (1997), and Yuan and Bentler (1997). The multidisciplinary journal *Structural Equation Modeling* has been published quarterly since 1994.

The following SEM software was found to be generally available in 2002. A comparative review of three of the most popular SEM programs (AMOS, EQS, and LISREL) is provided by Kline (1998a).

AMOS (Arbuckle, 1994, 1997) is a general-purpose SEM package (http://www.smallwaters.com/) also available as a component of SPSS statistical analysis software.

CALIS (Hartmann, 1992) is a procedure available with SAS statistical analysis software (http://www.sas.com/).

EQS (Bentler, 1989, 1995) is a well-known SEM package focusing on estimation with nonnormal data (http://www.mvsoft.com/).

EzPath (Steiger, 1989) provides SEM capability for SYSTAT statistical analysis software (http://www.spssscience.com/systat/).

LISCOMP (Muthén, 1988) pioneered estimation for nonnormal variables and is a predecessor of MPLUS.

LISREL (Jöreskog and Sörbom, 1993), with coupled modules PRELIS and SIMPLIS, is one of the oldest SEM software packages. It has been frequently upgraded to include alternative estimation methods and goodness-of-fit tests, as well as graphical interfaces (http://www.ssicentral.com/).

MPLUS (Muthén and Muthén, 1998) is a program suite for statistical analysis with latent variables that include SEM (http://www.statmodel.com/index2.html).

Mx (Neale, 1997), a matrix algebra interpreter and numerical optimizer for SEM, is available as freeware (http://views.vcu.edu/mx/).

SEPATH for STATISTICA software provides SEM with extensive Monte Carlo simulation facilities (http://www.statsoftinc.com/).

STREAMS (Structural Equation Modeling Made Simple) is a graphical model specification interface for AMOS, EQS, and LISREL (http://www.gamma.rug.nl).

TETRAD software (Scheines et al., 1994) provides tools for developing SEM by generating input files for CALIS, EQS, or LISREL (http://hss.cmu.edu/HTML/departments/philosophy/TETRAD/tetrad.html)

11.2.2   The Defining Features of SEM

A SEM with latent variables is composed of up to three sets of simultaneous equations, estimated concurrently: 1) a measurement model (or submodel) for the endogenous (dependent) variables, 2) a measurement (sub)model for the exogenous (independent) variables, and 3) a structural (sub)model, all of which are estimated simultaneously. This full model is seldom applied in practice. Generally, one or both of the measurement models are dropped. SEM with a measurement model and a structural model is known as SEM with latent variables. Alternatively, one can have a structural model without any measurement models (SEM with observed variables) or a measurement model alone (confirmatory factor analysis). In general, a SEM can have any number of endogenous and exogenous variables.

Suppose that we have a multivariate problem with p endogenous variables and q exogenous variables. For simplicity, we will assume that all variables are measured in terms of variations from their means. The first SEM component, the measurement model for the endogenous variables, is given by

$$y = \Lambda_y \eta + \varepsilon \tag{11.1}$$

This postulates that m latent (unobserved) endogenous variables, represented by the (m by 1) column vector η, are described (indicated) by the p observed endogenous variables, represented by the (p by 1) vector y. Typically, p > m. The vector ε of unexplained components (measurement errors) of the observed endogenous variables is defined to have a variance–covariance matrix Θ_ε. The parameters of this measurement model are the elements of the (p by m) matrix Λ_y and the (p by p) variance–covariance matrix Θ_ε. As usual, we require that ε is uncorrelated with η.

A SEM measurement model is used to specify latent (unobserved) variables as linear functions (weighted averages) of other variables in the system. When these other variables are observed, they take on the role of "indicators" of the latent constructs.[2] In this way, SEM measurement models are similar to factor analysis, but there is a basic difference. In exploratory factor analysis, such as principal components analysis, all elements of the Λ_y factor loadings matrix are estimated and will take on nonzero values. These values (factor loadings) generally measure the correlations between the factors and the observed variables, and rotations are routinely performed to aid in interpreting the factors by maximizing the number of loadings with high and low absolute values. In SEM, the modeler decides in advance which of the parameters defining the factors are restricted to be zero, and which are freely estimated or constrained to be equal to each other or to some nonzero constant. Specification of each parameter allows the modeler to conduct a rigorous series of hypothesis tests regarding the factor structure. Also, in SEM one can specify nonzero covariances among the unexplained portions of both the observed and latent variables. The ability to assign free SEM parameters is governed by rules pertaining to the identification of the entire SEM (Section 11.2.4). Since there can be a large number of possible combinations in a measurement model with more than just a few variables, exploratory factor analysis is sometimes used to guide construction of a SEM measurement model.

Second, a similar measurement model is available for the exogenous variables:

$$x = \Lambda_x \xi + \delta \tag{11.2}$$

where ξ denotes the (n by 1) vector of n latent exogenous variables, which we postulate are indicated by the (p by 1) vector x of observed exogenous variables. The vector δ of measurement errors, of the observed exogenous variables (uncorrelated with ξ), is defined to have a variance–covariance matrix Θ_δ. The parameters of Equation (11.2) are the elements of the (q by n) matrix Λ_x and the (q by q) symmetric

---

[2]In advanced applications, models can be specified in which latent variables are functions only of other latent variables. Such "phantom" latent variables allow the modeler to constrain parameters to be within certain ranges (e.g., greater than zero) and to construct other types of special effects, such as random effects and period-specific effects in dynamic data.

variance–covariance matrix Θ_δ. Unlike the endogenous variables, the variance–covariance matrix of the observed exogenous variables, denoted by Φ, is taken as given.

Finally, the structural model captures the causal relationships among the latent endogenous variables and the causal influences (regression effects) of the exogenous variables on the endogenous variables:

$$\eta = B\eta + \Gamma\xi + \zeta \tag{11.3}$$

where B is the (m by m) matrix of direct effects between pairs of latent endogenous variables, Γ is the (m by n) matrix of regression effects of the latent exogenous variables on the latent endogenous variables, and the (m by 1) vector of errors in equations, ζ, is defined to have the variance–covariance matrix Ψ. It is assumed that ζ is uncorrelated with ξ. The parameters here are the elements of B, Γ, and Ψ matrices. By definition, the main diagonal of B must be zero (no variable can effect itself directly), and identification requires that the matrix $(I - B)$ must be nonsingular. (Further identification conditions for the entire system are discussed in Section 11.2.4.) Recursive models are those in which the variables can be rearranged such that B has free elements only below the main diagonal.

A SEM can be viewed as always being comprised of these three equations, but if observed endogenous variables are used directly in the structural model, Equation (11.1) is trivialized by assuming m = p, Λ_θ is an identity matrix, and Θ_ε is a null matrix. Similarly, for a structural model with observed exogenous variables, we assume Equation (11.2) with n = q, Λ_δ as an identity matrix, and Θ_δ as a null matrix. SEM with observed variables then replaces η with x and ξ with x in Equation (11.3). However, SEM with observed variables still allows specification of error term covariances through the Ψ parameter matrix.

The general SEM system — consisting of Equations (11.1) through (11.3) — is estimated using covariance analysis (method of moments). In variance analysis methods, model parameters are determined such that the variances and covariances of the variables replicated by Equations (11.1) to (11.3) are as close as possible to the observed variances and covariances of the sample. It can be easily shown that the model-replicated combined variance–covariance (moments) matrix of the observed (p) endogenous and (q) exogenous variables, arranged so that the endogenous variables are first, is given by

$$\Sigma = \left[\begin{array}{c|c} \Lambda_y(I-B)^{-1}(\Gamma\Phi\Gamma'+\Psi)\left[(I-B)^{-1}\right]'\Lambda_y'+\Theta_\varepsilon & \Lambda_y(I-B)^{-1}\Gamma\Phi\Lambda_x' \\ \hline \Lambda_x\Phi\Gamma'\left[(I-B)^{-1}\right]'\Lambda_y' & \Lambda_x\Phi\Lambda_x'+\Theta_\delta \end{array} \right] \tag{11.4}$$

The (symmetric) submatrix in the upper left-hand quadrant of Equation (11.4) represents the SEM reproduction of the moments of the observed endogenous variables. Here the regression effects matrix Γ translates the given moments of the exogenous variables (Φ), which are then added to the moments of the errors in equations (Ψ). This combined inner term is then translated through the effects of the endogenous variables on one another and the measurement parameters Λ_y. Finally, errors in measurement are added on. The full rectangular submatrix in the upper right-hand quadrant (transposed in the lower left-hand quadrant) represents the covariances between the observed endogenous and exogenous variables. Here the given exogenous variable moments Φ are interpreted through the structural effects $(I - B)^{-1}\Gamma$ and the two measurement models (Λ_x and Λ_y). Finally, the symmetric submatrix in the lower right-hand quadrant represents the (factor analytic) measurement model for the exogenous variables.

For SEM with observed variables, Equation (11.4) reduces to

$$\Sigma(\theta) = \left[\begin{array}{c|c} \left[(I-B)^{-1}(\Gamma\Phi\Gamma'+\Psi)\left[(I-B)^{-1}\right]'\right] & (I-B)^{-1}\Gamma\Phi \\ \hline \Phi\Gamma'\left[(I-B)^{-1}\right]' & \Phi \end{array} \right] \tag{11.5}$$

This shows that the total effects of x on y are given by $(I - B)^{-1} \Gamma$. It can also be shown that the total effects of y on y are given by $(I - B)^{-1} - I$. If there is only one endogenous variable, Equation (11.5) reduces to the normal equations for multiple regression.

One of the most common SEM applications is with a single measurement model on the endogenous variable side. In this configuration, the model-replicated moments are

$$
\Sigma = \left[
\begin{array}{c|c}
\Lambda_y (I-B)^{-1} (\Gamma \Phi \Gamma' + \Psi) \left[(I-B)^{-1} \right]' \Lambda_y' + \Theta_\varepsilon & \Lambda_y (I-B)^{-1} \Gamma \Phi \\
\hline
\Phi \, \Gamma' \left[(I-B)^{-1} \right] \Lambda_y' & \Phi
\end{array}
\right]
\tag{11.6}
$$

An important distinction in SEM is that between direct effects and total effects. Direct effects are the links between a productive variable and the variable that is the target of the effect. These are the elements of the B and Γ matrices. Each direct effect corresponds to an arrow in a path (flow) diagram. A SEM is specified by defining which direct effects are present and which are absent. With most modern SEM software this can be done graphically by manipulating path diagrams. These direct effects embody the causal modeling aspect of SEM.[3] Total effects are defined to be the sum of direct effects and indirect effects, where indirect effects correspond to paths between the two variables that involve intervening variables. The total effects of the exogenous variables on the endogenous variables (given by $(I - B)^{-1} \Gamma$) are sometimes known as the coefficients of the reduced-form equations.

The general SEM system is estimated using covariance (structure) analysis, whereby model parameters are determined such that the variances and covariances of the variables implied by the model system are as close as possible to the observed variances and covariances of the sample. In other words, the estimated parameters are those that make the variance–covariance matrix predicted by the model as similar as possible to the observed variance–covariance matrix, while respecting the constraints of the model. Covariance analysis appears at first to be quite different from least square regression methods, but it is actually just an extension of least squares into the realm of latent variables, error term covariances, and nonrecursive models (i.e., models with feedback loops). In some simple cases, covariance analysis is identical to least squares. Estimation methodology is discussed in Section 11.2.5.

Advantages of SEM compared to most other linear-in-parameter statistical methods include the following capabilities:

1. Treatment of both endogenous and exogenous variables as random variables with errors of measurement
2. Latent variables with multiple indicators
3. Separation of measurement errors from specification errors
4. Testing of a model overall rather than coefficients individually
5. Modeling of mediating variables
6. Modeling of error term relationships
7. Testing of coefficients across multiple groups in a sample
8. Modeling of dynamic phenomena such as habit and inertia
9. Accounting for missing data
10. Handling of nonnormal data

These capabilities are demonstrated in many of the applications reviewed in Section 11.3.

---

[3]For discussions of SEM in the context of causal modeling see Berkane (1997), Pearl (2000), Shipley (2000), and Spirtes, Glymour, and Scheines (2001).

11.2.3 A Brief History of Structural Equation Models

It is generally agreed that no one invented SEM. One simple view is that SEM is the union of latent variable (factor analytic) approaches, developed primarily in psychology and sociology, and simultaneous equation methods of econometrics. Upon closer inspection, we see that modern SEM evolved out of the combined efforts of many scholars pursuing several analytical lines of research. Bollen (1989) proposed that SEM is founded on three primary analytical developments: (1) path analysis, (2) latent variable modeling, and (3) general covariance estimation methods. Here we will highlight the contributions of each of these three areas.[4]

Path analysis, developed almost exclusively by geneticist Sewall Wright (1921, 1934), introduced three concepts: (1) the first covariance structure equations, (2) the path diagram or causal graph, and (3) decomposition of total effects between any two variables into total, direct, and indirect effects. Shipley (2000) describes how and why path analysis was largely ignored in biology, psychology, and sociology until the 1960s. Prior to the 1960s, econometricians also pursued the testing of alternative causal relationships through the use of overidentifying constraints on partial correlations (e.g., Haavelmo, 1943), but for many years economics was also uninformed about the solutions inherent in path analysis (Epstein, 1987; Shipley, 2000). During the 1960s, sociologists in particular, led by Blalock (1961), Boudon (1965), and Duncan (1966), discovered the potential of path analysis and related partial correlation methods. Path analysis was then superseded by SEM, in which general covariance structure equations specify how alternative chains of effects between variables generate correlation patterns. Modern SEM still relies on path diagrams to express what the modeler postulates about the causal relationships that generate the correlations among variables.

The development of models in which inferences about latent variables could be derived from covariances among observed variables (indicators) was pursued in sociology during the 1960s. These latent variable models contributed significantly to the development of SEM by demonstrating how measurement errors (errors in variables) can be separated from specification errors (errors in equations). A seminal contribution was that of Blalock (1963). These models led directly to the first general SEM, developed by Jöreskog (1970, 1973), Keesling (1972), and Wiley (1973).

Wright's path analysis lacked the ability to test specific hypotheses regarding a postulated causal structure. Work by Lawley (1940), Anderson and Rubin (1956), and Jöreskog (1967, 1969) led to the development of maximum likelihood (ML) estimation methods for confirmatory factor analysis, which in turn led to the estimation of models in which confirmatory factor analysis was combined with path analysis (Jöreskog, 1970, 1973; Keesling, 1972). ML estimation allowed testing of individual direct effects and error term correlations, and it is still the most widely used estimation method for SEM (Section 11.2.5).

Modern SEM was originally known as the JKW (Jöreskog–Keesling–Wiley) model. SEM was initially popularized by the wide distribution of the LISREL (Linear Structural Relationships) program developed by Jöreskog (1970), Jöreskog et al. (1970), and Jöreskog and Sörbom (1979). For some time, SEM was synonymous with LISREL, but there are now many SEM programs available (Section 11.2.1).

11.2.4 Model Specification and Identification

Any SEM is constructed in terms of postulated direct effects between variables and optional error term covariances of several types. Each postulated effect usually corresponds to a free parameter. If the SEM has no measurement models (no latent variables), there are four types of potential free parameters: (1) regression effects of the exogenous variables on the endogenous variables, (2) effects of the endogenous variables on one another, (3) variances of the unique portions (error terms) of each endogenous variable,

---

[4]For more detailed perspectives on the genesis of SEM, see Aigner et al. (1984), Duncan (1975), Goldberger (1972), Bielby and Hauser (1977) and Bentler (1980). Historical background is also discussed in many of the SEM texts listed in Section 11.2.1.

and (4) covariances between the error terms of the endogenous variables. If the SEM contains a measurement submodel for the endogenous variables, the above error term variances and covariances pertain to the *latent* endogenous variables, and the potential list of free parameters is increased to include (5) effects (similar to factor loadings) of the latent variables on the observed indicators, (6) variances of the (measurement) error terms of the observed variables, and (7) covariances between the error terms of the observed variables. If the SEM contains a measurement submodel for the exogenous variables, there will be a similar opportunity for error term variances and covariances pertaining to exogenous latent variables. Modern SEM software allows specification of a model using one or more of three tools: matrix notation, symbolic equations, and graphs, by specifying arrows in a flow diagram.

We are usually in search of a parsimonious description of travel behavior. In SEM, the primary measure of parsimony is the *degrees of freedom* of the model, which is equal to the difference between the number of free parameters in the model and the number of known quantities. The number of known quantities in covariance analysis is equal to the number of free elements in the variance–covariance matrix of the variables. The art of constructing a SEM involves specifying an overidentified model in which only some of the possible parameters are free and many are restricted to zero, but the model is nevertheless a reasonable representation of the phenomena under study (criteria for assessing model fit are discussed in Section 11.2.6). Theory and good sense must guide model specification. A saturated, or just-identified, SEM has zero degrees of freedom and fits perfectly, but it is only of interest as a baseline for certain goodness-of-fit criteria and as a means of exploring candidate parameters for restriction to zero. The most common ways of reducing model complexity are to eliminate weak regression effects, to reduce the number of indicators of each latent variable, and to minimize weak covariances between error terms. For SEM with latent variables, it is recommended that the measurement model(s) be developed first, followed by the structural model (Anderson and Gerbing, 1988).

Estimation of a model is not possible if more than one combination of parameter values will reproduce the same data (covariances). Such an indeterminate model is termed to be unidentified or underidentified. In models of travel behavior with a single endogenous variable, identification is not generally a problem, except when caused by special patterns in the data (empirical underidentification). In SEM, empirical underidentification can also be a problem, but the cause of an indeterminate solution is usually the design of the model (structural underidentification). The flexibility of SEM makes it fairly easy to specify a model that is not identified.

Heuristics are available to guide the modeler. There are separate rules of thumb for the measurement model and structural model, but an entire system may be identified even if a rule of thumb indicates a problem with one of its submodels, because restrictions in one submodel can aid in identifying the other submodel. Rules of thumb for identification of measurement models are reviewed in Bollen (1989, pp. 238–254), Reilly (1995), and Shipley (2000, pp. 164–171). These rules involve the number of observed variables to which each latent variable is linked and whether or not the error terms of the latent variables are specified as being correlated.[5]

Rules of thumb for identification of structural models (and the only concern for SEM with observed variables) are reviewed in Bollen (1989, pp. 88–104), Rigdon (1995), and Shipley (2000, pp. 171–173). Basically, all recursive models, in which there are no feedback loops in the chains of direct effects, will be identified as long as there are no error term correlations. Nonrecursive models can be broken into blocks in which all feedbacks are contained within a block, so that the relationship between the blocks is recursive. If each block satisfies identification conditions, then the entire model is also identified (Fox, 1984; Rigdon, 1995). The modeler can also check the rank order of a composite matrix involving the exogenous variable effects and the effects among the endogenous variables to verify that a structural model will be identified even if there are unlimited error term correlations (Bollen, 1989).

---

[5]The "three measure rule" asserts that a measurement model will be identified if every latent variable is associated with at least three observed variables; and the "two measure rule" asserts that a measurement model will be identified if every latent variable is associated with at least two observed variables and the error term of every latent variable is correlated with at least one other latent variable error term.

Confronted with an underidentified model, SEM software might diagnose the identification problem. However, detection is not guaranteed, and the program might either produce peculiar estimates or fail to converge to a solution. Detection is generally based on interrogating the rank of the information matrix of second-order derivatives of the fitting function. Unfortunately, rank is almost always evaluated sequentially and pertains only to a local solution. Thus, when a deficiency is detected, only the first parameter involved in the problem is identified and there is no information about other parameters that are also involved in the indeterminacy (McDonald, 1982). Identification problems can also be uncovered by testing whether the same solution is obtained when reestimating the model with an alternative initial solution, or by substituting the model-reproduced variance–covariance matrix for the sample matrix. Also, by using methods of modern computer algebra, the rank of an augmented version of the Jacobian matrix of first derivatives of the fitting function can establish whether a model is structurally identified (Bekker et al., 1994). An abnormally large coefficient standard error or covariance is evidence of undetected identification problems.

11.2.5   Estimation Methods and Sample Size Requirements

The fundamental principle of covariance analysis is that every linear statistical model implies a variance–covariance matrix of its variables. The functional form of every element in the combined variance–covariance matrix of the endogenous and exogenous variables can be derived from the SEM equations using simple matrix algebra. Covariance analysis works by finding model parameters such that the variances and covariances implied by the model system are as close as possible to the observed variances and covariances of the sample. In simple multiple regression, this exercise leads to the normal equations of ordinary least squares. For SEM with multiple endogenous variables, especially SEM with latent variables, estimation becomes more challenging, and quite a few different methods have been developed. Selection of an appropriate SEM estimation method depends on the assumptions one is willing to make about the probability distribution, the scale properties of the variables, the complexity of the SEM, and the sample size.

The mostly commonly used SEM estimation methods today are normal theory ML, generalized least squares (GLS), weighted least squares (WLS), in forms such as asymptotically distribution-free weighted least squares (ADF or ADF-WLS), and elliptical reweighted least squares (EGLS or ELS).[6] These methods all involve a scalar fitting function that is minimized using numerical methods. Parameter standard errors and correlations are computed from the matrices of first and second derivatives of the fitting function. The product of the optimized fitting function and the sample size is asymptotically chi-square distributed with degrees of freedom equal to the difference between the number of free elements in the observed variance–covariance and the number of free parameters in the model.[7] In SEM group models, the variance–covariance data are stacked and hypothesis tests can be conducted to determine the extent to which each group differs from every other group.

ML is the method used most often. The ML solution maximizes the probability that the observed covariances are drawn from a population that has its variance–covariances generated by the process implied by the model, assuming a multivariate normal distribution. The properties of ML estimators have been thoroughly investigated with respect to the effect of violations from normality and sample size on biases of estimators, nonconvergence, and improper solutions (e.g., Boomsma, 1982; Bollen, 1989; Finch et al., 1997; Hoogland and Boomsma, 1998; Kline, 1998b). The bottom line is that ML estimation

---

[6]Lesser used methods include unweighted least squares (ULS), diagonally weighted least squares (DWLS), and instrumental variable (IV) methods, such as three-stage least squares. IV methods are sometimes used to establish initial values for ML, GLS, and WLS.

[7]Depending on the estimation method and whether the correlation or variance-covariance matrix is being analyzed, either the sample size or the sample size minus one is used in the chi-square calculation. Also, under certain assumptions, the chi-square distribution can be considered to be non-central, and some goodness-of-fit criteria (Section 11.2.6) correspond to how well a model reduces the noncentrality parameter of the distribution.

is fairly robust against violations of multivariate normality for sample sizes commonly encountered in transportation research. Excess kurtosis has been shown in simulation studies to be the main cause of biases in ML estimates, and some software packages provide measures of multivariate kurtosis (Mardia, 1970) as an aid in assessing the accuracy of ML estimates and goodness of fit. Skewness is less of a problem. Corrections have also been developed to adjust ML estimators to account for nonnormality. These include a robust ML (RML) standard error estimator (Browne, 1984; Bentler, 1995) and a scaled ML (SML) test statistic (Satorra and Bentler, 1988). In addition, Bayesian full-information ML estimators based on the expectation-maximization (EM) algorithm are now becoming available for use with missing and nonnormal data (Lee and Tsang, 1999; Lee and Shi, 2000).

The robustness of corrected ML estimation means that it can be used in many situations with discrete choice variables, ordinal scales used to collect data on feelings and perceptions (e.g., Likert scales), and truncated and censored variables.[8] In order to further reduce biases, ADF-WLS and related elliptical estimators have been specifically designed for limited endogenous variables. These estimators have been shown to be consistent and asymptotically efficient, with asymptotically correct measures of model goodness of fit, under a broad range of conditions (Bentler, 1983; Browne, 1982, 1984; Muthén, 1983, 1984; Bock and Gibbons, 1996). Comparisons of the performance of ADF-WLS vs. alternative methods are provided by Sugawara and MacCallum (1993), Fan et al. (1999), and Boomsma and Hoogland (2001). The major disadvantage of ADF-WLS and related estimators is that they require a larger sample size than ML, due to their heavy reliance on asymptotic assumptions and required computation and inversion of a matrix of fourth-order moments.[9]

Sample size issues have received considerable attention (e.g., Anderson and Gerbing, 1988; Bentler, 1990; Bentler and Yuan, 1999; Bollen, 1990; Hoogland and Boomsma, 1998). The consensus is that the minimum sample sizes for ADF-WLS estimation should be at least 1000 (Hoogland and Boomsma, 1998), some say as high as 2000 (Hoyle, 1995; Ullman, 1996; Boomsma and Hoogland, 2001). ML estimation also requires a sufficient sample size, particularly when nonnormal data are involved. Based on Monte Carlo studies of the performance of various estimation methods, several heuristics have been proposed:

1. A minimum sample size of 200 is needed to reduce biases to an acceptable level for any type of SEM estimation (Kline, 1998b; Loehlin, 1998; Boomsma and Hoogland, 2001).
2. Sample size for ML estimation should be at least 15 times the number of observed variables (Stevens, 1996).
3. Sample size for ML estimation should be at least five times the number of free parameters in the model, including error terms (Bentler and Chou, 1987; Bentler, 1995).
4. Finally, with strongly kurtotic data, the minimum sample size should be ten times the number of free parameters (Hoogland and Boomsma, 1998). Bootstrapping is an alternative for ML estimation with small samples (Shipley, 2000).

11.2.6 Assessing Goodness-of-Fit and Finding the Best Model

Many criteria have been developed for assessing overall goodness of fit of a structural equation model and measuring how well one model does vs. another model.[10] Most of these evaluation criteria are based on the chi-square statistic given by the product of the optimized fitting function and the sample size.

---

[8] A current limitation is that SEM estimation methods will only support dichotomous and ordered polychotomous categorical variables. This means that a multinomial discrete choice variable must be represented in terms of a multivariate choice model by breaking it down into component dichotomous variables linked by free error covariances (Muthén, 1979).

[9] A previous disadvantage of WLS and related methods, computational intensity, has been eliminated with the capabilities of modern personal computers.

[10] For overviews of SEM goodness-of-fit, see Bentler (1990), Bollen and Long (1992), Gerbing and Anderson (1992), Hu and Bentler (1999), and Mulaik et al. (1989).

The objective is to attain a *nonsignificant* model chi-square, since the statistic measures the difference between the observed variance–covariance matrix and the one reproduced by the model. The level of statistical significance indicates the probability that the differences between the two matrices are due to sampling variation. While it is generally important to attain a nonsignificant chi-square, most experts suggest that chi-square should be used as a measure of fit, not as a test statistic (Jöreskog and Sörbom, 1993). One rule of thumb for good fit is that the chi-square should be less than two times its degrees of freedom (Ullman, 1996).

There are problems associated with the use of fitting-function chi-square, mostly due to the influences of sample size and deviations from multinormality. For large samples it may be very difficult to find a model that cannot be rejected due to the direct influence of sample size. For such large samples, critical N (Hoetler, 1983) gives the sample size for which the chi-square value would correspond to $p = 0.05$; a rule of thumb is that critical N should be greater than 200 for an acceptable model (Tanaka, 1987). For small sample sizes, asymptotic assumptions become tenuous, and the chi-square value derived from the ML fitting function is particularly sensitive to violations from multinormality. Many of the following goodness-of-fit indices use normalizations to cancel out sample size in the chi-square functions, but the mean of the sampling distribution of these indices is still generally a function of sample size (Bollen, 1990; Bentler and Yuan, 1999).

Goodness-of-fit measures for a single model based on chi-square values include 1) root mean square error of approximation (RMSEA), which measures the discrepancy per degree of freedom (Steiger and Lind, 1980); 2) Z-test (McArdle, 1988); and 3) expected cross-validation index (ECVI) (Browne and Cudeck, 1992). Most SEM programs provide these measures together with their confidence intervals. It is generally accepted that the value of RMSEA for a good model should be less than 0.05 (Browne and Cudeck, 1992), but there are strong arguments that the entire 90% confidence interval for RMSEA should be less than 0.05 (MacCallum et al., 1996).

Several goodness-of-fit indices compare a proposed model to an independence model by measuring the proportional reduction in some criterion related to chi-square.[11] Most programs calculate several of these indices using a model with no restrictions whatsoever as the baseline model. Using such a naïve baseline, a rule of thumb for most of the indices is that a good model should exhibit a value greater than 0.90 (Mulaik et al., 1989; Bentler, 1990; McDonald and Marsh, 1990). Unfortunately, in many applications these indices will be very close to unity because of the very large chi-square values associated with such independence models. This renders them of little use when distinguishing between two well-fitting models. However, there is more than one interpretation of an independence model, so these indices should be recalculated using a baseline model that is appropriate for each specific application (Sobel and Bohrnstedt, 1985).

The performance of models with substantially different numbers of parameters can be compared using criteria based on the Bayesian theory. The Akaike Bayesian information criterion (variously abbreviated ABIC, BIC, or AIC) compares ML estimation goodness of fit and the dimensionality (parsimony) of each model (Akaike, 1974, 1987).[12] Modifications of the ABIC, the consistent Akaike information criterion (CAIC) (Bozdogan, 1987), and the Schwarz Bayesian criterion (SBC) (Schwarz, 1978) take into account the sample size as well as the model chi-square and number of free parameters. These criteria can be used to compare not only two alternative models of similar dimensionality, but also the models to the

---

[11]These indices, which mainly differ in terms of the normalization used to account for sample size and model parsimony, include: (1) normed fit index, which is variously designated in SEM software output as NFI, BBI, or D1 (Bentler and Bonett, 1980); (2) non-normed fit index (NNFI, TLI or RNI) (Tucker and Lewis, 1973; Bentler and Bonett, 1980); (3) comparative fit index (CFI) (Bentler, 1989; Steiger, 1989); (4) parsimonious normed fit index (PNFI) (James et al., 1982); (5) relative normed index (designated as RFI or r) (Bollen, 1986); and (6) incremental fit index (IFI or D2) (Bollen, 1989; Mulaik et al., 1989).

[12]Discussions of the role of parsimony in model evaluation and the effects of sample size and model complexity on criteria such as the three used here are provided by Bentler (1990), Bentler and Bonett (1980), McDonald and Marsh (1990), and Mulaik et al. (1989).

independence model at one extreme and the saturated model (perfect fit) at the other extreme. The model that yields the smallest value of each criterion is considered best.

Goodness-of-fit measures based on the direct comparison of the sample and model-implied variance–covariance matrices include (1) root mean square residual (RMR), or average residual value; (2) standardized RMR (SRMR), which ranges from zero to one, with values less than 0.05 considered a good fit (Byrne, 1989; Steiger, 1990); (3) goodness-of-fit index (GFI); (4) adjusted GFI (AGFI), which adjusts GFI for the degrees of freedom in the model; and (5) parsimony-adjusted GFI (PGFI) (Mulaik et al., 1989). R^2 values are also available by comparing estimated error term variances to observed variances. It is important to distinguish between R^2 values for reduced form equations and those for the structural equations.

Based on these goodness-of-fit tests for a model, a travel demand modeler can take one of three different courses of action:

1. Confirm or reject the model being tested based on the results. If a model is accepted, it should be recognized that other unexamined models might fit the data as well or better. Confirmation means only that a model is *not rejected*.
2. Two or more competing models can be tested against each other to determine which has the best fit. The candidate models would presumably be based on different theories or behavioral assumptions.
3. The modeler can also develop alternative models based on changes suggested by test results and diagnostics, such as first-order derivatives of the fitting function. Models confirmed in this manner are post hoc. They may not fit new data, having been created based on the uniqueness of an initial data set.

The availability of published results from previous studies affects the balance between a confirmatory or exploratory approach for a given application. Such results from structural equation modeling in travel behavior research are reviewed in the remainder of this paper. The following bibliography is organized by topic, and the citations within each section are generally in chronological order.

11.3 Transportation Research Applications

The earliest known applications of SEM to travel behavior are a joint model of vehicle ownership and usage (Den Boon (1980), reviewed in Section 11.3.1) and a dynamic model of mode choice and attitudes (Lyon (1981a, 1981b), Section 11.3.2). Tardiff (1976) and Dobson et al. (1978) (Section 11.3.4) developed simultaneous equation models of travel behavior and attitudes that are precursors of SEM applications. Finally, insightful early discussions of SEM as a potential tool in modeling travel demand can be found in Charles River Associates (1978) and Allaman et al. (1982).

11.3.1 Travel Demand Modeling Using Cross-Sectional Data

Models of vehicle ownership and usage are a natural application for SEM, through which it is possible to capture the mutual causal effects between vehicle ownership and distance traveled in a simultaneously estimated system, rather than through sequential estimation with selectivity corrections. Den Boon (1980) shows how this can be accomplished. Later, Golob (1998) modeled travel time, vehicle miles of travel, and car ownership together, using data for Portland, Oregon. A model of household vehicle usage and driver allocation was developed by Golob et al. (1996). WLS estimation is used with U.S. data for urban regions within California. Vehicle usage is expressed in reduced-form equations as a function of household and vehicle characteristics.

Pendyala (1998) investigates the dependence of SEM on the homogeneity of a causal travel behavior process across the population of interest. Results are presented from models estimated on simulated data generated from competing causal structures. These estimates are shown to perform poorly in the presence of structural heterogeneity.

Fujii and Kitamura (2000b) and Golob (2000) developed models of trip chain generation. As these models encompass activity duration in addition to conventional travel measures of trip generation and travel time, they are further discussed in Section 11.3.3.

Axhausen et al. (2001) tests causal hypotheses linking car ownership, season ticket ownership, and modal usage in Switzerland. The results confirm the dominance of car ownership, which drives the other variables. However, car usage was found to be complementary with public transport usage through direct positive links to season ticket ownership and public transport usage. Following up on this work, Simma and Axhausen (2001a) compared interrelationships between car ownership, season tickets, and travel and found consistent results in models using similar data from three countries (Germany, Great Britain, and Switzerland).

Simma and Axhausen (2001b) demonstrate a SEM that captures travel behavior relationships between male and female heads of household. The endogenous variables were car ownership, distances traveled by males and females, and male and female trips by two types of activities. Exogenous variables included the employment status of each head, family characteristics, and measures of residential accessibility and local land use.

Finally, Simma (2000) and Simma et al. (2001) investigated the effects of spatial structure on car ownership, trips by mode, and travel distance, using trip diary and environmental data for Austria. Household-based accessibility measures were found to be more influential than municipal and regional measures developed from gravity models and land use characteristics.

11.3.2 Dynamic Travel Demand Modeling

Panel data modeling is a natural application for SEM. Models can be specified with repeated variables joined by lagged causal effects and possibly autocorrelated error structures. Moreover, time-invariant individual-specific terms can be incorporated in error structures, and period effects can be isolated with certain types of panel data.

Lyon (1981a, 1981b, 1984) was the first to develop a dynamic SEM incorporating travel choices and attitudes. At the time of this work, the lack of available SEM estimation methods for nonnormal variables motivated the use of a sequential IV approach to parameter estimation. This work represents an important breakthrough in the application of SEM to the modeling of travel behavior and values. SEM allows the exploration of mutual causality between attitudes and behavior (Section 11.3.4).

In Golob and Meurs (1987, 1988) are early examples of SEM applied to (Dutch) panel trip diary data. These models suffer from a lack of exogenous variable effects. Golob and van Wissen (1989) unify explanation of car ownership and travel distances by mode, but the SEM is once again short on exogenous household characteristics, with the exception of household income. ML estimation is applied to Dutch data.

In joint dynamic models of car ownership and trip generation (Golob, 1989) and car ownership and travel time expenditures (Golob, 1990b) it is demonstrated that a SEM applied to (Dutch) panel data is able to capture both panel conditioning biases and period effects. The models also capture lags between car travel needs and vehicle transactions and incorporate autocorrelated errors. In a related discussion that is now outdated, Golob (1990a) explores use of SEM with panel data on travel choices.

Kitamura (1989) uses dynamic log-linear models (generalized linear model, or GLM) applied to Dutch panel survey data instead of SEM to test alternative causal postulates concerning travel behavior. In general, SEM and GLM are intimately related (McCullagh and Nelder, 1989), and van Wissen and Golob (1990) directly compare GLM and SEM on the same data. The authors conclude that SEM is more effective in distinguishing the performance of competing hypotheses.

Once again using panel data for The Netherlands, van Wissen and Golob (1992) present a dynamic SEM of car fuel type choice and mobility that captured influences of reduced vehicle operating costs on latent demand for car travel. The model incorporates individual-specific time-invariant effects. WLS estimation was used.

Using data from a two-wave panel survey of residents of Davis, California, Mokhtarian and Meenak-shisundaram (1998, 1999) develop dynamic models of travel and three communication activities: personal meetings, object transfer (e.g., mail), and electronic transfer (phone, fax, and e-mail). The authors found very little evidence of the substitution of electronic communication for trips. The relatively small sample size restricts model complexity. ML estimation was used.

Fujii and Kitamura (2000a) use multiday panel data from drivers in the Osaka–Kobe region of Japan to test hypotheses concerning how drivers collect and process information about anticipated travel time. Anticipated travel time is modeled as a function of lagged anticipated time, lagged actual time, and time forecasted by different sources (e.g., mass media and word of mouth). The relatively small sample size called for ML estimation.

Multiday travel is also modeled by Simma and Axhausen (2001c). Using a 6-week travel diary for areas in Germany and pooling the data by week, the authors present SEM results that shed light on the nature of linkages between travel on successive days of the week, for individuals and household couples, in terms of both travel distances and trip making.

11.3.3 Activity-Based Travel Demand Modeling

SEM has considerable potential here. Activity participation and travel can be modeled within a comprehensive framework that captures (1) the direct relationships between activity demand and the need to travel to get to activity sites, (2) interrelationships between participation in different activities, and (3) feedbacks from travel time to activity time (travel time budget effects), all conditional on personal and household characteristics. Kitamura (1997) and Pas (1997a, 1997b) provide comprehensive overviews of activity-based travel demand modeling that include discussions of the role of SEM.

Kitamura et al. (1992) and Golob et al. (1994) were the first to apply SEM in modeling joint demand for activity duration and travel. Results, estimated using ML applied to California time use survey data, confirm a negative feedback of commute time to nonwork activities; individuals with longer commutes have less time available for discretionary activities.

Lu and Pas (1999) present a SEM of in-home activities, out-of-home activities (by type), and travel (measured various ways), conditional on socioeconomic variables. Estimation is by ML, and the emphasis is on interpretation of the direct and indirect effects. The activity diary data are for the greater Portland, Oregon, metropolitan area.

Golob and McNally (1997) model the interactions of household heads in activity and travel demand. Activities are divided into three types, and SEM results are compared using ML and WLS estimation methods. The authors conclude that, where possible, WLS methods should be used to estimate SEM applied to activity participation data.

Gould and Golob (1997) and Gould et al. (1998) use SEM to explore how travel time saved by working at home or shopping close to home might be converted to other activities and other travel. Certain population segments were found to exhibit latent demand for activities. ML estimation is applied to Portland data.

Golob (1998) develops a joint SEM of vehicle ownership, activity participation (by activity type), travel time expenditure (by trip purpose), and household aggregate vehicle miles of travel. The major distinction of this work is that an ordered discrete choice household car ownership variable is included with time use and distance generation variables in a single SEM. WLS is used with data for Portland.

Two independent joint trip chain and time use models were also published in 2000. Fujii and Kitamura (2000b) studied the latent demand effects of the opening of new freeways. The authors used SEM to determine the effects of commute duration and scheduling variables on after-work discretionary activities and trips. They used sequential instrumental variables estimation, which they refer to as a measurement model. Data are for the Osaka–Kobe region of Japan. Similarly, Golob (2000) estimated a joint model of work and nonwork activity duration, four types of trip chains, and three measures of travel time expenditure. ML estimation was applied to Portland data, and the effects of in-home work and residential accessibility were explored using the model.

Finally, Kuppam and Pendyala (2001) present three separate models estimated using WLS applied to data for Washington, D.C. The models focused on relationships between (1) activity duration and trip generation, (2) durations of in-home and out-of-home activities, and (3) activity frequency and trip chain generation.

11.3.4   Attitudes, Perceptions, and Hypothetical Choices

Applied to data on attitudes, perceptions, stated behavioral intentions, and actual behavior, SEM can be used to specify and test alternative causal hypotheses. It has been found that, as might be expected, causality is often mutual. The assumption that behavior is influenced by attitudes, perceptions, and behavioral intentions — without feedback from behavior to these other variables — does not hold up when tested using SEM. These results challenge the assumption, held by some, that stated preference (SP) choices or ratings can be directly scaled into revealed preference (RP) choice models. SEM results show that, in most applications, SP data are a direct function of RP choice.

Tardiff (1976) uses path analysis to demonstrate empirical evidence that the causal link from choice behavior to attitudes is stronger than the link from attitudes to choice behavior. Subsequent studies using different forms of simultaneous equation modeling showed consistently that attitudes, especially perceptions, are conditioned by choices, while at the same time, attitudes affect choices (e.g., Dobson et al., 1978).

Golob et al. (1997b) developed models in which travel times, attitudes toward carpooling, mode, and route choice are modeled over time using panel survey data for San Diego, California. The SEM, which assumes ordinal scales and discrete choice variables, has individual-specific terms that take advantage of repeated measurements to account for population heterogeneity.

Golob and Hensher (1998) employ SEM to address the dichotomy between an individual's behavior and his or her support for policies that are promoted as benefiting the environment. Through the use of latent variables, attitudes are related to behavioral variables representing mode choice and choice of compressed work schedules, all of which are conditioned by a set of exogenous variables. The attitude scales are treated as ordinal, choices are treated as discrete, and the SEM is estimated using WLS applied to data for major Australian urban areas.

A SEM that combines SP and RP data from the same households in California to explain vehicle usage as a function of vehicle type, vintage, and fuel type to predict use of limited-range electric vehicles was developed by Golob et al. (1997a). Joint SP and RP estimation using SEM allows SP and RP error terms to be correlated while simultaneously testing for causal effects of RP (experiences) on SP (preferences).

Morikawa and Sasaki (1998) employ a SEM in concert with a discrete choice model to capture the influence of latent subjective indicators of the attributes of choice alternatives on choice. Using a Dutch survey of intercity travel and joint ML estimation, the authors concluded that models with causality only from attitudes to behavior perform more poorly than those that incorporate a causal feedback from behavior to attitudes. The preferred model involves estimation of the SEM and discrete choice equations simultaneously.

Levine et al. (1999) present two latent variable models that explain financial support for public transport and support for an institutional reform in public transit planning. The models, estimated using ML applied data collected in southeast Michigan, contain as many as six latent endogenous variables with observed ordinal and discrete indicators, and several sociodemographic variables.

A SEM with five latent variables is used by Jakobsson et al. (2000) to investigate causality among acceptance of road pricing, behavioral intention concerning reductions in car usage, and feelings related to fairness and infringement on personal freedom. ML is applied using data from a Swedish survey.

Stuart et al. (2000) used SEM to determine how a series of ratings of attributes of the New York subway (e.g., crowding, personal security, cleanliness, predictability of service) are related to customers' ratings of value and overall satisfaction with the system. ML estimation is applied using a sample of over 1000 transit panel participants.

In a combination of attitudinal and activity-based modeling, Fujii et al. (2000) used SP (budget allocation) and RP data collected in the Osaka–Kobe region to study joint activity engagement. Satisfaction with the activity pattern, discretionary trip frequency, and discretionary travel time are modeled as a function of in-home and out-of-home activity duration broken down by household activity participation. Sequential IV estimation is used.

Sakano and Benjamin (2000) developed a SEM that modeled SP responses concerning a new mode, together with attitudes and perceptions about the travel environment, and exogenous personal and modal characteristics. The data are for Winston-Salem and Greensboro, North Carolina, and ML estimation is used. An important contribution is that model forecasts are computed and interpreted.

Gärling et al. (2001) explores decision making involving driving choices by using a SEM with latent variables to test links among attitudes toward driving, frequency of choice of driving, and revealed presence of a certain type of decision process known as script based. ML estimation is applied to Swedish survey data. The authors followed up the SEM results with laboratory experiments.

The effects of negative critical incidents on cumulative satisfaction with public transport is determined by Friman et al. (2001) by applying a SEM with a measurement model to Swedish data on attitudes and experiences. Friman and Gärling (2001) extend the results of the first study by applying a SEM to stated preference data involving satisfaction under a variety of conditions involving treatment by public transport employees, service reliability, clarity of service information, and comfort.

Golob (2001) tested a series of joint models of attitude and behavior to explain how both mode choice and attitudes regarding a combined high-occupancy vehicle (HOV) and toll facility (HOT lanes) differ across the population. Applying WLS estimation to a data set from San Diego, the author demonstrates that choices appear to influence some opinions and perceptions, but other opinions and perceptions are independent of behavior and dependent only on exogenous personal and household variables. None of the models tested found any significant effects of attitudes on choice.

Finally, Sakano and Benjamin (2001) estimate a SEM comprised of (1) endogenous RP choices, (2) endogenous SP choices, (3) endogenous attitudes, in the form of attribute importance ratings, (4) exogenous mode characteristics, and (5) exogenous personal characteristics. ML estimation was applied to data collected in the Puget Sound region.

11.3.5 Organizational Behavior and Values

Golob and Regan (2000) applied SEM in the form of confirmatory factor analysis with regressor variables (estimated using WLS) to analyze the interrelationships among fleet managers' evaluations of 12 proposed congestion mitigation policies. The data are from a survey of managers of trucking companies operating in California.

Using these same data, Golob and Regan (2001a) used a SEM to determine how perceptions concerning five aspects of traffic congestion problems differ across sectors of the trucking industry. The model also simultaneously estimates how these five aspects combine to predict the perceived overall magnitude of the problem, and multigroup estimation is used to determine how results vary across industry sectors.

Finally, Golob and Regan (2001b) use SEM in the form of a multivariate probit model to captured the influences of each of 20 operational characteristics on the propensity of trucking company managers to adopt each of 7 different traveler information technologies. The authors discuss using SEM with WLS estimation as an alternative to simulated moments for estimating multivariate probit models.

11.3.6 Driver Behavior

Driver behavior (or more generally, user behavior) is a growing subject area for the application of SEM. Traffic safety is one potential focus, while another is the application of advanced technologies such as vehicle navigation systems and Advanced Traveler Information Systems (ATIS).

Donovan (1993) studied how driving under the influence of alcohol is related to other types of behavior using SEM. Using survey data collected in Colorado, the author concluded that problematic driving behaviors are related to more general lifestyle choices involving unconventional psychosocial behavior.

In a study of the behavior of long-distance truck drivers, Golob and Hensher (1996) tested alternative hypotheses concerning causal relations among drug taking, compliance with shipping schedules, and the propensity to speed, using data from an Australian survey and WLS estimation. The authors concluded that increasing speed is positively influenced by the propensity to take stay-awake pills, which in itself is influenced by the propensity to self-impose schedules. McCartt et al. (1999) present results from a similar application of SEM using data from a survey of long-distance truck drivers in New York State.

In a study of the user-interface of route guidance systems, Fujii et al. (1998) modeled experimental data to determine how comprehension of map displays are related to the attributes of the display and sociodemographic characteristics of the driver.

Finally, Ng and Mannering (1999) used SEM to analyze experimental data from a driving simulator on drivers' speed behavior as a function of different types of advisory information (in vehicle and out of vehicle). Speeds and speed variances were modeled using instrumental variables.

11.4 Summary

Historically, travel demand modeling has been grounded in econometric methods. SEM is used more in biometrics, sociology, and psychology. Consequently, until recently, SEM was relatively unknown among transportation researchers and planners. While microeconometric choice models are highly appropriate for many situations, there are many other situations where one needs a modeling tool that is more flexible. SEM is finally being used regularly in travel behavior research, as witnessed by the more than 50 applications cited in this review. Half of these applications have been published within the past 3 years.

SEM is relatively easy to understand and use, and there are myriad applications in transportation research and planning for which SEM is an appropriate analysis tool. SEM is particularly useful when the task is market research, activity-based travel demand modeling, and project evaluation. SEM can also be useful whenever one is faced with analyzing either panel survey data or attitudinal data on preferences and perceptions, or when the subject involves accessibility, activity participation, and shared use of household resources.

References

Aigner, D.J. et al., Latent variable models in econometrics, in *Handbook of Econometrics*, Vol. 2, Griliches, Z. and Intrilgator, M.D., Eds., North Holland, Amsterdam, 1984, pp. 1321–1393.

Akaike, H., A new look at the statistical identification model, *IEEE Trans. Autom. Control*, 19, 716–723, 1974.

Akaike, H., Factor analysis and AIC, *Psychometrika*, 52, 317–332, 1987.

Allaman, P.M., Tardiff, T.J., and Dunbar, F.C., New Approaches to Understanding Travel Behavior, NCHRP Report 250, Transportation Research Board, Washington, D.C., 1982.

Anderson, J.C. and Gerbing, D.W., Structural equation modeling in practice: A review and recommended two-step approach, *Psychol. Bull.*, 103, 411–423, 1988.

Anderson, T.W. and Rubin, H., Statistical inference in factor analysis, in *Proceedings of the Third Berkeley Symposium on Mathematical Statistics and Probability*, Vol. V, Neyman, J., Ed., University of California, Berkeley, 1956, pp. 111–150.

Arbukle, J.L., AMOS: Analysis of moment structures, *Psychometrika*, 59, 135–137, 1994.

Arbukle, J.L., *AMOS Users' Guide*, Version 3.6, Smallwaters Corp., Chicago, 1997.

Axhausen, K.W., Simma, A., and Golob, T.F., Pre-commitment and usage: Season tickets, cars and travel, *Eur. Res. Reg. Sci.*, 11, 101–110, 2001.

Bekker, P.A., Merckens, A., and Wansbeek, T.J., *Identification, Equivalent Models, and Computer Algebra*, Academic Press, New York, 1994.

Bentler, P.M., Multivariate analysis with latent variables: Causal modeling, *Annu. Rev. Psychol.*, 31, 419–456, 1980.

Bentler, P.M., Some contributions to efficient statistics in structural models: Specification and estimation of moment structures, *Psychometrika*, 48, 493–517, 1983.

Bentler, P.M., *EQS Structural Equations Program Manual*, BMDP Statistical Software, Los Angeles, 1989.

Bentler, P.M., Comparative fit indexes in structural models, *Psychometrika*, 107, 238–246, 1990.

Bentler, P.M., *EQS Structural Equations Program Manual*, Multivariate Software, Inc., Encino, CA, 1995.

Bentler, P.M. and Bonett, D.G., Significance tests and goodness of fit in the analysis of covariance structures, *Psychol. Bull.*, 88, 558–606, 1980.

Bentler, P.M. and Chou, C.P., Practical issues in structural modeling, *Sociol. Methods Res.*, 16, 78–117, 1987.

Bentler, P.M. and Yuan, K.-H., Structural equation modeling with small samples: Test statistics, *Multivariate Behav. Res.*, 34, 181–197, 1999.

Berkane, M., Ed., *Latent Variable Modeling and Applications to Causality*, Springer-Verlag, New York, 1997.

Bielby, W.T. and Hauser, R.M., Structural equation models, *Annu. Rev. Sociol.*, 3, 137–161, 1977.

Blalock, H.M., Correlation and causality: The multivariate case, *Soc. Forces*, 39, 246–251, 1961.

Blalock, H.M., Making causal inferences for unmeasured variables from correlations among indicators, *Am. J. Sociol.*, 69, 53–62, 1963.

Bock, R.D. and Gibbons, R.D., High-dimensional multivariate probit analysis, *Biometrics*, 52, 1183–1194, 1996.

Bollen, K.A., Sample size and Bentler and Bonett's nonnormed fit index, *Psychometrika*, 51, 375–377, 1986.

Bollen, K.A., *Structural Equations with Latent Variables*, Wiley, New York, 1989.

Bollen, K.A., Overall fit in covariance structure models: Two types of sample size effects, *Psychol. Bull.*, 107, 256–259, 1990.

Bollen, K.A. and Long, J.S., Eds., Tests for structural equation models: An introduction, *Sociol. Methods Res.*, 21, 123–131, 1992.

Boomsma, A., On the robustness of LISREL against small sample size in factor analysis models, in *Systems under Indirect Observation: Causality, Structure, Prediction*, Part 1, Jöreskog, K.G. and Wold, H., Eds., North-Holland, Amsterdam, 1982, pp. 149–173.

Boomsma, A. and Hoogland, J.J., The robustness of LISREL modeling revisited, in *Structural Equation Modeling: Present and Future*, Cudeck, R., du Toit, S., and Sörbom, D., Eds., Scientific Software International, Chicago, 2001, pp. 139–168.

Boudon, R., A method of linear causal analysis: Dependence analysis, *Am. Sociol. Rev.*, 30, 365–373, 1965.

Bozdogan, H., Model selection and Akaike's information criterion (AIC): The general theory and its analytical extensions, *Psychometrika*, 52, 345–370, 1987.

Browne, M.W., Covariance structures, in *Topics in Multivariate Analysis*, Hawkins, D.M., Ed., Cambridge University Press, Cambridge, U.K., 1982, pp. 72–141.

Browne, M.W., Asymptotic distribution free methods in analysis of covariance structures, *Br. J. Math. Stat. Psychol.*, 37, 62–83, 1984.

Browne, M.W. and Cudeck, R., Alternative ways of assessing model fit, *Sociol. Methods Res.*, 21, 230–258, 1992.

Byrne, B.M., *A Primer of LISREL: Basic Applications and Programming for Confirmatory Factor Analysis*, Springer-Verlag, New York, NY, 1989.

Byrne, B.M., *Structural Equation Modeling with AMOS: Basic Concepts, Applications, and Programming*, Lawrence Erlbaum Associates, Hillsdale, NJ, 2001.

Charles River Associates, *On the Development of a Theory of Traveler Attitude-Behavior Interrelationships*, Final Report DOT-TSC-RSPA-78-14, U.S. Department of Transportation, Washington, D.C., 1978.

Cudeck, R., du Toit, S., and Sörbom, D., Eds., *Structural Equation Modeling: Present and Future*, Scientific Software International, Chicago, 2001.

Den Boon, A.K., *Opvattingen over Autogebruik en Milieuvervuiling*, Baschwitz Institute for Public Opinion and Mass Psychology, University of Amsterdam, 1980.

Dobson, R. et al., Structural models for the analysis of traveler attitude-behavior relationships, *Transportation*, 7, 351–363, 1978.

Donovan, J.E., Young adult drinking–driving: Behavioral and psychological correlates, *J. Stud. Alcohol*, 54, 600–613, 1993.

Duncan, O.D., Path analysis: Sociological examples, *Am. J. Sociol.*, 72, 1–16, 1966.

Duncan, O.D., *Introduction to Structural Equation Models*, Academic Press, New York, 1975.

Epstein, R.J., *A History of Econometrics*, Elsevier, New York, 1987.

Fan, X., Thompson, B., and Wang, L., Effects of sample size, estimation methods, and model specification on structural equation modeling fit indexes, *Structural Equation Modeling*, 6, 56–83, 1999.

Finch, J.F., West, S.G., and MacKinnon, D., Effects of sample size and nonnormality on the estimation of mediated effects in latent variables models, *Structural Equation Modeling*, 4, 87–107, 1997.

Fox, J., *Linear Statistical Models and Related Methods*, Wiley, New York, 1984.

Friman, M., Edvardsson, B., and Gärling, T., Frequency of negative critical incidents and satisfaction with public transport services I, *J. Retailing Consumer Serv.*, 8, 95–104, 2001.

Friman, M. and Gärling, T., Frequency of negative critical incidents and satisfaction with public transport services II, *J. Retailing Consumer Serv.*, 8, 105–114, 2001.

Fujii, S. and Kitamura, R., Anticipated travel time, information acquisition and actual experience: The case of the Hanshin expressway route closure, *Transp. Res. Rec.*, 1725, 79–85, 2000a.

Fujii, S. and Kitamura, R., Evaluation of trip-inducing effects of new freeways using a structural equations model system of commuters' time use and travel, *Transp. Res. B*, 34, 339–354, 2000b.

Fujii, S., Kitamura, R., and Kishizawa, K., An analysis of individuals' joint activity engagement using a model system of activity–travel behavior and time use, *Transp. Res. Rec.*, 1676, 11–19, 2000.

Fujii, S.R. et al., An experimental analysis of intelligibility and efficiency of in-vehicle route guidance system displays, in *Proceedings of the Fifth International Conference on Applications of Advanced Technologies in Transportation Engineering*, Henderson, C.T. and Ritchie, S.G., Eds., American Society of Civil Engineers, Reston, VA, 1998, pp. 106–113.

Gärling, T., Fujii, S., and Boe, O., Empirical tests of a model of determinants of script-based driving choice, *Transp. Res. F*, 4, 89–102, 2001.

Gerbing, D.W. and Anderson, J.C., Monte Carlo simulations of goodness of fit indices for structural equation models, *Sociol. Methods Res.*, 21, 132–160, 1992.

Goldberger, A.S., *Structural Equation Methods in the Social Sciences*, North-Holland, Amsterdam, 1972.

Golob, T.F., Effects of income and car ownership on trip generation, *J. Transp. Econ. Policy*, 23, 141–162, 1989.

Golob, T.F., Structural equation modelling of travel choice dynamics, in *New Developments in Dynamic and Activity-Based Approaches to Travel Analysis*, Jones, P.M., Ed., Gower, Aldershot, England, 1990a, pp. 343–370.

Golob, T.F., The dynamics of travel time expenditures and car ownership decision, *Transp. Res. A*, 24, 443–463, 1990b.

Golob, T.F., A model of household demand for activity participation and mobility, in *Theoretical Foundations of Travel Choice Modeling*, Gärling, T., Laitilla, T., and Westin, K., Eds., Pergamon, Oxford, 1998, pp. 365–398.

Golob, T.F., A simultaneous model of household activity participation and trip chain generation, *Transp. Res. B*, 34, 355–376, 2000.

Golob, T.F., Joint models of attitudes and behavior in evaluation of the San Diego I-15 Congestion Pricing Project, *Transp. Res. A*, 35, 495–514, 2001.

Golob, T.F., Bunch, D.S., and Brownstone, D., A vehicle usage forecasting model based on revealed and stated vehicle type choice and utilization data, *J. Transp. Econ. Policy*, 31, 69–92, 1997a.

Golob, T.F. and Hensher, D.A., Driver behavior of long-distance truck drivers: Effects of schedule compliance on drug use and speeding citations, *Int. J. Transp. Econ.*, 23, 267–301, 1996.

Golob, T.F. and Hensher, D.A., Greenhouse gas emissions and Australian commuters' attitudes and behaviour concerning abatement policies and personal involvement, *Transp. Res. D*, 3, 1–18, 1998.

Golob, T.F., Kim, S., and Ren, W., How households use different types of vehicles: A structural driver allocation and usage model, *Transp. Res. A*, 30, 103–118, 1996.

Golob, T.F., Kitamura, R., and Lula, C., Modeling the Effects of Commuting Time on Activity Duration and Non-Work Travel, paper presented at Annual Meeting of Transportation Research Board, Washington, D.C., 1994.

Golob, T.F., Kitamura, R., and Supernak, J., A panel-based evaluation of the San Diego I-15 Carpool Lanes Project, in *Panels for Transportation Planning: Methods and Applications*, Golob, T.F., Kitamura, R., and Long, L., Eds., Kluwer Academic Publishers, Boston, 1997b, pp. 97–128.

Golob, T.F. and McNally, M.G., A model of household interactions in activity participation and the derived demand for travel, *Transp. Res. B*, 31, 177–194, 1997.

Golob, T.F. and Meurs, H., A structural model of temporal changes in multi-modal travel demand, *Transp. Res. A*, 21, 391–400, 1987.

Golob, T.F. and Meurs, H., Modeling the dynamics of passenger travel demand by using structural equations, *Environ. Plann. A*, 20, 1197–1218, 1998.

Golob, T.F. and Regan, A.C., Freight industry attitudes towards policies to reduce congestion, *Transp. Res. E*, 36, 55–77, 2000.

Golob, T.F. and Regan, A.C., Impacts of highway congestion on freight operations: Perceptions of trucking industry managers, *Transp. Res. A*, 35, 577–599, 2001a.

Golob, T.F. and Regan, A.C., Trucking industry adoption of information technology: A structural multivariate probit model, *Transp. Res. C*, 10, 205–228, 2001b.

Golob, T.F. and van Wissen, L.J., A joint household travel distance generation and car ownership model, *Transp. Res. B*, 23, 471–491, 1989.

Gould, J. and Golob, T.F., Shopping without travel or travel without shopping: An investigation of electronic home shopping, *Transp. Rev.*, 17, 355–376, 1997.

Gould, J., Golob, T.F., and Barwise, P., Why do people drive to shop? Future travel and telecommunications trade-offs, paper presented at Annual Meeting of Transportation Research Board, Washington, D.C., January 1998.

Haavelmo, T., The statistical implications of a system of simultaneous equations, *Econometrica*, 11, 1–12, 1943.

Hartmann, W.M., *The CALIS Procedure: Extended Users' Guide*, SAS Institute, Cary, NC, 1992.

Hayduk, L.A., *Structural Equation Modeling with LISREL: Essentials and Advances*, Johns Hopkins University Press, Baltimore, 1987.

Hoetler, J.W., The analysis of covariance structures: Goodness-of-fit indices, *Sociol. Methods Res.*, 11, 325–344, 1983.

Hoogland, J.J. and Boomsma, A., Robustness studies in covariance structure modeling: An overview and a meta-analysis, *Sociol. Methods Res.*, 26, 329–367, 1998.

Hoyle, R.H., Ed., *Structural Equation Modeling*, Sage, Thousand Oaks, CA, 1995.

Hu, L. and Bentler, P.M., Cutoff criteria for fit indexes in covariance structure analysis: Conventional criteria versus new alternatives, *Structural Equation Modeling*, 6, 1–55, 1999.

Jakobsson, C., Fujii, S., and Gärling, T., Determinants of private car users' acceptance of road pricing, *Transp. Policy*, 7, 153–158, 2000.

James, L.R., Mulaik, S.A., and Brett, J.M., *Causal Analysis: Assumptions, Models and Data*, Sage, Thousand Oaks, CA, 1982.

Jöreskog, K.G., Some contribution to maximum likelihood factor analysis, *Psychometrika*, 32, 443–482, 1967.

Jöreskog, K.G., A general approach to confirmatory maximum likelihood factor analysis, *Psychometrika*, 34, 183–202, 1969.

Jöreskog, K.G., A general method for analysis of covariance structures, *Biometrika*, 57, 239–251, 1970.

Jöreskog, K.G., A general method for estimating a linear structural equation system, in *Structural Equations Models in the Social Sciences*, Goldberger, A.S. and Duncan, O.D., Eds., Academic Press, New York, 1973, pp. 85–112.

Jöreskog, K.G., New developments in LISREL: Analysis of ordinal variables using polychoric correlations and weighted least squares, *Qual. Quantity*, 24, 387–404, 1990.

Jöreskog, K.G., Gruvaeus, G.T., and van Thillo, M., *ACOVS: A General Computer Program for Analysis of Covariance Structures*, Educational Testing Services, Princeton, NJ, 1970.

Jöreskog, K.G. and Sörbom, D., *Advances in Factor Analysis and Structural Equation Models*, Abt Books, Cambridge, MA, 1979.

Jöreskog, K.G. and Sörbom, D., *LISREL 8 User's Reference Guide; PRELIS 2 User's Reference Guide*, Scientific Software International, Chicago, 1993.

Kaplan, D., *Structural Equation Modeling: Foundations and Extensions*, Sage Publications, Newbury Park, CA, 2000.

Keesling, J.W., Maximum likelihood approaches to causal analysis, unpublished Ph.D. dissertation, Department of Education, University of Chicago, 1972.

Kitamura, R., A causal analysis of car ownership and transit use, *Transportation*, 16, 155–173, 1989.

Kitamura, R., Activity-based travel demand forecasting and policy analysis, report of the Travel Model Improvement Program, in *Activity-Based Travel Forecasting Conference: Summary, Recommendations and Compendium of Papers*, Engelke, L.J., Ed., Texas Transportation Institute, Arlington, TX, 1997.

Kitamura, R. et al., A comparative analysis of time use data in The Netherlands and California, in *Proceedings of the 20th PTRC Summer Annual Meeting: Transportation Planning Methods*, PTRC Education and Research Services, London, 1992, pp. 127–138.

Kline, R.B., Software programs for structural equation modeling: AMOS, EQS, and LISREL, *J. Psychoeduc. Assessment*, 16, 302–323, 1998a.

Kline, R.B., Ed., *Principles and Practice of Structural Equation Modeling*, Guilford Press, New York, 1998b.

Kuppam, A.R. and Pendyala, R.M., A structural equations analysis of commuter activity and travel patterns, *Transportation*, 28, 33–54, 2001.

Lawley, D.N., The estimation of factor loadings by the method of maximum likelihood, *Proc. R. Soc. Edinburgh*, 60, 64–82, 1940.

Lee, S.Y. and Shi, J.Q., Bayesian analysis of structural equation model with fixed covariates, *Structural Equation Modeling*, 7, 411–430, 2000.

Lee, S.Y. and Tsang, S.Y., Constrained maximum likelihood estimation of two-level covariance structure model via EM type algorithms, *Psychometrika*, 64, 435–450, 1999.

Levine, J. et al., Public choice in transit organization and finance: The structure of support, *Transp. Res. Rec.*, 1669, 87–95, 1999.

Loehlin, J.C., *Latent Variable Models: An Introduction to Factor, Path, and Structural Analysis*, Lawrence Erlbaum Associates, Mahwah, NJ, 1998.

Lu, X. and Pas, E.I., A structural equations model of the relationships among socio-demographics, activity participation and travel behavior, *Transp. Res. A*, 31, 1–18, 1999.

Lyon, P.K., Time-Dependent Structural Equations Modelling of the Relationships between Attitudes and Discrete Choices Behavior of Transportation Consumers, dissertation, Northwestern University, Evanston, IL, and Technical Report IL-11-0012, Office of Policy Research, Urban Mass Transportation Administration, Washington, D.C., 1981a.

Lyon, P.K., Dynamic Analysis of Attitude-Behavior Response to Transportation Service Innovation, Report 423-II-2, prepared for the Office of Policy Research, Urban Mass Transportation Administration, Transportation Center, Northwestern University, Evanston, IL, 1981b.

Lyon, P.K., Time-dependent structural equations modeling: A methodology for analyzing the dynamic attitude-behavior relationship, *Transp. Sci.*, 18, 395–414, 1984.

MacCallum, R.C., Browne, M.W., and Sugawara, H.M., Power analysis and determination of sample size for covariance structure modeling, *Psychol. Methods*, 1, 130–149, 1996.

Mardia, K.V., Measures of multivariate skewness and kurtosis with applications, *Biometrika*, 57, 519–530, 1970.

Maruyama, G., *Basics of Structural Equation Modeling*, Sage Publications, Newbury Park, CA, 1998.

McArdle, B.H., The structural relationship: Regression in biology, *Can. J. Zool.*, 66, 2329–2339, 1988.

McCullagh, P. and Nelder, J.A., *Generalized Linear Models*, 2nd ed., Chapman & Hall, London, 1989.

McCartt, A.T. et al., Causes of sleepiness-related driving among long-distance truck drivers including violations of the hours-of-service regulations, in *Proceedings of the Conference on Traffic Safety on Two Continents, Mälmo, Sweden*, Swedish National Road and Transport Research Institute, Linkoping, 1999, pp. 155–172.

McDonald, R.P., A note on the investigation of local and global identifiability, *Psychometrika*, 47, 101–103, 1982.

McDonald, R.P. and Marsh, H.W., Choosing a multivariate model: Noncentrality and goodness of fit, *Psychol. Bull.*, 107, 247–255, 1990.

Mokhtarian, P.L. and Meenakshisundaram, R., Beyond Tele-Substitution: A Broader Empirical Look at Communication Impacts, Working Paper UCB-ITS-PWP-98–33, Partners for Advanced Transit and Highways, University of California, Berkeley, CA, 1998.

Mokhtarian, P.L. and Meenakshisundaram, R., Beyond tele-substitution: Disaggregate longitudinal structural equations modeling of communication impacts, *Transp. Res. C*, 7, 33–52, 1999.

Morikawa, T. and Sasaki, K., Discrete choice models with latent variables using subjective data, in *Travel Behaviour Research: Updating the State of Play*, Ortuzar, J. de D., Hensher, D.A., and Jara-Diaz, S., Eds., Pergamon, Oxford, 1998, pp. 435–455.

Mueller, R., *Basic Principles of Structural Equation Modeling: An Introduction to LISREL and EQS*, Springer-Verlag, New York, 1996.

Mueller, R., Structural equation modeling: Back to basics, *Structural Equation Modeling*, 4, 353–369, 1997.

Mulaik, S.A. et al., Evaluation of goodness-of-fit indices for structural equation models, *Psychol. Bull.*, 105, 430–445, 1989.

Muthén, B., A structural probit model with latent variables, *J. Am. Stat. Assoc.*, 74, 807–811, 1979.

Muthén, B., Latent variable structural equation modeling with categorical data, *J. Econometrics*, 22, 43–65, 1983.

Muthén, B., A general structural equation model with dichotomous, ordered categorical and continuous latent variable indicators, *Psychometrika*, 49, 115–132, 1984.

Muthén, B., *LISCOMP: Analysis of Linear Structural Equations with a Comprehensive Measurement Model*, Scientific Software, Mooresville, IN, 1988.

Muthén, L. and Muthén, B., *Mplus User's Guide*, Muthén and Muthén, Los Angeles, 1998.

Neale, M.C., *Mx: Statistical Modeling*, 4th ed., Department of Psychiatry, Virginia Commonwealth University, Richmond, VA, 1997.

Ng, L. and Mannering, F., Statistical Analysis of the Impact of Traveler Advisory Systems on Driving Behavior, paper presented at Annual Meeting of the Transportation Research Board, Washington, D.C., Jan. 10–14, 1999.

Pas, E.I., Recent developments in activity-based travel demand modeling, report of the Travel Model Improvement Program, in *Activity-Based Travel Forecasting Conference: Summary, Recommendations and Compendium of Papers*, Engelke, L.J., Ed., Texas Transportation Institute, Arlington, TX, 1997a, pp. 79–102.

Pas, E.I., Time Use and Travel Demand Modeling: Recent Developments and Current Challenges, paper presented at 8th Meeting of the International Association for Travel Behavior Research, Austin, TX, Sept. 21–25, 1997b, in *Perpetual Motion: Travel Behavior Research Opportunities and Applications Challenges*, Mahmassani, H.S., Ed., Pergamon, Oxford, in press.

Pearl, J., *Causality: Models, Reasoning, and Inference*, Cambridge University Press, Cambridge, U.K., 2000.

Pendyala, R., Causal analysis in travel behaviour research: A cautionary note, in *Travel Behaviour Research: Updating the State of Play*, Ortuzar, J. de D., Hensher, D.A., and Jara-Diaz, S., Eds., Pergamon, Oxford, 1998, pp. 35–48.

Reilly, T., A necessary and sufficient condition for identification of confirmatory factor analysis models of complexity one, *Sociol. Methods Res.*, 23, 421–441, 1995.

Rigdon, E.E., A necessary and sufficient identification rule for structural models estimated in practice, *Multivariate Behav. Res.*, 30, 359–383, 1995.

Sakano, R. and Benjamin, J., A Structural Equation Analysis of Stated Travel by Commuter Rail, paper presented at Annual Meeting of the Transportation Research Board, Washington, D.C., Jan. 9–13, 2000.

Sakano, R. and Benjamin, J., A Structural Equation Analysis of Revealed and Stated Travel Mode and Activity Choices, paper presented at Annual Meeting of the Transportation Research Board, Washington, D.C., Jan. 7–11, 2001.

Satorra, A. and Bentler, P.M., Scaling corrections for chi-squared statistics in covariance structure analysis, in *Proceedings of the American Statistical Association*, American Statistical Association, Alexandria, VA, 1988, pp. 308–313.

Scheines, R. et al., *TETRAD II: Tools for Discovery*, Lawrence Erlbaum Associates, Hillsdale, NJ, 1994.

Schoenberg, R., Covariance structure models, *Annu. Rev. Sociol.*, 15, 425–440, 1989.

Schwarz, G., Estimating the dimension of a model, *Ann. Stat.*, 6, 461–464, 1978.

Shipley, B., *Cause and Correlation in Biology*, Cambridge University Press, Cambridge, U.K., 2000.

Simma, A., Verkehrsverhalten als eine Funktion soziodemografischer und räumlicher Faktoren, Working Paper 55, Institute of Transportation, Traffic, Highway- and Railway-Engineering (IVT), Swiss Federal Institute of Technology (ETHZ), Zurich, 2000.

Simma, A. and Axhausen, K.W., Structures of commitment in mode use: A comparison of Switzerland, Germany and Great Britain, *Transp. Policy*, 8, 279–288, 2001.

Simma, A. and Axhausen, K.W., Within-Household Allocation of Travel: The Case of Upper Austria, paper presented at Annual Meeting of the Transportation Research Board, Washington, D.C., Jan. 7–11, 2001b.

Simma, A. and Axhausen, K.W., Successive Days, Related Travel Behaviour? Working Paper 62, Institute of Transportation, Traffic, Highway- and Railway- Engineering (IVT), Swiss Federal Institute of Technology (ETHZ), Zurich, 2001c.

Simma, A., Vrtic, M., and Axhausen, K.W., Interactions of Travel Behaviour, Accessibility and Personal Characteristics: The Case of the Upper Austria Region, paper presented at European Transport Conference, Cambridge, England, Sept. 10–12, 2001.

Sobel, M.E. and Bohrnstedt, G.W., Use of null models in evaluating the fit of covariance structure models, in *Sociological Methodology, 1985*, Tuma, N.B., Ed., Jossey-Bass, San Francisco, 1985, pp. 152, 178.

Spirtes, P., Glymour, C., and Scheines, R., *Causation, Prediction, and Search*, 2nd ed., MIT Press, Cambridge, MA, 2001.

Steiger, J.H., *EzPath Causal Modeling: A Supplementary Module for SYSTAT and SYGRAPH*, SYSTAT, Inc., Evanston, IL, 1989.

Steiger, J.H., Structural model evaluation and modification: An interval estimation approach, *Multivariate Behav. Res.*, 25, 173–180, 1990.

Steiger, J.H. and Lind, J.C., Statistically Based Tests for the Number of Common Factors, paper presented at Annual Meeting of the Psychometric Society, Iowa City, IA, 1980.

Stevens, J., *Applied Multivariate Statistics for the Social Sciences*, Lawrence Erlbaum Associates, Mahwah, NJ, 1996.

Stuart, K.R., Mednick, M., and Bockman, J., A structural equation model of consumer satisfaction for the New York City subway system, *Transp. Res. Rec.*, 1735, 133–137, 2000.

Sugawara, H.M. and MacCallum, R.C., Effect of estimation method on incremental fit indexes for covariance structure models, *Appl. Psychol. Meas.*, 17, 365–377, 1993.

Tanaka, J.S., How big is big enough? Sample size and goodness of fit in structural equation models with latent variables, *Child Dev.*, 58, 134–146, 1987.

Tardiff, T.J., Causal inferences involving transportation attitudes and behavior, *Transp. Res.*, 11, 397–404, 1976.

Tucker, L.R. and Lewis, C., A reliability coefficient for maximum likelihood factor analysis, *Psychometrika* 38, 1–10, 1973.

Ullman, J.B., Structural equation modeling, in *Using Multivariate Statistics*, 3rd ed., Tabachnick, B.G. and Fidell, L.S., Eds., HarperCollins College Publishers, New York, 1996, pp. 709–819.

van Wissen, L.J. and Golob, T.F., Simultaneous equation systems involving binary choice variables, *Geogr. Anal.*, 22, 224–243, 1990.

van Wissen, L.J. and Golob, T.F., A dynamic model of car fuel type choice and mobility, *Transp. Res. B*, 26, 77–96, 1992.

Wiley, D.E., The identification problem for structural equations with unmeasured variables, in *Structural Equation Models in the Social Sciences*, Goldberger, A.S. and Duncan, O.D., Eds., Academic Press, New York, 1973, pp. 69–83.

Wright, S., Correlation and causation, *J. Agric. Res.*, 20, 557–585, 1921.

Wright, S., The method of path coefficients, *Ann. Math. Stud.*, 5, 161–215, 1934.

Yuan, K.-H. and Bentler, P.M., Mean and covariance structure analysis: Theoretical and practical improvements, *J. Am. Stat. Assoc.*, 92, 767–77, 1997.

III

Systems Simulation and Applications

12

Microsimulation

CONTENTS

Eric J. Miller
University of Toronto

12.1 Introduction

The purpose of this chapter is to provide an overview of microsimulation concepts and methods that may be used in travel-related forecasting applications. Including this very brief introductory section, the chapter is divided into eight sections. Section 12.2 defines the term *microsimulation*. Section 12.3 discusses the reasons why microsimulation may prove useful or even necessary for at least some types of activity-based travel forecasting applications. Section 12.4 defines the important concepts of objects, agents, and cellular automata, which represent fundamental organizing constructs in modern microsimulation models. Section 12.5 discusses a key step in the microsimulation process — synthesizing and updating the attributes of the population or sample of individuals whose behavior is being simulated. Section 12.6 then discusses some of the major issues associated with the development and application of operational microsimulation methods, while Section 12.7 provides representative examples of transportation-related microsimulation applications. Finally, Section 12.8 presents several microsimulation models drawn from a range of applications, including activity-based travel forecasting.

12.2 What Is Microsimulation?

While many current modeling efforts are microsimulation based, the term itself is rarely defined and tends to mean different things to different people. Perhaps due to the recent proliferation of network microsimulators, many people tend to think of microsimulation specifically in terms of network route choice and performance models. On the other hand, given the use of microsimulation methods to generate the disaggregate inputs required by their models, many activity modelers think of microsimulation as the procedures used in this input data synthesis and updating process. In this chapter we adopt a comprehensive definition of microsimulation as a method or approach (rather than a model per se) for exercising a disaggregate model over time.

Simulation generally refers to an approach to modeling systems that possess one or both of the following two key characteristics.

1. The system is a dynamic one, whose behavior must be explicitly modeled over time.
2. The system's behavior is complex. In addition to the dynamic nature of the system (which generally in itself introduces complexity), this complexity typically has many possible sources, including:
   a. Complex decision rules for the individual actors within the system
   b. Many different types of actors interacting in complex ways
   c. System processes that are path dependent (i.e., the future system state depends both on the current system state and explicitly on how the system evolves from this current state over time)
   d. A generally open system on which exogenous forces operate over time, thereby affecting the internal behavior of the system
   e. Significant probabilistic elements (uncertainties) that exist in the system, with respect to random variations in exogenous inputs to the system or the stochastic nature of endogenous processes at work within the system

Note that in speaking of complexity, we are not merely referring to the difficulty in dealing with very large models with large data sets defined over many attributes for hundreds if not thousands of zones. Rather, we are referring to the more fundamental notion of the difficulty in estimating likely future system states given the inherently complex nature of the system's behavioral processes.

Given the system's complexity, closed-form analytical representations of the system are generally not possible, in which case numerical, computer-based algorithms are the only feasible method for generating estimates of future system states. Similarly, given the system's path dependencies and openness to time-varying exogenous factors, system equilibrium often is not achieved, rendering equilibrium-based models inappropriate in such cases.[1] In the absence of explicit equilibrium conditions, the future state of the system again generally can only be estimated by explicitly tracing the evolutionary path of the system over time, beginning with current known conditions. Such numerical, computer-based models that trace a system's evolution over time are what we generally refer to as simulation models.

Note that conventional four-stage travel demand models most clearly are not simulation models under this definition. Conventional four-stage models are static equilibrium models that predict a path-independent future year end state without concern for either the initial (current) system state or the path traveled by the system from the current to the future year state.

The prefix micro simply indicates that the simulation model is formulated at the disaggregate or micro level of individual decision-making agents (or other relevant units), such as individual persons, households, and vehicles. A full discussion of the relative merits of disaggregate vs. more traditional aggregate modeling methods is beyond the scope of this chapter. It is fair to say that a broad consensus exists within the activity–travel demand modeling community that disaggregate modeling methods possess considerable advantages over more aggregate approaches (including minimization of model bias, maximization of model statistical efficiency, improved policy sensitivity, and improved model transferability — and hence usability within forecasting applications), and that they will continue to be the preferred modeling approach for the foreseeable future. With respect to microsimulation, the relevant question is to what extent does microsimulation represent a feasible and useful mechanism for using disaggregate models within various forecasting applications.

To answer this question, first consider the well-known short-run policy analysis or forecasting procedure known as sample enumeration. In this procedure, a disaggregate behavioral model of some form has been developed (say, for sake of illustration, a logit work trip mode choice model). A representative

---

[1] Although many examples of equilibrium-based simulation models exist.

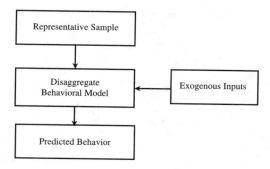

FIGURE 12.1 Forecasting with sample enumeration (short-run microsimulation).

sample of decision makers (in this case workers) typically exists, since such a sample is generally required for model development. This sample defines all relevant inputs to the model with respect to the attributes of all the individuals in the sample. The short-run impact of various policies that might be expected to affect work trip mode choice can then be tested by "implementing" a given policy, and then using the model to compute the response of each individual to this policy. Summing up the responses of the individuals provides an unbiased estimate of the aggregate system response to the policy in question.

Figure 12.1 summarizes this procedure. This figure can be taken as a very generic representation of a microsimulation process for the case of a short-run forecast, in which all model inputs except those relating to the policy tests of interest are fixed, and hence all that needs to be simulated are the behavioral responses of the sampled decision makers to the given policy stimuli. Thus, in such cases, sample enumeration and microsimulation are essentially synonymous, and use of the latter term simply emphasizes the disaggregate, dynamic[2] nature of the model. The majority of activity-based microsimulation models developed to date basically fall into this category of short-run sample enumeration-based models.

Sample enumeration is a very efficient and effective forecasting method providing that:

1. A representative sample is available
2. One is undertaking a short-run forecast (so that the sample can be assumed to remain representative over the time frame of the forecast)
3. The sample is appropriate for testing the policy of interest (i.e., the policy applies in a useful way to the sample in question)

Many forecasting situations, however, violate one or more of these conditions. Perhaps most commonly, one is often interested in forecasting over medium to long time periods, during which time the available sample will clearly become unrepresentative (people will age and even die; workers will change jobs or residential locations; new workers with different combinations of attributes will join the labor force; etc.). The question then becomes how to properly update the sample in order to maintain its representativeness. In other cases, the sample may not be adequate to test a given policy (e.g., it contains too few observations of a particularly important subpopulation for the given policy test). If this is the case, how does one extend the sample so that a statistically reliable test of the policy can be performed? Finally, there may be cases in which a suitable sample simply does not exist (e.g., perhaps the model has been transferred from another urban area). In such a case, how does one generate or synthesize a representative sample?

In all of these cases, microsimulation provides a means of overcoming the limitations of the available sample. In the case of the sample becoming less and less representative over time, Figure 12.2 presents a simple microsimulation framework in which the sample is explicitly updated over time. Updating may involve changing attributes of the households or persons in the sample (age of population, changes

---

[2]In such cases, the dynamics involved are usually quite short-run, particularly relative to the much longer-term demographic and socioeconomic dynamics that are discussed immediately below.

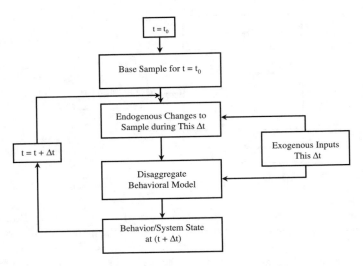

FIGURE 12.2 Typical microsimulation model design using an available sample.

in household structure, changes in income or employment, etc.), deleting households or persons from the sample (due to out-migration from the study area, death, etc.), or adding new households or persons (due to in-migration, births, etc.). The result is that the sample (hopefully) remains representative for each point in simulated time and so provides a valid basis for predicting behavior at each such point in time.

If the original sample is either inadequate or missing altogether, then, as shown in Figure 12.3, an additional step must be inserted into the model, involving synthesizing a representative sample from other available (typically more aggregate) data, such as census data. Procedures for doing this are discussed in greater detail below.

Note that these figures assume that the disaggregate behavioral model is itself a dynamic one that must be stepped through time (and hence its inclusion within the time loop). Many current models, however, are fairly static in nature. In such cases, the behavioral model can be removed from the time loop and executed only once, using the desired future year sample that has been estimated through the microsimulation procedure. Thus, one may distinguish between static microsimulation, in which a fixed, repre-

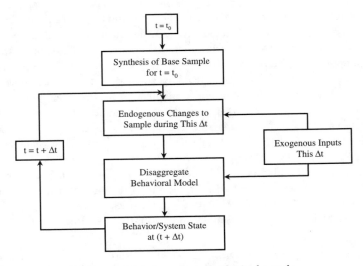

FIGURE 12.3 Typical microsimulation model design using a synthesized sample.

sentative sample is used to test various policy alternatives within a microanalytic framework, and dynamic microsimulation, wherein the representative sample changes or evolves over time as a function of endogenous or exogenous processes.

In summary, key features of the full microsimulation process include the following:

1. The model must have as its primary input a disaggregate list of actors (or entities or behavioral units) upon which it operates. This list often consists of a representative sample of individuals drawn from the relevant population. Alternatively, it is becoming more common to use a 100% sample, i.e., a list of the entire set of actors in the population. In general, two sources exist for this input list or base sample: a sample of actual individuals drawn from the population, obtained through conventional survey methods, or a list of synthetic individuals, statistically constructed from more aggregate data concerning the population being modeled (e.g., census data).

2. Over time, the demographic, social, and economic characteristics of the population being analyzed will change. Thus, the list of actors must be updated over time within the model so that it remains representative of the population at each point in time. Processes affecting the evolution of the population over time can be both endogenous (aging, births, deaths, etc.) and exogenous (in-migration, etc.).

3. Once the attributes of the list of actors are known for a given point in time, the behavior of these actors can be simulated using a relevant behavioral model. Depending on the application, this model may deal with a single process (e.g., activity–travel choices in response to travel demand management (TDM) measures) or many nested processes (residential location, employment location, activity–travel, etc.), involving a complex set of interconnected submodels. In general, this behavior will depend on past and current system states (endogenous factors) as well as exogenous factors.

4. Primary outputs from the microsimulation include both the attributes and behaviors of the actors over time. These outputs are generally expressible both in aggregate terms (link volumes, average modal splits, etc.) and in terms of disaggregate trajectories of the individuals being simulated (i.e., the historical record of the behavior of each individual over time).

Different microsimulation applications, of course, will involve different implementations of Figure 12.3. So-called static microsimulations, in particular, do not require updating of the base sample or population, and they do not require iterating the model through time. Similarly, microsimulations that are based on an observed sample of actors do not require a synthetic sample to be constructed.

12.3 Why Microsimulate?

As briefly discussed in the previous section, a primary motivation for adopting a microsimulation modeling approach is that it may well be the best (and in some cases perhaps the only) way to generate the detailed inputs required by disaggregate models. The strength of the disaggregate modeling approach is in being able to fix decision makers within explicit choice contexts with respect to:

1. Salient characteristics of the actors involved
2. Salient characteristics of the choice context (in terms of the options involved, the constraints faced by the actors, etc.)
3. Any context-specific rules of behavior that may apply

This inherent strength of the disaggregate approach is clearly compromised if one cannot provide adequately detailed inputs to the model. Such compromises occur in at least two forms. One involves using overly aggregate forecast inputs, resulting in likely aggregation biases in the forecasts. The other involves developing more aggregate models in the first place, so as to reduce the need for disaggregate forecast input data, thereby building the aggregation bias into the model itself. A strong case can be made that a primary reason for the relatively slow diffusion of disaggregate modeling methods into travel demand forecasting practice is due to the difficulty practitioners have in generating the disaggregate forecast inputs required by these methods. As described in the previous section, microsimulation in principle eliminates this problem by explicitly generating the detailed inputs required for each actor simulated.

A second driving force for using microsimulation relates to the outputs required from the travel demand model. As discussed further in Chapter 4, many emerging road network assignment procedures are themselves microsimulation based, and hence require quite microlevel inputs from the travel forecasting model.

In addition, the disaggregate nature of the model outputs, in which behavior is explicitly attached to individual actors with known attributes, permits very detailed analysis of model results. For example, the impacts of a given policy on specific subgroups (the elderly, the poor, suburban vs. central area inhabitants, etc.) can be readily identified, since the disaggregate model outputs can, in principle, be aggregated in almost any user-specified fashion. Thus, questions of equity, distribution of cost and benefits (both spatially and socioeconomically), details of the nature of the behavioral responses (e.g., who travels less or more, who changes modes, etc.), etc., can all be explored within a microsimulation framework in ways that are generally infeasible with more aggregate models.

A third point is that, despite the obviously large computational requirements of a large microsimulation model, in many cases microsimulation is computationally more efficient than conventional methods for dealing with large-scale forecasting problems. In particular, a micro list-based approach to storing large spatial databases is far more efficient than aggregate matrix-based approaches. To illustrate this, consider a simple example in which one might want to keep track of the number of workers by their place of residence, place of work, number of household automobiles, and total number of household members. Further assume that there are 1000 traffic zones, 3 auto ownership levels (e.g., 0, 1, 2+), and 5 household size categories (e.g., 1, 2, 3, 4, 5+). To save this information in matrix format would require a four-dimensional matrix with a total of $1000 \times 1000 \times 3 \times 5 = 15 \times 10^6$ data items. Also note that a large number of the cells in this matrix will have the value zero, either because they are infeasible (or at least extremely unlikely; e.g., 2+ autos in a one-person household) or because one simply does not observe nonzero values for many cells (as will be the case for many origin–destination (O-D) pairs).

In a list-based approach, one record is created for each worker, with each record containing the worker's residence zone, employment zone, number of household autos, and household size. Thus, four data storage locations are required per worker, meaning that as long as there are less than $(15 \times 10^6)/4 = 3.75 \times 10^6$ workers in this particular urban area, the list-based approach will require less memory (or disk space) than the matrix-based approach to store the same information. Obviously, as the number of worker attributes that need to be stored increase, the relative superiority of the list-based approach increases.

The advantages of list-based data structures for large-scale spatial applications have been recognized for at least 30 years. Aggregate urban simulation models such as NBER, for example, developed in the 1970s, used list-based data structures (see Ingram et al., 1972; Wilson and Pownall, 1976). The key point to be made here with respect to microsimulation is that once one begins to think in list-based terms, the conceptual leap to microsimulation model designs is a relatively small one. Or, turning it around, if one takes a microsimulation approach to model design, efficient list-based data structures quickly emerge as the natural way for storing information.

Whether microsimulation possesses other inherent computational advantages relative to more aggregate methods is less clear. Certainly one can advance the proposition that by working at the microlevel of the individual decision maker, relatively simple, clear, and computationally efficient models of process can generally be developed. Whether this efficiency in computing each actor's activities translates into overall computation time savings relative to other approaches given the large number of actors being simulated remains to be seen. It is, however, important to note that Harvey and Deakin (1996) report a major advantage of the STEP microsimulation model (discussed further below): it can provide model results in many applications much more quickly (both in terms of getting ready to do the model runs and in terms of the computational effort involved in running the model) than conventional four-stage modeling systems. Vovsha et al. (2002) similarly argue that microsimulation is far more efficient than conventional methods for modeling very large urban areas such as the New York metropolitan area.

A fourth argument in favor of microsimulation is that it raises the possibility of emergent behavior, that is, of predicting outcomes that are not "hardwired" into the model. Simple examples of emergent behavior of relevance to this discussion might include the generation of single-parent households by a demographic simulator as a result of more fundamental processes dealing with fertility and household formation and dissolution, or the prediction of unexpected activity–travel patterns by an activity-based model as a result of the occurrence within the simulation of certain combinations of household needs, constraints, etc.

The importance of emergent behavior within travel demand forecasting is at least twofold. First, it offers the potential for the development of parsimonious models in the sense that relatively simple (but fundamental) rules of behavior can generate very complex behavior. This is an attractive property of any model for two reasons. In practical terms, it implies computational efficiency. At a more theoretical level, parsimony is an important criteria in the evaluation of any model: it is generally assumed that a model that can satisfactorily explain behavior with fewer variables, parameters, rules, etc., is preferred, all else being equal, over more complicated formulations. Second, while all models are to at least some degree captive to past behavior through use of historical data to estimate model parameters, the potential for emergent behavior increases the likelihood of the model generating unanticipated outcomes, and hence for departures from the trend to occur.

Finally, it may well be the case that microsimulation models will ultimately prove easier to explain or "sell" to decision makers than more aggregate models. Since microsimulation models are formulated at the level of individual actors (workers, homeowners, parents, etc.), relatively clear and simple "stories" can be told concerning what the model is trying to accomplish (e.g., the model estimates the out-of-home activities that a given household will undertake on a typical weekday, and when and where these activities will occur) to which laypeople can readily relate. The technical details of the model's implementation typically will be very complex, but the fundamental conceptual design is, in most cases, surprisingly simple to convey to others.

12.4 Object, Agents, and Cellular Automata

The purpose of microsimulation is to model the behavior of actors, or objects, in the real world. In a travel-related microsimulation, these objects may include persons, households, vehicles, jobs, firms, dwelling units, etc. The real world consists of these objects evolving and interacting over time; microsimulation models attempt to emulate this evolution and interaction with as high a degree of fidelity as is required or feasible in the given application.

Object-oriented software systems employ object-oriented analysis, design, and programming techniques. These techniques were specifically developed to handle complex problems such as large-scale microsimulation. In an object-oriented system, the program consists of many objects that have their own states and behaviors. There is a one-to-one mapping between objects in the real world and objects in the simulated world. The conceptual benefits of the one-to-one mapping between the real world and objects in an object-oriented system should be clear. Every object in the real world is represented by a similar object in the simulated world — these objects are known as abstractions of their real-world counterparts. The behavior of objects in an object-oriented program is given by a set of methods or member functions. There is a member function corresponding to every behavior that the object exhibits. For example, the decision to change jobs is a behavior that is part of the person object.

Traditional programming methods (such as functional or procedural programming) are built on data flow diagrams and are not well suited for microsimulation applications. These methods use an approach known as top-down design with stepwise refinement and require that the program begin with a single problem statement that can be refined over time. Many complex real-world systems, however, do not consist of a single abstract problem; rather, they are comprised of a set of objects that interact in complex ways over time.

Object-oriented techniques provide a solution that is conceptually cleaner and easier for noncomputer scientists to understand and validate. By matching real-world objects with their synthetic counterparts, researchers can focus on understanding the problem rather than programming the solution. When dealing with a system as complex as a microsimulation of events in the real world, any technique that will help to reduce this inherent complexity will surely benefit the project. Object-oriented systems are generally better equipped than procedural systems in solving problems where there is significant complexity in the problem domain.

To give an example, we can think of the individual persons in the system. These persons have certain behaviors that can easily be described. In addition, there is information about each person that must be stored — this information might include age, gender, education level, marital status, job, etc. Object-oriented systems allow us to encapsulate behavior and data together in the object. The object's behavior is responsible for changing its state. For example, a person will have the behavior that he or she ages over time. This aging process is a behavior that is programmed into the person object that updates the data corresponding to the individual's age. Other behaviors are responsible for moving, finding jobs, making travel decisions, etc.

Agents are objects of particular interest in microsimulation models in that they exhibit autonomous behavior that is typically the primary focus of the microsimulation. Trip makers making travel decisions, households making residential location or auto ownership decisions, and firms hiring or firing workers are all examples of agents who independently make decisions as a function of their own attributes and of the state of the system that they find themselves within. These actions of the agents, in turn, change the system state over time (congestion levels, housing prices and vacancy rates, unemployment rates, etc.). Multiagent simulation models are simply microsimulation models that focus on modeling autonomous agent objects, often using specialized programming languages (e.g., SWARM) that have be specifically developed to facilitate this type of modeling.

Cellular automata are a very special form of object or agent, in which the agents exist within a regular spatial pattern. Traffic zones, roadway links, and grid cells are all examples of objects that exist in a regular spatial pattern (i.e., in which one object's relationship or interaction with another object is determined by their spatial relationship).[3] In such cases, each object or cell can be modeled as an autonomous agent that interacts with other cells in a highly localized way (e.g., usually only with its immediate neighbors). In such cases, very efficient (typically integer-based) algorithms can be developed to model cell (and hence system) behavior. The TRANSIMS network microsimulation model is an example of the cellular automata approach. In this model roads are divided into small segments, where each segment is approximately one car length in size. Each road segment is a cell. At each instant in time each cell is either occupied by a vehicle or not, and its interactions with other cells simply consist of receiving or not receiving a vehicle from the upstream cell, and of sending or not sending a vehicle to the downstream cell.

12.5 Agent Attribute Synthesis and Updating

Microsimulation models by definition operate on a set of individual actors whose combined simulated behaviors define the system state over time. As previously discussed, in short-run forecasting applications, a representative sample may often exist that can define the set of actors whose behavior is to be simulated (Figure 12.1). In medium- and long-term forecasting applications, however, even if such a sample exists for the base year of the simulation, this sample cannot generally be assumed to remain representative over the forecast time period. In such cases the microsimulation model must be extended to include methods for updating the attributes of the set of actors so that they continue to be representative at each point of time within the simulation (Figure 12.2). In addition, in many applications (particularly larger-scale, general-purpose regional modeling applications), the base year sample of actors either may not be available or may not be suitable for the task at hand. In such cases, the microsimulation model must also include a procedure for synthesizing a suitable base year set of actors as input to the dynamic behavioral

---

[3]Or, an equivalent interaction structure exists, even if it is not spatial in nature.

simulation portion of the model (Figure 12.3). Each of these two processes — synthesis and updating — are discussed in the following two subsections.

Before discussing synthesis and updating methods, however, one other important model design issue needs to be addressed. The discussion to this point has assumed that the set of actors being simulated is a sample drawn in an appropriate way from the overall population. Situations exist, however, in which it may be useful or even necessary to work with the entire population of actors within the microsimulation, rather than a representative sample. At least two major reasons exist for why one might prefer to work at the population level rather than with a sample.

First, situations exist in which computing population totals based on weighted sample results can be difficult to do properly. Consider, for example, the problem of simulating residential mobility. Assume that one is working with a 5% sample of households. Then, on average, each household in the sample will carry a weight of 20 in terms of its contribution to the calculation of population totals. If it is determined within the simulation that a given sample household will move from its current zone of residence i to another zone j, does this imply that 20 identical households make the same move? The answer is probably not. More complex weighting schemes can undoubtedly be devised, but it may prove to be conceptually simpler, more accurate, and perhaps even computationally more efficient to deal directly with the residential mobility decisions of every household and thereby avoid the weighting problem entirely.

All sample-based models inherently represent a form of aggregation in that each observation in the sample stands for or represents n actual population members (where, as illustrated above, 1/n is the average sample rate). These n population members will possess at least some heterogeneity and hence variability in behavior. In many applications (microsimulation or otherwise) this aggregation problem is negligible, and the efficiency in working with a (small) sample of actors rather than the entire population is obvious. In many other applications, such as the one described above, however, use of a sample may introduce aggregation bias into the forecast unless considerable care (and associated additional computational effort) is taken. In such cases, the relative advantages of the two approaches are far less clear.

Second, as one moves from short-run, small-scale, problem-specific applications to longer-run, larger-scale, general-purpose applications (e.g., testing a wide range of policies within a regional planning context — presumably an eventual goal of at least some modeling efforts), the definition of what constitutes a representative sample becomes more ambiguous. A sample that is well suited to one policy test or application may not be suitable for another. This is particularly the case when one requires adequate representation spatially (typically by place of residence and place of work) as well as socioeconomically. In such cases, a sufficiently generalized sample may be so large or sufficiently complex to generate that it might be just as easy to work with the entire population.

In trying to build a case for population-based microsimulations, one certainly cannot ignore the computational implications (in terms of processing time, memory, and data storage requirements) of such an approach. Nevertheless, it is important to note that the conceptual case for population-based microsimulation does exist, in at least some applications; computing capabilities and costs are continuously improving, and several population-based models are currently operational or under development.

The synthesis and updating methods discussed in the following subsections do not depend in any significant conceptual way on whether they are operating on a sample or the entire population. For simplicity of discussion, however, the presentations in these sections assume that it is a disaggregated representation of the entire population that is being either synthesized or updated.

12.5.1 Synthesis

All population synthesis methods start with the basic assumption that reliable aggregate information concerning the base year population is available, generally from census data. These data typically come in the form of one-, two-, or possibly multiway tables, as illustrated in Figure 12.4. Collectively, these tables define the marginal distributions of each attribute of the population of interest (age, sex, income, household size, etc.). In addition, any two-way or higher cross-tabulations provide information concerning the joint distribution of the variables involved. The full multiway distribution of the population

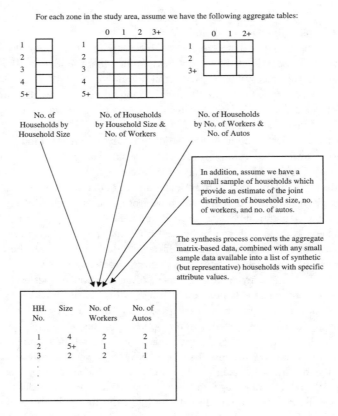

FIGURE 12.4 Population synthesis.

across the entire set of attributes, however, is not known, although it may be the case that sample data (which provide specific attributes for the observed individuals) obtained from sources such as activity–travel surveys, public use microdata sample (PUMS) files, etc., are available. In such cases, these data provide an estimate of the joint population attribute distribution. The synthesis task, as shown in Figure 12.4, is to generate a list of individual population units (in the case of Figure 12.4, households) that is statistically consistent with the available aggregate data.

All synthesis procedures developed to date use some form of Monte Carlo simulation to draw a realization of the disaggregate population from the aggregate data. At least two general procedures for doing this currently exist. The first appears to have been originally proposed by Wilson and Pownall (1976). In this method, the marginal and two-way aggregate distributions for a given zone (or census tract) are used sequentially to construct the specific attribute values for a given person (or household, etc.) living in this zone. For example, assume that we are synthesizing households with three attributes, X_1, X_2, and X_3. Also assume that we have the marginal distribution for X_1, which defines the marginal probabilities $P(X_1 = x_1)$ for the various valid values x_1 for this attribute. We also have the joint distributions for X_1 and X_2 and for X_2 and X_3, which can be used to define the conditional probabilities $P(X_2 = x_2|x_1)$ and $P(X_3 = x_3|x_2)$. An algorithm for generating specific values (x_{1h}, x_{2h}, x_{3h}) for household h is then:

1. Generate a uniform random number u_{1h} on the range $(0, 1)$. Given u_{1h}, determine x_{1h} from the distribution $P(X_1 = x_{1h})$.
2. Generate a uniform random number u_{h2}. Given x_{1h} and u_{2h}, determine x_{2h} from the distribution $P(X_2 = x_{2h}|x_{1h})$.
3. Generate a uniform random number u_{3h}. Given x_{2h} and u_{3h}, determine x_{3h} from the distribution $P(X_3 = x_{3h}|x_{2h})$.

This process is then repeated until all households, each with a specific set of attributes, have been generated.[4] This procedure is conceptually straightforward, easy to implement, and has been used in several models, including in Mackett (1990) and Miller et al. (1987).

As Wilson and Pownall (1976) note, this process implies a causal structure in terms of the order in which the conditional probabilities are computed (i.e., in the assumptions concerning which attributes are conditional upon which others). In practical applications it is not always clear to what extent this conditioning is guided by theoretical considerations, as opposed to the availability of a given set of cross-tabulations. Alternatively, sufficient redundancy often exists within available census tables so that multiple paths through these tables may exist, leaving it to the modeler to determine which path is best for computing the joint attribute sets (e.g., perhaps one has two-way tabulations of X_1 by X_3 as well as the other two-way tabulations previously assumed; in such a case, which order of conditioning is best?).

More fundamentally, this procedure ignores the potential for significant multiway correlations among the variables, except for the very limited two-way correlations permitted within the arbitrarily assumed conditional probability structure. This is a potentially serious problem. One approach for resolving this problem is a procedure developed by Beckman et al. (1996) for use in TRANSIMS.

The TRANSIMS procedure also starts with aggregate census tabulations for each census tract. In addition, however, it utilizes PUMS files that consist of 5% representative samples of almost complete census records for collections of census tracts. Adding up the records in a PUMS provides an estimate of the full multiway distribution across all attributes for the collection of census tracts. If one assumes that each census tract has the same correlation structure as its associated PUMS, then the PUMS multiway distribution provides important additional information to the synthesis process. Skipping over a number of important details, primary steps in the TRANSIMS procedure are as follows:

1. For each public use micro area (PUMA), construct the multiway distribution of attributes from the corresponding PUMS.
2. A two-step iterative proportional fitting (IPF) procedure is used to estimate simultaneously the multiway distributions for each census tract within a PUMA, such that each distribution satisfies the marginal distributions for the census tract (as defined by aggregate census tables) and has the same overall correlation structure as the PUMS-based multiway distribution. This IPF procedure can be interpreted as the constrained maximum entropy estimate of the multiway distribution given the known information and the available PUMS data.
3. Individual households are then randomly drawn from the full multiway distribution for each census tract.

The TRANSIMS procedure is relatively straightforward to implement and appears to perform well in validation tests to date (Beckman et al., 1996). In particular, it clearly performs better than either drawing households directly from the PUMS multiway distribution (i.e., without filtering this distribution through the census tract marginal distributions by means of the two-step IPF procedure) or drawing households directly from the tract marginals (i.e., a simplified version of the Wilson and Pownall procedure). While more operational experience is obviously required with population synthesis methods, the general thrust exemplified by the TRANSIMS approach appears to be well founded: use a full information approach that accounts for multiway correlation among the attributes being synthesized.

The discussion to this point has focussed on the synthesis of the resident population who will eventually be called upon within the model to make activity–travel decisions. For many (typically

---

[4]Wilson and Pownall proposed this algorithm for the case of generating a small sample. In this case "sampling with replacement" (as occurs in the algorithm outlined) is acceptable. If an entire population set is to be generated, then the algorithm shown must be altered so that it involves "sampling without replacement." That is, after each household is drawn, the aggregate household distributions should be modified to reflect the fact that this household has been removed from the distribution, thereby altering slightly the probability distributions for subsequent households.

short-run) applications, this may be sufficient. In many other (particularly longer-run) applications, however, many other system entities may also need to be known at a disaggregate level, and hence may also need to be synthesized. These entities might include firms, residential buildings, nonresidential buildings, vehicles, jobs, etc.

While the synthesis problem for such entities is, in principle, the same as for the population case, additional issues may arise in at least some cases. These may include:

1. In some cases, limitations may exist with respect to the depth, breadth, or reliability of data that are available to support disaggregate synthesis activities. In particular, data sources other than the census will almost certainly have to be drawn upon in order to synthesis many of these entities.
2. In at least the case of firms, it may well be that significantly greater heterogeneity exists than for people or households. Firms vary tremendously in terms of size, function, location preferences, needs, etc. Can we credibly synthesis a disaggregate population of firms for an urban area? If so, at what level of detail or specification? Do we need to work at a completely disaggregated level of analysis for firms, or is a more aggregate level acceptable for our purposes?
3. Many of the questions raised in number 2 are indicative of the lack of experience (and hence insight) that we have with modeling aspects of the urban system, other than the distribution of the resident population and their subsequent activity–travel patterns. We have a long tradition of assuming that the supply or attraction side of the process is fixed and given. Dynamic microsimulation (particularly when performed over medium- to long-term forecast periods) severely challenges this assumption, and ultimately requires us to model supply as well as demand.

The proliferation of commercial GIS-based databases undoubtedly represents a major resource to support the development of supply and attraction databases and models. The completeness, accuracy, etc., of these databases, however, may be a considerable problem for some time to come. Further, even given excellent databases, one should not underestimate both the importance and the difficulty of developing credible supply or attraction components for our activity–travel simulation models.

12.5.2   Updating

Once the base year population has been provided to the model, through either a survey sample or a synthesis procedure, this population must be updated at each time step within the simulation run. The nature of this updating obviously depends on the attributes involved, the processes being simulated, the size of the simulation time step, etc. Assuming, however, that one is simulating household processes over a number of years, in 1-year time steps, demographic and socioeconomic processes that need simulating as part of the updating process may well include:

- Aging
- Births and deaths
- Marriages and divorces
- Other changes in household structure (adult children leaving the home, etc.)
- Nonfamily household formation and dissolution
- Changes in education level
- Changes in employment status (entry to or exit from the labor market, change in job location or type, etc.)
- Changes in residential location
- Changes in automobile holdings (types and numbers of vehicles, etc.)

With the exception of aging, which is a completely deterministic accounting process, each of these processes requires a submodel of some sort. Demographic and household structure attributes are generally handled using very simple probability models: either fixed transition rates based on empirical data (e.g., fertility rates for women by age group) or simple parametric probability functions (e.g., simple logit models to determine household type transition probabilities). In all such cases, Monte Carlo simulation methods are used to generate household-specific events (birth of a child, etc.) on a household-by-household and year-by-year basis.

Treatment of employment status, residential location, and automobile holdings varies far more widely across models, depending on their application. Each of these can be a significant part (or even the primary focus) of the behavioral modeling component of the microsimulation (see below). Alternatively, if the application permits, one or more of these might be handled in terms of transition probabilities in the same way as the demographic variables discussed above.

As with synthesizing procedures, limited experience exists, at least within the travel demand forecasting community, with demographic and socioeconomic updating methods. For examples of specific methods used to date, see Miller et al. (1987), Kitamura and Goulias (1991), Goulias and Kitamura (1992), and Oskamp (1995). All of these examples should be treated as being illustrative and experimental in nature rather than in any way definitive in terms of *the* method to use. Considerable experience with demographic forecasting obviously exists among demographers. Traditional demographic forecasting, however, does not attempt to work at the fine spatial scale required by our travel demand forecasting applications. Our challenge is to adapt existing methods or develop new ones that can operate reliably at the census tract–traffic zone level required for travel demand forecasting.

Considerable experience with demographic forecasting obviously exists among demographers. In addition, a long and extensive tradition of microsimulation exists within the social sciences, involving a wide range of socioeconomic applications. Beginning with the seminal work of Orcutt (1957, 1976) and Orcutt et al. (1961, 1986), economists and other social scientists have developed, particularly over the last 10 years, very impressive policy-sensitive microsimulation models designed to inform politicians and other decision makers on a wide range of economic and social policies (see, for example, Lambert et al., 1994; Caldwell, 1996; and Troitzsch, 1996). Indeed, these models have reached the point of acceptance where they are routinely used within many countries to analyze the benefits and costs (and the winners and losers) associated with most major policy issues, including such high-profile issues as health care reform in the United States and taxation policy in Canada.[5] While much can be learned from the social science microsimulation experience, one should note the following:

1. Most operational models are static rather than dynamic in nature (i.e., they do not evolve or update the base population over time) and so provide little insight into how to deal with the dynamics of population evolution.
2. It is also the case that most social science microsimulations either are completely aspatial or, at best, operate at an extremely gross spatial scale (e.g., perhaps by state within a national model). Similarly, traditional demographic forecasting typically is undertaken at very gross spatial scales (e.g., state or province or perhaps at a regional level). Travel behavior forecasting ultimately requires a much finer spatial scale, typically down to the traffic zone level. The challenge for travel behavior researchers is to adapt existing methods or to develop new ones that can operate reliably at the census tract–traffic zone level required for travel demand forecasting.

All other system entities explicitly represented within the model (such as possible firms, buildings, etc.) that can be expected to change over time must also be updated. As with population updating, this may involve the use of simple transition probabilities (x% of all housing stock of a given type undergoes renovation each year), exogenous inputs, or behavioral models of varying complexity. As has already been discussed above concerning the synthesis problem associated with these entities, the development of credible, appropriate updating methods represents a significant research and development effort.

---

[5]Among many others. For a discussion of applications, see, for example, Citro and Hanushek (1991).

12.6 Issues in Microsimulation Modeling

Large-scale, operational applications of travel-related microsimulation models are a quite recent phenomenon. Considerable research and development work with respect to these models is still under way as part of the process of moving microsimulation out of the laboratory and into operational practice. This section briefly discusses some of the key issues that must be addressed as microsimulation modeling continues to evolve and improve as an operational planning tool.

1. Continued development and testing of population synthesizing and updating methods. Just as conventional four-stage models depend fundamentally on the population and employment inputs provided to them, microsimulation models depend on the population demographic and socioeconomic inputs to the behavioral components of the model. While the TRANSIMS procedure for population synthesis appears very attractive (and emerges out of at least 20 years of experience in the literature with related but simpler methods), clearly much more operational experience is required before such a method can be considered a proven tool. Updating methods similarly have clearly been demonstrated to be feasible, but require much further incremental experimentation, improvement, and optimization.

2. Determination of appropriate levels of aggregation. Even in a microsimulation model, aggregation inevitably occurs. Aggregation can occur in:

   a. Space: typically through the use of zones as the spatial unit of analysis, even when modeling individual decision makers within these zones.

   b. Time: primarily in terms of the time step used to move the model through simulated time. A model that operates on a 1-year time step is temporally more aggregate than one that steps through time on a month-by-month basis.

   c. Attributes: no matter how detailed the model's description of an individual, there is always some point beyond which two individuals will be considered identical. Individuals are, however, exactly that, and by treating them as identical we are, in fact, introducing some amount of aggregation into the analysis.

   d. Behavior: for example, perhaps in a given model all types of nongrocery shopping — everything from buying shoes to buying a new car — might be aggregated into a single activity category.

   A major rationale for the disaggregate modeling approach is the minimization of aggregation bias. In the theoretical development of our disaggregate models it is often easy to pretend that these models truly operate at the level of unique individuals acting within their actual individual choice contexts. It must be recognized, however, that any operational model will inevitably reach some finite limit of disaggregation (where this limit may be defined by data availability, theoretical insight, methodological capabilities, computational feasibility, or application requirements), beyond which aggregate homogeneity assumptions are inevitably required. This is neither good nor bad, but rather simply a fact of model building. The key point is to recognize this fact and to make intelligent decisions concerning where finer levels of disaggregation are both required and achievable, and where more aggregate representations either can be used because of the nature of the problem (relative homogeneity does exist, system state estimates are robust with respect to this component of the model, etc.) or must be used due to inherent limitations in our modeling capabilities.

   Over and above a general concern with finding appropriate levels of disaggregation in our microsimulations, specific issues include:

   a. Treatment of space. Microsimulation models have been developed at a wide variety of spatial scales, ranging from point-to-point travel to large zone-based systems. While the microsimulation approach is largely independent of spatial scale, issues of appropriate spatial representation are critical to accurate travel demand forecasting. Considerable uncertainty also exists about what level of spatial disaggregation is required to support forecasting requirements for emissions analysis, etc. It is not currently clear what level of spatial disaggregation is likely to be supportable with respect to data and computational capabilities, even given modern Geographic Information Systems (GIS), etc.

    b. Treatment of time. Different urban processes operate within very different time frames. Residential and employment location processes operate over periods of years, typically involving brief periods of intense activity (e.g., looking for a new home or job), followed possibly by decades of inactivity. Most demographic processes operate on approximately a yearly scale. Activity–travel decisions, however, occur more typically within daily or weekly time frames. Tailpipe emissions from a vehicle depend critically on the second-by-second decisions of the vehicle's driver. Within each of these components of the overall travel demand process decisions need to be made concerning the best time step to use in modeling the given component. Is second-by-second simulation of vehicle performance really necessary or can a longer time step (say 5 sec) be used? Is the day or the week the fundamental step in modeling household activity and travel dynamics (or is hour-by-hour or minute-by-minute simulation required)? Can 1-year time steps by used to simulate residential mobility decisions (and if so, how does one handle the microdynamics of the housing search process, which typically occurs over a period of a few weeks or, at most, months)? These questions become even more problematical as one attempts to bring these model components into a comprehensive modeling system. It is easy to speak about the need for integrated land use–transportation models, for example, but how does one actually integrate these models, given their very different time frames?

    c. Selection of attributes. Models vary in terms of the definition and detail of the attributes of persons, households, etc., being modeled. Decisions concerning these attributes obviously affect, among other components of the model, the nature of the population synthesis and procedures required to generate and update these attributes over time. Trade-offs may well often occur between the ability of the synthesis or updating procedures to reliably provide a given attribute and the relative importance of the attribute within the behavioral model.

3. Linkages among model components. As has been mentioned at various points throughout this chapter, linkages between location choice, activity–travel decisions, and network assignment and performance models represent both a trend and a desirable feature in microsimulation model development. In particular, analysis of the full range of possible impacts of a given policy may often require a relatively comprehensive modeling system, given the wide range of possible short-run and long-run responses available to individuals and households in many cases.

    While conceptually attractive, comprehensive microsimulation models obviously bring with them a host of model design issues, not the least of which is the computational feasibility of such models. It is to be expected that many modelers will continue to develop individual models for various components of the overall process, both as a means for best making progress in the development of these components and as a means for analyzing problems directly addressable by such models. At the same time, other modelers will continue with the task of developing comprehensive modeling systems, often with simplified versions of the current state-of-the-art component models. Both types of activities obviously are mutually reinforcing and are to be encouraged.

4. Demonstration of the statistical properties of microsimulation models. Almost all microsimulation models include stochastic elements. Surprisingly, little attention seems to have been paid to the statistical properties of these models. This may partially be due to the preliminary nature of most models: when one is busy trying to show that the thing simply works at all, one may be forgiven for not worrying what the average outcome of a hundred replications of the same model run might look like. It may also reflect a reluctance on the part of modelers to come to grips with the issue, given both the magnitude of the computational effort to generate a single model run and the complexity of the outcome of the simulation experiment — i.e., a massively multidimensional data structure defining the final system state.

    This issue, however, must be addressed, since the output of any single run of a stochastic model is simply one random draw from the unknown distribution of possible outcomes. The representativeness of this single outcome (and hence its usefulness for planning purposes) is also by definition unknown. In classical stochastic simulations, this problem is resolved by executing many replications of the run, each one of which generates additional information concerning the underlying unknown distribution of outcomes. This process continues until one has generated a sufficient number of observations to be able to say statistically meaningful things about the distribution of possible outcomes — in particular, to provide reliable estimates of the means and variances of the final system state.

Much work is required to address this issue in the case of activity-based travel demand microsimulation models. Considerable experimentation is needed to determine the statistical properties of both individual model components and overall modeling systems — in particular, to develop guidelines concerning when replications need to be undertaken and, if performed, how many are generally required. As Axhausen (1990) points out, many standard methods exist for reducing internal variation within simulation model runs, and the usefulness and appropriateness of using such methods must be investigated. Finally, thought must be given to how one does average over a set of simulated outcomes in cases of such complexity and high dimensionality as are typical of these applications.

5. Demonstration of computational feasibility. One should never make the mistake of underestimating the computational intensity of microsimulation models. In addition to requiring considerable amounts of CPU time, the memory and disk storage requirements of a large microsimulation model are enormous. Early microsimulation models quickly bumped up against computational limits or made significant design comprises in order to maintain computational feasibility. With the continuing rapid expansion of the computing power cost-effectively available to both researchers and planners, the definition of what is computationally feasible is being upgraded almost daily.

Nevertheless, the computational challenges associated with large-scale microsimulations are significant, to say the least. This is particularly the case for population-based (as opposed to sample-based) models. The magnitude of the problem also grows as we move toward more integrated, comprehensive models (e.g., combined models of residential and employment location choice, activity–travel, and network assignment).

Ultimately, all of the issues discussed above come together and interact with the issue of computational feasibility in a classic engineering design problem involving trade-offs between cost and performance. Every increase in model disaggregation, every extension of its comprehensiveness, every improvement in its statistical reliability comes at a cost in computer time, memory, and storage. Conversely, at any point in time, current computational capabilities establish upper bounds in terms of what is cost-effectively doable within the model.

12.7 Example Applications

Orcutt (1957, 1976) is generally credited with being the seminal developer of microsimulation as an operational policy analysis tool, with his applications occurring in the field of social welfare policies. A wide variety of travel-related microsimulation modeling applications exist, including auto ownership, residential mobility, dynamic network assignment, and activity–travel models. These are each discussed in turn in the following subsections.

12.7.1 Microsimulation of Auto Ownership

Some of the earliest applications of microsimulation in the transportation field involved dynamic modeling of auto ownership (e.g., Barnard and Hensher, 1987; Daly, 1982). Behavioral modeling of auto ownership has almost always occurred as a stand-alone activity, outside of the normal activity–travel demand modeling process. Within the travel demand modeling process, auto ownership has typically been treated as just one socioeconomic exogenous input to the demand process. For some purposes this may be adequate, in which case a transition probability treatment within a microsimulation modeling system would be adequate. However, many current policy issues (notably concerning emissions and energy use) relate in no small way to household decisions concerning the number and types of vehicles that they own, as well as the interactions between vehicle holdings and (auto) travel demand. Thus, a strong case exists for including explicit models of household automobile choice within the overall travel demand modeling process (Miller and Hassounah, 1993). A recent example of a household vehicle transaction model developed for use in a microsimulation forecasting system is provided by Mohammadian and Miller (2002).

12.7.2 Microsimulation of Housing Markets and Residential Mobility

Many of the microsimulation models developed to date fall into this general category. Early work includes that undertaken by Wegener (1983), Mackett (1985, 1990), and Miller et al. (1987). This continues to be an active area for research efforts, including work by Spiekermann and Wegener (1993), Oskamp (1995), and Miller et al. (2002a).

Given the central role that life cycle stage and household structure play in determining residential mobility, these models typically deal in detail with population and household synthesis and updating — issues of considerable importance to travel demand models. In addition to the technical issues relating to synthesis and updating already discussed, note that the discussion to this point has been relatively indifferent to the unit of analysis within microsimulation models. In residential mobility modeling it has long been recognized that both households and persons (with the latter being further subdivided into workers, nonworkers, etc.) must be maintained within the modeling system, given that some decisions are inherently household level in nature (e.g., residential choice), while others inherently occur at the level of the individual (e.g., changing jobs), with interactions between both levels continuously occurring (e.g., the decision to change jobs may have ramifications for household income levels and hence the suitability or affordability of the current residential location; the decision on whether or where to move may be influenced by the impact that the move would have on commuting times and costs). As a result, such models generally maintain both households and persons (and mappings between the two) as explicit elements of their database. This dual representation presumably will prove useful to activity-based models, both as they move to more household-level formulations and as they become more integrated with residential mobility models within more comprehensive microsimulation frameworks.

In addition, of course, housing market models are intended to forecast medium- to long-term evolution of the spatial distribution of the residential population, another key input into travel demand models. Modeling housing markets as part of an integrated approach to dealing with transportation–land use interactions is discussed in detail in Chapter 5.

12.7.3 Microsimulation of Auto Route Choice and Network Performance

As has already been briefly mentioned, many current and emerging road network assignment procedures are microsimulation based. From a more general travel demand modeling perspective, three points should be noted concerning such models:

1. As has already been discussed, the input requirements of these network microsimulation models may in some instances drive the design criteria for activity-based travel forecasting models. TRANSIMS is perhaps the best example of this point, in that the network performance and emissions modeling needs are clearly in this case driving the overall system design.
2. The interface between the activity-based models and the network models generally does not simply consist of the outputs from the one becoming the inputs to the other. Typically, dynamic route assignment procedures simultaneously determine route choice and trip departure time choice (given assumptions about desired arrival times). Thus, these models intrude into at least one component of the activity-based modeling domain: the microscheduling of trips. Again, this may well have design implications for activity-based models to the extent that they are intended to be integrated with network microsimulation models.

3. Most current network microsimulation models appear to have been developed with short-run (and, in some cases, real-time) forecasting applications in mind, often specifically relating to Intelligent Transportation Systems (ITS) applications. Whether these models are well suited for medium- to long-term forecasting applications is an unanswered question at this point in time. Issues include the level of detail of network representation often required by these models (e.g., are we able to specify the traffic signal settings and offsets 20 years into the future, as may be required by some models), as well as the match between network model precision (e.g., second-by-second calculations of vehicle performance) and the accuracy of the activity–travel demand model's predictions (even with microsimulation), given the inevitable uncertainties associated with medium- to long-term forecasting.

12.7.4 Activity-Based Microsimulation Models

Given the inherently disaggregate nature of activity-based models, as well as the fact that these models typically incorporate some level of dynamics, one might argue that a large portion of the extensive activity-based modeling literature should be included in this section.[6] This has not been attempted here. Rather, emphasis has been placed on including models that emphasize the connection between activity modeling and travel demand forecasting in at least a quasi-operational manner, and that do this within an explicit microsimulation framework.

Bonsall (1982) provides a very early example of the application of microsimulation to the problem of predicting commuters' participation in a proposed ridesharing program. Although very specialized in nature, Bonsall's report is noteworthy for the time period of development of the model and for the clarity with which general issues of microsimulation modeling are discussed.

Axhausen (1990) reports on a considerable tradition in Germany of activity-based microsimulation modeling of destination and mode choice, tracing back to Kreibich's (1978, 1979) initial work in the late 1970s. Much of this German work has been generally inaccessible to North American audiences since with the exception of Kreibich's papers, most of his work has only been published in German. Axhausen's contribution was to combine an activity chain simulation model (which had been the focus of the work of Kreibich) with a mesoscopic traffic flow simulator.[7] Axhausen's report is noteworthy in at least two respects: (1) it represents an early attempt to link an activity-based model directly to a network assignment model — clearly an essential step in developing a true activity-based travel demand forecasting capability; and (2) the decision to use a mesoscopic rather than microscopic traffic simulator provides a useful counterpoint to the general North American trend of leaping directly to the extreme microlevel for this later type of model.

The Microanalytic Integrated Demographic Accounting System (MIDAS) (Kitamura and Goulias, 1991; Goulias and Kitamura, 1992, 1996) represents an extremely important milestone in the development of transportation-related microsimulation models. Developed for the Dutch government, MIDAS is an operational microsimulation-based forecasting tool. Starting with a nationwide sample of households obtained from the Dutch Mobility Panel, the model has two main components: a socioeconomic and demographic component that simulates household transitions, including births, deaths, and household type changes, as well as changes in persons' employment status, personal income, driver's license possession, and education; and a mobility component that simulates auto ownership, trip generation, and modal split. Although the application is somewhat atypical (i.e., predicting overall national travel levels rather than intraurban trip making), the model contains most of the attributes of the activity-

---

[6]Very explicity simulation-based activity-based models such as STARCHILD (Recker et al., 1986a, 1986b) and SMASH, developed by Ettema et al. (1993), come to mind.

[7]Mesoscopic network models generally work at the level of the individual vehicle, but make use of much more simplified models of vehicle performance than the microscopic models discussed above. For a detailed discussion of the potential merits of mesoscopic models, see Miller and Hassounah (1993).

based travel forecasting microsimulation modeling paradigm presented earlier in this chapter. In particular, the model's treatment of the demographic and socioeconomic updating problem is very strong.

Another very important but not well-known microsimulation model is STEP, which has been evolving in its capabilities for more than 20 years (Harvey and Deakin, 1996). STEP is a descendent from the pioneering disaggregate modeling system developed by Cambridge Systematics and the University of California at Berkeley for the San Francisco Bay Area Metropolitan Transportation Commission (MTC) in the late 1970s. STEP can accept a representative input sample or synthesize a representative sample from a range of data sources. It contains a wide range of model components, including simple residential and workplace location choice models, along with a suite of travel demand models (including recently developed activity-based models). STEP has been used in a number of air quality and transportation control measure (TCM)-related studies in the San Francisco Bay Area, as well as in the analysis of pricing policies in Los Angeles, San Diego, Sacramento, the Puget Sound region, and Chicago.

In 1992 the Federal Highway Administration (FHWA) commissioned four groups — RDC, Inc., Caliper Corporation, MIT, and the Louisiana Transportation Research Center (LTRC) — to propose new modeling systems to replace the conventional four-stage system. It is noteworthy that two of the four groups (RDC and LTRC) proposed activity-based microsimulation designs, while a third (MIT) proposed a disaggregate activity-based approach that certainly could be implemented within a microsimulation framework (Spear, 1994). Further, both RDC's Sequenced Activity-Mobility System (SAMS) and LTRC's Simulation Model for Activities, Resources and Travel (SMART) postulated an integrated, comprehensive modeling system beginning with land use and flowing through activity–travel decisions to dynamic assignment of vehicles to networks (and hence calculation of congestion, emissions, etc.).

Since the FHWA study, a prototype of the Activity Mobility Simulator (AMOS), the central component of the proposed SAMS system, has been developed and used in Washington, D.C., to evaluate alternative TDM strategies (RDC, 1995). AMOS represents an example of an activity-based travel microsimulator. As currently implemented, it represents a stand-alone tool for analyzing a specific type of short-run transportation policies that is not currently tied to either a demographic simulator (as in the case of MIDAS) or a network simulator (as in the case of Axhausen's model). More generally, however, it represents a potential stepping-stone toward a more comprehensive microsimulation system, such as SAMS, that would include these other microsimulation components, among others.

TRANSIMS (Barrett et al., 1995) represents an ambitious attempt to develop a comprehensive microsimulation travel demand forecasting model. The TRANSIMS program is well documented in the literature, and so no attempt will be made in this chapter to provide a complete description of the model. In many respects, the TRANSIMS work has defined much of the current state of the art in microsimulation modeling, as well as challenged other researchers to develop their own thoughts and models. At the same time, the current operational component of TRANSIMS largely focuses on network modeling (i.e., route assignment and network performance) aspects of the overall modeling problem, with many of the higher-level travel demand components being treated in a much less definitive way.

In The Netherlands, a very comprehensive activity-based modeling system, known as ALBATROSS, has been developed (Arentze and Timmermans, 2000) that is microsimulation based. ALBATROSS is noteworthy for being totally rule based in nature, rather than based on random utility choice models (which tend to represent the typical North American approach to activity-based modeling). Microsimulation is essential to the development of rule-based models, since it represents the only effective computational strategy for implementing complex systems of rule-based decision making.

12.7.5 Fully Operational Models

In recent years, several fully operational microsimulation models that forecast the travel demand for entire urban regions have been developed and put into operational practice. These include activity-based models in Portland, Oregon (Bradley and Bowman, 1998), San Francisco (Bradley et al., 2001), and Toronto (Miller et al., 2002b), as well as a very large trip-based model for the New York metropolitan region (Vovsha et al., 2002). The Portland, San Francisco, and New York models all operate on 100% synthesized populations. The Toronto model currently uses a large (5%) sample of households that is exogenously updated to represent future year conditions, but the intention is to eventually run this model on a fully synthesized 100% population.

These models demonstrate the feasibility of the microsimulation approach for practical, large-scale urban transportation planning applications. They also confirm that the potential strengths of microsimulation models — most notably, computational efficiency, behavioral realism, and policy sensitivity — are all, indeed, achievable within current computational, data, and modeling methodological capabilities.

Although not yet fully operational, another very large-scale, largely (but not completely) microsimulation-based modeling effort is currently under way in the state of Oregon. The Oregon State Department of Transportation is well on its way to constructing an integrated transportation–land use modeling system (see Chapter 5 for a discussion of such models) for the entire state that includes microsimulations of personal travel and residential housing markets on a statewide basis (Hunt and Abraham, 2001). Once operational, this model will again confirm the computational feasibility of the approach. It will also represent the most complete implementation to date of a Figure 12.3 type model, in which the trip-making population and the travel-generating activity system (and the salient attributes of these entities) are endogenously updated within the modeling system over time.

12.8 Concluding Remarks

Microsimulation is quickly becoming the method of choice for implementing disaggregate behavioral models of travel-related processes. Models of arbitrarily large complexity can be developed within the microsimulation paradigm, limited only by computational capabilities, data requirements, and our theoretical insights into the processes modeled. As each of these constraints becomes less binding (especially computational limitations), the power and usefulness of microsimulation within operational planning applications increases.

This chapter has discussed the definition of and motivation for microsimulation, the strengths and weaknesses of the approach, and the key issues involved in developing and applying such models. It has also provided a brief description of a range of microsimulation models that have developed over the past 20 years.

References

Arentze, T. and Timmermans, H.J.P., *ALBATROSS: A Learning-Based Transportation Oriented Simulation System*, Eindhoven University of Technology, Netherlands, 2000.

Axhausen, K., A simultaneous simulation of activity chains and traffic flow, in *Developments in Dynamic and Activity-Based Approaches to Travel Analysis*, Jones, P., Ed., Avebury, Aldershot, U.K., 1990, pp. 206–225.

Barnard, P.O. and Hensher, D.A., Policy Simulation with Discrete/Continuous Choice Model Systems: A Discussion of the Issues in the Context of the Auto Market, Working Paper 30, Transport Research Group, Macquarie University, Australia, 1987.

Barrett, C. et al., An Operational Description of TRANSIMS, LA-UR-95-2393, Los Alamos National Laboratory, Los Alamos, New Mexico, 1995.

Beckman, R.J., Baggerly, K.A., and McKay, M.D., Creating synthetic baseline populations, *Transp. Res. A*, 30, 415–429, 1996.

Bonsall, P.W., Microsimulation: Its application to car sharing, *Transp. Res. A*, 15, 421–429, 1982.

Bradley, M. and Bowman, J.L., *A System of Activity-Based Models for Portland, Oregon*, Travel Model Improvement Program, U.S. Department of Transportation, Washington, D.C., May 1998.

Bradley, M. et al., Estimation of Activity-Based Microsimulation Model for San Francisco, paper presented at 80th Annual Meeting of the Transportation Research Board, Washington, D.C., January 2001.

Caldwell, S.B., Dynamic Microsimulation and the CORSIM 3.0 Model, Institute for Public Affairs, Department of Sociology, Cornell University, Ithaca, NY, 1996.

Citro, C.F. and Hanushek, E.A., Eds., *The Uses of Microsimulation Modelling, Vol. 1: Review and Recommendations*, National Academy Press, Washington, D.C., 1991.

Daly, A., Policy analysis using sample enumeration: An application to car ownership forecasting, in *Proceedings of the 10th PTRC Summer Annual Meeting, Transportation Planning Methods Seminar*, PTRC, London, 1982, pp. 1–13.

Ettema, D., Borgers, A., and Timmermans, H., Simulation model of activity scheduling behavior, *Transp. Res. Rec.*, 1413, 1–11, 1993.

Fowler, M., *UML Distilled: A Brief Guide to the Standard Object Modeling Language*, 2nd ed., Addison-Wesley, Boston, 2000.

Goulias, K.G. and Kitamura, R., Travel demand forecasting with dynamic microsimulation, *Transp. Res. Rec.*, 1357, 8–17, 1992.

Goulias, K.G. and Kitamura, R., A dynamic model system for regional travel demand forecasting, in *Panels for Transportation Planning: Methods and Applications*, Golob, T., Kitamura, R., and Long, L., Eds., Kluwer Academic Publishers, Boston, 1996, chap. 13, pp. 321–348.

Harvey, G. and Deakin, E., *Description of the Step Analysis Package*, draft manuscript, Deakin/Harvey/Skabardonis, Hillsborough, NH, 1996.

Hunt, J.D. and Abraham, J.E., Heterogeneous Agents in Land Use Transport Interaction Modelling, paper presented at the WEHIA 2001 Conference, Honolulu, 2001.

Ingram, G.K., Kain, J.F., and Ginn, J.R., *The Detroit Prototype of the NBER Urban Simulation Model*, National Bureau of Economic Research, New York, 1972.

Kitamura, R. and Goulias, K.G., *MIDAS: A Travel Demand Forecasting Tool Based on a Dynamic Model System of Household Demographics and Mobility*, Projectbureau Integrale Verkeer-en Vervoerstudies, Ministerie van Verkeer en Waterstaat, The Netherlands, 1991.

Kreibich, V., The successful transportation system and the regional planning problem: An evaluation of the Munich rapid transit system in the context of urban and regional planning policy, *Transportation*, 7, 137–145, 1978.

Kreibich, V., Modeling car availability, modal split and trip distribution by Monte-Carlo simulation: A short way to integrated models, *Transportation*, 8, 153–166, 1979.

Lambert, S. et al., An Introduction to STINMOD: A Static Microsimulation Model, STINMOD Technical Paper 1, National Centre for Social and Economic Modelling, Faculty of Management, University of Canberra, Australia, 1994.

Mackett, R.L., Micro-analytical simulation of locational and travel behaviour, in *Proceedings PTRC Summer Annual Meeting*, Seminar L: Transportation Planning Methods, PTRC, London, 1985, pp. 175–188.

Mackett, R.L., Exploratory analysis of long-term travel demand and policy impacts using micro-analytical simulation, in *Developments in Dynamic and Activity-Based Approaches to Travel Analysis*, Jones, P., Ed., Avebury, Aldershot, U.K., 1990, pp. 384–405.

Miller, E.J. and Hassounah, M.I., Quantitative Analysis of Urban Transportation Energy Use and Emissions: Phase I Final Report, report submitted to Energy, Mines and Resources Canada, Joint Program in Transportation, University of Toronto, 1993.

Miller, E.J. et al., Microsimulating urban systems, in *Computers, Environment and Urban Systems*, special issue: Geosimulation: Object-Based Modeling of Urban Phenomena, 2002a, forthcoming.

Miller, E.J., Noehammer, P.J., and Ross, D.R., A micro-simulation model of residential mobility, in *Proceedings of the International Symposium on Transport, Communications and Urban Form, Vol. 2: Analytical Techniques and Case Studies*, Monash University, Melbourne, 1987, pp. 217–234.

Miller, E.J. et al., A Microsimulation Model of Daily Household Activity and Travel, paper presented at Understanding and Modeling Travel Behavior session of the 98th Annual Meeting of the Association of American Geographers, Los Angeles, March 2002b.

Mohammadian, A. and Miller, E.J., Understanding and modeling dynamics of household automobile ownership decisions: A Canadian experience, *Proc. 7th Int. Conf. Appl. Adv. Technol. Transp.*, American Society of Civil Engineering, Atlanta, GA, 2002, pp. 672–679.

Orcutt, G., A new type of socio-economic system, *Rev. Econ. Stat.*, 58, 773–797, 1957.

Orcutt, G., *Policy Evaluation through Discrete Microsimulation*, 2nd ed., Brookings Institute, Washington, D.C., 1976.

Orcutt, G. et al., *Microanalysis of Socioeconomic Systems: A Simulation Study*, Harper and Row, New York, 1961.

Orcutt, G., Merz, J., and Quinke, H., *Microanalytic Simulation Models to Support Social and Financial Policy*, New-Holland, New York, 1986.

Oskamp, A., LocSim: A Microsimulation Approach to Household and Housing Market Modelling, PDOD Paper 29, presented at 1995 Annual Meeting of the American Association of Geographers, Chicago, March 15–18, 1995, Department of Planning and Demography, AME — Amsterdam Study Centre for the Metropolitan Environment, University of Amsterdam, The Netherlands.

RDC, Inc., Activity-Based Modeling System for Travel Demand Forecasting, DOT-T-96-02, U.S. Department of Transportation, Washington, D.C., 1995.

Recker, W.W., McNally, M.G., and Root, G.S., A model of complex travel behavior. Part I: Theoretical development, *Transp. Res. A*, 20, 307–318, 1986a.

Recker, W.W., McNally, M.G., and Root, G.S., A model of complex travel behavior. Part II: An operational model, *Transp. Res. A*, 20, 319–330, 1986b.

Spear, B.D., New Approaches to Travel Demand Forecasting Models: A Synthesis of Four Research Reports, DOT-T-94-15, U.S. Department of Transportation, Washington, D.C., 1994.

Spiekermann, K. and Wegener, M., Microsimulation and GIS: Prospects and First Experience, paper presented at Third International Conference on Computers in Urban Planning and Urban Management, Atlanta, Georgia, July 23–25, 1993.

Taylor, D.A., *Object-Oriented Technology: A Manager's Guide*, Addison-Wesley, Reading, MA, 1990.

Troitzsch, K.G. et al., *Social Science Microsimulation*, Springer-Verlag, Heidelberg, 1996.

Vovsha, P., Peterson, E., and Donnelly, R., Micro-Simulation in Travel Demand Modeling: Lessons from the New York "Best Practices" Model, paper presented at 81st Annual Meeting of the Transportation Research Board, January 2002.

Wegener, M., The Dortmund Housing Market Model: A Monte Carlo Simulation of a Regional Housing Market, Arbeits Paper 7, Institut fuer Raumplanung, Universitaet Dortmund, Netherlands, 1983.

Wilson, A.G. and Pownall, C.E., A new representation of the urban system for modelling and for the study of micro-level interdependence, *Area*, 8, 246–254, 1976.

Further Reading

Bradley and Bowman (1998) provide a detailed description of the Portland, Oregon, activity-based microsimulation model, which defines the current state of best practice in random utility-based microsimulation modeling on a citywide scale. Arentze and Timmermans (2000) similarly provide an excellent and extensive discussion of their rule-based modeling system, which defines a credible alternative to the nested logit methods of Bradley and Bowman. Although now somewhat dated, Miller et al. (1987) provide a detailed presentation of the application of microsimulation to an integrated land use–transportation modeling system.

The International Association of Travel Behaviour Research (IATBR) has held conference workshops on microsimulation modeling at both the 1997 (Austin, Texas) and 2000 (Gold Coast, Australia) conferences. The proceedings from these two conferences contain summaries of the workshop findings plus a number of papers dealing with specific microsimulation modeling issues and applications.

A useful newsgroup dedicated to microsimulation in the social sciences is found at simsoc@mailbase.ac.uk.

13

Mobile Source Emissions: An Overview of the Regulatory and Modeling Framework

CONTENTS

Debbie A. Niemeier
University of California

13.1 Introduction

According to the latest report of the Environmental Protection Agency (EPA) documenting air quality trends in the United States (EPA, 1999b), mobile sources accounted for 51% of carbon monoxide (CO) emissions, 34% of nitrogen oxide (NO_x) emissions, and 29% of volatile organic compound (VOC) emissions; NO_x and VOCs react in sunlight to form ozone. In the California South Coast Air Basin, running stabilized emissions account for between 60% (organic gases) and 90% (nitrogen oxides) of estimated total mobile source emissions inventories.

The health effects associated with high levels of pollutant concentrations for at-risk populations such as the elderly, children, and those suffering respiratory problems like asthma have been well established in the literature. For example, ozone has been shown to lead to coughing, nausea, and long-term lung impairment. Partially as a result of these risks, the Clean Air Act (CAA) has been used to regulate mobile source emissions since the 1970s. Over the years the mobile source-related provisions of the Act have increasingly been strengthened, particularly with the passage of the 1990 amendments, as more information on the associated health risks has become available.

To address contemporary mobile emissions regulatory requirements, it is necessary to link two important but distinct modeling practices: travel demand forecasting and air quality modeling. And for nearly a decade, the transportation and air quality research and professional communities have worked closely together to improve the interface between these models. At the same time, however, regional governments are required to utilize these same models for demonstrating compliance with federal air quality regulations. The net result is that improvements to the mobile source modeling process not only are subject to the technical scrutiny of model developers and researchers, but also are assessed in terms of how model modifications will impact state or regional progress toward meeting air quality goals.

This chapter begins with an overview of the legislative framework, which defines the need for mobile source modeling and outlines broad rules for how the modeling is to be undertaken. This discussion is followed by an overview of contemporary mobile source modeling practices. Here, the focus is on a review of the basic foundation underpinning the models used to prepare on-road mobile source emissions inventories. There are also a number of key underlying concepts and practices highlighted during this review that are designed to help transportation researchers better understand the foundation of the current modeling practice. The chapter ends with reflections on future research needs.

13.2 Legislative Framework of Transportation Conformity

The link created between transportation and air quality was the result of three important events: passage of the Clean Air Act Amendments in 1990 (CAAA90), passage of the Intermodal Surface Transportation Efficiency Act (ISTEA) in 1991, and implementation of the 1993 Conformity Rule (EPA, 1993b).[1] Together, these provided the legal framework that formally expanded the traditional mobility-oriented goals of regional and statewide transportation planning to include those associated with improving air quality.

13.2.1 CAAA90

The CAAA90 required states with nonattainment areas for ozone, carbon monoxide, nitrogen dioxide, sulfur dioxide, or particulate matter with an aerodynamic diameter of less than 10 μm (PM_{10}) to prepare state implementation plans (SIPs). SIPs describe how the state will meet the National Ambient Air Quality Standards (NAAQS), including discussion of any control measures that will be required to achieve attainment for ozone, CO, and PM_{10}. The SIPs also establish the regional mobile emission budgets, which represent the ceiling on total allowable emissions for the region's transportation plan (RTP)[2] and the region's transportation improvement program (TIP), which is a multiyear prioritized list of federally funded or approved transportation improvements. The RTP and TIP must conform to the SIP in that planned emissions must not exceed the budgets prescribed in the SIP when conducting a regional emissions analysis. The regional emission analyses typically include total emissions generated by travel on the regional transportation network and all proposed regionally significant transportation projects minus any benefits associated with adopted emission control programs.

The SIPs also describe a minimum rate of progress toward attainment by specifying emissions targets for both the attainment year and every third year until the attainment year has been reached. In total, there are 13 specific provisions of CAA with which SIPs must comply. These can be found in CAA, §110(a)(2), and 42 U.S.C., §7410(a)(2); these provisions cover a range of issues, including monitoring, enforcement, reporting responsibilities, and permitting of new sources, among others.

---

[1]While the Clean Air Act Amendments of 1977 also included a conformity requirement (Section 176(c)), the 1990 Amendments dramatically expanded the statutory framework by further defining conformity and by requiring the U.S. EPA to "...promulgate criteria and procedures for demonstrating and assuring conformity..."

[2]The RTPs provide a 20-year vision of transportation investments.

In terms of jurisdiction, the EPA is the federal agency responsible for creating, implementing, and enforcing federal air quality regulations. Its jurisdictional authority includes establishing regulations, setting vehicle emission standards, supervising state air quality programs, and approving SIPs. State agencies share the responsibility of setting mobile emission standards, preparing the SIPs, and creating, implementing, and enforcing air quality regulations that will bring states into compliance with the state and federal requirements (CARB, 2001a).

In many states, such as California, there are county or regional governmental entities charged with regional oversight. These agencies develop and enforce regulations and control measures that will reduce industrial and area-wide pollutants emissions from their jurisdictional sources. In California these governmental entities are known as either air pollution control districts (APCDs) or air quality management districts (AQMDs). The districts are responsible for establishing and maintaining monitoring networks and preparing air basin emissions inventories (CARB, 2001a).

Air districts in nonattainment air basins are required to produce attainment demonstration plans, which describe the methods and dates for attainment. Local air districts work with the state agencies to design attainment plans and with the local planning agencies to ensure that RTPs do not exacerbate air quality problems. The air districts submit plans to the state agencies for approval. The district plans are then aggregated into the SIP. The state agency in charge of the air quality process is then charged with submitting attainment plans (and updates and revisions) to the EPA for approval.

In preparing the SIP, the CAAA90 specifies that each metropolitan planning organization and the respective departments of transportation "must demonstrate that the applicable criteria and procedures" in 40 Code of Federal Regulations (CFR), Parts 93.110–119, are satisfied. The applicable criteria and procedures vary depending on the action being considered (e.g., a conformity lapse vs. a conformity update); however, all actions must use the latest planning assumptions (40 CFR, Part 93.110) and latest emissions models (40 CFR, Part 93.111), and the SIP must have emerged as part of a interagency consultative process (40 CFR, Part 93.112).

13.2.2   ISTEA

The 1991 Intermodal Surface Transportation Efficiency Act complemented the CAAA90 in two ways. First, it legislatively supported the CAAA90's provisions associated with mobile emissions by providing the flexibility to use transportation funding to improve air quality (Larson, 1992). The ISTEA also created new funding categories, such as the Surface Transportation Program and the National Highway System within the Highway Program, and allowed flexibility to allocate funds between program categories and across transportation modes. Newly created programs, such as the Congestion Mitigation and Air Quality Improvement (CMAQ) Program, specifically provided funding to state and local governments for transportation projects and programs that would assist regions in attaining the requirements specified by the CAAA90.

Perhaps most important, ISTEA fundamentally changed the transportation planning process. New requirements for establishing transportation planning boundaries were specified. In particular, for nonattainment areas planning boundaries were expected to match air quality boundaries. For those metropolitan planning organizations (MPOs) in ozone and carbon monoxide nonattainment areas, long-range transportation plans had to be coordinated with the transportation control measures specified in the SIP. The financially constrained transportation improvement programs, whose planning horizons and priorities had to complement the CAAA90 3-year emissions reduction requirements for the more serious nonattainments areas (Larson, 1992), were required to be consistent with the long-range transportation plans.

The basic framework of ISTEA was maintained with the passage of the Transportation Equity Act for the 21st Century (TEA-21) in 1998. TEA-21 continued ISTEA's legislative support of the CAAA90 by reauthorizing the CMAQ Program and placing continued emphasis on the coordination of transportation planning with air quality goals. Titles 23 and 49 of the U.S.C. condensed the 23 planning factors identified in ISTEA to 7 broad planning factors designed to ensure that a range of planning alternatives were considered. With respect to air quality, the planning process must consider projects and strategies that will *protect and enhance the environment, promote energy conservation,* and *improve quality of life.* The requirement to formally integrate this planning factor into the planning process reinforces the link between TEA-21 and the Clean Air Act. Finally, in response to the revised and new NAAQS promulgated in 1997 for ozone, PM_{10}, and $PM_{2.5}$, TEA-21 ensured that the newly required $PM_{2.5}$ monitoring networks would be established and financed by EPA's administrator. TEA-21 also codified timetables for designating whether areas were in attainment for the new $PM_{2.5}$ NAAQS and the revised ozone NAAQS (U.S. DOT, 1998).

While the CAAA90 ensured that air quality improvements were achieved by requiring development of implementation plans that specified dates for meeting prescribed ambient standards, the ISTEA and TEA-21 reinforced coordination between transportation planning and the state implementation plans. The body of rules and procedures by which the CAAA90 conformity provisions are interpreted is known as the transportation conformity rule (40 CFR, Parts 51 and 93, as amended by 62 FR 43780, August 1997).

13.2.3 Transportation Conformity Rule

The transportation conformity rule requires that planners make certain that any federally funded or approved transportation projects in their region are consistent with statewide air quality goals. This means that transportation plans, programs, and projects cannot result in new NAAQS violations, increase the frequency or severity of existing violations, or delay attainment. Under the conformity rule, regions must demonstrate that all federally funded transportation plans, programs, and projects are consistent with the mobile source emissions budgets established in the SIPs (EPA, 1993a).

The Federal Highway Administration (FHWA) makes conformity determinations for regional plans at least every 3 years or as plans change. The CAA also requires that transportation control measures (TCMs) must be considered and adopted to offset any emission increases that result from increased vehicle travel for ozone severe or extreme nonattainment areas. TCMs are generally expected to reduce inventory emissions by reducing vehicle use or improving traffic flow (CAA, Section 108(f)(1)(A)). Events that impact the mobile emission budget, such as a SIP revision that adds or deletes a TCM, can trigger a conformity determination. Not all TCMs are legally enforceable; however, those that are not legally enforceable cannot accrue emission credits in either attainment or maintenance SIPs. However, in the case of conformity, emission credits are often generated by TCMs that are not credited in the SIPs. For example, in the recent Puget Sound conformity analysis, one of the TCMs included a public smog awareness program in which alerts were triggered by potentially high ozone weather conditions. The TCM was designed to encourage voluntary behavioral changes and, while implemented, no emissions reductions were credited in the maintenance plan inventory. In another example, Denver municipalities agreed to a street sanding and sweeping program designed to reduce PM_{10} that was subsequently credited in the conformity analysis, but not in the SIP (Howitt and Moore, 1999).

By law, conformity determinations rely on modeling practices from both travel demand forecasting and air quality modeling. Travel demand models must be used to estimate the vehicle activity in most nonattainment and maintenance areas. The vehicle activity is, in turn, multiplied by emission factors that must be derived in federally approved vehicle emission models. Since there is no direct way to measure regional mobile source emissions, the application of the models becomes very important when demonstrating conformity (Stephenson and Dresser, 1995). To date, the modeling practices have typically followed each other sequentially more or less as shown in Figure 13.1, which represents the conformity modeling process of the Puget Sound Regional Council (PSRC).

PSRC utilizes a standard four-step model to forecast future travel volumes with the standard modeling steps of trip generation, trip distribution, mode choice, and trip assignment. Note that after the trip assignment step, there is a feedback loop to both mode choice and trip distribution to reconcile the output travel speeds implied by assigned volumes to the input speeds assumed at earlier stages of the process. The travel demand modeling output, volumes, distances, and speeds serve as inputs to vehicle

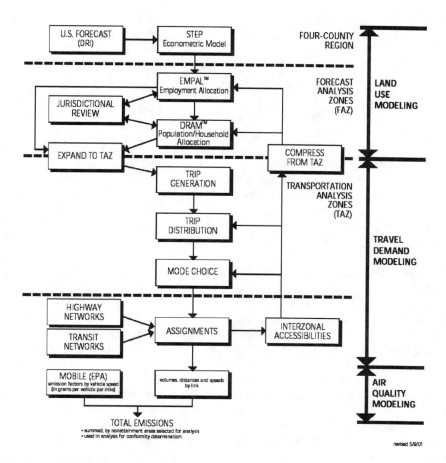

FIGURE 13.1 Puget Sound mobile source modeling process. Overview of models used in PSRC transportation planning to prepare mobile source emissions.

emissions inventory models. In general, four-step models are considered to have relatively low accuracy, particularly with respect to the speed estimates. For example, UTPS has an accuracy range of 5 to ~30% error in overall vehicle miles traveled (VMT) estimates, and 5 to ~20% mi/h error in terms of average speeds (UMTA, 1977; Levinsohn, 1985), both of which are key inputs to the mobile source models.

While most transportation analysts have a firm understanding of the processes shown in the upper two thirds of Figure 13.1, there is far less understanding of the components of the mobile emissions modeling — shown as the single box, MOBILE (EPA), in the lower left-hand corner — or the types of off-model manipulations that are performed in deriving total emissions. In the remainder of this chapter, the focus will be on elaborating these modeling components, particularly with respect to various spatial and temporal uncertainties and assumptions.

13.3 Motor Vehicle Emissions Modeling Processes

Compared to travel demand models, motor vehicle emission models have a relatively shorter history; most were developed in response to CAA requirements beginning in the mid-1970s. In the early 1990s major improvements were undertaken to enhance the modeling capabilities specifically for the purpose of developing SIP emissions inventories and for conducting conformity demonstrations for SIP emission budgets.

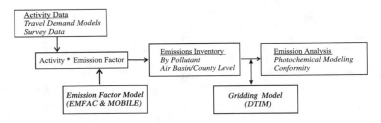

FIGURE 13.2 Mobile emissions inventory modeling process.

Both conformity and SIP mobile emission budgets are currently prepared using emission rates produced by one of two models: the MOBILE series developed by the EPA or the Motor Vehicle Emission Inventory model series developed for California by the California Air Resources Board (CARB). The latest releases of these models are MOBILE6 and EMFAC2000, respectively. The basic methodological steps used to derive emission rates in MOBILE6 and EMFAC are relatively similar for both models (Figure 13.2).

Regional mobile emissions are calculated by multiplying an emission factor (EF) by an associated travel activity. Travel activity includes data from both the travel demand models (speed, miles) and surveys (number of vehicle starts and vehicle soak time). Very generally, speed and miles are used to compute speed–VMT distributions for estimating on-road running emissions; the number of vehicle starts is used to quantify the increased emissions that occur when a vehicle is started, and soak time, the time a vehicle is not operating, is used to characterize evaporative emissions, which occur when fuel vapors escape through the tank or fuel delivery systems.

The emissions inventory is produced by combining these estimates into a single total for a range of pollutants, including hydrocarbons (HC), carbon monoxide (CO), nitrogen oxides (NO_x), particulates (PM), lead (Pb), sulfur oxides (SO_x), and carbon dioxide (CO_2). The primary emphasis for estimating mobile source inventories is typically placed on the first four pollutants, HC, CO, NO_x, and PM. Hydrocarbon emissions result when unburned fuel moves through the exhaust system, which is a function of the types and condition of vehicle emission controls, or through diurnal, hot soak, and resting loss evaporative processes. Diurnal emissions arise when ambient temperatures rise and fuel evaporates while a vehicle is sitting; hot soak emissions occur immediately after the engine is turned off; and resting loss emissions are a function of permeation through plastic or rubber fittings, which takes place as the vehicle sits for long periods of time.

Carbon monoxide is produced mostly by gasoline-powered engines and is created as a by-product of incomplete combustion when carbon in fuel is only partially oxidized, rather than fully oxidized to CO_2. Carbon monoxide reduced the flow of oxygen in the bloodstream. Nitrogen oxides are also formed during combustion under high pressure and temperature when oxygen reacts with nitrogen; diesel vehicles tend to produce greater amounts of NO_x because of their high air–fuel ratios (which creates excess oxygen in the combustion process). Both NO_x and HC are ozone precursors. Finally, exhaust particulate matter emissions are small carbon and sulfur particles that are produced mainly by diesel vehicles.

Once regional mobile emissions have been estimated for each of these pollutants, the inventory is typically used for one of two purposes, preparing SIP updates or evaluating conformity. There are a number of technical difficulties that arise in using the mobile emission inventories for either purpose. For example, to prepare the SIP, the inventories must be converted to gridded emissions suitable for photochemical modeling. That is, the period-based (e.g., A.M. and P.M. periods) link emissions created using the travel demand modeling data must be converted to gridded hourly vehicle emissions. For conformity, difficulties arise because the geographic boundaries for creating air basin inventories (encompassing whole counties) are typically not the same as the boundaries used in regional transportation planning (sometimes encompassing only partial counties) and the scale of regional inventories makes it difficult to conduct regional transportation alternatives evaluations.

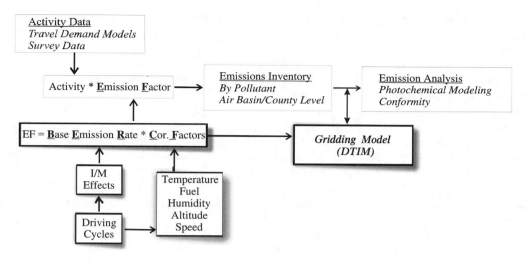

FIGURE 13.3 Major components of emission factor models.

In California, a model known as the Direct Travel Impact Model (DTIM) was developed by the California Department of Transportation to help overcome some of the difficulties in converting inventories to the gridded inputs needed for photochemical modeling. The model is used as a postprocessing step before photochemical modeling. More recently, the model use has also been extended to help conduct conformity determinations by allowing the modeling of transportation systems alternatives at a regional level. There are a few problems with DTIM, and a new, updated gridding model was recently developed at the University of California–Davis (UC Davis) that, as will be discussed, could help to mitigate some of the problems associated with using DTIM to perform conformity determinations.

Looking at Figure 13.3, it can be seen that creating the emission factors and conducting the postprocessing of the inventory includes a number of steps. Understanding the basics associated with these steps is important for understanding how the interface between transportation and mobile emissions modeling can be improved. From Figure 13.3, the process of creating an emission factor begins with the development of the basic emission rates (BERs).

13.3.1   Base Emission Rates

BERs are the fundamental building blocks used in deriving emission factors. BERs are established using vehicle testing data for carbon monoxide (CO), hydrocarbons (HC), and nitrogen oxides (NO_x), and are adjusted for deterioration in vehicle emission control over time. The vehicle testing data are collected during laboratory dynamometer experiments that are conducted by driving a vehicle over an established speed–time trace, known as a driving cycle and bagging emissions during the test. The bagged emissions, after being adjusted to reflect inspection and maintenance (I/M) control programs and nontest conditions, are used to estimate the BERs.

13.3.2   Driving Cycles

In addition to being used for developing emission inventories, driving cycles are also used to ensure that light-duty vehicles and trucks comply with mandated emission standards. Three programs were designed to accomplish the regulatory intent contained in the CAA: certification, assembly line testing (known as selective enforcement audit), and recall. Under the Clean Air Act (Section 203(a)(1)), a motor vehicle manufacturer must obtain a certificate of conformity demonstrating compliance with emission standards prior to selling new cars in the United States. A manufacturer submits information to the EPA, including test data demonstrating that its new motor vehicles will comply with the applicable emission standards.

Since it is a preproduction program, a manufacturer collects dynamometer test data from low-mileage, production-intended vehicles, that is, vehicles assembled as closely as possible to those that are planned for production. The test results from the vehicles are adjusted to project useful life emission levels (called certification levels) by the emission deterioration factors specifically for the vehicle technology. If the certification levels are below the standard and the manufacturer has demonstrated that the vehicle meets all emission requirements, a certificate of conformity can be issued.

Section 206(h) of the Clean Air Act authorizes the EPA to conduct testing of new motor vehicles or engines at the time they are produced to determine whether they comply with the applicable emission standards. This assembly line testing may be conducted by the EPA or by the manufacturer under conditions specified by the EPA. If the EPA determines that the vehicles or engines do not comply with the regulations, the EPA may suspend or revoke the applicable certificate.

Driving cycles are used for dynamometer testing of in-use vehicles under the recall program; the EPA uses test data to evaluate the emission performance of vehicles in actual use. If it is determined that a class or category of vehicles or engines does not conform with the applicable regulations when in actual use throughout its useful life, the manufacturer is required to submit a plan to remedy the nonconformity at the manufacturer's expense (CAAA, Section 207(c)).

Thus, driving cycles serve multiple functions, ranging from conducting vehicle certification to preparing emission inventories. Defining a representative driving cycle for dynamometer testing for this broad range of purposes is surely one of the most difficult tasks in deriving the BERs. And for years the primary focus in terms of the driving cycle was on meeting the needs of vehicle certification and recall, rather than those associated with building emission inventories. Perhaps the most well-known cycle, the Federal Test Procedure (FTP), was created in the early 1970s primarily to comply with federal vehicle certification standards (Austin et al., 1993). To create the cycle, six drivers from EPA's West Coast Laboratory drove a 1969 Chevrolet over a single route in Los Angeles, which at the time was chosen to reflect the typical home-to-work journey. A range of operating parameters was computed for the six traces, including idle time, average speed, maximum speed, and number of stops per trip. After discarding one of the six traces, the trace with the actual driving time closest to the average was selected as the most representative rush-hour driving behavior trace. The selected trace included 28 "hills" of nonzero speed activity separated by idle periods.

After slight modifications to accommodate the limitations of the belt-driven chassis dynamometers in use at the time, the final cycle, also known as the Urban Dynamometer Driving Schedule (UDDS or FTP), was finalized in the early 1970s. The FTP is 7.46 mi in length, has an average speed of 19.6 mi/h and a maximum speed of 56.7 mi/h, and is 1372 sec long. It includes 505 sec of cold start and 867 sec of running hot stabilized (see Figure 13.4). The cycle has been the standard driving cycle for emissions

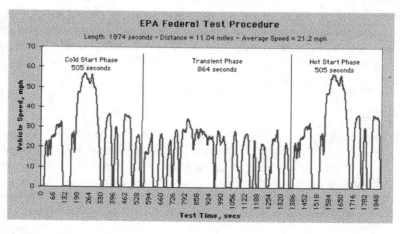

FIGURE 13.4 The federal test procedure.

certification of light-duty vehicles beginning with the 1972 model year. Since the passage of the 1990 Clean Air Act Amendments, the FTP has also served as the primary means by which BERs are established for the MOBILE models used to prepare mobile source emission inventory models.

Almost from the beginning concerns were raised about the representativeness of the FTP. In driving studies conducted in Baltimore and Spokane, Washington, in the early 1990s (which was after the passage of the CAAA90, when inventory preparation became critically important), speed and acceleration rates were observed that were far in excess of those simulated by the FTP (EPA, 1993b). For example, the maximum and average speeds represented in the FTP are 56.7 and 19.5 mi/h, respectively, while in Baltimore speeds as high as 96 mi/h, with averages around 25 mi/h, were observed. The use of driving data collected solely in the Los Angeles region was also criticized when driving patterns in other regions showed that the FTP overrepresented time at stop and cruise modes between 25 to 35 mi/h, and underrepresented acceleration rates and cruise conditions between 40 and 50 mi/h and above 60 mi/h (St. Denis et al., 1994).

Partially in response to concerns raised about the FTP, CARB created a second standard cycle, the unified cycle (UC), which is currently used to set the BER for estimating mobile source inventories in California. The UC was constructed with chase car data collected in the early 1990s, also in the greater metropolitan Los Angeles area (Austin et al., 1993). The UC is slightly longer than the FTP (see Table 13.1), with an average cycle speed of 24.6 mi/h. Note also that the UC encompasses higher speeds and greater acceleration rates (see Figure 13.5). Positive kinetic energy (PKE), which is a measure of acceleration engine work (Watson and Milkins, 1983), is also higher in the UC than in the FTP.

The method used to construct driving cycles is fairly similar for both the EPA cycles and the CARB cycles. The standard practice has been to first collect chase car data, which involves using an instrumented vehicle (the chase vehicle) to follow a randomly selected vehicle (the target) in traffic. In addition to a range of variables (e.g., traffic conditions, roadway type, grade, etc.), the target vehicle's speed is recorded using a laser range finder mounted on the chase vehicle. This technique yields data on hundreds of drivers across many routes, types of roadways, and congestion levels. (For an overview and discussion on chase car sampling design and data collection efforts, see Morey et al. (2000)).

TABLE 13.1 Characteristics of the FTP and UC

Characteristic	FTP	UC
Duration (sec)	1372	1435
Distance (mi)	7.5	9.8
Average speed (mi/h)	19.5	24.6
Maximum speed (mi/h)	56.7	66.4
Maximum acceleration (mi/h/sec)	3.3	6.8
PKE (ft/sec^2)	1.2	1.6

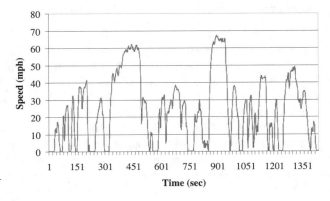

FIGURE 13.5 The unified test cycle.

Using a combination of chase and target vehicle data, the driving cycles are constructed by dividing the collected speed–time traces into smaller segments known as trip snippets or microtrips, depending on the protocol used to define segments (Lin and Niemeier, 2002b). A microtrip is defined as a segment of the speed–time trace that is bound by an idle mode (zero speed) at either end, while a trip snippet can have end points either bound by an idle mode (zero speed) or reflecting a change in traffic conditions such as facility type or level of service. For example, a trip snippet might be a portion of the speed time trace collected in the field with one end defined by an idle speed and the other end defined by a change in level of service.

The microtrips (or snippets) are then classified into collections of similar traffic conditions (e.g., average speed) or driving patterns (e.g., percent idle time). Although statistical clustering methods have been used sporadically for classifying microtrips into groups (e.g., Effa and Larsen, 1993), it is important to note that none of the current regulatory cycles were constructed using these techniques. Both the types of categories and the assignment of snippets or microtrips to the categories have arbitrarily delineated. Once data segments have been assigned to groups, snippets (or microtrips) are randomly chosen and linked together to form a driving cycle. As each microtrip is randomly selected, it is compared to one or more performance criteria, the most common being the speed–acceleration–frequency distribution (SAFD) of the complete sample data or a particular subset of the data. Thus, the cycle is built by iterative random selection of each segment (and subsequent segments) such that the addition of the segment to the cycle improves the match to the desired SAFD. The cycle construction is completed when the desired cycle length or duration is reached. Additional details on cycle construction can be found in Austin et al. (1993).

Literally thousands of driving cycles are generated using this procedure, and from that collection a single cycle is selected based on a set of target statistics. In theoretical work, target statistics have included such measures as average and maximum speed and acceleration, percent idle, PKE, and engine power, formulated as a function of vehicle speed, acceleration, vehicle mass, and road grade angle, which influences vehicle emissions and fuel consumption (An et al., 1997). In practice, one criterion dominates the selection of the final cycle: the difference between a particular driving cycle's joint SAFD, sometimes referred to as a Watson plot, and the SAFD of the sample data. The cycle with the minimum difference is selected as the final cycle.

In a recent National Academy of Science (NAS) review of MOBILE (National Research Council, 2000), the need for improving the spatial and temporal resolution associated with estimating mobile source emissions was clearly articulated. Although the issues were not directly connected to the driving cycles in the report, the cycles underlie many of issues discussed. The spatial representativeness of a driving cycle is a function of both the spatial nature of the underlying data and the method used to construct it. Thus, it is important to be able to identify the limits of the spatial generalizability of the cycle given the underlying data. The cycle construction method must also be reproducible and ideally stochastic. The temporal representation of the current cycles is 24 h, that is, the cycles represent average travel over 24 h. This raises the issue of how many cycles are needed and what each cycle should represent.

13.3.2.1 Spatial Representativeness

If we look at a comparison between the CARB UC, which is used to produce BERs for the California EMFAC model, and the FTP, which is used to establish MOBILE BERs, it is sensible to assume that the UC emission rates will be more representative of mobile emissions in Los Angeles than the FTP, simply because the UC is based on more recently collected data. In addition, underlying the UC rates are data that reflect significantly greater spatial coverage, which in turn incorporates a wider variety of operating conditions and roadway types. (Recall that the FTP ultimately reflects the average driving of six drivers over a single route). To carry this notion of spatial representativeness a bit further, would emission rates from the Los Angeles-based UC be equally representative of emissions from urban driving in, for example, Sacramento? In other words, can emission rates derived from a standard driving cycle, or even a set of driving cycles, be used to reasonably characterize emissions from driving conditions and behavior across different cities?

Implicit in the driving cycles used in both the California and EPA models is the assumption that driving variability across regions can be controlled for by using average speed (EMFAC) or congestion level and facility type (MOBILE). The EPA's MOBILE6 cycles were constructed using chase car data collected in three cities: Los Angeles, Spokane, and Baltimore (the data are known colloquially as the three-city data). The cycles are supposed to represent a variety of travel conditions by operational characteristics (i.e., congested and uncongested) and facility types (e.g., arterial, freeway, etc.) (EPA, 1997, 1999a). By controlling for operation and facility type, the EPA has assumed that the driving behavior and conditions in these three cities "are not dependent on the city in which the driving was performed" (EPA, 1997, p. 10).

Early research indicated that driving patterns in different U.S. cities were dissimilar enough to suggest significantly different emission rates (e.g., Milkins, 1983; LeBlanc et al., 1995). However, one key limitation to these studies was that they were based on examining driving data alone. While driving differences are important, they are not by themselves conclusive with regard to the actual emissions generated; a difference in average modal activity (i.e., accelerate, cruise, and decelerate) over the course of a trip does not necessarily translate to significant differences in overall tailpipe emission rates. In a recent study, however, new driving cycles were created with the explicit purpose of testing the hypothesis of spatial representativeness (Niemeier, 2002).

In this study, new cycles were created using the same method used to create the UC, except that the underlying data used to create the cycles were from a chase car study conducted in Spokane, Washington. The dynamometer findings suggested that the UC-generated emission rates should not be considered spatially representative. In other words, while emissions rates generated by the UC may be reflective of driving in Los Angeles, they are probably not good emission indicators for other regions in which driving patterns may be significantly different. These findings have been echoed by another recent study that looked specifically at California regional driving variability and the potential impact differences in driving would have on the final driving cycle form (Lin and Niemeier, 2002b).

Using recent driving data collected in the Bay Area, Sacramento, and Stanislaus regions of California, this study found that steady-state and acceleration driving events had significantly different frequencies, durations, and intensities (speed and acceleration) between the three California areas. Interestingly, the study also found that the California driving data reflected very different driving characteristics with respect to modal events when compared to EPA's three-city data. Using each region's driving data, two sets of congested and uncongested driving cycles were created. The differences in driving had a notable influence on the shape and form of the resulting driving cycles. For example, the Sacramento and Stanislaus uncongested freeway driving cycles reflected higher steady speeds with similar duration characteristics than did the cycle created using the Bay Area data. The Bay Area congested freeway cycle generally reflected more frequent low-speed cruises and idles than represented in the cycles created for the other regions. Both of these recent studies suggest strong evidence of driver behavior differences related to the particular spatial layout of the highway network or the culture of driving that might exist in the region, which almost certainly reflects at least some perceptions related to congestion.

13.3.2.2 Cycle Construction Methodology

Another of the oft-cited criticisms of the current driving cycles is that they do not fully reflect the range of a vehicle's operating conditions (National Research Council, 2000), typically expressed in terms of the frequency, intensity, and duration of modal events (Holmén and Niemeier, 1998). Certainly, under the current U.S. models, the segmenting of a driving trace into microtrips is based on fairly arbitrary criteria, such as average speed, the beginning or end of a particular facility type, or change in congestion level. The dependence of emission factors on average speed in the regulatory models has been identified as particularly problematic by a number of researchers (e.g., de Haan and Keller, 2000; Joumard et al., 2000; Ntziachristos and Samaras, 2000).

Many have argued that the solution to the modal event representation problem in the cycles is to develop a *modal emission model*. These are models that typically use one or more parameters in addition to average speed as a way of characterizing (or deriving) emission factors. The recent NAS report on

MOBILE suggested that the modal emission approach could form a very reliable basis for improving the emission factors in the inventory models (National Research Council, 2000). For example, when a new emission rate is needed (e.g., off-ramps), a driving cycle would be created and input into a modal emission model, and the resulting emission rate derived. This is essentially the same approach used in producing emission factors contained in the *German Handbook on Emission Factors for Transport* (SAEFL, 1995). As cited in de Haan and Keller (2000), the method uses emissions matrices of dynamometer testing results organized into cells defined by speed and speed–acceleration combinations.

Regardless of the modeling approaches, a driving cycle is required, and clearly the better the cycle is at replicating real-world driving, the more accurate the estimated emissions (e.g., de Haan and Keller, 2000). In most research, it has been argued that cycles would be vastly improved if driving variability in terms of modal activities (i.e., acceleration, deceleration, idle, and constant speed) was better replicated in the cycles (e.g., Lyons et al., 1986; Andre, 1996).

A method for stochastically constructing a driving cycle was recently proposed where the sequencing of modal events (i.e., cruise, idle, acceleration, or deceleration) is described using Markov process theory (Lin and Niemeier, 2002a). The Markov process approach allows the driving cycle to replicate the average (or global) driving characteristics while still preserving microtransient events (i.e., small timescale speed fluctuations) that contain the information related to driving variability. Instead of being divided into the traditional microtrips, the speed trace is divided into segments representing modal events (i.e., acceleration, deceleration, cruise, and idle) using a maximum likelihood approach for mixture decomposition (Symons, 1981).

The cycle construction method begins by denoting the length of a route-based speed trace as n and each of the observed data points (acceleration and deceleration rates) contained within n as y_i, i = 1, …, n. The vector of n observations of y_i is denoted as $\vec{y}$, with $\vec{\theta}$ as the corresponding vector of parameters such as the mean and variance of the observation (i.e., acceleration and deceleration rates). Under the assumptions of multivariate normality and equal covariance, the likelihood function can be defined to describe the likelihood that a realization of parameter set $\vec{\theta}$ will occur given the observed data:

$$L(\vec{\theta}\,|\,\vec{y}) = \prod_{g=1}^{G} (\pi_g^{n_g}\sigma_g^{-n_g}) \exp\left\{ -\frac{1}{2}\sum_{g=1}^{G}\sum_{C_g} \frac{(y_i - \mu_g)^2}{\sigma_g^2} \right\}$$

where G equals the total number of partitioning groups, where each group is designated g = 1, …, G; C_g is the collection of observations, y_i values; and n_g is the number of observations in C_g. The probability of y_i being in group g ($\Sigma_g \pi_g = 1$) is π_g. The above equation implies that the likelihood of some realization of parameter set $\vec{\theta}$ occurring is subject to how observations are divided into groups. By maximizing the likelihood function, we can obtain both the maximum likelihood estimates of the unknown parameters $\vec{\theta}$ and the group membership (C_g, n_g).

At the completion of the partitioning, a route-based speed trace is then divided into groups of segments that are organized by modal event bins. The modal event bins represent collections of modal events with similar average speed and acceleration–deceleration characteristics, such as average speed, maximum and minimum speeds, average acceleration, maximum acceleration, and maximum deceleration. This grouping is accomplished again using the maximum likelihood partitioning technique. A modal event bin, labeled by speed and acceleration, can contain hundreds, even thousands, of modal events of differing event durations.

To construct the actual driving cycle, transition probabilities are computed for each modal event bin. That is, the modal event bins represent the state space in the Markov process, where each transition probability, p_{ij}, is the probability of the next modal event chosen from modal event bin j, given that the current modal event is from bin i. As a segment is chosen (at random) from each successive event, the driving cycle is progressively extended. The selection process is repeated until the cycle length reaches some predefined measure (e.g., the average trip length). Any number of candidate cycles can be generated, and at the end of the procedure a single cycle is selected based on some set of predetermined performance characteristics. Additional details on the cycle construction method can be found in Lin and Niemeier (2002a). While there are some nuanced limitations with the new method, it represents an important advancement in constructing driving cycles that are theoretically defensible.

13.3.2.3 Number of Cycles

The final issue to be discussed in this section relates to the number of cycles needed to accurately and fully represent real-world conditions in the regulatory models. Since the costs associated with vehicle testing are significant, it is important to limit the number of cycles each vehicle is tested on to the minimum number required to adequately capture the range of real-world operating conditions. Both CARB and EPA use a relatively limited number of cycles to represent the range of real-world conditions.

In the EMFAC models, the cycles have traditionally been represented as trips. Emission rates derived using the UC (with an average speed of 27.4 mi/h) have to be adjusted to reflect a range of average trip speeds represented in the real world. This is accomplished using correction factors established for trips with average speeds from 5 to 65 mi/h, categorized into 5 mi/h speed bins. The correction factors are derived by dynamometer testing of 13 cycles, representing trips with average speeds representative of each speed bin. In contrast, EPA's new MOBILE model relies on six freeway cycles, three arterials, and one local road cycle for adjusting the basic emission rates derived from the FTP. Each of the cycles represents an operating condition. So, for example, the six freeway cycles include a high-speed cycle and respective levels of service A through F, and a low-speed high-congestion cycle denoted as G. Thus, in total, EPA and CARB use a relatively limited number of cycles to represent the range of real-world operating conditions.

Recall that CARB's cycles are created using data collected in the early 1990s, while EPA's cycles are derived using the three-city data. Given the range of concern related to the spatial representativeness of the cycles, there is a real need for understanding how many cycles are necessary to adequately represent real-world conditions, including regional variability. Recent exploratory work suggests that a limited number of cycles may be sufficient if each region develops and applies linear adjustments that reflect driving variability or the range of facilities found in that particular region (Niemeier, 2002).

13.3.3 Adjustments to the BERs: Inspection and Maintenance

Once the base emission rates have been derived from the driving cycles, and before they can be used as emission factors, they are modified to reflect the emission benefits associated with different inspection and maintenance programs, such as a smog check. Different adjusted basic emission rates are produced for various I/M scenarios and vehicle model years, classes, and technology (CARB, 1996; EPA, 2001a). The adjusted rates are then combined with fleet data to calculate composite emission rates for each vehicle class. The I/M benefits have long been controversial and the issues associated with these programs are thoroughly discussed in the recent NAS report (National Research Council, 2000) and an earlier General Accounting Office (GAO) report (GAO, 1997).

The basic methodology for computing the I/M benefits assumes that a proportion of the vehicle fleet can be identified as high emitters and that by repairing a proportion of these high emitters emissions benefits are achieved. For the proportion of repaired vehicles, the emission rates are adjusted based on a relationship between normal emissions and repaired emissions, which are a function of I/M program repair cost limits and the level of mechanic repair effectiveness.

13.3.4 Adjustments to the BERs: Correction Factors

Once composite basic emission rates have been produced by technology, model year, and I/M scenario, there are several corrections applied to the composite BERs in order to finally compute an emission factor, which can be multiplied by the respective travel activity to develop inventory totals. The correction factors are adjustments made to reflect off-test environments (i.e., outside of driving cycle conditions) for speed, fuel, humidity, temperature, and altitude. Details associated with how each of the factors is computed can be found in the respective technical manuals for CARB and EPA. However, it is worth considering the speed correction factors in more detail since these factors are combined with VMT–speed distribution output from the travel models, and running emissions are a significant portion of the mobile source emission inventories.

The two models differ in a very important philosophical way that affects how the final mobile source inventories are used in conformity. MOBILE6 now produces emissions factors using a facility congestion, link-based approach. That is, the emission factors reflect best estimates of the emissions generated by an average pass on a link segment for a given facility and level-of-service combination. In contrast, EMFAC produces trip-based emission factors that rely on average trip speeds (be they link or trip). To produce the requisite emission factors, the BERs must be corrected for speed because real-world emissions are generated at average speeds other than the 27.4 mi/h reflected in the UC-generated BERs, or 19.7 mi/h in the FTP BERs.

The new EMFAC and MOBILE models handle speed adjustments slightly differently. In EMFAC2000, 13 new cycles, known as the unified correction cycles (UCCs), with differing average trip speeds were developed using subsets of the UC chase car data to develop what is referred to as the cycle correction factors (CCFs). UCC45 and UCC60 are shown in Figure 13.6. The CCF equations are estimated using vehicle test data from the UC and UCC using what is typically referred to as the ratio of the mean method (CARB, 2000a). In the new MOBILE model, the ratio of the means (ROM) method is also used, but the adjustment is relative to the FTP.

The ROM is computed as the mean cycle emissions (by pollutant) divided by the mean baseline emissions for groups of vehicles categorized by model year and technology type. For each group a least squares curve is fit to the ratio of means as a function of speed, usually after applying a transformation such as the natural log to reduce the impact of nonnormality (CARB, 1996; EPA, 1997).

For example, in the latest version of EMFAC the CCF equations are modeled as second order for each emission category and technology group and are normalized to the bag 2 UC mean speed (27.4 mi/h) emission rates. The general CCF equation for any given emission category and technology grouping is (CARB, 2000a)

$$CCF(S)_{s,p,t,m} = EXP(A(S - 27.4) + B(S - 27.4)^2)$$

where $CCF_{s,p,t,m}$ represents the cycle correction factor for a given speed (s), pollutant type (p), technology group (t), and vehicle model year (m); S denotes the mean speed of the trip and can range from 2.5 to 65 mi/h; and A and B are estimated coefficients. For cycles on which vehicles have been tested by the CARB, the ratio of means variable ROM_s is defined as

$$ROM_{s,g} = \frac{\text{Average emissions on the UCCs cycle in g/mile}}{\text{Average bag 2 emissions on the UC cycle in g/mile}}$$

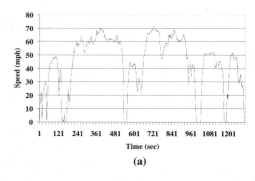

(a)

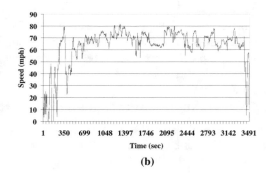

(b)

FIGURE 13.6 The UCC45 and UCC60 driving cycles.

where UCCs is CARB's unified correction cycle with average speed s and UC is CARB's baseline unified cycle for each gas, g. Note that for these models ROM_s will always equal 1 when the speed is equal to 27.4 mi/h. The models are designed for this to be the case, since the average speed of bag 2 (which is used to collect running emissions) of the UC is 27.4 mi/h, and thus when s = 27.4, the ratio of the numerator to the denominator will be equal to 1.

Using the ratio of means to fit a curve in this context is statistically awkward for two reasons (Utts et al., 2000). First, there is the loss of variability related to the vehicles when the data are averaged. Variability is particularly important for creating confidence intervals for the curves. In the past, the speed correction factors (SCFs) were produced without confidence intervals and there were few ways to gauge the accuracy of the SCFs. It is well known that emissions data tend to have a large variance, and thus the estimated curves are not likely to be accurate even for moderate sample sizes.

Second, when the SCFs (or CCFs) are created, the same vehicles are run on all cycles, resulting in correlated observations. Ignoring these correlations in fitting the SCF curves will result in biased estimates of the equation parameters (e.g., A and B in CARB's case). In traditional linear regression we fit the model $y = X\beta + \varepsilon$, where $\varepsilon \sim N(0, \sigma^2 I)$, and in the context of speed correction factors, y is a random vector of emissions, X is a matrix of explanatory variables, β is a vector of estimated coefficients, and ε is a vector of errors. The errors are assumed to independent and normally distributed with a constant variance σ^2, which corresponds to assuming that the vector ε has a multivariate normal distribution with a diagonal covariance matrix and constant values on the diagonal. If we make the usual assumption that the values in X are nonrandom, under this model $y \sim N(X\beta, \sigma^2 I)$. Believing that the errors are independent implies that there is no correlation between emissions measured on the same or different cycles — an assumption that makes sense in this context only if different vehicles are used for each measurement.

In a recent study that examined the effect of taking into consideration the correlated observations by using a repeated measures model to estimate SCF curves (Utts et al., 2000), it was clear that while the differences in the SCF estimates obtained using the two methods were not large, the ratio of means curve was almost always below the repeated measures curve (see Figure 13.7). This is because, in general, dynamometer emissions measurements are skewed toward higher values, but taking the mean of the emissions and then the ratio severely dampens the influence of the extreme high values. Note also that the authors found that the differences between the two curves were consistently largest at the lowest speed. That is partly an artifact caused by forcing the curves to be identically 1 at the baseline speed of 27.4 mi/h, but it also seems to be related to the fact that there is considerably more variability at the low speeds than at the higher speeds, and thus the dampening effect of taking the means before taking the ratio would be most acute at the lowest speeds.

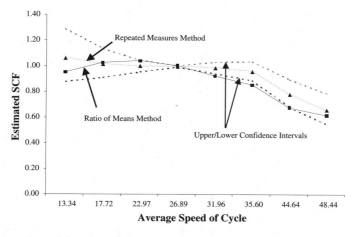

FIGURE 13.7 A comparison of the ratio of means curves and the repeated measures model for estimating speed-related correction factors. (Adapted from Utts, J.N. and Ring, D., *Transp. Res. D*, 5, 103–120, 2000.)

There is a real issue associated with how base emission rates are adjusted. In theory, the base emission rates reflect clean newer-model vehicles, rather than the inventory fleet. Rates developed from these vehicles are then adjusted to correct for speed variations, among many others, as will be discussed in the next section. Thus, the issue of whether the speed adjustments should dampen or accentuate the outliers is an important consideration, and one that remains largely unresolved.

13.3.5 Fleet Characterization

Both the EMFAC and MOBILE models require certain fleet information to properly weight BERs and to incorporate future fleet characteristics into inventory estimates (EPA, 2001b). The fleet information that is typically gathered includes a distribution of vehicles by age, the average annual mileage accumulation rates by age and vehicle category, and estimates of the future vehicle fleets. The vehicle population estimates are used for calculating tons per day emission estimates for exhaust and evaporative emissions, while vehicle age distributions are important for calculating reliable average fleet emission factors (CARB, 2000b). The information included in the newest models (MOBILE6 and EMFAC2000) provides updated fleet assumptions from the prior model versions, although in some cases, the underlying methodologies for estimating travel or population fractions have also changed slightly (EPA, 2001b).

Recall that CAA (42 U.S.C., §7511(a)) requires that each nonattainment state must submit periodic emissions inventories every 3 years until attainment. The CAA also states that following completion of the emissions inventory forecasts of future emissions and identification of budget shortfalls, the state can allocate specific mobile emission budgets for on-road sources. And as part of the *applicable criteria and procedures* required in the preparation of the SIP, the modeling and analysis must be conducted using the latest planning assumptions (40 CFR, Part 93.110) and the latest emissions models (40 CFR, Part 93.111).

In both the newest releases of EPA's and CARB's models there has been a conscious effort to update the basic planning assumptions. For example, in MOBILE6, among other changes, the fleet information was updated from 1990 to 1996 and the number of vehicle classes was expanded from 8 to 28 vehicle categories, with fleet aging increased to 25 years. In contrast, in MOBILE5 there were eight vehicle categories, including light-duty gasoline vehicles, diesel vehicles, gasoline trucks category 1 (0 to 6000 lb gross vehicle weight rolling [GVWR]), gasoline trucks category 2 (6001 to 8500 lb GVWR), and diesel trucks (0 to 8500 lb GVWR), and heavy-duty gasoline vehicles, diesel vehicles, and motorcycles. The light-duty gasoline trucks categories 1 and 2 reflected the different emission standards for two gross vehicle weight categories. However, starting with a phase-in period in 1994, the EPA expanded its regulatory classifications to include four light-duty truck categories. MOBILE6 reflects the original two truck categories plus the additional four in the category of gasoline-fueled trucks, effectively increasing the number of light-duty truck categories in the model from two to six (EPA, 2001b).

The EPA has indicated that the increase in vehicle categories allows for greater fleet representation of class-specific trends (e.g., differences in mileage accumulations in certain heavy-duty vehicle categories) and facilitates future fleet calculations (EPA, 2001b). However, the EPA also notes that data were not easily available for a number of the 28 categories and that, in some instances, the data that were available were applied across multiple vehicle classes.

There are also a number of significant changes in the characterization of the fleet in CARB's new EMFAC model. The new fleet data represent the years 1997 to 1999, updated from the early 1990s, and for the first time the data are county specific. The number of vehicle categories has increased from 7 to 13, and the distribution of fleet aging has been extended from 35 to 45 years. Approximately 30 million Department of Motor Vehicles (DMV) registration records provided by the California Energy Commission (CEC) were used to develop the new EMFAC vehicle age distributions (CARB, 2000b).

Both EMFAC and MOBILE use similar strategies for weighting final emission rates. First, as noted earlier, an emission rate is computed for each model year. The emission rates are then weighted to reflect each model year's contribution to the annual VMT by vehicle class. The weighted (or composite) emission rates are then summed to form weighted emission factors by vehicle class. The fleet characterization updates have been shown to be particularly important to consider for conformity. The NAS report (National Research Council, 2000) recommended that the EPA update the fleet characteristics at regular intervals, "every 2 years or so" (p. 130). In theory this is an important effort that, particularly given the recent history of significant increases in sport utility vehicle purchases, seems both useful and relevant. However, the efficacy of this recommendation has to be evaluated in light of the methods used to construct the weighted emission rates and the SIP and conformity rules.

An example of the kinds of problems that can occur with updated fleet characteristics can be seen in the San Joaquin Valley. On February 1, 2002, the Federal Highway Administration notified the state of California of the potential for a conformity lapse for the Fresno County 1998 Regional Transportation Plan (Rogers and Ritchie, 2002); like SIPs, the federal regulations require that an RTP be found conforming every 3 years. The potential conformity lapse was generated because the Council of Fresno County Government developed the 2002 RTP using EMFAC7f, California's previous mobile emission model, which had been replaced by a newer model. EMFAC7f included fleet assumptions from the early 1990s, while the newest model version (EMFAC2000) incorporates fleet assumptions from 1997 to 1999. The letter states that to avoid a conformity lapse, the most recent fleet data must be incorporated into the RTP conformity determination pursuant to guidance issued in a January 18, 2001, EPA–U.S. Department of Transportation (DOT) joint memorandum.

This presents a very serious problem with respect to making a conformity determination because the mobile emissions budget included in the 2005 attainment SIP was developed using the fleet characteristics included in EMFAC7f; the updated fleet in the latest version of EMFAC includes a greater number of older heavy-duty vehicles (recall that the new fleet has an age distribution of 45 years rather than the 35 years represented in the previous version of EMFAC) and a significant redistribution of light-duty vehicles based on the recent upsurge in sport utility vehicle purchases. In addition, assuming that CARB and presumably EPA have incorporated information on use of the so-called defeat devices (to reduce control measure effectiveness and increase NO_x emissions), which were recently made available as part of the EPA–Department of Justice settlement with industry, the heavy-duty vehicle fleet is likely to be represented as higher emitting in EMFAC2000. To be technically correct, the older version of EMFAC (7g) should be run with the new fleet data. However, this necessitates binning the new fleet data at age 35 and reweighting all of the base emission rates to reflect the new fleet fractions. A sketch plan level of analysis of off-model adjustments, which take into account the updated fleet characteristics, shows as much as a 50% increase in NO_x emissions and close to a doubling of HC emissions.

The California SIP on-road motor vehicle emission inventories, the conformity budgets based on those inventories, and modeled attainment and rate-of-progress demonstrations are all based on earlier versions of EMFAC. While it is important to understand the mobile source impacts of contemporary fleet characteristics, using these fleet characteristics to develop conformity inventories that must be compared to budgets derived with older data clearly results in incompatibilities between the conformity inventory and the SIP-prescribed mobile emission budgets. Technically, the best approach to eliminating these inconsistencies and improving SIP and conformity comparisons is to update SIPs to include new budgets derived with the latest models and the latest planning assumptions. However, this is both a time-consuming (as long as 3 years) and expensive process. If the CARB and EPA are to follow the NAS report recommendations, substantial thought has to go into how to handle these kinds of issues, which are likely to have a significant impact on conformity determinations.

13.3.6   The Mobile Emissions Inventory

The final step in preparing a mobile emissions inventory is to combine the emission factors with the travel activity data. MOBILE6 produces emission rates in grams of pollutant per vehicle mile traveled, which can also be reported as grams per mile or grams per vehicle per unit time (day or hour) (EPA, 2002). As can be seen, emission rates will change over time when fleets turn over, leading to new emission factors that are then combined with estimates of travel activity (e.g., total VMT), which also change over time, to produce the final mobile vehicle emission inventories expressed in terms of tons per hour, day, month, season, or year. MOBILE6 also computes subcomponents of the total inventory emissions, including evaporative emissions, running losses, resting losses, and refueling emissions.

EMFAC2000 computes inventory estimates of the total emissions for the entire state of California, subtotals for each of the 15 air basins, 35 air pollution control districts, and 58 counties (Gao and Niemeier, 2001). The model produces emission rates and inventories of exhaust and evaporative hydrocarbons, carbon monoxide, oxides of nitrogen, and particulate matter associated with exhaust, tire wear, and brake wear. Hydrocarbon emissions estimates are produced for total hydrocarbon, total organic gases, and reactive organic gases. The model also produces emissions of oxides of sulfur, lead, and carbon dioxide, which are used to estimate fuel consumption.

EMFAC2000 also calculates the emissions inventory for every hour and every month of the year. After selecting a specific month for analysis, the model provides the area-specific hourly temperatures and relative humidity, and properly adjusts the properties of dispensed fuel. Emissions inventories can be backcast to 1970 or forecast to the year 2040. EMFAC2000 produces a number of seasonal inventories for different purposes. The annual average inventories are derived by weighting each month of emissions for the year equally for a specific area. An extensive inspection and maintenance simulation program in EMFAC2000 also allows users to determine the incremental effects of adding or deleting certain programmatic elements.

Because MOBILE6 uses facility-based emission factors, the computation of conformity inventories is fairly straightforward. For running stabilized it is the multiplication of link-based VMT by the appropriate emission factor. It is important to note that the units match in this case; factors are produced and used by facility segment (or modeling link). In contrast, recall that EMFAC produces emission factors that are trip based. This has led to the technically incorrect use of emission factors at the link level when preparing conformity inventories. That is, trip-based emission factors are being applied to the link-based VMT coming from the travel demand models.

In both models, one of the key inputs is obviously the speed–VMT distributions produced by the MPOs. For most MPOs, regional VMT is divided into speed (EMFAC) or facility level of service bins (MOBILE) and multiplied by the emission factor. In California, for regions without a travel model default, speed distributions[3] have typically been generated using Caltrans traffic counts on urban freeways and the Highway Performance Monitoring System (HPMS)[4] (CARB, 1996). In California, a second model is frequently used to produce the emission inventories, the Direct Travel Impact Model.

13.3.6.1 The Direct Travel Impact Model

The DTIM model (the latest version is DTIM4) was developed in California by Caltrans to enhance vehicle emission modeling tools. The model was originally developed to provide detailed emissions input for photochemical grid models such as the Urban Airshed Model, but has since been used to estimate regional vehicle emissions (Caltrans, 1999). DTIM calculates traffic analysis zone (TAZ) emissions that are gridded at the TAZ centroid based on information produced by the travel forecast models for each link in a network. This approach produces an emissions inventory that (1) provides useful information on the spatial and temporal distribution of emissions, with a few caveats described below, and (2) provides

---

[3]Default VMT-by-speed distributions use HPMS VMT estimates with traffic speed counts from Caltrans. HPMS estimates a "typical" VMT distribution for each of 6 facility types in 3 geographical area types by taking the proportion of travel on each facility type. Vehicle speeds by facility are obtained from traffic counts by speed collected by Caltrans for the California Highway Patrol (CARB, 1996). The sum of the speed distributions for each facility, weighted by the proportion of travel on each facility type from HPMS, results in the default VMT-by-speed distribution.

[4]HPMS is regarded as a benchmark for VMT estimates. During the calibration of the regional travel models, VMT estimates derived from the regional models are compared to the HPMS VMT estimates for that region. The difference between the two must not be statistically significant. If it is, corrections to the travel model must be made before it can be considered "calibrated." Once a regional model has been calibrated to HPMS, however, the regional data obtained from the model is generally regarded as more accurate than DOT-derived data. The logic behind this reasoning is that while HPMS provides sound VMT data, the region-specific nature of the travel demand models allows for more accuracy in the additional activity data they provide.

a means of approximating differences in emissions for varying transportation alternatives (Systems Application Inc., 1998).

DTIM uses output from commonly used travel demand models (e.g., MINUTP and TRANPLAN) to perform the emissions inventory calculations. A speed postprocessor algorithm, which can calculate hourly speeds by roadway link, is also available as an option in DTIM. These speeds can then be used in place of speed output from the travel demand models, which often do not reflect very accurate levels of congestion on a roadway link.

DTIM's speed postprocessor algorithm has three major modules: one for unsignalized facilities, the second for signalized facilities, and the third for queuing when the volume–capacity ratio exceeds 1 (Systems Application Inc., 1998). The unsignalized facility module uses a Bureau of Public Roads (BPR) type function taken from the *1985 Highway Capacity Manual*. The signalized facility module is from the *1997 Highway Capacity Manual*. The queuing module of the speed postprocessor methodology is a slightly modified version of the queuing algorithm proposed by Dowling and Skabardonis (1992). Research has clearly shown the need for better postprocessing of speed estimates produced by the travel demand models for the purposes of air quality modeling (Dowling and Skabardonis, 1992; Helali and Hutchinson, 1994).

After transportation data are imported, grid cells are defined by transforming node coordinates to the Universal Transverse Mercator (UTM) coordinates. Emissions are calculated at the grid level for three types of travel activity data: link, intrazonal, and trip end. Link emissions are computed for each grid cell based on the network characteristics contained within that grid cell. Intrazonal and trip-end emissions are computed for each TAZ at the TAZ centroid. Although TAZ-level emissions can be distributed into grid cells contained within the TAZ (which are needed for subsequent photochemical air quality modeling), this is rarely done; it requires that the user define activity ratios for each grid cell within the TAZ. Since travel demand modelers rarely have this level of information, emissions are usually assigned to the grid cell containing the TAZ centroid.

Although originally developed to prepare gridded emissions input for the airshed photochemical models because it has a convenient connection to travel demand models, DTIM is often used to develop emission inventories for conformity purposes. Theoretically, DTIM inventories and those generated with EMFAC should be comparable when aggregated to the same level, such as county totals. However, large gaps between their respective estimates are always found. One reason for gaps between the model inventory estimates is that link-based transportation data are applied to match trip-based emission rates from EMFAC. DTIM ignores this problem and uses the trip-based emission rates as a proxy for link-based rates. It is not at all clear that EMFAC rates are as valid for the homogeneous link speeds assumed in transportation models as they are for average trip speeds (Ito et al., 2002).

13.4 Travel Inputs from the Transportation Models

Although the discussion thus far has focused mostly on the impact of changes in modeling practices from the emission model perspective, there are also significant issues that can arise when changes are incorporated into the travel demand forecasting practice. The passage of the conformity rule necessitated the defining of common parameters to be passed between the air quality and transportation forecasting models (Loudon and Quint, 1992). As noted earlier, to date the key parameters have been VMT and vehicle speeds.

13.4.1 Vehicle Miles Traveled

VMT is used for many different applications, and in many different formats. Some of the most common VMT-related mobile emissions data needs include VMT by speed for estimating running stabilized emissions; VMT for use in year-by-year forecasting, tracking, and comparison; and VMT by grid square and hour of the day for photochemical modeling (National Cooperative Highway Research Program (NCHRP), 1997a). The 1990 CAAA states that the HPMS estimates of VMT should be the primary means by which total travel is calculated in nonattainment areas (NCHRP, 1997a). The EPA allows only two data sources for VMT estimation: the HPMS and the network-based travel demand model estimates, which cannot be more than 20% different from HPMS estimates.

HPMS is a state-specific database containing extensive VMT and travel information that includes location, lanes, average annual daily traffic, and highway mileage. For data collection, HPMS divided the roadway system into sections, with traffic counts conducted on different sections each year. Since not every section is counted every year; interim updates are accomplished by applying growth factors to the most recent estimates. The HPMS also incorporates correction factors for the number of vehicle axles, the day of the week, and season of the year. The HPMS database sections may differ from the modeling domain of a particular region's travel model (Fleet and DeCorla-Souza, 1992) and methods have to be developed to account for differences in spatial coverage.

13.4.2 Vehicle Speed

Traditionally, speeds in travel demand models have been used as tools to calibrate traffic volumes on the network with the primary focus to obtain valid link volumes. Emission models, on the other hand, were developed with an acute sensitivity to vehicle speeds since pollutant emissions are dependent on vehicle speed (de Haan and Keller, 2000; Ireson et al., 1992; EPA, 1999a). Tests have demonstrated that vehicle emissions can vary dramatically, with higher levels of emissions generally occurring at the upper and lower speed ranges (NCHRP, 1997a).

A variety of speed estimation techniques have been developed over the years (NCHRP, 1997b). In practice, the two primary estimation techniques in use rely on estimates of volume-to-capacity (V/C) or the Highway Capacity Manual (HCM) curves. The most common V/C technique is the Bureau of Public Roads (BPR) curve developed in the late 1960s. The V/C curves are monotonically increasing functions that require a minimum amount of data, are extremely easy to use, and are spreadsheet friendly. Because of these attributes, they are ideal for regional forecasting even though they are not as accurate or reliable as the other estimation techniques (NCHRP, 1997b).

Conversely, the HCM method requires some training, has large data requirements, and involves using off-model software. In 1997, NCHRP (1997b) reported on the state of the practice for travel speed estimation techniques among state departments of transportation, regional and local agencies, and private firms (Table 13.2). The research focused on the BPR and HCM methods, and concluded that each approach has advantages and weaknesses, although the HCM methods are generally acknowledged to be of greater quality.

Speed input for air quality modeling can also be derived using simulation models. These models are often described as the solution to the problems facing the standard travel demand models (Bachman, 1998). Simulation models can be deterministic or stochastic and generally come in one of two forms:

TABLE 13.2 NCHRP Survey Findings on Speed Estimation Techniques

Bureau of Public Roads	Highway Capacity Manual
Used in travel models for long-range regional transportation plans	Used most often for congested speed estimation
Insensitive to the impact of signal spacing, timing, and coordination	Complex and difficult to apply without specialized software
Initially fit to 1965 HCM data; requires updating to the 1994 HCM	Requires extensive data for reliable results
Reflects national averages, but not local conditions	Underestimates speeds on urban arterials by 19%, while other facility types are within 10% of observed speeds
Generally underestimates speeds when demand exceeds capacity	Limited to volumes that are greater than or equal to capacity
Concern about low speed estimates that occur at very high V/C ratios	

Source: NCHRP, Planning Techniques to Estimate Speeds and Service Volumes for Planning Applications, NCHRP Report 387, Transportation Research Board, National Research Council, National Academy Press, Washington, D.C., 1997.

macroscopic or microscopic. Macroscopic models approximate traffic flow as a fluid and use a road segment as a base unit. A major limitation to the macroscopic model for mobile emissions modeling, however, is that the time spent in each driving mode (cruise, acceleration, and deceleration) is based on average flow rates and certain simplified assumptions (e.g., constant rates of acceleration and deceleration) instead of a detailed simulation of each vehicle's travel path (Skabardonis, 1994). Alternatively, microscopic models, also called microsimulation models, track individual vehicles as well as their relationship to other vehicles. Microscopic simulations are capable of producing second-by-second vehicle movement as the vehicle travels in the network.

Simulation models have the theoretical and computational capability to predict regional facility-level data at a resolution needed to predict emission-specific activity, particularly as the mobile emissions models begin to require greater resolution. However, most models are developed to answer specific problems in a local network, such as traffic congestion around a shopping mall, instead of describing complete system activity (Reynolds and Broderick, 2000).

For example, the INTRAS model (Wick and Liebermann, 1980) was used to simulate vehicle movement on freeways and ramps based on car-following, lane-changing, and queue discharge algorithms. FRESIM (Halati and Torres, 1990), a model that succeeded INTRAS, improved the representation of driver behavior, the logic for merging and lane changing, and the modeling of real-time ramp metering. ATMS (Junchaya et al., 1992), a traffic simulation model, employed parallel processing to simulate vehicle movement based on real-time link travel time by small time slices. Finally, the INTEGRATION model (Van Aerde, 1992; Berkum et al., 1996) simulated individual vehicle movements, including those with route guidance systems. The focus of these models is mainly on the solution of various traffic flow problems using limited, localized networks.

A newer generation of microsimulation models, with an expanded scope designed around regional systems, may prove exceptionally useful for generating detailed regional travel data for emission estimates. These models include, among others, MICE, a noncommercial dynamic traffic assignment simulation model (Adamo et al., 1996); TRAF-NETSIM (Chatterjee et al., 1997); TRANSIMS, an activity-based microsimulation model still under development (Los Alamos National Laboratory, 1998); and DYNASMART, a dynamic traffic assignment simulation model (University of Texas at Austin, 2000).

Of these models, all are capable of providing individual vehicle travel speed at second-by-second resolution for emissions estimates. The differences between the models lie in the type of trip–activity assignment algorithm used and the underlying traffic simulation rules that the models embed in their algorithms. Among the dynamic traffic assignment microsimulation models, only DYNASMART has the capability for simulating fairly large networks. The model is designed to replicate most real-world traffic situations and provides the detailed vehicle activities required by emissions inventory models. It also achieves a balance between representational detail, computational efficiency, and input and output data sizes.

Individual vehicle activity is modeled at a resolution of 6 sec. The effects of Advanced Traffic Management Systems (ATMS) strategies, trip chaining, driver classes, geometric and operational restrictions, mode fixed schedule, and capacity changes can be explicitly simulated, and satisfies most key physical properties and spatial and temporal constraints pertaining to vehicles, traffic, and highway networks, such as the link flow conservation equations, the first in, first out (FIFO) rule, and the vehicle speed–density relationship. The model is currently being expanded to better handle larger networks.

As we begin to better understand the emissions generation processes and as the ability to model these processes improves, more detailed vehicle activities, such as highly resolved link volumes and link speeds, and activity indicators, such as durations, will be desired. In the meantime, a number of studies have been conducted to improve the results coming from the standard travel demand models specifically for emissions inventory modeling. Postprocessing techniques have been developed to improve the prediction of estimated speeds produced by the standard travel demand models (Dowling and Skabardonis, 1992; Helali and Hutchinson, 1994; Skabardonis, 1997); to disaggregate daily (or other time period) link volumes into hourly volumes (Benson et al., 1994; Quint and Loudon, 1994; Niemeier and Utts, 1999); to adjust daily volumes into season-specific volumes (Benson et al., 1994; Quint and Loudon, 1994); and to disaggregate trips into TCM-related improvements, such as carpools (Everett, 1998).

Other efforts at improving the resolution of travel demand outputs thought to directly affect mobile emissions have included expanding the number of periods estimated by the travel demand models, to include prepeak/off-peak and postpeak/off-peak period assignments in addition to the standard peak and off-peak period assignments (Eash, 1998); using traffic systems to better determine off-network (local) VMT (Flood, 1998); directly estimating VMT by running mode (e.g., cold start, hot start, and hot stabilized mode) using a specialized equilibrium assignment model (Venigalla et al., 1999); and dividing the standard travel demand model output into emission-homogeneous speed flow regions, which are associated with different sets of disaggregated emissions rates (Roberts et al., 1999).

Most standard travel demand model results can be converted into the kinds of detailed data required by the emissions inventory models using postprocessing or expanded modeling techniques. However, because most postprocessors are based on surveys or traffic monitoring where traffic conditions vary, they are also specific to a certain region and time of year. Different sets of data have to be prepared and frequently updated for application in different regions or to reflect important seasonal fluctuations needed for air quality modeling.

13.5 Importance of Modeling Tools for Transportation Conformity

Because federal highway funding can be severely impacted when conformity determinations fail, each time elements of the modeling process are changed metropolitan areas may face challenges that could threaten successful conformity determinations. This was made particularly clear in a recent analysis that examined the timing associated with conformity in case studies of the Los Angeles and Sacramento, California, and Houston, Texas, metropolitan areas (Eisinger et al., 2002). The analysis explored how conformity determinations are placed at risk as EPA approves new mobile source emissions modeling tools because of the rigid timing of scheduled updates to regional transportation plans, transportation improvement programs, and state implementation plans.

Conformity regulations require the use of EPA-approved mobile source emissions models within 2 years of the model's approval date. When air quality management plans are based on older versions of EMFAC or MOBILE, the mobile source emissions budgets are often smaller than emissions estimates produced by new modeling tools. Thus, conformity findings become difficult when new emissions estimates are compared to budgets based on older model versions. The alternative is for the states and metropolitan areas to avoid conformity issues by undertaking a SIP update using the new modeling tools, a process that is very expensive and can take as long as 3 years to complete. Either way, the modeling tools become a critical element in determining whether or not, and how, areas achieve successful conformity determinations, which ensures continued receipt of federal transportation funding.

Conformity modeling, as it is currently interpreted statutorily, requires use of the latest planning assumptions for VMT, travel speeds, and, according to the joint EPA–FHWA guidance document issued in January 2001, the vehicle fleet age and distribution. Thus, conformity determinations, which take place at least once every 3 years and sometimes more frequently, may end up using substantially different planning assumptions or emissions models from those used to set the emissions budgets contained in the SIP.

This was the heart of a recent discussion related to Fresno's RTP approval. Here, the vehicle fleet assumptions were incorporated into the most recent version of EMFAC (EMFAC2000) — a version that produced inventory estimates substantially higher than the EMFAC version used to prepare the original emission budgets (EMFAC7g). In follow-up actions, the CARB responded by stating that the fleet information was embedded too tightly in the mobile emissions factor software (EMFAC) for fleet assumptions to be readily updated (Marvin, 2000). Thus, CARB's position was that the latest fleet assumptions reflect those contained with the latest approved model (EMFAC7g). However, the updated fleet represented in the new EMFAC model clearly resulted in higher emissions than estimated using the previous EMFAC7g model and reflected current knowledge of the vehicle fleet. While this does not, in and of itself, imply that conformity will fail, it is an important distinction with which to be reckoned. As our understanding of mobile source modeling continues to grow, the need for better methods of incorporating new knowledge and for extending the capabilities of both the modeling and policy frameworks will become increasingly critical in order to effectively measure actual progress toward attainment.

13.6 Reflections on the Future

The political and institutional dynamics of both the establishment and implementation of air quality regulations and the conformity process will continue to play out over many years. Currently, the conformity and SIP processes are fairly rigid. The mobile source inventory used in the SIPs is based on emissions estimated using MOBILE or EMFAC (California only). For on-road emissions, inputs to these models, namely speed and VMT, are usually provided by MPOs in urban areas and the county or state transportation agencies for nonurban areas. Inventories are then adjusted to reflect implementation of various control strategies; the adjusted inventories become the emissions budgets contained in the SIPs. The budgets are finalized in a SIP submittal that must be approved by the EPA. Changing an approved emissions budget requires development of a revised SIP, which must then be resubmitted and approved by the EPA.

In short, there is very little flexibility to handle or evaluate marginal changes without an extensive SIP review, a process that can take up to 3 years. This limits our ability to readily incorporate new scientific and technological developments into the process. It will be important, particularly over the next 4 to 5 years, to develop thoughtful alternatives that accelerate our ability to integrate new developments that affect our assumptions about future air quality while still meeting the spirit and intent of the conformity regulations.

With the February 2001 Supreme Court decision that upheld EPA's authority to act under the Clean Air Act, and the March 2002 U.S. Court of Appeals decision, which upheld the 1997 promulgated 8-h ozone and new $PM_{2.5}$ standards, there will be increased need for better understanding of fine particulate matter pollution. We know that particulate matter in vehicle exhaust occurs chiefly in the ultrafine mode (aerodynamic diameter $\leq$ 100 nm), and because these soot particles are small, they can remain airborne for days to weeks and can significantly affect human health (Pope et al., 2002), as well as global climate through light absorption and scattering effects (Chughtai et al., 1991).

We also understand very little about how alternative fuels and engine technologies affect particulate matter and NO_x emissions. Recent studies have suggested that diesel particulate filter (DPF) technology, a potentially lower-cost short-term solution to reduce PM emissions by outfitting existing diesel buses with particle traps, may significantly reduce particle emissions (Holmén and Ayala, 2002), but more data are needed on both the driving patterns and technology. There has been increased interest in the air quality community in the development of real-time onboard vehicle emissions systems for heavy-duty vehicles. Understanding how to effectively model the activity patterns using this data is something the transportation community can do well. With the important role diesel vehicles play in the U.S. economy for low-cost, long-distance transport of goods, these vehicles will have a significant impact on air quality for the foreseeable future (Lloyd and Cackette, 2001).

Concerns have also been raised that particle *number* concentration, rather than *mass* concentration, may have a more direct relationship with adverse human health effects (Donaldson et al., 1998). Currently, no reliable models exist to predict the ultrafine particle size distributions from vehicles, especially under the various modes of vehicle operation (speed, acceleration rate) and the wide range of atmospheric meteorological conditions (temperature, relative humidity) that occur in real-world driving.

Taking advantage of increasing computational capabilities, a major research thrust in the transportation–mobile emissions interface over the next few years will be directed toward producing highly resolved spatial and temporal models for both transportation activities and mobile emissions. Already some of this work has begun with the weekday–weekend effect, where a number of studies have shown a day-of-week variation in ozone concentrations at most monitoring sites in the South Coast Air Basin (SoCAB) and the San Francisco Bay Area Air Basin (SFBAAB), as well as many other areas. Surprisingly, as ozone concentrations have declined over time, the frequency of high ozone concentrations occurring on Sundays went from being the lowest of the week to the highest of the week by the late 1990s (CARB, 2001b).

These trends are important because they reflect the types of emission control strategies used, where ambient ozone trends are governed by changes in the temporal and spatial patterns of precursor emissions (Fujita et al., 2000). One of the major limitations to better understanding the weekday–weekend effect is the lack of good transportation modeling estimates for weekends. In order to achieve the improved spatial and temporal resolutions, there will be a significant increase in not only the amount and type of data required (National Research Council, 2001), but also the need for better methodological techniques, perhaps, for example, new simulation methods that can be used in conjunction with existing travel models to better estimate weekend travel activity.

There have been many modeling techniques that would be of great use in the transportation–mobile emissions modeling interface and are now routinely applied to many types of transportation problems, including, for example, duration models and vehicle transaction models. The use of these techniques for problems associated with mobile emissions modeling could span such topics as better prediction of I/M effectiveness, improving vehicle fleet representation, and simulating activity data for large-scale estimation of evaporative emissions. In particular, human behavior lies at the heart of the effectiveness of the I/M programs. We need better models to understand how decisions are made on repair costs and repair effectiveness, and which I/M programs actually work best for reducing fleet emissions. For example, the California Bureau of Automotive Repair has shown that I/M effectiveness is higher at test-only stations (as opposed to test and repair stations), yet on-road failures are still quite high and fleet turnover is not well understood.

The separation between the transportation modeling and the air quality research communities is still quite vast. One result of this gap is that the mobile emissions researchers often spend time trying to solve problems or develop methods that already have a long history of being successfully tackled in the transportation community. Likewise, those in the transportation research community have sometimes been applied to mobile emissions problems that are portrayed as fairly simple when in fact they are quite complex. It is almost certain that the next generation of new knowledge will come from those individuals able to be innovative in the interface between the two fields.

Acknowledgments

I thank my students for their thoughtful input and ideas and the discussions we have had through the years about transportation and air quality. In particular, Yi Zheng, Kathy Nanzetta, Trish Hendren, Jen Morey, Oliver Gao, and Erin Foresman all offered wonderful suggestions and comments. I also thank Prof. Britt Holmén at the University of Connecticut and Doug Eisinger and Tom Kear of the UC Davis–Caltrans Air Quality Project for generously lending their knowledge and insight.

References

Adamo, V. et al., A dynamic network loading model for simulation of queue effects and while trip re-routing, in *Proceedings of the Planning and Transport Research and Computation (International) Co. Meeting*, Seminars D and E: Transportation Planning Methods, Part 1, PTRC Education and Research Services, London, 1996.

An, F. et al., Development of a comprehensive modal emissions model: Operating under hot-stabilized conditions, *Transp. Res. Rec.*, 1587, 73–84, 1997.

Andre, M., Driving Cycles Development: Characterization of the Methods, SAE Technical Paper, Series 961112, Society of Automotive Engineers, Warrendale, PA, 1996, pp. 1–13.

Austin, T.C. et al., Characterization of Driving Patterns and Emissions from Light-Duty Vehicles in California, A932-185, California Air Resources Board, Sacramento, CA, 1993.

Bachman, W., A GIS-Based Modal Model of Automobile Exhaust Emissions, CR823020, prepared for EPA, Office of Air Quality Planning and Standards Research, Triangle Park, NC, 1998.

Benson, J.D. et al., Development of Emissions Estimates for the Conformity Analysis of the El Paso FY-94 TIP, FHWA/TX-94/1375–2, Texas Transportation Institute, College Station, 1994.

Berkum, E. et al., Applications of dynamic assignment in The Netherlands using the INTEGRATION model, in *Proceedings of the Planning and Transport Research and Computation (International) Co. Meeting*, Seminars D and E: Transportation Planning Methods, Part 1, PTRC Education and Research Services, London, 1996.

California Air Resources Board, Methodology for Estimating Emissions from On-Road Motor Vehicles, Vol. II: EMFAC7G, Technical Support Division, Mobile Source Emission Inventory Branch, CARB, California EPA, Sacramento, 1996.

California Air Resources Board, Revisions to the State's On-Road Motor Vehicle Emissions Inventory, Technical Support Document, Correction Factors, CARB, Sacramento, 2000a, chap. 6.

California Air Resources Board, Revisions to the State's On-Road Motor Vehicle Emissions Inventory, Technical Support Document, County Specific Vehicle Age Distribution and Population Matrices, CARB, Sacramento, 2000b, chap. 7.

California Air Resources Board, The 2001 California Almanac of Emissions and Air Quality, California EPA, Sacramento, 2001a.

California Air Resources Board, The Ozone Weekend Effect in California, Planning and Technical Support Division, The Research Division, CARB, Sacramento, November 2001b.

Caltrans, *Direct Travel Impact Model 3.1: User's Guide*, California Department of Transportation, Sacramento, 1999.

Chatterjee, A. et al., *Improving Transportation Data for Mobile Source Emission Estimates*, Transportation Research Board, National Research Council, Washington, D.C., 1997.

Chughtai, A.R. et al., Spectroscopic and solubility characteristics of oxidized soots, *Aerosol Sci. Technol.*, 15, 112–126, 1991.

de Haan, P. and Keller, M., Emission factors for passenger cars: Application of instantaneous emission modeling, *Atmos. Environ.*, 35, 4629–4638, 2000.

Donaldson, K. et al., Ultrafine (nanometre) particle mediated lung injury, *J. Aerosol Sci.*, 29, 553–560, 1998.

Dowling, R. and Skabardonis, A., Improving average travel speeds estimated by planning models, *Transp. Res. Rec.*, 1366, 68–74, 1992.

Eash, R., Time period and vehicle class highway assignment for air quality conformity evaluation, *Transp. Res. Rec.*, 164, 66–72, 1998.

Effa, R.C. and Larsen, L.C., Development of real-world driving cycles for estimating facility-specific emissions from light-duty vehicles, in *Air and Waste Management Association Specialty Conference on the Emission Inventory: Perception and Reality*, Air and Waste Management Association, Pasadena, California, 1993.

Eisinger, D. et al., Conformity: The New Force behind SIP Deadlines, Association of Air and Waste Management, January 2002.

Environment Protection Agency, Criteria and procedures for determining conformity to state or federal implementation plans of transportation plans, programs, and projects funded or approved under Title 23, U.S.C., or the Federal Transit Act, as amended August 1995, November 1995, and August 1997, Fed. Regist., 58, 62188, 1993a.

Environmental Protection Agency, Review of Federal Test Procedure: Modification Status Report, EPA 420-R-93-007, Office of Mobile Sources, Office of Air and Radiation, U.S. EPA, Ann Arbor, MI, 1993b.

Environmental Protection Agency, Development of Speed Correction Cycles, Draft M6.SPD.001, U.S. EPA Assessment and Modeling Division, National Vehicle Fuel and Emissions Laboratory, Ann Arbor, MI, 1997.

Environmental Protection Agency, Facility Specific Speed Correction Factors, Draft EPA420-P-99-002 M6.SPD.002, Office of Mobile Sources, Office of Air and Radiation, U.S. EPA, Ann Arbor, MI, 1999a.

Environmental Protection Agency, National Air Quality and Emissions Trends Report, Office of Air Quality Planning and Standards Research, Triangle Park, NC, 1999b.

Environmental Protection Agency, Determination of CO Basic Emission Rates: OBD and I/M Effects for Tier 1 and Later LDVs and LDTs, M6.EXH.009, Assessment and Modeling Division, Office of Transportation and Air Quality, U.S. EPA, Ann Arbor, 2001a.

Environmental Protection Agency, Fleet Characterization Data for MOBILE6: Development and Use of Age Distributions, Average Annual Mileage Accumulation Rates, and Projected Vehicle Counts for Use in MOBILE6, M6.FLT.007, Assessment and Modeling Division, Office of Transportation and Air Quality, U.S. EPA, Ann Arbor, 2001b.

Environmental Protection Agency, User Guide to MOBILE6.0: Mobile Source Emission Factor Model, EPA420-R-02-001, Assessment and Modeling Division, Office of Transportation and Air Quality, U.S. EPA, Ann Arbor, 2002.

Everett, J., Emission impacts of utilizing vehicle class distributions by mode for TDM analysis, in *Transportation, Land Use, and Air Quality: Making the Connection*, Conference Proceedings, American Society of Civil Engineers, Portland, OR, 1998.

Fleet, C.R. and DeCorla-Souza, P., VMT for air quality purposes, in *Transportation Planning and Air Quality*, Proceedings of the National Conference, American Society of Civil Engineers Board, Portland, OR, 1992, pp. 126–141.

Flood, M.J., Redetermining highway performance monitoring system derived off-network VMT calculations using GIS, in *Transportation, Land Use, and Air Quality: Making the Connection*, Conference Proceedings, American Society of Civil Engineers, Portland, OR, 1998.

Fujita, E. et al., Weekend/Weekday Ozone Observations in the South Coast Basin, ACI-0-29086-01, National Renewable Energy Laboratory, Golden, CO, 2000, pp. 165–172.

Gao, O. and Niemeier, D., Mobile Emissions Inventories: EMFAC7g and EMFAC2000 Comparison by Air Basin, prepared as part of the UC Davis–Caltrans Air Quality Project (http:/ AQP.engr.ucdavis.edu), University of California, Davis, 2001, p. 32.

General Accounting Office, Air Pollution: Limitations of EPA's Motor Vehicle Emissions Model and Plans to Address Them, report to the Chairman, Subcommittee on Oversight and Investigations, House of Representatives, Washington, D.C., 1997.

Halati, A. and Torres, J., *Freeway Simulation Model Enhancement and Integration: FRESIM User's Guide*, Federal Highway Administration, Washington, D.C., 1990.

Helali, K. and Hutchinson, B., Improving road link speeds for air quality models, *Transp. Res. Rec.*, 1444, 71–77, 1994.

Holmén, B. and Ayala, A., Ultrafine PM emissions from natural gas, oxidation-catalyst diesel and particle-trap diesel heavy-duty transit buses, *Environ. Sci. Technol.*, 2002, submitted.

Holmén, B.N., and Niemeier, D., Characterizing the effects of driver variability on real-world vehicle emissions, *Transp. Res. D*, 3, 117–128, 1998.

Howitt, A.M. and Moore, E.M., Linking Transportation and Air Quality Planning: Implementation of the Transportation Conformity Regulations in 15 Nonattainment Areas, U.S. EPA, FHWA, U.S. DOT, Washington, D.C., 1999.

Ireson, R.G., Generating detailed emissions forecasts using regional transportation models: current capabilities and issues, in Ireson, R.G., Fieber, J.L., and Causley, M.C., in *Transportation Planning and Air Quality*, American Society of Civil Engineers, New York, NY, 1992, pp. 142–160.

Ito, D. et al., How VMT–speed distributions can affect mobile emissions inventory modeling, *Transportation*, 28, 409–425, 2002.

Joumard, M.A. et al., Influence of driving cycles on unit emissions from passenger cars, *Atmos. Environ.*, 34, 4621–4628, 2000.

Junchaya, T. et al., ATMS: Real-Time Network Traffic Simulation Methodology with a Massively Parallel Computing Architecture, paper presented at 71st Annual Meeting of the Transportation Research Board, Washington, D.C., National Research Council, 1992.

Larson, T.D., A Summary: Air Quality Programs and Provisions of the Intermodal Surface Transportation Efficiency Act of 1991, U.S. DOT, FHWA, Washington, D.C., 1992.

LeBlanc, D.C. et al., Driving pattern variability and impacts on vehicle carbon monoxide emissions, *Transp. Res. Rec.*, 1472, 45–52, 1995.

Leslie, T. et al., Notification of Potential Conformity Lapse, Government of California, The Honorable Gray Davis, Sacramento, February 1, 2002.

Levinsohn, D.M., Use of UTPS for Subarea Highway Analysis: A Case Study, U.S. Department of Transportation, Federal Highway Administration, 1985.

Lin, J. and Niemeier, D., Estimating regional air quality vehicle emission inventories: Constructing robust driving cycles, *Transp. Sci.*, 2002a, forthcoming.

Lin, J. and Niemeier, D., Development of regional driving cycles using regional driving characteristics, *Transp. Res. D*, 2002b, submitted.

Lloyd, A.C. and Cackette, T.A., Diesel engines: Environmental impact and control, *J. Air Waste Manage. Assoc.*, 51, 809–847, 2001.

Los Alamos National Laboratory, Transportation Analysis Simulation System (TRANSIMS): The Dallas Case Study, prepared for the U.S. DOT and the U.S. EPA, Los Alamos National Laboratory, Texas Transportation Institute, 1998.

Loudon, W.R. and Quint, M., Integrated Software for Transportation Emissions Analysis: Transportation Planning and Air Quality, paper presented at Proceedings of the National Conference, American Society of Civil Engineers, Portland, OR, 1992, pp. 161–176.

Lyons, T.J. et al., The development of a driving cycle for fuel consumption and emissions evaluation, *Transp. Res. A*, 20, 447–462, 1986.

Marvin, C., Letter clarifying information available for use in conformity determinations, M. Ritchie, California Division Administrator, FHWA, and L. Rogers, Regional Administrator, FTA, Sacramento, CARB, Feb. 20, 2000.

Milkins, E. and Watson, H., Comparison of Urban Driving Patterns, SAE Technical Paper, Series 830939, Society of Automotive Engineers, Warrendale, PA, 1983.

Morey, J.E. et al., Validity of chase car data used in emissions cycles, *J. Transp. Stat.*, 3, 15–28, 2000.

National Cooperative Highway Research Program, Improving Transportation Data for Mobile Source Emission Estimates, NCHRP Report 394, Transportation Research Board, National Research Council, National Academy Press, Washington, D.C., 1997a.

National Cooperative Highway Research Program, Planning Techniques to Estimate Speeds and Service Volumes for Planning Applications, NCHRP Report 387, Transportation Research Board, National Research Council, National Academy Press, Washington, D.C., 1997b.

National Research Council, *Modeling Mobile Source Emissions*, National Academy Press, Washington, D.C., 2000.

Niemeier, D., Spatial applicability of emissions factors for modeling mobile emissions, *Environ. Sci. Technol.*, 36, 736–741, 2002.

Niemeier, D.L. and Utts, J., Using observed traffic volumes to improve fine-grained regional emissions estimation, *Transp. Res. D Transport Environ.*, 4, 313–332, 1999.

Ntziachristos, L. and Samaras, Z., Speed dependent representative emissions factors for catalyst passenger cars and influencing parameters, *Atmos. Environ.*, 34, 4611–4619, 2000.

Pope, C.A. et al., Lung cancer, cardiopulmonary mortality, and long-term exposure to fine particulate air pollution, *J. Amer. Med. Assoc.*, 287, 1132–1141, 2002.

Quint, M.M. and Loudon, W.R., Transportation activity modeling for the San Joaquin emissions inventory, *Transp. Res. Rec.*, 1444, 109–117, 1994.

Reynolds, A.W. and Broderick, B.M., Development of an emissions inventory model for mobile sources, *Transp. Res. D*, 5, 77–101, 2000.

Roberts, C.A. et al., Forecasting dynamic vehicular activity on freeways: Bridging the gap between travel demand and emerging emissions models, *Transp. Res. Rec.*, 1664, 31–39, 1999.

SAEFL, Luftschadstoff-Emissionsen des SStrassenverkehrs 1950–2010, SRU 255, can be obtained from SAEFL, Dokummentationsdienst, Bern, Switzerland, 1995, p. 420 (in German; English abstract).

Skabardonis, A., Feasibility and Demonstration of Network Simulation Techniques for Estimation of Emissions in a Large Urban Area, Final Report, prepared for CARB, Sacramento, 1994.

Skabardonis, A., Modeling framework for estimating emissions in large urban areas, *Transp. Res. Rec.*, 1587, 85–95, 1997.

St. Denis, M.J. et al., Effects of in-use driving conditions and vehicle/engine operating parameters on "off-cycle" events: Comparison with federal test procedure conditions, *J. Air Waste Manage. Assoc.*, 4, 31–38, 1994.

Stephenson, A. and Dresser, G.B., State-of-the-Practice Report on Mobile Source Emissions Models (revised), 1279-3, 0-1279, prepared for the Texas Department of Transportation in cooperation with the U.S. DOT/FHWA College Station, Texas Transportation Institute, 1995, pp. 1–50.

Symons, M.J., Clustering criteria and multivariate normal mixtures, *Biometrics*, 37, 35–43, 1981.

Systems Application Inc., DTIM3 User's Guide, prepared for the California Department of Transportation, Sacramento, 1998.

University of Texas at Austin, DYNASMART-P, Vol. I: Capabilities, Algorithmic Aspects, Software Design and Implementation, Technical Report STO67-85-PI, prepared for the FHWA, U.S. DOT, Washington, D.C., 2000.

Urban Mass Transportation Administration (UMTA), An Introduction to Urban Travel Demand Forecasting: A Self-Instructional Text, U.S. Department of Transportation, Federal Highway Administration, Washington, D.C., 1977.

U.S. Department of Transportation, A Summary: Transportation Equity Act for the 21st Century, U.S. DOT, Washington, D.C., 1998.

Utts, J.N., Niemeier, D., and Ring, D., Statistical methods for estimating speed correction factors with confidence intervals for mobile source emissions models, *Transp. Res. D*, 5, 103–120, 2000.

Van Aerde, M., INTEGRATION: A Model for Simulating Integrated Traffic Networks: User's Guide for Model Version 1.4d, prepared by TT&E Branch, Ontario Ministry of Transportation, and Transportation Systems Research Group, Department of Civil Engineering, Queen's University, Canada, 1992.

Venigalla, M.M. et al., A specialized equilibrium assignment algorithm for air quality modeling, *Transp. Res. D*, 4, 29–44, 1999.

Watson, H. and Milkins, E., Comparison of Urban Driving Patterns, SAE Paper 830939, Society of Automotive Engineers, Warrendale, PA, 1983.

Wick, D.A. and Liebermann, E.B., Development and Testing of INTRAS: A Microscopic Freeway Simulation Model, Final Report Vol. 1, FHWA/RD-80/106, FHWA, Washington, D.C., 1980.

14

Demographic Microsimulation with DEMOS 2000: Design, Validation, and Forecasting

CONTENTS

Ashok Sundararajan
*AECOM Consulting
Transportation Group*

Konstadinos G. Goulias
Pennsylvania State University

14.1 Introduction

As illustrated earlier in this handbook, new concepts in travel demand modeling capture and predict travel behavior more realistically than ever before. At the heart of any forecasting model, however, are social, economic, and demographic data. In addition, the overwhelming majority of these new travel demand models need this type of data at the person or household levels. Moreover, predictions from these forecasting models are very sensitive to the accuracy of the information provided as input. Although this has been recognized for more than 20 years, sociodemographic forecasting for travel demand systems is progressing at a much slower pace than needed to support new policy initiatives and associated model-building efforts (Goulias, 1997).

Two of the reasons for this gap are the detailed microinput required by these models and the complexity involved in designing model systems that produce them. With the advent of new techniques in survey methods like activity-based surveys and panel surveys, precise person and household level information is now available to build demographic simulators. Also, tremendous advancements have been made in the development of microanalytic simulation models and related programming languages, which can effectively handle the complexities involved in the human life cycle evolution. Miller, in Chapter 12, presents the theoretical background and history of these models. In addition, Sundararajan (2001), Kazimi (1995), and Goulias (1991) provide reviews of demographic microsimulation methods. This chapter describes an application in demographic microsimulation using data from the United States that can be extensively applied to travel forecasting. The chapter also provides validation examples that are useful for other applications too.

14.2 DEMOS 2000

DEMOS is a microsimulator of social, economic, and demographic attributes describing an individual and a household. During the process of simulation an individual will be born and progressed through different life cycle stages. While progressing through these life cycle stages the individual is exposed to different events in the form of death, giving birth to a child, leaving the nest and living elsewhere, marrying or divorcing, acquiring a license and a job, buying a new vehicle, and so on. All these changes are simulated probabilistically in DEMOS. Most of the transition probabilities are obtained by cross-classification of the variables between two successive years of the Puget Sound Transportation Panel (PSTP) data. In the process, different person and household attributes are internally generated in a conditional way. For example, the age of the mother affects the age and number of children living in the household, and the income group in which a household lives affects the lifestyle and the number of vehicles owned by the household. DEMOS captures all these correlations by specifying the probability of change as a function of the household and person characteristics.

The model system combines the unique concepts of microsimulation and object-oriented programming (OOP) to develop a highly modular simulation model in which the submodels can be added, altered, or replaced without affecting the other components of the model system. The concepts of OOP and microsimulation go hand in hand because OOP was designed to handle complex problems for large-scale microsimulation types of applications. The purpose of OOP is to model the behavior of the real-world objects (Mandava, 1999). These objects can be persons, households, vehicles, firms, highways, intersections, and so on. The real world consists of these objects that are interacting and evolving over time. Each object has its own state and behavior. There is a one-to-one mapping of the objects in the real world and the simulated or virtual world.

An object in a simulated world, however, is considered an abstraction of the real-world object. The object has a number of methods that explain the behavior of the object simulated. For example, in DEMOS each individual is simulated based on two classes, namely, PERSON and HOUSEHOLD. As the names suggest, the PERSON class has all the methods describing the individual and the HOUSEHOLD class contains all the methods required to explain the household. Then every individual living in a household is an object of the PERSON and HOUSEHOLD classes. Any changes in the HOUSEHOLD attributes are updated to all the members of the household, reflecting the changes taking place at the PERSON level. For example, if an individual dies, the household attribute and type for every person is changed appropriately. Part of this more basic research work and the computer code were developed in Mandava (1999). Due to difficulties in internal validation of that earlier work, it was decided that a new computer code should be developed to incorporate some of the original ideas, but not the computer code from the earlier work; the results using the new code (DEMOS 2000) are described in this chapter. Next, a description of the DEMOS simulation process and structure of the software are given. Then the data used and the validation of select variables are presented. The chapter concludes with a few forecasting examples and a brief summary.

14.2.1 DEMOS Process

DEMOS uses longitudinal simulation for evolving the persons during the simulation. In longitudinal simulation an individual is first simulated through the entire simulation period, and then the next person is simulated. By doing this, the attributes describing the individual need not be output after every period. The variables describing the individual can be updated after every year. In this way the demand for external mass media is much lower than in other methods. By eliminating the input file at every simulation period, the computation time is reduced, as the model does not have to read in and read out data all the time. However, difficulties arise due to longitudinal simulation: it may be difficult to validate the model,

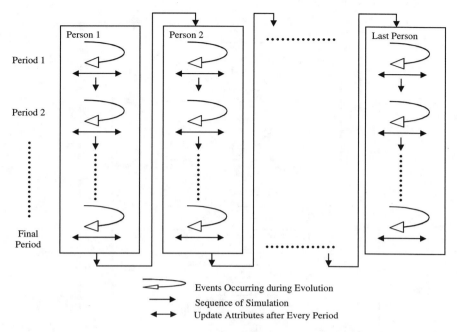

FIGURE 14.1 Longitudinal simulation of individuals. (From Hain, W. and Helberger, C., *Microanalytic Simulation Models to Support Social and Financial Policy*, Orcutt, G.H., Merz, J., and Quinke, H., Eds., North Holland, Amersterdam, 1986. With permission.)

and it is difficult to design interdependent processes between persons and households. A diagram of longitudinal simulation is shown in Figure 14.1.

Figure 14.2 shows the flowchart of the simulation process. First the input data for a particular individual are read and then he or she is progressed through the first year. In DEMOS the individual's household attributes are determined first. Then the person attributes are simulated, followed by simulation of the information and communication technologies (ICTs) owned and used, and then activity–travel duration models are applied to simulate activity and travel behavior. In case a child is born during the simulation, the children are simulated after the mother is simulated. Also, if an eligible single person gets married during the simulation period, then the new person is simulated based on data about the member in the original database. A user can provide the number of years and number of simulations (replications for each person to be simulated) as inputs.

The order in which an individual is exposed to different events is shown in Figure 14.3. First the individual is checked for the event "death." If he or she dies, then the individual is removed from the simulation. Following death, based on gender, the individual is exposed to "birth." The next event is "child leave nest." If the person is below 25 years age, then he or she is eligible to leave the parents' household. Based on marital status, the individual is then exposed to either "divorce" or "marriage." In all these cases changes are made to other members' household attributes as required. Then the income group of the household is simulated, followed by the total number of vehicles in the household. After the household characteristics are simulated, the person characteristics are estimated. The chances of the individual holding a driver's license are estimated, followed by the employment status and occupation type. Detailed descriptions of each of these events and the data used are provided in Sundararajan (2001).

14.2.2 Data Used as Input Population

The data used in the analysis are from waves 1 (1989), 2 (1990), 3 (1991), 4 (1992), 5 (1993), and 7 (1997) of the PSTP data. PSTP is the first general-purpose travel panel survey in an urban area in the

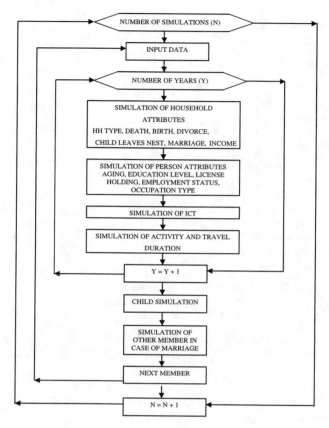

FIGURE 14.2 Flowchart of simulation.

United States. The survey was conducted in the Seattle metropolitan area by the Puget Sound Regional Council in partnership with the transit agencies in the region. It is a longitudinal survey in which similar measurements are made on the same sample at different times. Each measurement conducted during a time point is called a wave. The first survey was initiated in the fall of 1989. Murakami and Watterson (1990) provide more information regarding the origins of this panel survey.

PSTP's three components are household demographics, person socioeconomics, and travel behavior. Trip information was collected using a travel diary as an instrument. The travel diary consisted of every trip a person made during two consecutive weekdays, which remained approximately the same during the panel years. Each trip was characterized by trip purpose, type, mode, start and end times, origin and destination, and distance. In DEMOS, the first wave serves as the input population. Transition probabilities were estimated from *waves 1 and 2*, which determine the probability of a particular event to occur or not occur for an individual. *Waves 2, 3, and 4* are used to validate the model predictions. In waves 1 through 4 there was a total of 1621 respondents (928 households), and in wave 5 there were 1383 respondents or 801 households. Finally, wave 7 was also used to develop the information and communication technology ownership and use models (Sundararajan, 2001).

In addition to the PSTP, some additional information was used from the U.S. Census Bureau and the National Center for Health Statistics (NCHS). The U.S Census Bureau provides detailed data about the people and economy of the United States. NCHS is the federal government's principal vital and health statistics agency. NCHS data systems include data on vital events as well as information on health status, lifestyle, and health care.

In the simulation, the first wave of the PSTP data was used as the input population to DEMOS. The reasons for using the first wave as the input population are that (1) the short-term forecasting ability of

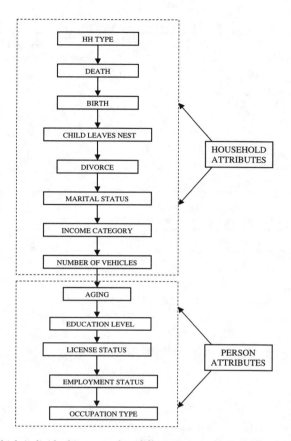

FIGURE 14.3 Order in which individual is exposed to different events during evolution.

the software can then be tested to the four remaining waves, thus allowing sufficient data for validation, and (2) Ma (1997) has done considerable research in developing the activity and travel indicators using the same data. These models can be directly embedded in DEMOS and can be used to study the activity and travel pattern of individuals in the future. In addition, Ma (1997) has also developed models for daily time allocation and models for daily activity and travel scheduling using the PSTP data. All these models can be incorporated in DEMOS, and then the microsimulator can be extended to predict the daily activity and travel budget of the individuals in the sample. Finally, the predicted activity and travel durations for different activities can be validated to the PSTP data.

Initially, the model was designed to simulate 1621 respondents. The PSTP data do not provide detailed information about the children in the household, but contain information on the total number of children between the ages of 1 to 5 and 6 to 17. Based on this information, the characteristics of the children were simulated (synthetically generated) separately. This resulted in a total of 2157 respondents, including the children. Finally, the model database was expanded to 8628 respondents by replicating the same characteristics of the individuals and households. The model can simulate a maximum of 25 years and 100 simulations. However, by changing the size of arrays at appropriate places, the simulation period can be expanded. During an average DEMOS run it takes about 10 min to simulate 10 years over 100 times for 1621 respondents, and about 60 min to simulate 20 years over 100 times for 8628 respondents, using a personal computer with 384 MB of RAM, Pentium III processor.

The summaries of the socioeconomic and demographic characteristics of the persons and households for wave 1 are provided in Tables 14.1a and b, respectively. The sample has more women than men and is relatively older, considering the fact that the average age for both men and women is around 47 years.

TABLE 14.1A Summary of Household Characteristics (Wave 1)

Household (HH) Characteristics	Percent (Number of Respondents = 1621)
HH income less than $30000	29.8
HH income $30,000–$70,000	59.2
HH income greater than $70,000	11.0
0 vehicles in HH	1.7
1 vehicle in HH	17.5
2 vehicles in HH	46.3
3 vehicles in HH	22.8
More than 4 vehicles	11.6
Average household size = 2.74	

TABLE 14.1B Summary of Person Characteristics (Wave 1)

Person Characteristics	Male	Female
Gender	46.3%	53.7%
Mean age	47.7[a]	46.9[b]
Have driver's license	96.9%	93.1%
Do not have driver's license	3.1%	6.9%
Employed	74.7%	57.1%
Not employed	25.3%	42.9%
Occupation (percent out of employed)		
Professional	28.3%	25.4%
Manager	16.9%	16.3%
Secretary	4.1%	28.2%
Sales	6.2%	7.8%
Other	44.4%	22.3%

[a] Minimum = 15; maximum = 89; standard deviation = 14.3.
[b] Minimum = 15; maximum = 90; standard deviation = 14.3.

Most of the respondents have a driver's license. About 75% of the men are employed, while only 57% of the women are employed. Out of those employed, about 44% of the men are employed as production workers or foremen, vehicle operators, service workers, and so forth. About 28% of the men are professionals and about 17% are managers. Among women, the majority are employed as secretaries or professionals. About 16% of the women are managers.

At the household level, the majority of the households have a total household income between $30,000 and $70,000. Only 1.7% of the households do not have a vehicle. About 46% of the households have at least two vehicles. The sample has a total of 928 households, and the average household size is 2.74 per household. Other wave data are documented in Sundararajan (2001).

DEMOS was developed using Microsoft Visual C++ (VC++) Version 6.0. One of the main reasons for choosing VC++ was its visual capabilities and its OOP approach. VC++ allows the data to be read from the Microsoft Access database. DEMOS relies on the input population from the first wave of the PSTP data, which is stored as an MS Access file. All the variables are stored in two large tables.

The OOP approach allows the use of classes and objects. DEMOS is based on three important classes and a source file:

- CDATA: This class holds all the variables from the database. The variables are established automatically once the input file is specified, while creating a project file initially. It is important to note that the variables in this class should be exactly the same as the variables in the database. If any modifications such as adding or deleting a variable or changing a variable name are made to the database later, then a new class has to be created.

- PERSON: This class holds all the methods or functions that are relevant to the individual.
- HOUSEHOLD: This class has all the methods or functions that are related to each household in the data.
- CDEMOSVIEW: This is a source file that can be considered the heart of DEMOS. Objects are created from the PERSON, HOUSEHOLD, and CDATA classes and the functions are called from this file in the specified order. Also, this file contains the relevant code used to aggregate the information and provide results.

An object of these classes is created for processing the information. For example, every individual is considered an object of the PERSON and HOUSEHOLD classes, identified by a row of characteristics explaining the individual.

The following is a complete list of input parameters fed into the model:

- Age and gender of the individual
- Employment status and occupation type of the individual
- License-holding status of the individual
- Total number of adults and the number of children between the ages of 1 and 5 and 6 and 17 in the household
- Income category and the number of vehicles in the household

The following is a complete list of the output from DEMOS for every year:

- Number of people alive and dead by age groups
- Number of women giving birth to a child by age group and total number of children in the household before the current birth
- Number of married people divorcing in the year
- Number of children leaving the household
- Number of singles or single parents getting married
- Number of people in the respective household types
- Number of people employed and not employed, by gender
- Number of people having and not having a license, by gender
- Number of people in respective income groups and number of vehicles
- Number in respective occupation types

The events that can occur during the evolution of an individual are represented by member functions or methods in the software. The programming methodology adopted to build each method and the probabilities used are explained in Sundararajan (2001). In the majority of these methods a Monte Carlo experiment is performed. A random number is drawn from a uniform distribution, and the random number is compared with the probability of the event. If the random number is less than the probability, then the event occurs; otherwise, the event does not occur. Also, the events are designed to occur in discrete times. So any event can occur to an individual during any time period based on his or her eligibility to the particular event. Additional details about the probabilities of occurrence for each event and the source data are reported in Sundararajan (2001).

14.2.3   Validation Method and Results

Validation involves testing the model's predictive capabilities by comparing model predictions and external data. In this section the forecasting ability of DEMOS is provided based on the comparison between the observed data in PSTP and the predicted results from DEMOS. Comparison data are from later waves, census data, and other external information. Usually, measures that check for forecasting accuracy are computed and inferences are drawn regarding the model's predictions. The main objective of this exercise is to check how synthetic evolution through DEMOS matches the real-world evolution. Validation also gives the opportunity to check if the external probabilities used from the U.S. Census, and other sources are applicable to the sample from the Puget Sound region that has been used here.

In DEMOS two different sets of probabilities are used. First, the directly observable parameters in PSTP data, like license holding, employment status, number of vehicles, income groups, household types, and occupation types, are estimated using the transition probabilities from waves 1 and 2 (this is in contrast to other simulators, such as MIDAS, that estimate probability models instead). The second set of probabilities is for the events that bring significant changes in the household attributes. These are birth, death, divorce, marriage, and children leaving the nest. The probabilities for these parameters have been estimated from U.S. Census, NCHS, and other panel surveys. So two different validation methods have been used.

In the first case, where sufficient data are available to validate at disaggregate levels, the predictions from DEMOS were compared with waves 2, 3, 4, and 5 of the PSTP data. Validation is made from the results obtained after simulating 1621 respondents 100 times. The following parameters are computed to test the forecasting accuracy, where error is the difference between observed and predicted values: For every year t,

- Absolute difference: $|\bar{P} - O|$

- Percent error of the predicted average: $100 * \left(\dfrac{\bar{P} - O}{O} \right)$

- Mean absolute percent error: $\dfrac{1}{n} \displaystyle\sum_{i=1}^{n} 100 * \left| \left(\dfrac{P_i - O}{O} \right) \right|$

- Mean square error: $\dfrac{\displaystyle\sum_{i=1}^{n} (P_i - O)^2}{n}$

- Theil's inequality coefficient (Theil, 1971): $U^2 = \dfrac{\displaystyle\sum_{i=1}^{n} (P_i - O)^2}{n * O^2}$

- Standard deviation: $\dfrac{\left(\displaystyle\sum_{i=1}^{n} P_i^2 - n * \bar{P}^2 \right)}{n - 1}$

where

$$\bar{P} = \frac{\displaystyle\sum_{i=1}^{n} P_i}{n}$$

$$P_i = \left(\sum_{j=1}^{k} p_j \right)_i$$

$$O = \sum_{m=1}^{l} o_m$$

k is the number of persons alive in each year; j = 1, 2, ..., k; p_j is the predicted value for person j from DEMOS; n is the number of simulations; i = 1, 2, ..., n; l is the number of persons in the PSTP sample (l = 1621 for waves 1, 2, 3, and 4; l = 1383 for wave 5); m = 1, 2, 3, ..., l; and o_m is the observed value for person m in the PSTP sample.

The mean absolute percent error (MAPE) has the observed value in the denominator. So when the observed values are large, MAPE gives a small value, even for a relatively large absolute difference. The mean square error (MSE) measures the average squared distance between the prediction and the observed values. It penalizes a large value more than a small value. Theil's inequality coefficient, U, is another measure that is obtained by dividing the MSE by the sum of the squares of the observed mean. It can be seen immediately that U = 0 occurs only when there is a perfect match between predictions and observations, and U = 1 results in predictions worse than no-change extrapolation. Also, U does not have any upper bound.

In the second case, where there are no disaggregate data to compare the predictions with observations, the predicted probabilities for the occurrence of the event are compared with the observed probabilities computed from the external data. In such cases, only the absolute difference and the percent error of the predicted average are calculated. Validation is made from the results obtained by simulating 8628 respondents 100 times to increase the sample size of predictions and allow the algorithms to produce "rare" events.

In the following discussion, a small selection from the validation results (that include birth, death, marriage, divorce, and child leaving nest) is provided. Death is based on the age and gender of the individual. Since the probabilities were estimated from the year 1998, DEMOS forecasts from the year 1998 were used to validate this event. Table 14.2 shows the validation results. The observed and predicted probabilities were converted to rates per 1000 persons, and the absolute error and percent prediction error were calculated. The predictions for male children less than or equal to 1 year age is underpredicted by almost 52%, while the predictions for female children are almost perfect. For male children between the ages of 2 and 14 the predictions are less than the observed number of deaths, while for female children they are more than the observed number of deaths. The model predictions for both males and females between 25 and 34 years old differ from observed rates by more than 20%.

The predictions fit the original rates for both males and females between 15 and 24 years old and above 35 years old. The prediction errors in these cohorts range from 0.02 to 6.44%. It can be observed that the predictions are more precise for age cohorts above 35 years old than for age cohorts less than 35 years old. In order to determine whether the proportion of different age groups is the same across both census and PSTP data, a marginal chi-square test was conducted. The chi-square statistic was 1404.14 for men and 1338.84 for women, while the critical value with $\alpha = 0.5$ and 7 degrees of freedom (df) is 14.06. Tables 14.3 and 14.4 show the chi-square calculations for men and women, respectively. The chi-square results provide more insights about the distribution of people in different age cohorts. There are about 90% less children in the age groups less than 1 and 2–14 than expected. Similarly, there are about 80% less men and women in the age group 25 to 34. Since there are not enough observations in the age group, the prediction error for these cohorts is very high.

The probability of a woman giving birth to a child is associated with the total number of children in the household and the age of the potential mother. Also, we assumed that a mother can give birth to up to three children. Table 14.5 provides the comparisons between the observed and the predicted data. It can be observed that the model predictions are always less than the observed rates. In fact, the prediction error for all the cases varies from –45 to –99%. This results in significant underpredictions for the number of births occurring every year. In order to determine whether there is a significant difference between the population distributions of PSTP and the U.S. Census, a chi-square test was conducted. The chi-square value was 520.49 with df = 3, while the critical value was 7.81. The chi-square calculations are provided in Table 14.6. So this again proves that there is a significant difference between the PSTP and the U.S. Census. It can be observed that there are about 67% more women in the age group of 40 to 49 in the PSTP data. Since the increase in age is characterized by decrease in the probability of birth for women, this significantly affects the number of births in the simulation year. Also, the number of births is a means through which new members are added into the simulation; this underprediction may have major effects on other results. Additionally, it is an indication of potential major differences between PSTP and the overall Seattle population.

TABLE 14.2 Validation of Death: Year 1998

Age Group	Observed Probability of Death (A)		Death Rate per 1000 People in Population (B = A*1000)		Predicted Probability of Death (C)		Death Rate per 1000 People in PSTP (D = B*1000)		Absolute Error for Death Rate (B – D)		Percent Error for Death Rate (D – B)/B	
	Male	Female	Male	Female	Male	Female	Male	Female	Male	Female	Male	Female
=1 year	0.0078	0.0065	7.8296	6.5365	0.0038	0.0067	3.7807	6.7340	4.0488	0.1975	−51.71%	3.02%
> 2–14	0.0003	0.0002	0.2741	0.2054	0.0002	0.0004	0.1711	0.3593	0.1030	0.1540	−37.59%	74.99%
15–24	0.0012	0.0004	1.1929	0.4352	0.0012	0.0004	1.1605	0.4072	0.0324	0.0280	−2.72%	−6.44%
25–34	0.0015	0.0007	1.5184	0.6820	0.0019	0.0004	1.8734	0.3898	0.3550	0.2922	23.38%	−42.84%
35–44	0.0026	0.0014	2.5865	1.4159	0.0030	0.0014	2.9679	1.4002	0.3814	0.0158	14.75%	−1.12%
45–54	0.0054	0.0031	5.4292	3.0968	0.0052	0.0029	5.2124	2.8852	0.2168	0.2116	−3.99%	−6.83%
55–64	0.0130	0.0079	12.9730	7.8886	0.0135	0.0074	13.4580	7.3736	0.4851	0.5149	3.74%	−6.53%
Above 65	0.0558	0.0476	55.8374	47.5760	0.0558	0.0467	55.8263	46.7002	0.0111	0.8759	−0.02%	−1.84%

TABLE 14.3 Chi-Square Calculations for Validation of Death: Men

Age Group	=1	2–14	15–24	25–34	35–44	45–54	55–64	Over 65	Row Marginal Total
				Men					
Census	2016205	27746795	19044000	19241000	22091000	16895000	10802000	14195000	132031000
PSTP	5.29	467.62	465.33	101.42	525.62	817.28	665.03	662.77	3710.36
Column marginal total	2016210.3	27747263	19044465	19241101.42	22091525.62	16895817.28	10802665.03	14195663	132034710.4
Column % of total	1.53	21.02	14.42	14.57	16.73	12.80	8.18	10.75	100.00
			Observed and Expected Counts (Men)						
Census: observed	2016205	27746795	19044000	19241000	22091000	16895000	10802000	14195000	132031000
Census: expected	2016153.6	27746483	19043930	19240560.72	22090904.82	16895342.48	10802361.46	14195264	132031000
PSTP: observed	5.29	467.62	465.33	101.42	525.62	817.28	665.03	662.77	3710.36
PSTP: expected	56.66	779.74	535.18	540.70	620.80	474.80	303.57	398.92	3710.36
Chi-square	46.57	124.94	9.12	356.90	14.59	247.05	430.40	174.52	1404.09

TABLE 14.4 Chi-Square Calculations for Validation of Death: Women

Age Group	=1	2–14	15–24	25–34	35–44	45–54	55–64	Over 65	Row Marginal Total
				Women					
Census	1925348	26471652	18176000	19503000	22407000	17680000	11864000	20191000	138218000
PSTP	5.94	445.25	491.21	128.27	685.64	925.4	763.53	726.55	4171.79
Column marginal total	1925353.9	26472097	18176491	19503128.27	22407685.64	17680925.4	11864763.53	20191727	138222171.8
Column % of total	1.39	19.15	13.15	14.11	16.21	12.79	8.58	14.61	100.00
			Observed and Expected Counts (Women)						
Census: observed	1925348	26471652	18176000	19503000	22407000	17680000	11864000	20191000	138218000
Census: expected	1925295.8	26471298	18175943	19502539.63	22407009.34	17680391.76	11864405.43	20191117	138218000
PSTP: observed	5.94	445.25	491.21	128.27	685.64	925.4	763.53	726.55	4171.79
PSTP: expected	58.11	798.97	548.60	588.64	676.30	533.64	358.10	609.42	4171.79
Chi-square	46.84	156.61	6.00	360.06	0.13	287.61	459.03	22.51	1338.79

TABLE 14.5 Validation of Birth Year 1998

No. of Children	Age Group	Observed Probability for Births (A)	For 1000 Persons (B = A × 1000)	Predicted Probability of Birth (C)	For 1000 Persons (D = C × 1000)	Absolute Difference	Systematic Difference
1st child	15–19	0.0395	39.5172	0.00058	0.58445	38.9328	−98.52%
	20–29	0.0462	46.1782	0.01429	14.28655	31.8916	−69.06%
	30–39	0.0159	15.9026	0.00563	5.63246	10.2701	−64.58%
	40–49	0.0009	0.8582	0.00030	0.29898	0.5593	−65.16%
2nd child	15–19	0.0092	9.2484	0.00222	2.22092	7.0275	−75.99%
	20–29	0.0395	39.4782	0.00890	8.89982	30.5784	−77.46%
	30–39	0.0213	21.2962	0.00629	6.28845	15.0077	−70.47%
	40–49	0.0011	1.1160	0.00027	0.26694	0.8490	−76.08%
3rd child	15–19	0.0016	1.6077	0.00088	0.87668	0.7310	−45.47%
	20–29	0.0182	18.1879	0.00650	6.49921	11.6887	−64.27%
	30–39	0.0133	13.2890	0.00310	3.09899	10.1900	−76.68%
	40–49	0.0008	0.8449	0.00028	0.27762	0.5673	−67.14%

TABLE 14.6 Chi-Square Calculations for Validation of Birth

Group	15–19	20–29	30–39	40–49	Row Marginal Total
Census	9595000	18015000	21532000	20648000	69790000
PSTP	342.2	170.79	442.08	936.53	1891.6
Column marginal total	9595342.2	18015170.8	21532442.1	20648936.5	69791891.6
Column % of total	13.75	25.81	30.85	29.59	100.00
Census: observed	9595000	18015000	21532000	20648000	69790000
Census: expected	9595082.1	18014682.5	21531858.5	20648376.9	69790000
PSTP: observed	342.2	170.79	442.08	936.53	1891.6
PSTP: expected	260.07	488.27	583.60	559.66	1891.60
Chi-square	25.94	206.44	34.32	253.79	520.49

TABLE 14.7 Validation of Marriage

Age Groups	Observed Probability	Predicted Probability	Absolute Difference	Percent Error
		Year 1990		
20–24	0.1892	0.1703	0.0189	−9.98
25–34	0.3718	0.3734	0.0016	0.44
35–44	0.1240	0.1262	0.0022	1.74
45–54	0.0283	0.0318	0.0035	12.30
55–64	0.0029	0.0029	0.0000	−1.22
		Year 1991		
20–24	0.1892	0.1569	0.0323	−17.07
25–34	0.3718	0.3827	0.0109	2.94
35–44	0.1240	0.1253	0.0013	1.04
45–54	0.0283	0.0293	0.0010	3.63
55–64	0.0029	0.0022	0.0007	−24.74
		Year 1992		
20–24	0.1892	0.1964	0.0072	3.83
25–34	0.3718	0.3734	0.0016	0.42
35–44	0.1240	0.1225	0.0015	−1.21
45–54	0.0283	0.0286	0.0003	0.90
55–64	0.0029	0.0022	0.0007	−24.21
		Year 1993		
20–24	0.1892	0.2001	0.0109	1.09
25–34	0.3718	0.3784	0.0066	0.66
35–44	0.1240	0.1247	0.0007	0.07
45–54	0.0283	0.0269	0.0014	−0.14
55–64	0.0029	0.0036	0.0007	0.07
		Year 1994		
20–24	0.1892	0.1902	0.0010	0.54
25–34	0.3718	0.3741	0.0023	0.63
35–44	0.1240	0.1212	0.0028	−2.24
45–54	0.0283	0.0275	0.0008	−2.95
55–64	0.0029	0.0030	0.0001	4.63

In the simulation, adult members living in single-person and single-parent households are eligible for marriage. Again, the marriage depends on the age of the individual, and any individual between the ages of 20 and 64 is eligible for marriage. Although the probabilities for marriage were computed from the year 1998, it was decided to validate for a 10-year period, 1990 to 1999. Tables 14.7 and 14.8 show the marital rate comparisons between the observed and the predicted probabilities. For individuals between the ages of 20 and 24 the percent error ranges from 0.01 to 17%. Similarly, for other age groups the prediction errors are very much below 20%, which leads to the conclusion that the forecasts are stable, except for persons between the ages of 55 and 64. In 1998, the prediction error ranges from 0.01 to 4.20% for individuals between the ages of 20 and 54. For the individuals between the ages of 55 and 64 the prediction error is about 28%.

The employment status at time t depends on the gender, age, and employment status at t − 1. Tables 14.9 and 14.10 show the results and comparisons.

In 1990, all the predictions for both men and women are fairly accurate, which is characterized by low values of the MAPE and U. In 1991, employed men and women are underpredicted, and unemployed men and women are overpredicted. For the employed men and women the MAPEs are around 4 and 8%, respectively. The value of U for both of these cases is less than 0.1. However, for unemployed men

TABLE 14.8 Validation of Marriage

Age Groups	Observed Probability	Predicted Probability	Absolute Difference	Percent Error
		Year 1995		
20–24	0.1892	0.1962	0.0070	3.68
25–34	0.3718	0.3605	0.0113	−3.05
35–44	0.1240	0.1228	0.0012	−0.99
45–54	0.0283	0.0300	0.0017	6.04
55–64	0.0029	0.0024	0.0005	−16.48
		Year 1996		
20–24	0.1892	0.1927	0.0035	0.35
25–34	0.3718	0.3779	0.0061	0.61
35–44	0.1240	0.1212	0.0028	−0.28
45–54	0.0283	0.0294	0.0011	0.11
55–64	0.0029	0.0034	0.0005	0.05
		Year 1997		
20–24	0.1892	0.1922	0.0030	1.58
25–34	0.3718	0.3693	0.0025	−0.68
35–44	0.1240	0.1195	0.0045	−3.59
45–54	0.0283	0.0286	0.0003	0.98
55–64	0.0029	0.0024	0.0005	−17.44
		Year 1998		
20–24	0.1892	0.1892	0.0000	0.01
25–34	0.3718	0.3604	0.0114	−3.06
35–44	0.1240	0.1216	0.0024	−1.95
45–54	0.0283	0.0295	0.0012	4.20
55–64	0.0029	0.0037	0.0008	27.53
		Year 1999		
20–24	0.1892	0.1894	0.0002	0.10
25–34	0.3718	0.3496	0.0222	−5.98
35–44	0.1240	0.1308	0.0068	5.50
45–54	0.0283	0.0271	0.0012	−4.22
55–64	0.0029	0.0036	0.0007	24.41

TABLE 14.9 Validation of Employment Status for Men

Year	1990 Yes	1990 No	1991 Yes	1991 No	1992 Yes	1992 No	1993 Yes	1993 No
Observed	552	196	575	173	523	224	435	194
Predicted	559.31	197.37	551.77	206.58	542.71	212.59	534.57	212.59
Absolute difference	7.31	1.37	23.23	33.58	19.71	11.41	99.57	18.59
Percent error	1.32	0.70	−4.04	19.41	3.77	−5.09	22.89	9.58
MAPE	1.53	3.31	4.04	19.41	3.77	5.75	22.89	11.24
MSE	101.13	65.83	617.71	1222.78	475.21	233.05	10017.90	578.89
Theil's U	1.82e-02	0.0414	0.0432	0.2021	0.0417	0.0682	0.2301	0.1240
Standard deviation	6.9408	8.0374	8.8806	9.8043	9.3596	10.1932	10.2368	10.6724

TABLE 14.10 Validation of Employment Status for Women

Year	1990 Yes	1990 No	1991 Yes	1991 No	1992 Yes	1992 No	1993 Yes	1993 No
Observed	501	372	553	320	470	400	393	356
Predicted	503	368.92	507.95	362.15	509.36	356.24	508.19	352.53
Absolute difference	2.00	3.08	45.05	42.15	39.36	43.76	115.19	3.47
Percent error	0.40	−0.83	−8.15	13.17	8.37	−10.94	29.31	2.9185
MAPE	1.49	2.09	8.15	13.17	8.37	10.94	29.31	2.92
MSE	88.22	92.84	2133.77	1896.47	1681.62	2074.04	13401.90	181.69
Theil's U	0.0187	0.0259	0.0835	0.1361	0.0873	0.1139	0.2946	0.0379
Standard deviation	9.2234	9.1758	10.2626	11.0026	11.5649	12.6771	11.5982	13.0906

and women the prediction errors are 19 and 13%, respectively. In 1992, the prediction error for all cases ranges between 3.77 and 10.9%, with relatively low values for U. Even if the MAPEs for all cases range from 0.97 to 29%, the absolute differences are higher. In 1993, the absolute differences for employed men and women are around 99.6 and 115.19, respectively. However, the unemployed women have a prediction error of 2.9% and a U value of 0.03, indicating fairly accurate prediction. In 1993, unemployed men and women are predicted reasonably well.

Comparing across years, the number employed increases from 1990 to 1991 and then decreases in 1992. This trend is common for both men and women. The model predictions for the first year of simulation, i.e., 1990, are more accurate than the other three years, as expected.

The number of vehicles in the household depends on the income group of the household and the number of vehicles in the previous time point. Table 14.11 provides validation results for zero-car households and one-car households. All other groups are similar. In 1990, the MAPE is between 2.33 and 10.88%. The values of U are less than 0.1, except for the households with more than four vehicles. For households that have more than four vehicles, U = 0.13. In 1991, the households with more than four vehicles are underestimated by about 22%. The MAPEs for no-vehicle and single-vehicle households are 16 and 10%, respectively. In 1992, the zero-vehicle households are underestimated by 26%. For all other households the MAPE ranges between 3.7 and 11.72%. The households that have two or three vehicles are predicted fairly accurately, compared to other households. The MSE for three-vehicle households is 2341.97, and in this case, the absolute difference is about 43.5. In the same year it was observed that about 24 respondents failed to provide information on the number of vehicles in the household. This might have been one of the contributing factors to the discrepancies. Finally, in 1993, the predictions are not as good as they are in the other years. The MAPE for households with more than four vehicles is 69.58%, and the inequality coefficient is 0.712. This shows that the root mean square (RMS) prediction error is 71% of the RMS error obtained by no-change extrapolation. The single-vehicle households are still predicted well, with U = 0.078. The MAPEs for no-vehicle, two-vehicle, and three-vehicle households range from 11.5 to 25%. The number of vehicles is correlated to the income group of the household. In the validation of income groups for the year 1993, it was observed that high-income groups had a high MAPE. Similarly, the households having more than four vehicles also had a high MAPE. This might suggest that high-income households have a tendency to drop out of the panel survey as the panel progresses. However, this cannot be verified because of the lack of sufficient information.

Occupations have been grouped into five different types: professionals, managerial, secretarial, sales, and other. Table 14.12 shows a selection of the validation results. As observed in all other variables, the predictions in the first year match the real-world results almost perfectly. This is evident from the MAPE,

TABLE 14.11 Validation of Number of Vehicles

	Number of Vehicles	1990	1991	1992	1993
0	Observed	26	25	31	26
	Predicted	25.67	24.5	22.86	20.58
	Absolute difference	0.33	0.5	8.14	5.42
	Percent error	−1.27	−2.00	−26.26	−20.85
	MAPE	10.88	16.00	27.81	25.15
	MSE	12.37	25.60	94.60	58.24
	Theil's U	0.0135	0.2024	0.3138	0.2935
	Standard deviation	3.5192	5.0602	5.3504	5.3996
1	Observed	286	253	279	234
	Predicted	291.41	277.71	257.87	240.14
	Absolute difference	5.41	24.71	21.13	6.14
	Percent error	1.89	9.77	−7.57	2.62
	MAPE	3.77	10.04	8.44	6.49
	MSE	1.72	271.39	751.71	333.76
	Theil's U	0.0046	0.1167	0.0983	0.0781
	Standard deviation	12.0228	16.2308	17.5589	17.2931

TABLE 14.12 Validation of Occupation Type A

Occupation Type		1990	1991	1992	1993
Professional	Observed	352	357	358	322
	Predicted	355.9	393.35	417.59	432.23
	Absolute difference	3.9	36.35	59.59	110.23
	Percent error	1.11	10.18	16.65	34.23
	MAPE	2.74	10.18	16.65	34.85
	MSE	153.02	1515.07	3764.23	12852.00
	Theil's U	0.0351	0.1090	0.1714	0.3521
	Standard deviation	11.7984	13.9894	14.6771	16.0950
Manager	Observed	211	102	137	91
	Predicted	211.12	220.28	220.36	217.71
	Absolute difference	0.12	118.28	83.36	126.71
	Percent error	0.06	115.96	60.85	139.24
	MAPE	4.02	115.96	60.85	139.24
	MSE	111.32	14175.90	7134.76	16249.20
	Theil's U	0.0500	1.1673	0.6166	1.4008
	Standard deviation	10.6033	13.6966	13.7021	13.9916

which ranges from 2.7 to 7.8%, and the value of U is very small (less than 0.1 in all cases). The managerial, secretarial, and sales occupations were accurate because the absolute difference in all three occupation types was less than 1. There is an increase in MAPE in the second year for all the occupation types. The managerial occupation type is overestimated by 116%, and the MSE is 14175.9. This is because the number of persons employed in the managerial profession decreased from 211 in 1990 to 102 in 1991. The model does not capture this drastic decline. The value of inequality coefficient for the managerial occupation type is greater than 1. The MAPEs for all other occupation types range between 10.2 and 22%. In 1992, the number of managers is still overpredicted by 61%. For all other occupation types the MAPEs range between 12 and 21%. Finally, in 1993, the managers are overestimated by about 139%. The secretaries, sales, and other types are still predicted well because the MAPE ranges from 7 to 12%. Individuals employed in the professional field have a U value of 0.35.

In the following discussion, validation results for number of individuals alive in different age cohorts are provided. Every individual has been classified into one of eight age groups. Since the PSTP does not provide information about children below 15 years old, it is not possible to validate this particular case. Table 14.13 shows a validation example for men. In the first year, the predictions are fairly accurate, as the MAPEs for all cases range between 0.05 and 4.27%. The age group 15 to 24 has an absolute error of only 0.01. The value of Theil's U for all the age groups is less than 0.06, indicating that predicted values very closely match the observed values. For women, the trend remains the same for all age groups, having fairly low values of MAPE and U values of less than 0.05. For young women between 15 and 24 years of age the absolute error is 0.98.

When compared with the first year, the predictions in 1991 have higher errors. The young men between the ages of 15 and 24 are still predicted well, with an MAPE of 1.12% and absolute difference of only 0.2. The middle-aged men between 45 and 64 years old are also well predicted as shown by the very low values of inequality coefficient. Men above 65 years old are not predicted well when compared to other age groups. The value of U for this category is 0.145. Middle-aged women between 35 and 64 years old have MAPEs ranging from 1.5 to 7.7%. The other three age groups have MAPEs around 17%. Young women aged between 15 and 24 years have a U value of 0.21, indicating that the RMS of the prediction is 21% of the RMS error of no-change extrapolation.

In 1992, men aged between 25 and 34 and older than 65 years old have U values of 0.226 and 0.213, respectively. The MSE for men above 65 years old was also high, showing that predictions are away from the observed values. The other age groups were predicted with reasonable accuracy, as the MAPE is less

TABLE 14.13 Validation of Number of Men between Ages 15 and 24

	Age	1990	1991	1992	1993
15–24	Observed	18.99	17.8	16.84	12.83
	Predicted	19	18	16	8
	Absolute difference	0.01	0.2	0.84	4.83
	Percent error	0.05	1.12	−4.99	−37.65
	MAPE	0.05	1.67	6.88	60.38
	MSE	0.01	0.34	1.78	24.81
	Theil's U	0.0053	0.0324	0.0834	0.6226
	Standard deviation	0.1000	0.5505	1.0418	1.2231
25–34	Observed	101.27	87.03	71.06	57.58
	Predicted	100	74	58	35
	Absolute difference	1.27	13.03	13.06	22.58
	Percent error	−1.25	−14.97	−18.38	−39.22
	MAPE	1.33	17.61	22.52	64.51
	MSE	2.83	171.67	172.44	513.24
	Theil's U	0.0168	0.1771	0.2264	0.6473
	Standard deviation	1.1088	1.3814	1.3767	1.8487

than 11%. Young men between 15 and 24 years old had the least absolute difference. The same trend also applies to women because women aged between 25 and 34 and older than 65 years are not predicted well. The other age groups had MAPEs ranging from 5 to 7%.

In the year 1993 there are considerable differences between the observed and the predicted values for men between the ages of 15 and 44 years. The men aged between 15 and 24 years and 25 and 34 years have MAPEs greater than 60%, which is again proved by high values of Theil's inequality coefficient. It can be observed that older men are still predicted well, in spite of reduction of the sample size in 1993. Similarly, young women aged between 25 and 34 years are underpredicted, with U = 0.5715. One of the reasons for the discrepancy might be due to the assumption that 95% of the individuals reaching age 25 leave the household, resulting in fewer individuals greater than 25 years in the sample. Women above 45 years of age are predicted well, with reasonable accuracy as the MAPE ranges between 4.5 and 12.4%.

During the first three years, men between 15 and 24 and 45 and 64 years of age are well predicted compared to other age groups. Comparing across years, young women between 15 and 24 years of age are the most underpredicted. Similarly, women between 25 and 34 years of age are underestimated in all four years. Women between 25 and 64 years of age are very well predicted in all four years.

14.2.4   DEMOS Simulation Forecasts

In order to demonstrate the long-term forecasting capability of DEMOS, the model was executed for 20 years with 2157 respondents (including children) for 100 simulations. Figure 14.4 shows the PSTP population evolution for 20 years. It declines because it departs from a somewhat older initial sample and does not incorporate immigration. This is also confirmed by Figure 14.5 (births and deaths). The amount of detail we simulate is also illustrated by Figure 14.6, which shows the relative share of household types in this synthetic population.

14.2.5   DEMOS Information and Communication Technology Market Penetration Forecasts

Many recent articles describe the importance of information and (tele)communication technology for transportation (Golob, 2000; Mokhtarian, 1990, 1997, 2000; among others). Using PSTP data, Viswanathan et al. (2000, 2001) and Viswanathan and Goulias (2001) have estimated probability models of ICT ownership and use. The DEMOS-generated demographic information can be used as input to these probability models of market penetration of ICT and use. A new set of these models was estimated (Sundararajan, 2001) and then embedded into DEMOS. Then the entire simulation model ran, including

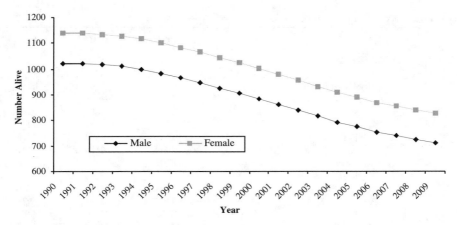

FIGURE 14.4   PSTP population in next 20 years.

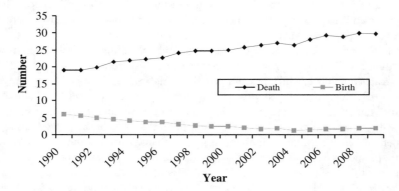

FIGURE 14.5   Deaths and births.

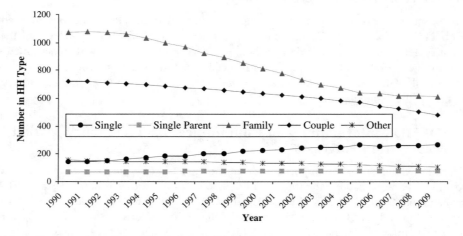

FIGURE 14.6   Number of people in different household (HH) types.

the children in the simulation. This resulted in a total of 2157 respondents. The average probabilities for computer usage at work and home, and Internet usage at work and home were calculated for both genders and different age groups. Also, the average probability for computer and Internet usage at home for different age groups was calculated. Finally, the average probabilities for the ownership of different mobile communication devices were estimated. Since wave 7 of the PSTP data had the information on ICT usage, it was decided that the model predictions would be compared to the actual data for the year 1997. To this end, the absolute error and percent error were computed for all average probabilities.

When comparing men and women, the model predicts that men use the computer and Internet more than women. This is consistent with the observed data, in which men were found to use the computer and Internet more than women both at work and home. When the predicted probabilities for computer usage for men at work and home were compared to the observed proportion of computer usage, it was found that the model underpredicts by about 14 and 20%, respectively. Also, the Internet usage for men from both places is underpredicted by DEMOS. The trend is similar for women, in which the predicted probabilities were less than the observed probabilities. Overall, for both men and women the percent errors are between 8 and 26%. Table 14.14 shows the comparisons between predicted and observed data.

DEMOS also calculates the average probability of computer and Internet usage at home for people in different age groups. The individuals were divided into four categories based on their age: less than 25, between 25 and 44, between 45 and 64, and greater than 65. Table 14.15 summarizes the results and also provides the comparisons between the predicted and observed data. The computer and Internet usage for individuals less than 25 years old is predicted well by DEMOS because the percentage prediction errors are around 1 and 5%, respectively. However, for individuals between ages 25 and 44 computer and Internet usage is underpredicted by about 22%. Similarly, the model underpredicts computer and Internet usage for individuals between 45 and 64 years old by about 13%. Finally, the model overestimates ICT usage for persons older than 65 years age.

The market penetration models (binary probit models) for computer usage show that it is highest among individuals between 25 and 44 years old and it decreases as age increases. Also, for young people below 25 years of age the probability of computer usage is less than those between 25 and 44. However, DEMOS predictions show that young people below 25 years tend to use the computer more than people between 25 and 44 years of age. This is due to the fact that the model underpredicts computer usage for people between 25 and 44 years of age by about 22%. DEMOS predictions were consistent with the trend

TABLE 14.14 Computer and Internet Usage at Home and Work between Men and Women

	Male				Female			
	Computer		Internet		Computer		Internet	
1997	Work	Home	Work	Home	Work	Home	Work	Home
Observed probability	0.5913	0.5946	0.4018	0.3944	0.5174	0.5509	0.2974	0.3047
Predicted probability	0.5107	0.4795	0.3400	0.2910	0.4381	0.4557	0.2711	0.2426
Absolute difference	0.0805	0.1151	0.0618	0.1034	0.0793	0.0952	0.0262	0.0621
Percent error	−13.62%	−19.36%	−15.37%	−26.22%	−15.33%	−17.28%	−8.83%	−20.37%

TABLE 14.15 Computer and Internet Usage at Home and Work across Different Ages

	Across Ages							
	Computer Usage at Home				Internet Usage at Home			
1997	< 25	25–44	45–64	= 65	<25	25–44	45–64	= 65
Observed probability	0.5509	0.6401	0.5277	0.3183	0.3047	0.4175	0.3230	0.1071
Predicted probability	0.5575	0.5006	0.4589	0.3754	0.2888	0.3230	0.2794	0.1475
Absolute difference	0.0065	0.1395	0.069	0.0571	0.0159	0.0945	0.0436	0.0403
Percent error	1.19%	−21.79%	−13.04%	17.94%	−5.22%	−22.63%	−13.51%	37.65%

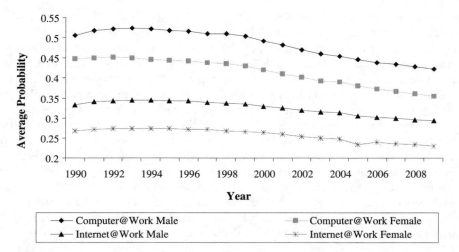

FIGURE 14.7 Computer and Internet usage at work.

for the other two age groups. The binary probit models for Internet usage at home show that middle-age people between 25 and 34 and 35 and 44 years old use the Internet more than other age groups. Internet usage is lower for young people than for middle-aged people. DEMOS predictions are consistent with the observed trend that Internet usage is high for middle-aged people and decreases as age increases. Using the sociodemographics predicted by DEMOS and the binary probit models developed for ICT usage, it is possible to forecast computer and Internet usage in the future. DEMOS was run for 100 simulations with 2157 respondents, and the ICT usage in the future is predicted for 20 years and provided in Figures 14.7 and 14.8.

Figure 14.7 shows computer and Internet usage at work from 1990 to 2009. Computer usage and Internet usage for both men and women seem to increase until 1994, and then follow a downward trend. There is a slight drop in both computer and Internet usage after the year 2000. This may be contradictory to reality, but the model predictions seem to be consistent because the sample is aging over time. By the year 2008 the probability of ownership or usage decreases significantly because the sample is still aging and the number of people in the sample is decreasing because of death, children moving out, etc. Although new members are added into the simulation in the form of birth and marriage, the amount of replacement is less than the amount leaving the sample.

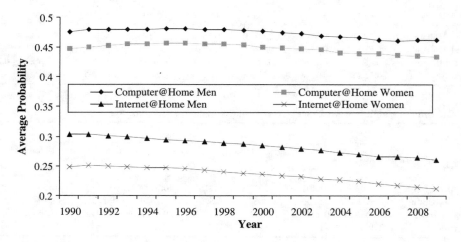

FIGURE 14.8 Computer and Internet usage at home.

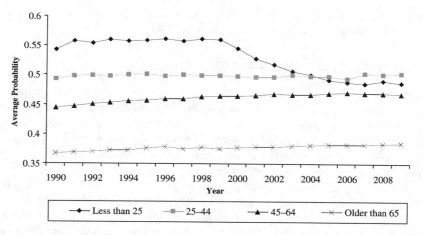

FIGURE 14.9 Computer usage at home across ages.

Figure 14.8 shows computer and Internet usage at home for both men and women from 1990 to 2009. The computer usage for both sexes seems to be relatively flat until 2002. After 2002 there seems to be a slight reduction in usage. However, the slope of reduction seems to be flatter than that for computer usage at work. Overall, there does not seem to be a big reduction in computer usage at home compared to usage at work. However, Internet usage at home for men and women decreases consistently from 1989. Again, the reason for the reduction in usage can be attributed to the aging of the population and the decreasing sample size over time. It can also be seen that men use the technologies more than women.

Figure 14.9 shows computer usage at home across different age groups from 1990 to 2009. As mentioned earlier, DEMOS predicts that young persons below 25 years of age use the computer more than middle-aged people 25 to 44 years of age. There seems to be a big drop in computer usage for people below 25 years old after 1999. The probability of usage then decreases considerably, by about 10% from 1990 to 2009, and becomes less than the probability of computer usage of 25 to 44 year old people. Computer usage for other age groups seems to increase slowly as time progresses. The increase in the probability of computer usage is clearly visible for the age group 45 to 64.

Figure 14.10 shows Internet usage across different age groups. It seems to be higher for middle-aged people between 25 and 44 years of age. This is consistent with the observed data. Internet usage for young people below 25 years of age is very similar to people between 45 and 64 years of age. In fact, after 2002

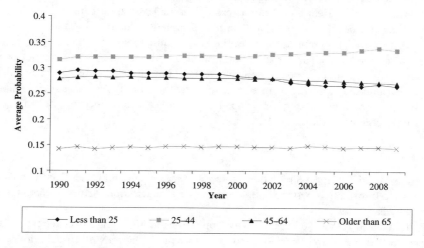

FIGURE 14.10 Internet usage at home across ages.

Internet usage for young individuals seems to decrease below the level of those 45 to 64 years old. For persons above 65 years Internet usage remains almost constant.

14.3 Summary and Conclusions

The DEMOS 2002 experiments presented in this chapter are motivated by the fact that no external demographic and socioeconomic forecasts are provided at levels that meet the data requirements of discrete choice models currently being developed in transportation planning (Goulias, 1991). This research combines the unique concepts of microsimulation and object-oriented programming to develop a dynamic, complex, realistic, and stochastic microsimulator that provides a quantitative description of the processes that determine population structure.

When object-oriented framework is used in software development, entities such as persons, households, highways, firms, intersections, and any other artifact can all be considered as objects that interact and evolve over time. Object attributes and associated functions are used to depict interactions among objects employing design and programming techniques that are specifically developed to handle complex problems such as this large-scale microsimulation.

Using data from the Puget Sound Transportation Panel survey, a demonstration of the predictive accuracy of the model system developed was also presented. This chapter shows one way for incorporating the effects of life cycle, lifestyle, and household structure on travel behavior and one way to incorporate the influence of telecommunication technologies in transportation. In addition, DEMOS effectively handles the problem of missing segments by synthetically simulating the children based on the information provided by PSTP.

The model system developed during this research is only the first step toward building a complete demographic microsimulator. The model system's predictions are reasonable for short-term forecasts. However, long-term forecasts require that a few changes be made to the model. For example, the sample size decreases every year as time progresses, and this decrease is very significant after 10 years of simulation. This is primarily due to the fact that many individuals leave the sample because of children leaving their parent's household or death. The number of people entering the simulation in the form of birth or marriage is not sufficient. In addition, by introducing regional immigration into the system every year, new households can be generated in the system. The characteristics of these new households can be generated in a probabilistic way based on in-migrant data that can be recruited in PSTP. In this way, the possibly different behavior of the new migrants can be accounted for. Regional out-migration is another aspect that requires careful scrutiny because out-migrants may be significantly different than in-migrants and long-time residents. Closely related to the concept of migration is residential relocation. It is possible, for a variety of reasons, that the household might move out of its present location. Combining migration and household relocation will result in forecasts that are more realistic than the current model (for an example using U.S. Census data, see Chung (1997)).

The current version of DEMOS does not compute sample weights (as in Ma and Goulias, 1996) to make the synthetic population representative of the Puget Sound region. So the effects of panel attrition (systematic refusal to continue participating in the survey) and refreshment (systematic recruitment to counter attrition and account for population changes) are not modeled. This can be rectified by computing sample weights and applying them every year. After these limitations have been rectified, models of activity and travel pattern indicators developed by Ma (1997) can be incorporated in the microsimulator to perform regional forecasting.

References

Chung, J., Transportation Impact Simulation of Access Management and Other Land Use Policies Using Geographic Information Systems, unpublished Ph.D. dissertation, Pennsylvania State University, University Park, 1997.

Golob, T.F., TRAVELBEHAVIOR.COM: Activity Approaches to Modeling Effects of Information Technology on Personal Travel Behavior, resource paper presented at Ninth International Association for Travel Behavior Research (IATBR), Gold Coast, Australia, 2000.

Goulias, K.G., Long-Term Forecasting with Dynamic Microsimulation, unpublished Ph.D. dissertation, University of California, Davis, 1991.

Goulias, K.G., Activity Based Travel Forecasting: What Are Some Issues? paper presented at Activity Based Travel Forecasting Conference Proceedings, Travel Model Improvement Program (TMIP), 1997 (http://www.bts.gov/tmip/papers/tmip/abtf/goulias.htm).

Hain, W. and Helberger, C., Longitudinal microsimulation of life income, in *Microanalytic Simulation Models to Support Social and Financial Policy*, Orcutt, G.H., Merz, J., and Quinke, H., Eds., North Holland, Amersterdam, 1986, pp. 251–266.

Kazimi, C., A Microsimulation Model for Evaluating the Environmental Impact of Alternative-Fuel Vehicles, unpublished Ph.D. dissertation, University of California, Irvine, 1995.

Ma, J., Activity Based and Microsimulated Travel Forecasting System: A Pragmatic Synthetic Scheduling Approach, unpublished Ph.D. dissertation, Pennsylvania State University, University Park, 1997.

Ma, J. and Goulias, K.G., Systematic self-selection and sample weight creation in panel surveys: The Puget Sound Transportation Panel case, *Transp. Res. A*, 31, 365–377, 1996.

Mandava, G., Sociodemographic Microsimulator for Transportation Forecasting (DEMOS), unpublished M.S. thesis, Pennsylvania State University, University Park, 1999.

Mokhtarian, P.L., A Typology of Relationships between Telecommunications and Transportation, *Transp. Res. A*, 24, 231–242, 1990.

Mokhtarian, P.L., The transportation impacts of telecommuting: Recent empirical findings, in *Understanding Travel Behavior in an Era of Change*, Stopher, P. and Lee-Gosselin, M., Eds., Elsevier, Oxford, 1997, pp. 91–106.

Mokhtarian, P.L., Telecommunications and Travel, Transportation in the New Millennium, Transportation Research Board, Washington, D.C., 2000 (http://www4.nationalacademies.org/trb/homepage.nsf/web/millennium_papers).

Murakami, E. and Watterson, W.T., Developing a household travel panel survey for the Puget Sound region, *Transp. Res. Rec.*, 1285, 40–46, 1990.

Sundararajan, A., Design and Validation of a Sociodemographic Microsimulator (DEMOS 2000) for Travel Forecasting, unpublished M.S. thesis, Pennsylvania State University, University Park, 2001.

Theil, W., *Applied Econometric Forecasting*, North-Holland, Amersterdam, 1971, pp. 26–36.

Viswanathan, K. and Goulias, K.G., Travel Behavior Implications of Information and Communications Technology (ICT) in the Puget Sound Region, paper presented at 80th Annual Meeting of the Transportation Research Board, Washington, D.C., 2001.

Viswanathan, K., Goulias, K.G., and Jovanis, P.P., Use of Traveler Information in the Puget Sound Region: A Preliminary Multivariate Analysis, paper presented at 79th Annual Meeting of the Transportation Research Board, Washington, D.C., 2000.

Viswanathan, K., Goulias, K.G., and Kim, T., On the Relationship between Travel Behavior and Information and Communications Technology (ICT): What Do the Travel Diaries Show? paper presented at 7th International Conference on Urban Transport and Environment for 21st Century, Lemnos Island, Greece, 2001.

15

Assessing the Effects of Constrained and Unconstrained Policy Scenarios on Activity–Travel Patterns Using a Learning-Based Simulation System[1]

Theo Arentze
Eindhoven University of Technology

Frank Hofman
Ministry of Transportation, Public Works and Water Management

Henk van Mourik
Ministry of Transportation, Public Works and Water Management

Harry Timmermans
Eindhoven University of Technology

CONTENTS

---

[1] Paper presented at the WCTR Conference, Seoul 2001.

15.1 Introduction

Over the last decade, it has been realized that the reduction of car use and mobility requires a multitude of policy instruments. Capacity-oriented measures, transport demand management, land use planning initiatives, and institutional approaches in principle all influence people's activity and hence travel patterns. Unfortunately, such policies are not always formulated and implemented with this goal in mind. Moreover, some policies may turn out to be counterproductive because not all factors and implications were taken into consideration. For example, the policy of stimulating to leave the car at home for the work trip has often ended up increasing travel, as other members of the household had an opportunity to use the car for other activities. Likewise, the stimulation of teleworking may have adverse affects, as time is now used for becoming involved in social and recreational activities (Harvey et al., 2000). To reduce the risk of stimulating the wrong set of policies, a rather advanced tool that incorporates the mechanisms described above may be helpful.

It was against this background that the Dutch Ministry of Transportation, Public Works and Water Management commissioned European Insitute of Retailing and Services Studies (EIRASS)/Urban Planning Group to develop a prototype of a rule-based model of activity behavior. The system developed for this purpose received the acronym Albatross. The system, which has been described in detail elsewhere (Arentze et al., 2000), predicts which activities are conducted where, when, for how long, with whom, and the travel involved. Household interaction and scheduling are modeled. The predictive ability of the system for a sample of activity diaries, plus a validation set, turned out to be relatively good.

This chapter reports the results of an application of the system to a set of scenarios. These simulations were conducted primarily to assess the face validity of the system, but may also have some substantive significance. Four scenarios were developed: (1) decrease in two-adult households, (2) change of work start times, (3) increase in part-time workers, and (4) Friday afternoon off. Note that some of these policies are unconstrained and relate to demographic change, while other scenarios constrain activity scheduling options. All these scenarios are related to ongoing trends in The Netherlands.

These policies affect the conditions and properties of activity programs. The policies were translated into the relevant attributes and conditions of the Albatross system. The model was then used to simulate the impact of the policy scenarios on activity–travel patterns. The impact of the scenarios was assessed in terms of such performance indicators as total travel distance, travel distance by car, number of tours, trip–tour ratio, car driver–total distance ratio, car passenger–total distance ratio, and public transport–total distance ratio.

This chapter will report the main results of the effects of these policy scenarios on activity–travel patterns. It will first briefly summarize Albatross. Next, the data collection process will be summarized. This will be followed by a description of the various policy scenarios, and the outcomes of the application of the model system to each of these scenarios. Interpretations will be given. The chapter will be completed with a summary of the main findings.

15.2 Albatross

Albatross is the latest, most comprehensive, and only operational computational process model of transportation demand (Arentze et al., 2000). It can be considered a multiagent rule-based system that predicts activity patterns. Essential to the system is that choice heuristics are used to simulate behavior. The system consists of a series of agents that together handle the (consistency of the) data, the derivation of choice heuristics from activity diary data, the simulation or prediction of activity patterns, the assessment and reporting of model performance, the calculation of various system performance indicators, and the evaluation of alternative model scenarios.

A series of rules have been derived from empirical activity diary data to identify the mechanisms underlying the organization of activities in time and space. It is assumed that individuals and households will try to meet particular criteria, subject to personal, household, temporal, spatial, institutional, and space–time constraints. The simulation process involves the application of the choice rules, subject to this wide range of constraints.

Central to the system is the scheduling agent. It controls the scheduling processes in terms of a sequence of steps. In each step, this agent identifies the condition information required for making principal scheduling decisions, sends appropriate calls to agents for the required analyses, passes the obtained information to the rule-based system, and translates returned decisions into appropriate operations on the current schedule.

The simulation starts with an activity program. This defines the set of activities that need to be completed on any given day. This program consists of mandatory activities (e.g., work, school) and discretionary activities. The mandatory activities constitute the skeleton of the activity program. Scheduling then involves selecting the discretionary activities, adding these to the skeleton, and determining the schedule position and profile of each added discretionary activity. A series of decision tables, used to represent the choice heuristics, are consulted to simulate this process in a series of steps.

The first step simulates the selection of discretionary activities, travel party, and the duration of the discretionary activities. The selection decision is modeled as a yes or no decision for each of a predefined set of optional activities. If a particular activity is selected, the model system simulates whether another instance of the same activity should be added to the activity program. If this is not the case, the selection of possible other activities is further modeled. The order in which activities are evaluated is predefined based on assumptions about activity priorities. Temporal, household, and institutional constraints (facility opening times) are systematically considered to assess the feasibility of selection and duration decisions.

The next two steps involve the identification of the scheduling position of each selected activity. The set of possible positions is first reduced to positions that comply with a specific episode of the day (e.g., early morning, around noon, etc.) by selecting a start time interval. Next, trip chaining decisions simulate whether the activity is to be linked to a previous or future activity. The outcome of this step simulates whether the conduct of that activity involves a single-stop tour from home or a more complex pattern. The combination of start time and trip chaining decisions often determines the schedule position. If more options still prevail, the system selects the position that leaves maximum freedom of choice for subsequent scheduling steps. Temporal constraints, available locations, and normative travel times determine the feasibility of choice options in this step.

The next steps in the scheduling process involve the choice of transport mode and location. Mode decisions are made at the level of tours. Possible interactions between mode and location choices are explicitly taken into account by using location information as conditions of mode selection rules. The location of fixed activities is assumed given. For nonfixed activities, the system dynamically defines the location choice set while accounting for available locations, mode-specific travel times, opening times of available facilities, and the time window and minimum duration for the activity. The temporal constraints follow from the earlier duration, schedule position, start time, trip chaining, and mode decisions. A location decision involves the choice of a heuristic that uniquely identifies a location from the choice set in terms of a noninferior combination of attractiveness and travel time. The option "other" is included to cover alternative heuristics. If this option is chosen, the system selects a location at random from a drive time band, if the choice set includes more than one location in that band.

15.3 Data

The simulations reported in this chapter are based on an activity diary survey that was administered in two municipalities in the Rotterdam region in The Netherlands in 1997. These diaries were also used to derive the rules underlying the Albatross model system. The diary involved two consecutive days. Days were designated for households such that the sample was balanced across the days of the week. Respondents were invited to report the sequence of activities that they conducted during these days, and to detail the start and end times, the location where the activity took place, the transport mode (chain), the travel time per mode, and the travel party. A precoded scheme was used for activity reporting. Several modes of administration were used. Response rates ranged between 64 and 82%, dependent on mode of administration for those households that indicated during a prescreening that they would be willing to participate in the survey. This resulted in a total of 2198 household days that were used for the analysis. The diaries were cleaned using the special-purpose program Sylvia (Arentze et al., 1999).

15.4 Model Performance

The activity diary data described above were used to derive a decision table for each decision step in the scheduling model. To allow validity testing, the data set was split into two parts. The first part includes 75% of the cases and was used to induce the decision tables, whereas the other cases were used to validate the derived decision rules. A CHAID-based tree induction algorithm was used to derive the decision rules.

It goes beyond the scope of the present chapter to discuss the induced decision tables and model performance in any detail. Readers are referred to Arentze and Timmermans (2000) for such details. Suffice to say that, although performance varied by step in the scheduling model, the overall performance was good compared to other model types (Arentze et al., 2000).

15.5 The Scenario Agent

One of the agents in the Albatross system is the scenario builder agent, called Agnes. This agent allows users to change the attributes of the land use and transportation environment, household characteristics, and the schedule skeleton. In addition, users can change the composition of the sample or population. The agent can be activated in three modes: population composition, behavioral change, and impact assessment.

In the *population composition* mode, population scenarios can be implemented by selecting one or more segments and specifying a multiplication factor. The factor determines the number of times or, if the factor is smaller than 1, the probability that each case of that segment will enter the prediction cycle.

In the *behavioral change* mode, users can simulate the effects of policy measures that are not incorporated as explanatory variables in the model. In that case, policy effects can only be predicted if the user can change directly some or all of the behavioral exogenous variables. For example, the current system does not have an adjustable road pricing variable. Yet, one may assume that variable road pricing may induce some people to leave home early. Users can modify the skeleton of the input schedules. In the current version, change to the structure–response pattern is not possible, unless one changes the decision rule base outside the system.

Scenarios about such behavioral change can be built as follows. First, a three-dimensional cross-tabulation can be created to select the target segment. This three-dimensional functionality allows users to vary either anticipated behavioral change by zone and two sociodemographic variables simultaneously or three sociodemographic variables. A drop-down list allows the user to select any combination of activity program facets considered relevant. Next, users are requested to indicate (1) the degree of adaptation penetration (i.e., the percentage of the selected segment or each category that will adapt), and (2) the size and direction of behavioral change. The degree of adaptation penetration may be set for all cases or a target segment.

In case of continuous variables, users can specify the size of changes in terms of an exact value, a minimum, a maximum, a proportion, a constant change, or a range. For nominal variables, the user should specify percentage change for all categories minus 1, by specifying proportions in a matrix of given and new categories of the dimension. Furthermore, users can define a standard deviation dependent on the assumed heterogeneity under change. A truncated normal distribution is assumed to avoid predicted change in the wrong direction. Based on these specifications, Albatross uses Monte Carlo simulation to identify the specific respondents that will experience the change of interest.

In the *impact assessment* mode, users can change the exogenous variables of the system. The same principles apply. It should be noted that in addition to these individual scenarios, accumulative scenarios can be built.

In the following sections, we will report the development of the scenarios and the changes in activity–travel patterns that the system predicts should these scenarios be implemented. The formulation of these scenarios used the scenario-building agent.

15.6 Scenario 1: Decrease in Two-Adult Households

15.6.1 Scenario Definition

The reduction of household size has been a constant trend over the last decades in The Netherlands as well as other postindustrial countries. Demographic forecasts suggest that this trend will continue to be influential in the foreseeable future. To explore the impact of such demographic change on activity–travel patterns, this first scenario assumes that the number of two-adult households will be reduced by 10% and that a proportional increase in the number of single-adult households will occur.

This scenario was implemented by decreasing the number of two-adult-households and increasing the number of single-adult households in the sample. This was done by setting the multiplication factor in the scenario builder to 0.90 for the two-adult cases and 1.58 for the single-adult cases. Due to these settings, each two-adult case has a chance of 0.90 of being selected in the new sample and each single-adult case has a chance of 1 of being selected once and a chance of 0.58 of being selected twice in the new sample. The multiplier for the single-adult cases was determined such that the number of person days keeps constant. Thus, the scenario has no impact on population size, but only on the composition of the sample.

15.6.2 Results

The number of prediction runs is set to N = 20 for both the scenario and the null scenario. As it appears, this number results in sufficient power to be able to measure all effects of interest. Table 15.1 represents a selection of the most relevant indicators.

In the after case, the total distance traveled equals 83,054 km. This means a decrease in 1255 km (−1.5%) compared to the before case. The distance traveled by car is estimated by a weighted summation of kilometers traveled as car passenger and kilometers traveled as car driver, to prevent double counting of kilometers that are traveled by partners on the same trip (as car passenger and car driver). That is, a relative weight of 0.5 is used for car passenger, assuming that in 50% of the car passenger cases the same trip is already represented in the partner's schedule. Measured this way, the total distance traveled by car decreases approximately proportionally with the decrease in total travel distance (−1.3%). The shares of other modes in total kilometers traveled remain the same. The decrease in total travel distance is not caused by a decrease in the number of tours (the decrease is not significant). This implies that the effect is caused by a decrease in average distance traveled per tour. The increase in ratio between trips and tours indicates that activities are more often combined on the same tour.

Several other observations are relevant. First, although the total number of out-of-home activities has stayed constant, the distribution across activities has changed. An increase in frequencies of service activities (+3.8%) and "other" activities (+2.8%) compensates for a decrease in the frequency of bring and get activities (−4.0%).[2] The distribution of out-of-home activities across days of the week has stayed the same except for Tuesday. On Tuesday, the frequencies of nondaily shopping, social, and leisure activities have increased. The distribution across times of the day has undergone a change too. The share of activities starting before 10 A.M. has decreased, and the proportion of activities initiated during 4 to 6 P.M. has increased. The relative frequency of single-stop trips is smaller in the after case. This is consistent

---

[2] The "other" category includes medical visits, personal business and activities labeled as "other" in the questionnaire.

TABLE 15.1 Comparison of Means between Scenario 1 and the Null Scenario on a Selection of Indicators

Indicator	Mean	s	Δm	%Δm	t-value	df
Total travel distance	83054	1316	−1255[a]	−1.5107	−3.187	41
Car distance (driver + passenger)	60176	1083	−808[a]	−1.3434	−2.567	39
Number of tours	5306	75	−39	−0.7398	−1.990	33
Trips–tours ratio	2.420	0.007	0.0077[a]	0.3199	3.301	42
Car driver–total distance ratio	0.677	0.011	0.0032	0.4797	0.968	41
Car passenger–total distance ratio	0.095	0.007	−0.0041	−4.3711	−1.870	43
Public–total distance ratio	0.128	0.012	0.0011	0.8182	0.320	37
Slow–total distance ratio	0.100	0.004	−0.0002	−0.1678	−0.135	39

[a] Significant at the $\alpha = 0.05$ level in a two-sided test.

with the earlier observation that the trip–tour ratio has increased. The choice of transport mode shows an increase in car passenger share in out-of-home activities.

15.6.3 Interpretation of Results

The simulation results suggest that this scenario leads to a reduction of total travel distance caused by a decrease in the average distance traveled per tour. There are at least three possible explanations for this decrease. First, the choice of more local destinations may be related to the change in distribution across activity types, which is occurring simultaneously. However, given the nature of the redistribution — fewer bring and get activities and more service and "other" activities — this is not very likely to be the case. Second, the preference for nearest-location choice heuristics has increased or location choice sets have been narrowed down. This explanation is not very plausible either, considering the location selection decision tables (DTs). Third, individuals from single-adult households tend to live closer to the place where they work, because in choosing the residence location they do not have to take the workplace of a partner into account. Thus, according to this last explanation, the average tour length has decreased due to shorter commuting trips.

None of the above explanations take into account that other population attributes have changed simultaneously with the manipulation of household size. In the after case, older-age households, no-child households, and females are more strongly represented in the sample. The decrease in average trip length may be related to differences in behavior (i.e., lifestyles) of these groups. Such differences may be reflected in input schedule skeletons as well as the DTs. With respect to the latter, the age index, child index, and gender variable indeed recur in many decision rules. Given the fact that other sample characteristics co-vary with the manipulation, a critical question is whether the whole package of changes is realistic in the context of the scenario. If the scenario assumes that age index, child index, and gender co-vary with the decrease in household size as they do in the present sample, then the outcomes are still interpretable as consequences of the scenario. Nevertheless, the present method does not allow users to control the correlation structure.

15.7 Scenario 2: Change of Work Start Times

15.7.1 Scenario Definition

To simulate the effects of changing work start times, the second scenario assumes that 15% of the trips from home to work that currently take place during morning rush hour start earlier or later to avoid rush hour. This behavioral change is implemented as a change in the schedule skeleton using the scenario builder. Trips are considered to take place during morning rush hour if the work start time falls in the time interval of 7:00 to 9:30 A.M. In a random selection of 15% of these work activities, the work start time is set to 7:00 or 9.30 A.M., whichever manipulation implies the smallest change. End times are changed simultaneously in such a way that the duration of the work activity stays the same. The scenario further assumes that every sleep activity and every other work activity (e.g., the afternoon work episode) simultaneously shift on the timescale such that the intervals between the activities remain constant. A change in start and end times of an activity

TABLE 15.2 Comparison of Means between Scenario 2 and Null Scenario on a Selection of Indicators

	Mean	s	Δm	%Δm	t	df
Total travel distance	84545	1423	237	0.280	0.574	39
Car distance (driver + passenger)	61420	802	436	0.709	1.616	43
Number of tours	5359	39	14	0.258	1.015	43
Trips–tours ratio	2.409	0.006	−0.003	−0.128	−1.466	42
Car driver–total distance ratio	0.678	0.009	0.004	0.577	1.274	43
Car passenger–total distance ratio	0.097	0.007	−0.002	−1.572	−0.697	43
Public–total distance ratio	0.127	0.010	0.000	−0.345	−0.150	41
Slow–total distance ratio	0.098	0.003	−0.002	−1.990	−1.793	43

may give rise to conflicts with the timing of other activities in the schedule skeleton. Conflicts were consistently solved by means of moving activities on the timescale while keeping the duration of activities constant.

15.7.2 Results

The prediction of schedules under scenario conditions was repeated 20 times. Table 15.2 shows differences in means between after and before runs for a selection of travel demand indicators. There are no significant differences in terms of the total distance traveled, the total car distance traveled, or the other indicators. There are also no significant differences in distributions of relevant characteristics of schedules, tours, or activities. Thus, the model predicts that this scenario does not lead to a rescheduling of activities to an extent that has measurable impacts on output variables.

15.7.3 Interpretation of Results

The scenario involves a change in the timing of a small proportion (15%) of activities of the input schedule skeletons. The amount of time spent on the activities and the travel times do not change. The change may have an impact on at least three types of conditions that play a role in the DTs. First, it may lead to more time available in preferred time periods of the day for specific optional activities (e.g., more time in the afternoon for shopping). More time generally means higher probabilities of selecting the activity and choosing a long duration for the activity. Second, shifts in work start times may impact relationships with the partner's schedule. In particular, the characteristics of tours of the partner overlapping in time play a role in decision rules related to mode and location selection. Obviously, the extent to which tours of the partner overlap in time may increase or decrease if the work start time changes. Third, a changed start time of the work activities may impact opportunities for travel party and trip chaining choices. Theoretically, therefore, the scenario may lead to changes in activity schedules. However, the results indicate that if it does, the effects are not measurable at the aggregate level.

Finally, we should note that the system is not sensitive to all theoretically possible impacts of the scenario. A 15% reduction of work trips during the morning rush hour probably leads to a meaningful decrease in congestion. Faster travel times leave more room in schedules for other activities. The present system uses predicted travel times under free-floating conditions only and, therefore, is not sensitive to such changes. Furthermore, coupling constraints may lead to rescheduling of the activities of the partner. Such decision rules are only partly represented in the current model.

15.8 Scenario 3: Increase in Part-Time Workers

15.8.1 Scenario Definition

Shortening the work hours of full-time workers is a national policy in The Netherlands, aimed at reducing the unemployment rate and stimulating flexible work hours. The third scenario assumes a 10% reduction of full-time workers and a proportional increase in part-time workers.

TABLE 15.3 Multipliers Used to Change the Sample Composition According to Scenario 3

Household Type	Weekly Work Hours	No. of Cases	Multiplier
Single, 1 worker	17–32	44	1.5
Single, 1 worker	> 32	220	0.9
Double, 1 worker	17–32	36	2.78
Double, 1 worker	> 32	632	0.9
Double, 2 worker	17–32	156	1.32
Double, 2 worker	> 32	496	0.9

This scenario could be implemented by editing the schedule skeletons and the job status of individuals and households. However, schedule skeletons may change in more ways than just work hours. For example, fixed activities may be reallocated across household members. Therefore, the scenario was implemented by increasing the share of part-time workers in the current sample and decreasing the share of full-time workers. This was done such that the number of person days remains constant. Full-time and part-time workers were identified based on weekly work hours of individuals. More than 32 h of work in a week was identified as full-time and 17 to 32 h as part time.[3]

The change was applied to individuals belonging to a dual-earner household with the same probability as to those belonging to a single-earner household (although one may argue that the first group is more prone to change to a part-time contract than the second group). Table 15.3 shows the multipliers that were used to implement the scenario for the different groups.

15.8.2 Results

As in the previous cases, the impacts of the scenario were analyzed based on 20 prediction runs. Table 15.4 shows the results for a selection of travel demand indicators. This scenario leads to a reduction of total distance traveled of 3.0%. The percentage reduction of total distance traveled by car is of the same magnitude (i.e., 2.8%). The share of distance traveled by slow mode has increased (3.7%), while the shares for the other modes have remained the same. This suggests that the extra share of slow mode is drawn from no specific fast mode in particular.

The number of tours has remained the same, implying that the decrease in distance traveled is exclusively caused by a decrease in average distance by tour (as was the case in scenario 1). Furthermore, the absence of a change in the trip–tour ratio suggests that the average number of activities per tour has stayed the same. We conclude therefore that the reduction of travel demand is caused by a decrease in travel required per activity.

The analysis of activity patterns confirms and details this picture. Although the total number of out-of-home activities has remained the same, the distributions across activity type, day of week, and time

TABLE 15.4 Comparison of Means between Scenario 3 and Null Scenario on a Selection of Indicators

	Mean	s	Δm	%Δm	t	df
Total travel distance	81862	1682	−2446[a]	−2.988	−5.340	35
Car distance (driver + passenger)	59350	1012	−1634[a]	−2.753	−5.397	41
Number of tours	5334	41	−11	−0.210	−0.802	43
Trips–tours ratio	2.415	0.008	0.003	0.132	1.266	40
Car driver–total distance ratio	0.674	0.009	0.000	0.031	0.068	43
Car passenger–total distance ratio	0.102	0.011	0.003	2.939	1.017	33
Public–total distance ratio	0.120	0.013	−0.007	−5.843	−2.002	35
Slow–total distance ratio	0.104	0.004	0.004[a]	3.691	3.007	39

[a] Significant at the $\alpha = .05$ level in a two-sided test.

---

[3]For dual-earner households the same cut-off points of 17 and 32 hours work at the household level were used. As it appears, these cut-off points result in the desired size of the shift from full-time workers to part-time workers.

of day show changes. We see a decrease in the number of work activities and increases in both bring–get and voluntary work activities. In addition, we see a decrease in out-of-home leisure activities and an increase in service activities.

As for day of the week, there are more activities on the early days (Monday and Tuesday) and fewer on Saturdays in the after case. Finally, with respect to time of day, the shares of activities starting before 10 A.M. and after 6 P.M. decrease, and the share of activities starting between 10 and 12 A.M. increase. Furthermore, the composition of tours has undergone changes. The percentage of tours consisting of only one activity has decreased, whereas the share of two-activity tours has increased.[4] Finally, we mention that the share of car mode in tours has decreased and the percentage of slow modes has increased.

The scenario involved a redistribution of cases across part-time and full-time workers. The third section of the report allows us to assess the extent to which other dimensions of the sample composition have changed simultaneously. Redistributions have occurred on almost all other dimensions. Only for age group, child index, and gender was the size of the differences meaningful. With regard to age, a shift has occurred from the 25 to 45 years old group (−1.7%) to the older than 45 years old group (+2.4%). As for child index, the percentage of households without children has increased (+1.4%), while households with young children decreased (−2.0%). With regard to gender, the after case comprises fewer males (−3.3%) and more females (+2.4%).

15.8.3 Interpretation of Results

As in the case of scenario 1, the increase in part-time work leads to a reduction of total travel distance. The reduction is less than one would expect based on the reduction of work activities implied by the scenario. The decrease in work activities is compensated by an increase in other out-of-home mandatory activities (bring–get, voluntary work). There is a slight increase in the average number of activities conducted on tours, but this cannot explain the reduction in the total distance traveled. Instead, the reduction of travel demand is caused by a decrease in the average distance traveled per trip. This is probably related to the shift of activities from work to other mandatory activities. Locations of work tend to be more distant from the home location than destinations of other activities. Therefore, replacing work activities by other mandatory activities entails that more distant locations are replaced by destinations closer to home, resulting in less travel demand per trip. Although activities have a more local character in the after case, approximately the same proportion of the distance is traveled by car as before.

Besides the reduction of total travel distance, the scenario has impacts on the spread of activities across days of the week and times of the day. In the after case, there are more activities during the early days of the week (Monday and Tuesday) and fewer activities on Saturday. At the same time, a smaller proportion of activities is initiated during morning rush hour (as well as in the evening). These changes probably lead to a slightly more even distribution of traffic on a daily and weekly basis.

Just as in the case of scenario 1, it is not possible to attribute the impacts to the variable of interest alone. Changing the sample composition with respect to work status of households has led to changes on other dimensions correlated with work status. Specifically, the shares of households belonging to older age groups, households without children, and females have increased in the after case. The present analysis does not allow us to unravel the separate impacts of the work status variable and these covariates.

15.9 Scenario 4: Friday Afternoon Off

15.9.1 Scenario Definition

This last scenario assumes a 50% reduction of work hours on Friday. The scenario is implemented by editing the schedule skeleton (as in scenario 2) rather than changing the sample composition (as in scenarios 1 and 3). Only Friday schedule skeletons involving more than 360 min work time were selected.

---

[4]The decrease of single-stop tours has not led to a significant increase of the trip/tour ratio.

In each selected skeleton, the duration of each work episode was reduced 50% by changing the end time while keeping the start time constant. In most cases, this means canceling the afternoon work part. Shortening the workweek in this way is a realistic scenario in the Dutch context.

15.9.2 Results

As in the foregoing scenarios, predictions were repeated 20 times. Sample composition is the same in the before and after cases, because the scenario involves changes in schedule skeletons only. A selection of most relevant travel demand indicators is represented in Table 15.5. As it appears, there is no significant difference in total travel distance between the before and after cases. However, the distance traveled by car and the number of tours have increased 1.3 and 1.6%, respectively. The increase in car kilometers is accompanied by a decrease in the public transport share in the total distance traveled, whereas the share of slow mode has not changed. Finally, we note that the average number of trips per tour has remained the same.

The increase in number of tours is caused by an increase in flexible activities conducted on Friday (+9.5%). These concern shopping, service, and social activities. The number of out-of-home leisure activities remains constant, but the duration of these activities tends to be longer in the after case. Furthermore, we see a shift in the start time of activities. The share of activities initiated between 10 A.M. and 4 P.M. increases at the expense of the before 10 A.M. and after 6 P.M. periods of the day. A bigger proportion of activities is conducted with other members of the same household and on multiple-stop tours.

15.9.3 Interpretation of Results

In a substantial number of cases, part of the time that has become available because of the 50% work time reduction is used for conducting flexible activities such as shopping, service, and social activities. In other cases, the extra time leads to decisions to substitute short-duration leisure activities with longer-duration leisure activities. These effects could be expected considering the underlying activity selection and activity duration DTs. Available time in preferred time slots for the concerned activity is an important condition variable in both DTs (dependent on household and individual characteristics). The effect is not limited to activity selection and duration decisions. Start time decisions also tend to change. The observed shift in start time of activities can be interpreted as a shift toward the more preferred times of day for conducting activities such as shopping and social activities. In summary, all these effects can be understood as consequences of relaxing temporal constraints on activity schedules.

The increase in the number of out-of-home activities leads to an increase in the total number of trips and tours undertaken by the individuals. It is noteworthy that the total distance traveled does not increase substantially. The distance traveled by car does increase significantly, accompanied by a decrease in kilometers traveled by public transport. Probably, there is a substitution between car and public transport directly related to work activity duration (on Friday). As the DTs for mode choice suggest, the preference for public transport decreases and the preference for car increases when the duration of work activities

TABLE 15.5 Comparison of Means between Scenario 4 and Null Scenario on a Selection of Indicators

	Mean	s	Δm	%Δm	t	df
Total travel distance	84613	997	305	0.3602	0.887	43
Car distance (driver + passenger)	61765	1028	781[a]	1.2645	2.557	40
Number of tours	5429	45	84[a]	1.5537	5.811	43
Trips–tours ratio	2.414	0.007	0.002	0.0973	1.011	43
Car driver–total distance ratio	0.677	0.010	0.003	0.4020	0.847	43
Car passenger–total distance ratio	0.106	0.007	0.008[a]	7.2188	3.364	42
Public–total distance ratio	0.117	0.006	−0.010[a]	−8.8994	−4.339	41
Slow–total distance ratio	0.100	0.004	0.000	0.0040	0.003	39

[a] Significant at the $\alpha = 0.05$ level in a two-sided test.

(strongly) decreases. Another effect that might be influential is that shortening of work hours leads to an increase in the time the car is available at home in cases where the car is used for work. Possibly, in the after case, the partners of workers may make more use of the car for their activities than they did before. This may explain at least partly the predicted increase in car use.

Finally, we should point out a potential source of bias in the predictions. The present model predicts individuals' decisions to select flexible activities and to choose the duration of activities exclusively on a daily basis. Thus, the system does not account for the possible substitution of activities between days of the week. In this case, this is apparent in that the increase in activities on Friday has no consequences for the probability of selecting the same activities on other days of the week. Probably, the present model overestimates the impact of the scenario in the sense that in reality extra activities on Friday will at least partly be compensated by fewer activities on other days of the week. To be able to accommodate such substitution effects, the present DTs must be extended with conditions representing the history of activities (e.g., the time past since the last time the activity was executed).

15.10 Conclusions

The purpose of this chapter was to illustrate the use of Albatross for impact analysis using four realistic scenarios as examples. The scenarios concerned population, behavioral or institutional developments, and trends. Additionally, scenarios with respect to land uses, transportation networks, and opening hours are supported by the model system, but were not illustrated in this chapter. Two methods of implementing scenarios were used: editing input variables and changing the composition of the sample through differential selection of cases. Both methods assume (small) changes in an existing sample of households and corresponding activity skeletons.

The system proves to be sensitive for impacts on the entire spectrum of situational and decision dimensions of activities, such as day of the week, activity type, duration, start time, trip chaining, etc. By considering decisions on these dimensions in interaction, the system is able to predict rescheduling of activities in response to a change. For example, the change in start time of work activities in scenario 2 did not give rise to rescheduling of activities according to the model, while the shortening of work hours in scenario 4 did.

Substantially, the results of these simulations suggest that demographic change, related to an increase in single-person households and an increase in part-time workers, will reduce mobility. Another interesting result is that changing departure times in the morning, with a larger percentage of people avoiding peak hour traffic, will not significantly affect activity–travel patterns, in the sense that it will not have a major impact on travel distances, mode, and destination choice. In contrast, a compressed workweek, with Friday afternoon off, will increase distance traveled.

References

Arentze, T.A., Borgers, A., Hofman, F., Fugi, S., Joh, C., Kikuchi, A., Kitamura, R., Timermans, H., and van der Waerden, P., Rule-based versus utility-maximizing models of activity–travel patterns: A comparison of empirical performance, in Hensher, D., Ed., *Travel Behaviour Research: The Leading Edge*, Pergamon, Amsterdam, 2001, pp. 569–584.

Arentze, T.A., Hofman, F., van Mourik, H., and Timmermans, H., Albatross: A multi-agent rule-based model of activity pattern decisions, *Transp. Res. Rec.*, 1706, 136–144, 2000.

Arentze, T.A., Hofman, F., Kalfs, N., and Timmermans, H., System for the logical verification and inference of activity (SYLVIA) diaries, *Transp. Res. Rec.*, 1660, 156–163, 1999.

Arentze, T.A. and Timmermans, H., *Albatross: A Learning-Based Transportation Oriented Simulation System*, European Institute of Retailing and Services Studies, Eindhoven, The Netherlands, 2000.

Harvey, A., Holler, B., and Spinney, J., Flexibility and Mobility: A Time Use Perspective, Paper presented at the 9th IATBR Conference, Gold Cost, Australia, July 2000.

16

Centre SIM: First-Generation Model Design, Pragmatic Implementation, and Scenarios

JoNette Kuhnau
Pennsylvania State University

Konstadinos G. Goulias
Pennsylvania State University

CONTENTS

16.1 Introduction

Regional transport system simulation is most often based on a computerized model system that contains the resident population's social, demographic, economic, and location characteristics. The system also contains statistical models of travel behavior that use as input the population characteristics to predict the number of trips people make and the places among which these trips are made, the means used to travel among these places, and the routes chosen to get from one place to another. One system designed to do this in sequence is called the four-step model system, and it has been under continuous scrutiny for the past 30 years (e.g., see the resource papers from the 1972 conference on Urban Travel Demand Forecasting, published by the Highway Research Board, now named the Transportation Research Board, in Special Report 143).

In order to depict and compute the routes chosen, the roadway and public transportation network with its characteristics (e.g., roadways and terminals) are also needed. In addition, many model systems also incorporate urban and rural spatial attributes, representing locations where persons pursue their everyday activities (e.g., work, education, eating meals, shopping). All this information is arranged in an electronic map to represent the spatial separation of places and the spatial organization of the transportation network connecting these places.

For modeling purposes, a study area is usually identified based on administrative and institutional criteria (e.g., a county, a city), but very often functionality is also taken into account (e.g., the region that contains Philadelphia, Pennsylvania, contains geographic portions in the state of New Jersey). This electronic map with embedded databases and behavioral equations constitutes software in which scenarios of change can be imposed (e.g., building a new road and therefore modifying the road network, or building new residences and therefore changing the spatial distribution of activity locations) and their effects computed by the embedded models. Outputs of the system include the number of vehicles per hour (volumes) traveling on each element of the network (links and nodes) and a variety of performance measures, such as vehicle kilometers of travel and travel speeds. In addition, postprocessing of this information can be used to derive air pollutant emissions. Forecasting is performed by estimating expected changes in the use of land, demographics, and network characteristics, incorporating these changes into the digital map, and running the entire software application to predict future traffic volumes.

Model building like the above for long-range transportation planning, has been very important, but it is becoming increasingly more important for cities and metropolitan planning organizations (MPOs) for short-range operational analysis. For example, cities also need tools that help them to maintain building facility and asset inventories and to study their policies a year ahead (e.g., parking changes and signal coordination), and provide data to study operational improvements in public transportation.

To do this, the state of the practice in planning for transportation engineers is still the four-step urban transportation planning system (UTPS) model consisting of trip generation, trip distribution, mode choice, and traffic assignment. These sequential models require large quantities of data, significant time investment, and considerable computer resources, even for small urban areas. Yet, they are loaded with well-known deficiencies such as lack of behavioral consideration, untested assumptions about uniform trip-making behavior within artificial zones, lack of time-of-day considerations, and so forth. Because these models have little or no behavioral basis, they cannot be used to evaluate the effects of transportation demand management strategies and other programs and small improvements in the transportation network that may have effects that are smaller than the model's error. Further, the models have little or no temporal resolution, making time-dependent issues, such as emissions estimates, impossible. This capability has become critical for regions with severe transportation problems, such as areas classified as air quality nonattainment areas (see Chapter 13 on emissions issues and Chapter 3 on behavioral issues).

Even as researchers develop improved travel demand methods, in practice there is not sufficient experience or resources for new data collection, analysis, and model development. This leaves academia and practicing transportation planners with an ever-increasing gap between theory and practice. Person- and household-based activity surveys are becoming more widely used, but small metropolitan planning organizations are not prepared to discard their current models and start over with activity-based models. Even as activity-based models become more accepted in practice, the resources for data collection, model development, and calibration are not available for the striking majority of small metropolitan areas. To assist planners in moving from the traditional models to behavior-based methods, transitional approaches are needed that can incorporate activity and behavioral data without abandoning the models that are well understood and available. This is particularly critical for urban areas that do not have the resources for data collection and for which a complete transition to a new model is not yet possible.

In this chapter the first version of a model system (Centre SIM) incorporating activity information in a traditional UTPS model is presented. Centre SIM improves the accuracy and predictive capabilities of the traditionally applied four-step model. This is based on a simple and practical methodology that effectively incorporates person-based activity behavior within the structure of a four-step model.

The remainder of the chapter is organized as follows. First an overview of the study area, issues, and background is provided. This is followed by a summary of the Centre SIM model effort and a set of scenarios studied with the model. The chapter concludes with a brief summary and a discussion on the next steps.

16.2 Study Area, Issues, and Background

The study area for the Centre SIM model is Centre County, Pennsylvania, a region of approximately 136,000 persons. Centre County includes the main campus of the Pennsylvania State University (Penn State), with over 50,000 students, faculty, and staff. The presence of Penn State dominates the time-of-day travel patterns of the region because a large number of residents (faculty and students) have flexible activity and travel schedules, creating an observed pattern of peak traffic flow in the evening rather than in the morning. It is therefore important to consider the special characteristics of the Penn State community when simulating the Centre County population.

The University Park Campus Master Plan was approved by the Pennsylvania State University in 1999. The plan outlines the development and redevelopment of the University Park campus (in Centre County) over the next 25 years. The master plan contains provisions for substantial construction of new classroom and laboratory facilities, but it is fundamentally based on the concept of an accessible campus with a student-oriented campus core. The Penn State Master Plan Transportation Committee (MPTC) was subsequently formed to study ways in which to achieve the environment- and pedestrian-friendly campus described in the master plan. One of the visions is exploring the effects of projects aimed at making the core of the campus pedestrian friendly (e.g., see the intermodal transportation concept at http://www.opp.psu.edu/divisions/cpd/trans/ITChome.htm (accessed May 2002)).

In addition, university growth requires a rearrangement of the parking structures. The effects of moving parking lots and construction of additional parking need to be computed and added to the traffic flow impacts. In parallel, Centre County is experiencing major additions to its roadway network and significant growth and land use changes. Therefore, the objective of this application example is to study and assess the effects of the campus projects, while at the same time accounting for all other changes taking place in the county. Figure 16.1 shows the study area's location in Pennsylvania, its jurisdictional composition (the southernmost municipalities are the most populated), and identifies one major roadway construction project that is under way with the objective of creating a four-lane controlled access (freeway) connection between the region and Interstate 80 as part of Interstate 99.

16.3 Centre SIM Background

In the past, a transportation demand model for Centre County, within an initiative called Access Management Impact Simulation (AMIS), was developed to aid in operational studies of the region, such as impact fee assessment, signal coordination, and traffic calming (Chung, 1997). At that time, the model was intended as a more detailed representation of development to help public agencies assign fees based on traffic contribution by each development (Goulias and Marker, 1998) and to study traffic engineering highway improvements and demonstrate their impacts by exploiting the capabilities of Geographic Information Systems (GIS) (Chung and Goulias, 1996).

In 2000, this model was further updated, improved, and calibrated with the specific objective to support the Master Plan Transportation Committee in its deliberations. This model encompasses all of Centre County with 1067 traffic analysis zones (TAZs), 3458 roadway links, and 3073 nodes. The digital map is divided into TAZs in its more densely populated region that are based on census blocks (equivalent to city blocks), while other TAZs are based on census block groups (in rural areas with low residential densities). In AMIS the amount of travel each household produces is estimated at the household level using other surveys (Chung, 1997). Cluster analysis was used to create groups of households with similar trip-making patterns in a day. The day was divided into ten time periods, and the number of trips for each household type in each period was estimated. The evolution of households was also simulated in a sociodemographic prediction model from 1980 to 1990 (for validation purposes), and then simulated again to 1997. The number of trips attracted by each business site in the county was calculated at the individual business level based on 1997 data. The data included number of employees for each business and type of business, with trip rate equations extracted from the widely used Institute of Transportation Engineers (ITE) *Trip Generation Manual*. The time-of-day profiles for

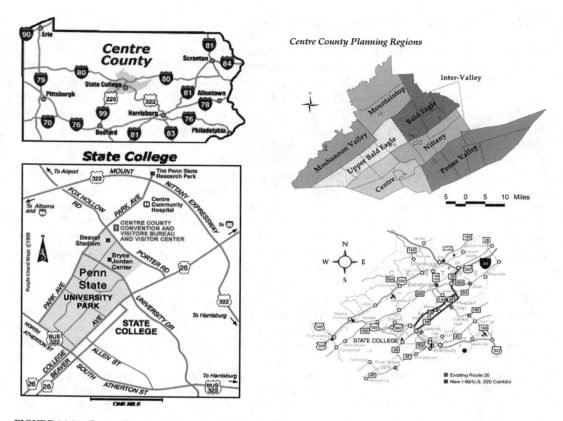

FIGURE 16.1   Centre County in its geographical context.

customer arrivals were derived from a telephone survey of the businesses in Centre County. In this way, the amount of traveling (trip generation rates) is estimated at the household and business levels, which is better than zonal trip generation because it does not require the assumption of uniform characteristics throughout the TAZ.

In a parallel study, the Penn State campus was used to demonstrate the feasibility of an activity-based evacuation management model system, and for this reason, more detailed data were used to study time allocation on campus (Alam, 1998). To this end, an activity survey was conducted in October and November 1996 and was used to determine the number of Penn State faculty, staff, and students traveling during different periods in the day. An example of this time allocation in a time slice of the 24 periods in a day and for each activity type is provided in Figure 16.2. These activity data were used in conjunction with other parking data to determine the number of vehicle trips originating and ending in the zones on campus.

Alam's study made the key connection between activity modeling and the four-step framework using a *building presence model* (this is the key idea used in developing Centre SIM to a zone presence model covering the entire county). Activity patterns for the different population segments provide probabilities for activity participation at time points (buildings on campus) throughout the day. By combining this activity participation information (in Figure 16.2) with land use data (e.g., the detailed information about the use of buildings on campus), the geographic distribution of persons in the study area is produced. Figure 16.3 provides an example of time-of-day allocation to activity locations in Alam's model. Each cubical solid shape in the map of Figure 16.3 is the amount of persons in that specific building. In this way, Alam built 24 maps (1 for each hour in a day) that depict campus life (Alam, 1998).

As part of the Penn State Master Plan project, the original AMIS model was updated by adding network (roadway) and development changes that occurred between 1997 and 2000. Development changes

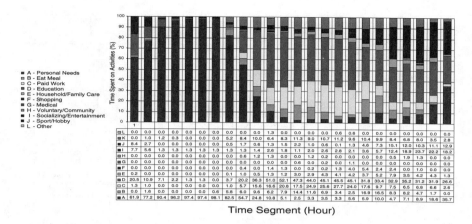

	1	2	3	4	5	6	7	8	9	10	11	12	13	14	15	16	17	18	19	20	21	22	23	24
L	0.0	0.0	0.0	0.0	0.0	0.0	0.0	0.0	0.0	0.0	1.3	0.0	0.0	0.0	0.0	0.6	0.6	0.0	0.0	0.0	0.0	0.0	0.0	0.0
K	0.0	1.0	1.2	0.3	0.0	0.0	0.0	5.2	8.4	10.0	6.4	8.3	11.3	9.0	10.7	11.2	9.6	10.4	9.9	8.4	6.8	8.0	3.5	2.8
J	8.4	2.7	0.0	0.0	0.0	0.0	0.0	0.5	1.7	0.6	1.3	1.5	2.2	1.0	0.6	0.1	1.3	4.6	7.3	15.1	12.0	10.3	11.1	12.9
I	7.7	5.6	1.3	1.3	1.3	1.3	1.3	1.3	1.3	1.4	2.6	1.8	1.1	2.0	2.6	2.8	2.1	3.6	5.7	12.4	18.9	23.7	22.2	18.2
H	0.0	0.0	0.0	0.0	0.0	0.0	0.0	0.6	1.2	1.3	0.0	1.2	0.0	0.0	0.0	0.0	0.0	0.5	1.9	1.3	0.0	0.0		
G	0.0	0.0	0.0	0.0	0.0	0.0	0.0	0.0	0.0	0.0	0.0	1.2	0.2	0.0	0.0	0.0	0.0	0.5	1.9	1.3	0.0	0.0		
F	0.0	0.0	0.0	0.0	0.0	0.0	0.0	0.0	0.4	0.0	1.4	1.3	0.0	0.2	1.3	4.0	5.4	2.4	2.4	0.0	1.0	0.0	0.0	
E	0.2	0.0	0.0	0.0	0.0	0.0	0.0	0.1	1.0	0.5	1.3	1.2	3.0	2.9	4.3	4.1	4.2	3.7	5.2	7.9	3.5	4.2	4.3	1.3
D	20.5	10.9	7.1	2.2	1.3	1.3	0.0	3.7	20.2	36.3	51.0	52.1	47.3	44.0	45.1	45.5	45.1	31.8	33.4	32.9	35.2	31.2	31.9	26.6
C	1.3	1.0	0.0	0.0	0.0	0.0	0.0	1.0	5.7	15.6	16.6	20.8	17.5	24.9	25.6	27.7	24.0	17.6	9.7	7.5	6.5	6.8	6.6	2.6
B	0.0	1.6	0.0	0.0	0.0	0.0	0.6	5.8	6.0	9.6	6.2	7.9	14.4	11.6	6.9	3.4	2.5	16.9	16.5	8.3	8.2	4.7	1.7	0.0
A	61.9	77.2	90.4	96.2	97.4	97.4	98.1	82.5	54.7	24.8	10.8	5.1	2.5	3.3	3.5	3.3	5.6	5.9	10.0	4.7	7.1	8.9	18.6	35.7

Time Segment (Hour)

Legend: A - Personal Needs; B - Eat Meal; C - Paid Work; D - Education; E - Household/Family Care; F - Shopping; G - Medical; H - Voluntary/Community; I - Socializing/Entertainment; J - Sport/Hobby; L - Other. Vertical axis: Time Spent on Activities (%).

FIGURE 16.2 Students' time allocation in the Penn State survey.

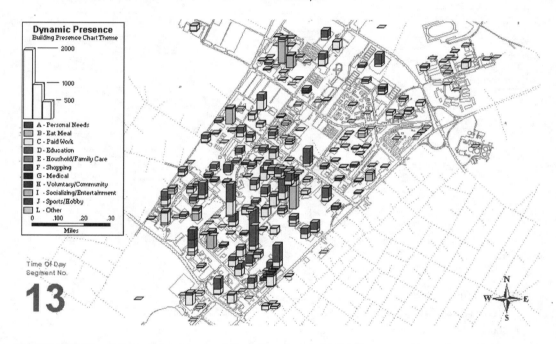

FIGURE 16.3 The Alam spatial and temporal time allocation model (12:00 noon to 1:00 P.M.).

included construction of commercial areas and major residential areas. The calibration of one model version was performed by assigning traffic to the network and comparing the resulting volumes to peak period traffic counts. For each traffic assignment run, differences between volumes predicted by the model and actual volumes were noted. Adjustments were subsequently made to improve the model system and minimize these differences. This was deemed insufficient for studies that require detailed information about time-of-day travel and network volumes for minor roadways. Different options for improvements were explored, targeting incorporation of Alam's approach in a county-wide frame and increasing the level of detail usually found in four-step models.

Kuhnau (2001) developed this new framework for simulating the entire county for each of the 24 daily segments by adding some key information and simplifying the Alam activity model. This model system is named Centre County Simulation (Centre SIM). Figure 16.4 provides a flowchart of all the information sources and modeling steps that are a combination of the Alam procedures to assign persons at geographic

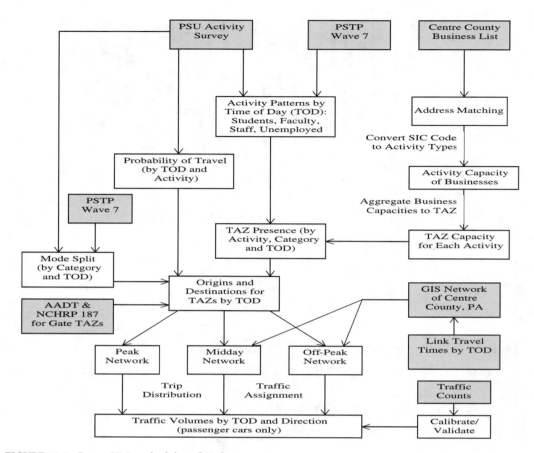

FIGURE 16.4 Centre SIM methodology flowchart.

locations pursuing activities, and the four-step model system. Below we identify a few key aspects of this model effort.

The primary data elements used in this Centre SIM are:

- Penn State activity survey
- Puget Sound Transportation Panel (PSTP) wave 7 (data from year 1997)
- 2000 U.S. Census population statistics
- 2000 business list for Centre County
- Hourly and daily traffic counts
- GIS network of Centre County
- Link travel times for three traffic conditions (peak, midday, off-peak)

An activity survey of a representative sample of the Centre County population, not only the Penn State campus, would have been more appropriate for this application. However, this was not available during model design and testing, so a much smaller survey designed by Alam, together with data from other regions, was used. This has some important implications, as discussed in the summary and conclusions. For simulation of activity patterns, the population of Centre County greater than 18 years of age was divided into the following six categories:

- Penn State students
- Penn State faculty

- Penn State staff
- Unemployed persons
- Professionals
- Workers (Non-Penn State staff)

Each of these categories was assumed to have different activity and travel patterns affecting the overall travel demand observed. It is important to note that persons under 18 years of age were not included in this model due to lack of information about activity patterns and employment rates. This may not be a serious deficiency because we focus on passenger travel by car to demonstrate feasibility and identify the next steps. From 2000 U.S. Census data, only about 2500 persons are between 16 and 18 years old, less then 2% of the total population of the county. As mentioned above, the person-based activity patterns used in this model were derived from a 2-day activity diary of 102 Penn State students, faculty, and staff. These activity patterns were modified slightly for persons employed outside the Penn State campus. This was done because of the unique characteristics of the location and constraints of the Penn State campus in relation to State College. Figure 16.5 shows the time allocation of students by time of day.

Kuhnau (2001) developed an algorithm to convert these time allocations by each population segment to time allocations for the entire population. Then these population time allocation shares by activity type are converted to time allocation at destinations. To do this, another algorithm is needed that is in essence a distribution model based on capacity and type of activity that each business location allows. Travel is then computed using a series of steps that are similar to the four-step procedures. A key difference here, however, is that the spatial and temporal distribution of activity types is used to derive the propensity to travel and modal share by each of the population segments above. Figure 16.6 is one of the outputs obtained from the model, and it contains both the spatial distribution of persons in activities and the traffic volumes predicted. Model validation and calibration were also needed for this type of model, in a way similar to that of other four-step applications. As Kuhnau (2001) shows, however, a model that departs from activity participation and contains within it the time-of-day variation in travel is closer to observed counts and requires less drastic calibration steps.

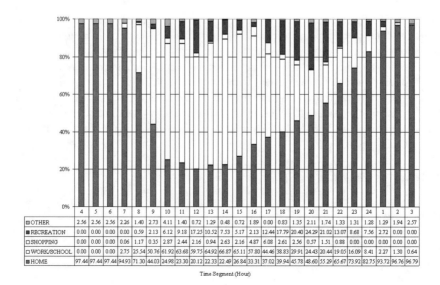

	4	5	6	7	8	9	10	11	12	13	14	15	16	17	18	19	20	21	22	23	24	1	2	3
OTHER	2.56	2.56	2.56	2.26	1.40	2.73	4.11	1.40	0.72	1.29	0.48	0.72	1.89	0.00	0.83	1.35	2.11	1.74	1.33	1.31	1.28	1.29	1.94	2.57
RECREATION	0.00	0.00	0.00	0.00	0.59	2.13	6.12	9.18	17.25	10.52	7.53	5.17	2.13	12.44	17.79	20.40	24.29	21.02	13.07	8.68	7.56	2.72	0.00	0.00
SHOPPING	0.00	0.00	0.00	0.06	1.17	0.35	2.87	2.44	2.16	0.94	2.63	2.16	4.87	6.08	2.61	2.56	0.57	1.51	0.88	0.00	0.00	0.00	0.00	0.00
WORK/SCHOOL	0.00	0.00	0.00	2.75	25.54	50.76	61.92	63.68	59.75	64.92	66.87	65.11	57.80	44.46	38.83	29.91	24.43	20.44	19.05	16.09	8.41	2.27	1.30	0.64
HOME	97.44	97.44	97.44	94.93	71.30	44.03	24.98	23.30	20.12	22.33	22.49	26.84	33.31	37.02	39.94	45.78	48.60	55.29	65.67	73.92	82.75	93.72	96.76	96.79

Time Segment (Hour)

FIGURE 16.5 Time use pattern for students. (From Kahnau, J.L., Master's thesis, Pennsylvania State University, University Park, 2001. With permission.)

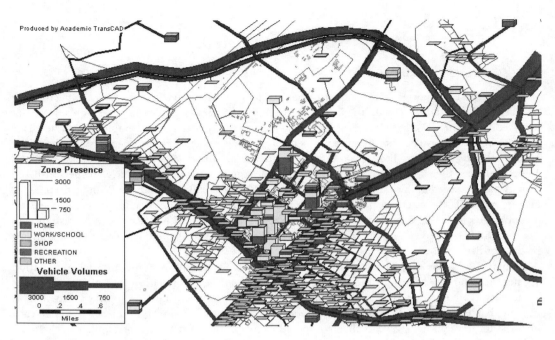

FIGURE 16.6 Centre SIM activity participation (zonal presence) and travel in the period 5:00 P.M. to 6:00 P.M.

16.4 Centre SIM Scenario Testing

This section describes how the Centre SIM model was used to assess proposed transportation projects and plans on the Penn State University Park campus. The scenarios include network (roadway) changes, as well as a travel demand management (TDM) proposal to spread the Penn State peak hour travel demand. The results of these scenarios are discussed from both the modeling and planning perspectives.

The objectives of scenario testing with the Centre SIM model were to demonstrate the capabilities of the model, evaluate alternative plans proposed by Penn State University, and analyze possible solutions to the localized congestion problems around the Penn State campus. First, a base scenario of existing conditions was simulated to provide a benchmark against which the simulated alternatives were evaluated. The changes proposed in each scenario were then created, simulated, analyzed, and compared to determine the effects of different proposals on the transportation network on the Penn State campus and in the surrounding State College area. The versatility of the model was demonstrated by its ability to model a wide variety of proposals and produce interpretable results necessary to make recommendations about the feasibility of each project.

16.4.1 Description of Scenarios

As mentioned in the first section, many projects are envisioned in the Penn State Master Plan. Specifically, several proposed parking and roadway changes required simulation to determine their potential impacts on traffic volumes and vehicle circulation patterns on campus. The simulations were designed to provide a quantitative basis for decision making about the feasibility of further planning for each alternative. Road closures and construction obviously will have an effect on traffic flows on campus. In addition, the construction of new buildings brings increased demand for parking facilities, while at the same time it removes surface parking capacity. The three proposed roadway changes evaluated in this research are described below and shown in Figure 16.7:

FIGURE 16.7 Penn State roadway scenarios.

Network scenario 1: Closure of Shortlidge Road. Involves closing one block of the only direct north–south route through the core campus, connecting downtown State College to Park Avenue (a major east–west route bordering the Penn State campus).

Network scenario 2: Extension of Bigler Road. Extends an existing route to College Avenue, creating a new direct north–south link through campus from Park Avenue to downtown State College.

Network scenario 3: Conversion of Curtin Road to transit only. Includes prohibiting all but transit vehicles from a section of roadway heavily used by transit. The route connects the easternmost parts of the campus (including major residence and parking facilities) with the westernmost areas (consisting of classrooms, laboratories, and offices).

Another scenario tested was defined to examine the model capabilities when the time-of-day aspects of the model are used. The proposed TDM policy scenario involved staggering the work starting and ending times for Penn State staff in an attempt to mitigate the severe demands placed on portions of the transportation network at 8:00 A.M. and 5:00 P.M. For this scenario, it was assumed that one third of the staff would start at the hours of 7:00, 8:00, and 9:00 A.M. Consequently, the corresponding work end times would be approximately 4:00, 5:00, and 6:00 P.M. No assumptions were made about changes in travel during the period around 12:00 P.M. (lunch) because it was unclear how staff would change their midday behavior due to different work starting times. In addition, no changes were made to the faculty or student activity patterns because they do not have fixed schedules dictated by a single employer. Thus, the hours simulated for the staggered hours policy were 6:00, 7:00, and 8:00 A.M. for the work arrival patterns because, for example, travel must occur in the 6:00 to 7:00 A.M. segment for a staff member to be engaged in work at 7:00 A.M. The 4:00, 5:00, and 6:00 P.M. hours were simulated for the work departure pattern since if work ends at 4:00 P.M., a trip to the next activity will occur in the 4:00 to 5:00 P.M. time period (the same time segment as the end of the workday).

16.4.2 Network Changes

Modifications to the transportation network were easily modeled in the GIS software TransCAD, using its algorithms to recalculate the shortest path between an origin–destination pair and reassign the trip route. First, three different networks were created to test each of the scenarios individually before examining the effects of combinations of alternatives.

FIGURE 16.8 Sample output from Shortlidge Road scenario.

For network scenario 1, the roadway closure was simulated by removing the appropriate section of Shortlidge Road from the link layer in the GIS map of the model. In scenarios 1 to 3 that involved only network changes, it was assumed that the activity patterns, zone presence, travel patterns, and mode split would remain constant. Employing these assumptions, the base origin–destination patterns (i.e., the origin–destination matrix) can be used for simulation of all three roadway scenarios. This assumption was reasonable because, for example, a road closure of one block changes the route choices available for a trip and may slightly change the travel times between some origin–destination pairs, but would be unlikely to influence the decisions whether to make a trip, what mode to use, and where to go (trip generation, mode split, and trip distribution).

A new link was added to connect the existing section of Bigler Road to College Avenue, simulating scenario 2. Characteristics such as travel time, number and capacity of lanes, and directions of flow were assigned to the link to make it usable for traffic in the GIS network. Next, the link designated as transit only was removed to simulate scenario 3. Personal vehicles only, not transit vehicles, were simulated in the Centre SIM model; therefore, creating the transit-only section in essence removes it from the non-transit (personal vehicle) network. A new network file was created for each of these scenarios for use in the reassignment of traffic flows on the modified transportation system.

After completing the network changes described, the origin–destination matrix resulting from trip distribution in the base model was assigned to each of the three new networks. TransCAD then produces the typical traffic assignment graphic and database output for each scenario. An example of the traffic assignment output from the Shortlidge Road scenario (scenario 1) is shown in Figure 16.8 for the 5:00 to 6:00 P.M. time segment. Although other time segments were simulated, the P.M. peak in State College exhibits the most significant impacts because the overall travel demand is highest during this time. The result in the P.M. peak from the simulation of the Shortlidge Road scenario, shown here, is thus the worst-case outcome of the proposed alternative.

In this form, however, impacts of the scenario can only be seen through manual link-by-link comparisons between the base and scenario models. To create a clearer demonstration of the scenario impacts, Figure 16.9 shows the differences between baseline and the Shortlidge scenario, again for the 5:00 to 6:00 P.M. segment (the peak hour). The positive values indicate increases in traffic volumes due to the road closure, while the negative values represent decreases in vehicle volumes. The major volume decrease in traffic all along Shortlidge Road was expected because it no longer provides a direct route through campus. In addition, one parking deck structure lies on each side of the closure, and as a result, these facilities

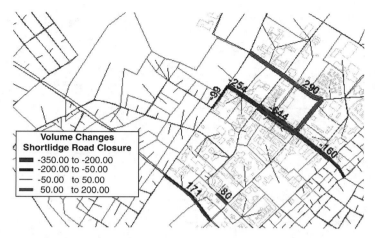

FIGURE 16.9 Sample output from scenario impact evaluation.

can only be entered and exited from one direction if the scenario is implemented. This type of comparative analysis was performed for the other two individual alternatives (Kuhnau, 2001).

16.4.3 Staggered Work Hours Policy Scenario Evaluation

The staggered work hours scenario, unlike the other three alternatives described, involved behavioral (activity pattern) changes as opposed to physical (network) changes. Thus, the same network used for the base condition was also used to evaluate this policy scenario. To simulate the changes in arrival and departure times to and from work, a modified activity distribution was created. Whereas formerly the distribution of Penn State staff arrivals at work had a concentration of 50% arriving in the 7:00 to 8:00 A.M. time segment, the new activity distribution was designed with 30% of the staff category engaged in work at 7:00 A.M., 60% at 8:00 A.M., and 90% at 9:00 A.M. It should be noted that at no time during the day were 100% of the staff engaged in work, as might be expected. A similar staggered distribution of departure times was used for the 4:00, 5:00, and 6:00 P.M. time segments. These two work arrival–departure patterns are summarized in Figure 16.10. Note that the staggered hours pattern has more gradual arrival and departure patterns than the relatively rapid transition patterns in the existing work pattern.

Several of the critical corridors of concern in the peak hours, with significant congestion and delay, are Park Avenue, Atherton Street, College Avenue, and Beaver Avenue (Figure 16.1). To evaluate the effectiveness of the staggered work hours scenario, the volumes on these corridors were compared before and after policy implementation. However, because these corridors at times have traffic volumes exceeding capacity in the peak hours, the volume changes due only to the staggered hours policy are difficult to determine. User equilibrium traffic assignment was used for the base and TDM scenarios. This means an equilibrium point is found such that an alternate route will not decrease the travel time between any origin–destination pair. However, the relationship between the base and TDM equilibrium points are not linear. As less Penn State traffic use congested routes, delays and travel times may improve to the point where other traffic will shift to these roadways from less attractive and congested routes. As a result, the volume reductions in the peak hour do not equal the volume increases in the "shoulder" hours (6:00 a.m., 8:00 A.M., 4:00 P.M., and 6:00 p.m.). These nonlinear effects are not obvious in reviewing the proposed policy, but the model can be used to predict how traffic patterns will change as the result of a policy scenario such as this one:

> The changes for the eastbound (EB) and northbound (NB) directions of the corridors in the morning hours (6, 7, and 8 A.M.) are summarized in Figure 16.11, and the changes for the westbound (WB) and southbound (SB) directions of the same roadways for the same period are shown in Figure 16.12.

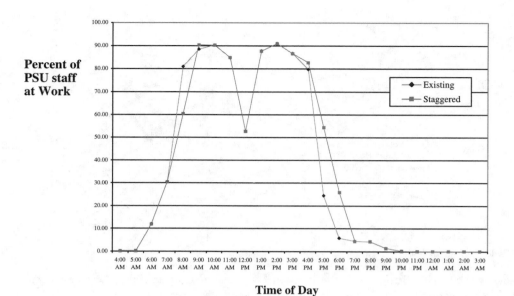

FIGURE 16.10 Existing and staggered hours of staff work patterns.

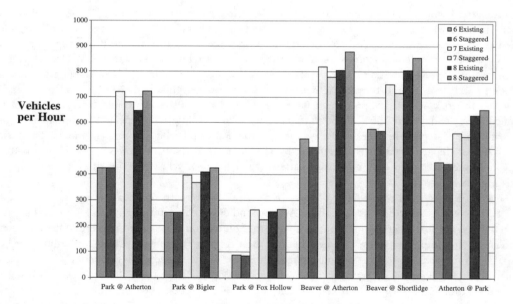

FIGURE 16.11 EB and NB volume changes due to staggered hours policy.

It should be noted that on each of these links, the volumes under the staggered hours scenario fell within the 2 standard deviation range of the mean volume for that hour. The impacts of the staggered work hours scenario were not as great as expected, especially in the Park Avenue corridor (see Figure 16.1), which is of particular interest to Penn State because of its proximity to the campus and the long delays and queues in peak hours. The peak hour traffic volume decreases are probably not worth the major effort required to implement the policy. Significant resistance to this change by the staff members affected would presumably be encountered, and without significantly better travel conditions, the policy would be viewed as ineffective and unnecessary.

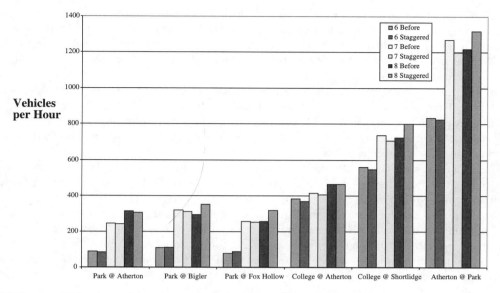

FIGURE 16.12 WB and SB volume changes due to staggered hours policy.

16.5 Summary and Next Steps

The Centre SIM model represents the first application in which spatial and temporal information about activity participation is used in a more traditional four-step modeling framework. The results of this simulation demonstrate the feasibility of incorporating activity data into the traditional UTPS travel demand model. The Centre SIM methodology provides the capabilities of more complex models without the need for significant investments in model development and data collection, a critical feature for small regions with limited resources. The effectiveness of this model was demonstrated by validating the output to traffic counts in the study area. It was also compared to previous models of Centre County. The Centre SIM model has expanded abilities in terms of simulating travel throughout the day and testing network and policy scenarios. The model has also been shown to better replicate existing traffic patterns than a previous model (AMIS). Therefore, the Centre SIM model is more accurate, in addition to moving toward a more theoretically sound approach.

Further, the Centre SIM methodology is a unique application of a geography-based travel demand model using activity data, starting first from the spatial distribution of activities. This is in contrast to the traditional approach in activity-based modeling, where an individual is assigned a series of activities and later these activities are assigned to locations. In addition, the geographic basis of the model provides for simulation of land use policies and scenarios, although these were not demonstrated here. The Centre SIM methodology can also accommodate travel viewed either as a derived demand or as an activity to itself. This characteristic is important as travel behavior theory continues to be refined. The Centre SIM model will not become outdated in the near future due to advances in travel research, so its capabilities can continue to be enhanced and developed.

Several issues were identified during development of the Centre SIM model that were not addressed or solved in this application. These elements would improve the foundation of the model or its modeling capabilities. For the Centre SIM model specifically, a representative activity survey of the Centre County population (not just the Penn State population) would provide a better foundation for the activity patterns. Additionally, a new survey would support the definition of population segments and the separation of the population into the segments. With a more complete data set, persons between 16 and 18 years of age, discounted in the current model, can be simulated in the Centre SIM model. This is an activity currently under way for the entire county that will also provide input to a new long-range plan.

Another use of the survey data would be to analyze the trip duration data to develop region-specific friction factors. Factors based on local characteristics and travel patterns would be a distinct improvement over the generic factors used in this research. The temporal distribution of trip durations can also be used to produce friction factors that vary by time of day, similar to the varied network travel times already used in the model.

For long-range planning purposes, two components may be added to the Centre SIM model to aid in forecasting. First, the addition of a sociodemographic microsimulator would provide the ability to simulate the population through 20 or 30 years (see Chapter 14 of this handbook and the Chung (1997) application). The microsimulator can provide the input needed for Centre SIM to model activity and travel demands for future time points.

Second, although the Centre SIM model already has capabilities for analyzing land use scenarios, an explicit land use simulation tool can be added as a feature to the model. Land use changes and development are critical to the growth of a region over time and are directly tied to travel demand and traffic patterns. However, current practice in projecting land use consists of trend analysis and the opinions of local planning experts about where developments of various types may occur. These methods are subjective and possibly highly unreliable for long-range planning horizons. Examples of new simulation models exist (see Chapter 12 on land use models) and should be used for this application. Several smaller-scale improvements may also be made to the Centre SIM methodology to improve its current accuracy and simulation capabilities; these are reported elsewhere (Kuhnau, 2001).

References

Alam, S.B., Dynamic Emergency Evacuation Management System Using GIS and Spatio-Temporal Models of Behavior, M.S. thesis, Department of Civil and Environmental Engineering, Pennsylvania State University, University Park, 1998.

Chung, J.H., Transportation Impact Simulation of Access Management and Other Land Use Policies Using Geographic Information Systems, doctoral dissertation, Department of Civil and Environmental Engineering, Pennsylvania State University, University Park, 1997.

Chung, J.H. and Goulias, K.G., Access management using GIS and traffic management tools in Pennsylvania, *Transp. Res. Rec.*, 1551, 114–122, 1996.

Goulias, K.G. and Marker, J.T., Jr., Procedure for Using the Access Management Impact Simulation (AMIS) Model within the Context of Act 209 and Act 47, final report submitted to PennDOT, PTI 9819, University Park, PA, 1998.

Kuhnau, J.L., Activity-Based Travel Demand Modeling within the Urban Transportation Planning System, M.S. thesis, Department of Civil and Environmental Engineering, Pennsylvania State University, University Park, 2001.

Index

N

O